The IMA Volumes in Mathematics and its Applications

Volume 123

Series Editor
Willard Miller, Jr.

Springer Science+Business Media, LLC

Institute for Mathematics and its Applications
IMA

The **Institute for Mathematics and its Applications** was established by a grant from the National Science Foundation to the University of Minnesota in 1982. The IMA seeks to encourage the development and study of fresh mathematical concepts and questions of concern to the other sciences by bringing together mathematicians and scientists from diverse fields in an atmosphere that will stimulate discussion and collaboration.

The IMA Volumes are intended to involve the broader scientific community in this process.

Willard Miller, Jr., Professor and Director

* * * * * * * * * *

IMA ANNUAL PROGRAMS

Continued at the back

Brian Marcus   Joachim Rosenthal
Editors

Codes, Systems, and Graphical Models

With 93 Illustrations

Springer

Brian Marcus
IBM Almaden Research Center, K65-802
650 Harry Rd.
San Jose, CA 95120
USA
e-mail: marcus@almaden.ibm.com

Joachim Rosenthal
Department of Mathematics
University of Notre Dame
Notre Dame, IN 46556-5683
USA
e-mail: rosen@nd.edu

Mathematics Subject Classification (2000): 11T, 37B, 65F, 68Q, 93B, 93C, 94A, 94B

Library of Congress Cataloging-in-Publication Data
Codes, systems, and graphical models / editors, Brian Marcus, Joachim Rosenthal.
    p. cm. — (The IMA volumes in mathematics and its applications ; 123)
    Includes bibliographical references and index.
    ISBN 978-1-4612-6534-4   ISBN 978-1-4613-0165-3 (eBook)
    DOI 10.1007/978-1-4613-0165-3
    1. Coding theory—Congresses. 2. System theory—Congresses. 3. Symbolic
dynamics—Congresses. I. Marcus, Brian, 1949– II. Rosenthal, Joachim, 1961– III. IMA
Workshop on Codes, Systems, and Graphical Models (1999) IV. Series.
QA268 .C63 2001
003'.54—dc21   00-052274

Printed on acid-free paper.

Production managed by A. Orrantia; manufacturing supervised by Joe Quatela.
Camera-ready copy prepared by the IMA.
Printed and bound by Sheridan Books, Inc., Ann Arbor, MI.

9 8 7 6 5 4 3 2 1

SPIN 10789096

FOREWORD

This IMA Volume in Mathematics and its Applications

CODES, SYSTEMS, AND GRAPHICAL MODELS

is based on the proceedings of a very successful 1999 IMA Summer Program with the same title.

I would like to thank G. David Forney, Jr. (Massachusetts Institute of Technology), Brian Marcus (IBM Almaden Research Center), Joachim Rosenthal (University of Notre Dame), and Alexander Vardy (University of California, San Diego) for their excellent work as organizers of the two weeks summer program. Special thanks to Brian Marcus and Joachim Rosenthal for their role as editors of the proceedings.

I also take this opportunity to thank the National Science Foundation (NSF) and the National Security Agency (NSA), whose financial support made the workshop possible.

Willard Miller, Jr., Professor and Director
Institute for Mathematics and its Applications
University of Minnesota
400 Lind Hall, 207 Church St. SE
Minneapolis, MN 55455-0436
612-624-6066, FAX 612-626-7370
miller@ima.umn.edu
World Wide Web: http://www.ima.umn.edu

PREFACE

Codes and systems based on graphical models form a prominent area of research in each of the following subjects: coding theory, systems theory, symbolic dynamics and automata theory. The aim of the IMA Workshop on Codes, Systems and Graphical Models (August 2-13, 1999) was to bring together mathematicians, computer scientists, and electrical engineers in these subjects to learn how techniques from one area might be applied to problems in the other areas.

The workshop was divided into two weeks, each with a different focus. The first week, organized by Dave Forney and Alex Vardy, focused on codes on graphs and iterative decoding. The second week, organized by Brian Marcus and Joachim Rosenthal, turned to connections among coding theory, system theory and symbolic dynamics.

A major goal of coding theory is to find a class of codes, together with corresponding decoding algorithms, that realize in a practical way the promise of Shannon's coding theorems. Nearly forty years ago, Gallager introduced low density parity check codes (LDPC's), with iterative decoding algorithms, for this purpose. Recently, the subject of iterative decoding has enjoyed a burst of intense activity with the discovery of turbo codes and the realization that both turbo codes and LDPC's perform remarkably well. Also, it has recently been recognized that the iterative decoding algorithms used for these codes are instances of belief propagation, an algorithm originally developed by Pearl in artificial intelligence. This has led to much work devoted to understanding and exploiting this connection. While Pearl's algorithm is guaranteed to work well on graphs without cycles, it has been demonstrated empirically to work well on graphs with cycles. A central impetus of current work on iterative decoding is to understand why this is so. The first week of the workshop focused on these issues.

Coding theory, system theory and symbolic dynamics have much in common. Among the central themes in each of these subjects are the construction of state space representations, understanding of fundamental structural properties of sequence spaces, construction of input/output systems, and understanding the special role played by algebraic structure. The second week of the workshop was devoted to understanding how these themes are treated in each of the subjects and how ideas from one subject may be useful in the others. This continued the interactions initiated in an IEEE workshop on codes, systems and symbolic dynamics in 1993.

This volume is a collection of papers on the subjects of the workshop. Our intention was to collect in one place many pieces of work on these subjects; some were presented at the workshop, while others were not.

Part 1 contains papers with significant tutorial or overview content. Parts 2 and 3 contain papers on the main themes of week 1, 'Codes on Graphs' and 'Iterative Decoding'(although Part 3 contains some system-theoretic papers on non-iterative decoding). Parts 4 and 5 contain papers on the main themes of week 2, 'Convolutional Codes' and 'Symbolic Dynamics and Automata Theory'.

In addition to the material collected in this volume, the reader can find copies of slides of many of the workshop talks as well as an exceptional trove of related bibliographic material on the IMA Web site at:

http://www.ima.umn.edu/csg

For a collection of related papers, the reader may also wish to consult the Special Issue on 'Codes on Graphs and Iterative Algorithms' of the IEEE Transactions on Information Theory to be published in 2001.

Finally, we are happy to acknowledge the excellent organization and facilities provided by the IMA as well as financial support from the IMA, the IEEE Information Theory Society and the U. S. National Security Agency.

Brian Marcus (IBM Almaden Research Center)

Joachim Rosenthal (University of Notre Dame)

CONTENTS

Part 1. Overviews

Part I. Overviews

AN INTRODUCTION TO THE ANALYSIS OF ITERATIVE CODING SYSTEMS

TOM RICHARDSON* AND RÜDIGER URBANKE†

Abstract. This paper is a tutorial on recent advances in the analysis of iterative coding systems as exemplified by low-density parity-check codes and turbo codes.

The theory described herein is composed of various pieces. The main components are concentration of system performance over the ensemble of codes and inputs, the existence of threshold phenomena in decoding performance, and the computational and/or analytical determination of thresholds and its implications in system design.

We present and motivate the fundamental ideas and indicate some technical aspects but proofs and many technical details have been omitted in deference to accessibility to the concepts. Low-density parity-check codes and parallel concatenated codes serve as contrasting examples and as vehicles for the development.

Key words. turbo codes, low-density parity-check codes, belief propagation, iterative decoding, stability condition, threshold, output-symmetric channels, Azuma's inequality, support tree.

AMS(MOS) subject classifications. 94B05.

1. Introduction. This paper is a tutorial on recent advances in the analysis and design of iterative coding systems as exemplified by low-density parity-check (LDPC) codes and turbo codes. We will outline a mathematical framework within which both of the above mentioned coding systems may be analyzed. Certain aspects of the theory have important practical implications while other aspects are more academic. Provable statements concerning the asymptotic performance of these iterative coding systems can be made. Here, asymptotic refers to the length of the code. For codes of short length the theory we shall present does not yield accurate predictions of the performance. Nevertheless, the ordering of coding systems which is implied by the asymptotic case tends to hold even for fairly short lengths. The *bit error probability* is the natural measure of performance in the theory of iterative coding systems but the theory also offers insights and guidance for various other criteria, e.g., the block error probability.

We will here briefly describe an example and formulate some claims that represent what we consider to be the apex of the theory: Let us consider the class of (3,6)-regular LDPC codes (see Section 2.1.1 for a definition) of length n for use over an additive white Gaussian noise channel, i.e., we transmit a codeword consisting of n bits $x_i \in \{\pm 1\}$ and receive n values $y_i = x_i + z_i$ where the z_i are i.i.d. zero mean Gaussian random

---

*Bell Labs, Lucent Technologies, Murray Hill, NJ 07974, USA; email: tjr@lucent.com.

†Swiss Federal Institute of Technology – Lausanne, LTHC-DSC, CH-1015 Lausanne, Switzerland; email: Rudiger.Urbanke@epfl.ch.

variables with variance σ^2. We choose the particular code, as determined by its associated particular graph, at random (see Section 2). We transmit one codeword over the channel and decode using the sum-product algorithm for ℓ iterations. The following statements are consequences of the theory:

1. The expected bit error rate approaches some number $\epsilon = \epsilon(\ell)$ as n tends to infinity and this number is computable by a deterministic algorithm.

2. For any $\delta > 0$, the probability that the actual fraction of bit errors lies outside the range $(\epsilon - \delta, \epsilon + \delta)$ converges to zero exponentially fast in n.

3. There exists a maximum channel parameter σ^* (in this case $\sigma^* \simeq 0.88$), the *threshold*, such that $\lim_{\ell \to \infty} \epsilon(\ell) = 0$ if $\sigma < \sigma^*$ and $\lim_{\ell \to \infty} \epsilon(\ell) > 0$ if $\sigma > \sigma^*$.

Each of these statements generalize to some extent to a wide variety of codes, channels, and decoders. The existence of $\epsilon(\ell)$ in statement 1 is very general, holding for all cases of interest. In general there may not be any efficient algorithm to compute this number $\epsilon(\ell)$ but, fortunately, for the case of most interest, namely the sum-product algorithm, an efficient algorithm is known. Statement 2 holds in essentially all cases of interest and depends only on the asymptotics of the structure of the graphs which define the codes. Statement 3 depends on both the decoding algorithm used and the class of channels considered (AWGN). It depends on the fact that the channels are ordered by *physical degradation* (see section 7) and that the asymptotic decoding performance respects this ordering.

Although the threshold, as introduced above in 3, is an asymptotic parameter of the codes, it has proven to have tremendous practical significance. It is only a slight exaggeration to assert that, comparing coding systems with the same rates, the system with the higher threshold will perform better for nearly all n. The larger n is, the more valid the assertion is. Even though the assertion is not entirely true for small n, one can still significantly improve designs by looking for system parameters that exhibit larger thresholds.

To design for large threshold one needs to be able to determine it or at least accurately estimate it. Therefore an important facet of the theory of these coding systems deals with the calculation of the threshold.

In his Ph.D. thesis of 1961, Gallager [13] invented both LDPC codes and iterative decoding. With the notable exceptions of Zyablov and Pinsker [45] and Tanner [40] (see also Sourlas [39]), iterative coding systems were all but forgotten until the introduction of turbo codes by Berrou, Glavieux and Thitimajshima [4] (see also [19]). In the wake of the discovery of turbo codes LDPC codes were rediscovered by MacKay and Neal [24], completing the cycle which had started some thirty years earlier.

For some simple cases Gallager was able to determine thresholds for the systems he considered. In the work of Luby *et. al.* [22] the authors used threshold calculations for the binary erasure channel (BEC) and the binary symmetric channel (BSC) under hard decision decoding to *optimize* the parameters of *irregular* LDPC codes (Gallager had considered only regular LDPC codes) with respect to the threshold. By this approach they showed that very significant improvements in performance were possible. Indeed, they explicitly exhibited a sequence of LDPC code ensembles which, under iterative decoding, are capable of achieving the (Shannon) capacity of the BEC. To date, the BEC is the only non-trivial channel for which capacity achieving iterative coding schemes are explicitly known.

Another important aspect of the work presented in [22] is the method of analysis. The approach differed from that taken by Gallager in certain key respects. The approach of Gallager allowed statements to be made concerning the asymptotics of certain special *constructions* of his codes. The approach of Luby *et. al.*, allowed similar and, in some ways, stronger statements (such as the ones given in our example above) to be made concerning the asymptotics of *random ensembles*, i.e., randomly chosen codes from a given class. This placed the emphasis clearly on the threshold and away from particular constructions and opened the door to irregular codes for which constructions are generally very difficult to find.

In [30] the approach taken in [22] was generalized to cover a very broad class of channels and decoders. Also in [30], an algorithm was introduced for determining thresholds of LDPC codes for the same broad class of channels and the most powerful and important iterative decoder, the sum-product algorithm, also called belief propagation, which is the name for a generalization of the algorithm as independently developed in the AI community by Pearl [28].[1] In [29] the full power of these results was revealed by producing classes of LDPC codes that perform extremely close to the best possible as determined by the Shannon capacity formula. For the additive white Gaussian noise channel (AWGNC) the best code of rate one-half presented there has a threshold within 0.06dB of capacity, and simulation results demonstrate a LDPC code of length 10^6 which achieves a bit error probability of 10^{-6}, less than 0.13dB away from capacity. Recent improvements have demonstrated thresholds within 0.012dB of capacity [5]. These results strongly indicate that LDPC codes can approach Shannon capacity. As pointed out above, only in the case of the BEC has such a result actually been proved [23]. Resolving the question for more general channels remains one of the most challenging open problems in the field.

In the case of turbo codes the same general theory applies, although there are some additional technical problems to be overcome. In the setting of turbo codes belief propagation corresponds to the use of the BCJR algorithm for the decoding of the component codes together with an exchange of *extrinsic information*. This is generally known as "turbo decoding".[2]

---

[1]The recognition of the sum-product algorithm as an instance of belief propagation was made by Frey and Kschischang [12] and also by McEliece, Rodemich, and Cheng [26].

[2]The original incarnation of turbo decoding in [4] was not belief propagation, Robert-

Thus, one can similarly explore turbo codes and generalizations of turbo codes from the perspective of threshold behavior. It is possible to describe an algorithm for computing thresholds for turbo codes but such an algorithm appears computationally infeasible except in the simplest cases. Fortunately, it is possible to determine thresholds to any desired degree of accuracy using Monte-Carlo methods. The putative deterministic algorithm used to compute thresholds can be mimicked by random sampling. One can prove that certain ergodic properties of the computation guarantee convergence of the Monte-Carlo approach (assuming true randomness) to the answer that would have been determined by the exact algorithm. Moreover, all of the information used to optimize LDPC codes is available from the Monte-Carlo approach and, therefore, it is possible to optimize thresholds for various extensions of turbo codes. Generalizations of turbo codes appear to exhibit thresholds approaching Shannon capacity. The work described here appears in [31].

In a very precise sense, determining the threshold by the methods indicated above corresponds to modeling how decoding would proceed on an infinitely long code. One determines not merely the threshold, but the statistics of the entire decoding process. Decoding proceeds in discrete time by passing messages along edges in a graph. In the infinite limit one considers the distribution of the messages (pick an edge uniformly at random, what message is it carrying?) These distributions are parameterized in a suitable fashion and the algorithms mentioned above iteratively update the distributions in correspondence with iterations of the decoding process. The sequence of distributions so obtained and the method used to obtain them are referred to as *density evolution*. Clearly, density evolution is key to understanding the decoding behavior of such systems and the study of density evolution is a fundamental outgrowth of the general theory.

Our purpose in this paper is to provide the reader with a vehicle for quickly grasping the key features, assumptions, and results of the general theory. The paper consists of further details and examples intended to be sufficient to equip the reader with a practical overview of the current state of knowledge as embodied in this theory. The paper is loosely organized along the lines of assumptions and conclusions: Each section is devoted to stating assumptions required for some part of the theory, usually some examples, to stating what conclusions can be drawn, and indicating what mathematical techniques are used to draw the conclusions. We have ordered material from most general to most specific. Rather that attempting to present the most general, all-encompassing, form of the theory, we have tried to make the main ideas as accessible as possible.

2. Graphical representations of codes. The mathematical framework described in this paper applies to various code constructions based on graphs. To keep notation to a minimum, we will restrict our attention in

---

son [33] refined the algorithm into a form equivalent to belief propagation.

this paper to the standard parallel concatenated code with two component codes and to LDPC codes. For a more detailed treatment of turbo codes which includes also serially concatenated codes and generalized turbo codes we refer the reader to [31]. The theory developed here also extends to much more general graphical representations such as those developed by, e.g., N. Wiberg and H.-A. Loeliger and R. Kötter [44], Kschischang, Frey, Loeliger [21], Kschischang and Frey [20], or Forney [11]. LDPC codes as well as turbo codes can be represented using quite simple graphs and the decoders of interest operate directly and locally on these graphs. (In the case of turbo codes one should bear in mind *windowed* decoding. For a description of windowed decoding see Section 5.2. The theory extends to standard, i.e., non-windowed, turbo decoding by taking limits but the analogy between LDPC codes and turbo codes is closer in the case of windowed decoding.)

The theory addresses *ensembles of codes* and their associated *ensembles of graphs*. For block codes the idea of looking at *ensembles* of codes rather than individual codes is as old as coding theory itself and originated with Shannon. In the setting of turbo codes this idea was introduced by Benedetto and Montorsi [3] and it was motivated by the desire to bound the maximum likelihood performance of turbo codes.[3] The ensembles are characterized by certain fixed parameters which are independent of the length of the code. The graphs associated to LDPC codes, for example, are parameterized by their degree sequences (λ, ρ) (see below for details). The graphs associated to turbo codes are parameterized by the polynomials determining the constituent codes, their interconnection structure, i.e., parallel vrs. serial, and puncturing patterns. Given the size of the code n, these fixed parameters determine the number of the various node types in the graph. The ensemble of codes is defined in terms of the possible edges which complete the specification of the graph. In both of the above cases, the graphs are bipartite: One set of nodes, the *variable nodes*, corresponds to variables (bits) and the other set, the *check nodes*, corresponds to the linear constraints defining the code.

In general, the fixed parameters and the length n determine the nodes of the graph and a collection of permissible edges. One then considers the set of all permissible edge assignments and places on them a suitable probability distribution. The theory addresses properties of the ensemble as n gets large.

2.1. Ensembles of codes and graphs. We shall now present more detailed definitions for code ensembles of LDPC codes and parallel concatenated codes. These examples, although important, are not exhaustive. Note that the local connectivity of the graphs remains bounded independent of the size of the graph. This is the critical property supporting the

---

[3]A considerable amount of additional research has been done in the direction of bounding the maximum likelihood performance of a code from information on its spectrum, see e.g. [10, 42, 34, 35, 9].

asymptotic analysis, i.e., the concentration theorem.

2.1.1. LDPC codes. As described above, low-density parity-check codes are well represented by bipartite graphs in which the variable nodes corresponds to elements of the codeword and the check nodes correspond to the set of parity-check constraints satisfied by codewords of the code. *Regular* low-density parity-check codes are those for which all nodes of the same type have the same degree. Thus, a (3,6)-regular low-density parity-check code has a graphical representation in which all variable nodes have degree three and all check nodes have degree six. The bipartite graph determining such a code is shown in Fig. 1.

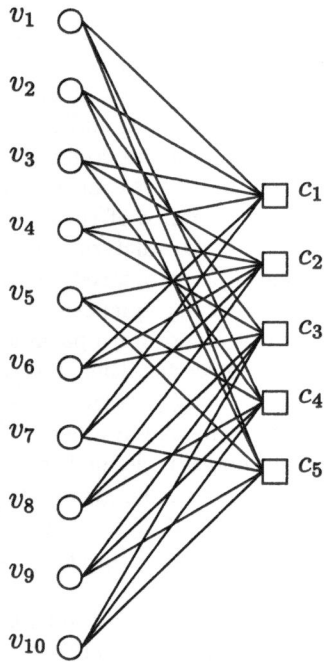

FIG. 1. *A* (3,6)-*regular code of length 10 and rate one-half. There are 10 variable nodes and 5 check nodes. For each check node c_i the sum (over GF(2)) of all adjacent variable nodes is equal to zero.*

For an *irregular* low-density parity-check code the degrees of each set of nodes are chosen according to some distribution. Thus, an irregular low-density parity-check code might have a graphical representation in which half the variable nodes have degree three and half have degree four, while half the constraint nodes have degree six and half have degree eight. For a given length and a given degree sequence (finite distribution) we define an *ensemble* of codes by choosing the edges, i.e., the connections between variable and check nodes, randomly. More precisely, we enumerate the edges emanating from the variable nodes in some arbitrary order and proceed in the same way with the edges emanating from the check nodes. Assume that the number of edges is E. Then a code (a particular instance of this ensemble) can be identified with a permutation on E letters. Note that all elements in this ensemble are equiprobable. In practice the edges are never chosen entirely randomly since, e.g., certain potentially unfortunate events, such as double edges and very short loops, in the graph construction can be easily avoided.

Hence, for a given length n the ensemble of codes will be determined once the various fractions of variable and check node degrees have been specified. Although this specification could be done in various ways the following notation introduced in [22] leads to particularly elegant statements of many of the most fundamental results. Let d_l and d_r denote the maximum variable node and check node degrees, respectively, and let $\lambda(x) := \sum_{i=1}^{d_l} \lambda_i x^{i-1}$ and $\rho(x) := \sum_{i=1}^{d_r} \rho_i x^{i-1}$ denote polynomials with non-negative coefficients such that $\lambda(1) = \rho(1) = 1$. More precisely, let the coefficients, λ_i (ρ_i) represent the fraction of *edges* emanating from variable (check) nodes of degree i. Then, clearly, this *degree sequence pair* (λ, ρ) completely specifies the distribution of the node degrees. The alert reader may have noticed several curious points about this notation. First, we do not specify the fraction of *nodes* of various degrees but rather the fraction of *edges* that emanate from nodes of various degrees. Clearly, it is easy to convert back and forth between this *edge perspective* and a *node perspective*. E.g., assume that half the variable nodes have degree three and half have degree four and that there is a total of n nodes. Since every degree three node has three edges emanating from it, whereas every degree four nodes has four edges emanating to it we see that there are in total $1/2 \cdot 3n$ edges which emanate from degree three nodes and that there are in total $1/2 \cdot 4n$ edges which emanate from degree four nodes. Therefore $\lambda_3 = \frac{1/2 \cdot 3}{1/2 \cdot 3 + 1/2 \cdot 4} = 3/7$ and $\lambda_4 = \frac{1/2 \cdot 4}{1/2 \cdot 3 + 1/2 \cdot 4} = 4/7$ so that in this case $\lambda(x) = 3/7x^2 + 4/7x^3$. Second, the fraction of edges which emanate from a degree i node is the coefficient of x^{i-1} rather than x^i as one might expect at first. The ultimate justification for this choice comes from the fact that, as we will see later, simple quantities like $\lambda'(0)$ or $\rho'(1)$ take on an operational meaning. A particular striking example of the elegance of this notation is given by the *stability condition* which we will discuss in Section 8.3. This condition takes on the form $\lambda'(0)\rho'(1) < g(\sigma)$ where g is a function of the channel parameter σ only.

2.1.2. Turbo codes. For every integer n we define an ensemble of standard parallel concatenated codes in the following manner. We first fix the two rational functions $G_1(D) = \frac{p_1(D)}{q_1(D)}$ and $G_2(D) = \frac{p_2(D)}{q_2(D)}$ which describe the recursive convolutional encoding functions. Although the general case does not pose any technical difficulties, in order to simplify notation, we will assume that all codewords of a convolutional encoder start in the zero state but are not terminated. For $x \in \{\pm 1\}^n$ let $\gamma_i(x)$, $i = 1, 2$, denote the corresponding encoding functions. Then for fixed component codes and a given permutation π on n letters the unpunctured codewords of a standard parallel concatenated code have the form $(x, \gamma_1(x), \gamma_2(\pi(x)))$. Therefore, for fixed component codes and a fixed puncturing pattern there is a one-to-one correspondence between permutations on n letters and codes in the ensemble. We will assume a uniform probability distribution on the set of such permutations. This is the same ensemble considered in [3] but the present focus is on the analysis of the performance of turbo codes under iterative decoding rather than under maximum likelihood decoding.

The graphical representation of the code contains variable nodes, as in the LDPC code case, and check nodes, which, in this case, represent a large number of linear constraints on the bits associated to the variable nodes. A check node represents the linear constraints imposed by an entire constituent code. Equivalently, it represents the trellis which in turn represents the constituent code. Hence, for standard parallel concatenated codes with two component codes, there are only two check nodes.

3. Decoding: Symmetries of the channel and the decoder. In this paper we will limit ourselves to the case of binary codes and transmission over memoryless channels since in this setting all fundamental ideas can be represented with a minimum of notational overhead. The generalization of the theory to larger alphabets [7] or channels with memory [14, 16, 15, 18, 25] is quite straightforward and does not require any new fundamental concepts. As usual for this case, we will assume antipodal signalling, i.e., the channel input alphabet is equal to $\{\pm 1\}$.

The decoding algorithms of interest operate directly on the graph, described in Section 2.1, that represents the code. The algorithms are localized and distributed: edges carry messages between nodes and nodes process the incoming messages received via their adjacent edges in order to determine the outgoing messages. The algorithms proceed in discrete steps, each step consisting of a cycle of information passing followed by processing. Generally speaking, computation is memoryless so that, in a given step, processing depends only on the most recent information received from neighboring nodes. It is possible to analyze decoders with memory but we will not consider such decoders here. Given the above setup we distinguish between two types of processing which may occur according to the dependency of the outgoing information. When the outgoing information along an edge depends only on information which has come in along *other* edges, then we say that the algorithm is a *message-passing* algorithm. The sum-product algorithm is the most important example of such an algorithm. The most important example of a non message-passing algorithm is the *flipping* algorithm. This is a very low complexity hard-decision decoder of LDPC codes in which bits at the variable nodes are 'flipped' in a given round depending on the number of *unsatisfied* and *satisfied* constraints they

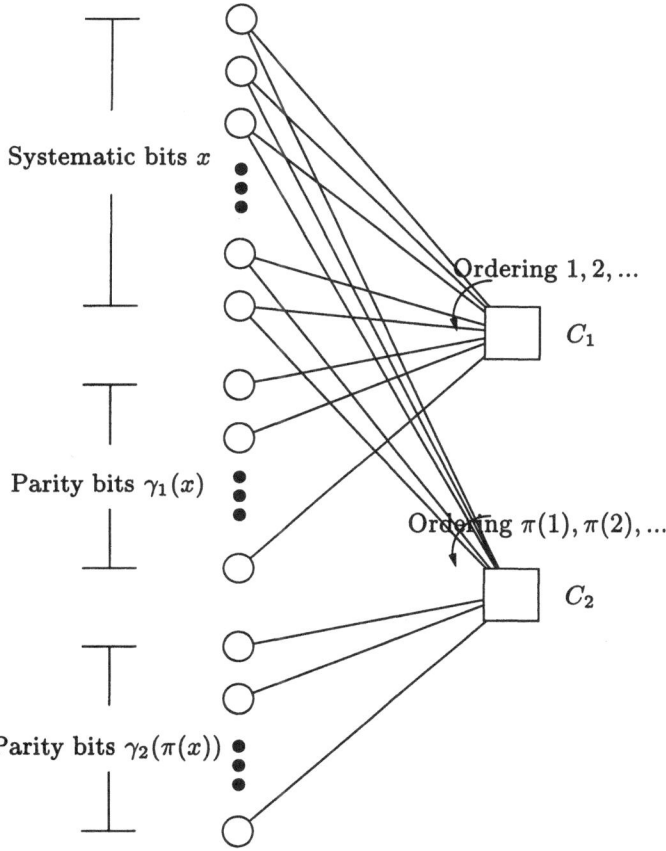

FIG. 2. *A graphical representation of a standard parallel concatenated code analogous to the bipartite graph of a LDPC code.*

are connected to. Here, we say that a constraint is satisfied if and only if the modulo two sum of its neighbor bits is 0. We note that the techniques used to analyze these two types of decoders are quite distinct and the nature of statements which can be made tend to differ significantly. (Nevertheless, certain aspects of the theory outlined here, in particular the concentration results, carry over to many variants of flipping.) The statements usually made for the flipping algorithm are more reminiscent of traditional coding theory in that they assert that for all error patterns of weight less than a given threshold the decoder will decode correctly [45, 38]. For message-passing decoders the overall focus is on the resulting bit error probability. A large fraction of high weight error patterns might be correctable but occasionally even low weight error patterns may lead to errors. Currently the best coding schemes are based on message-passing decoders but the flipping algorithm is often of interest because of its inherent extremely low complexity.

It is helpful to think of the messages (and the received values) in the following way. Each message which traverses an edge (v, c), in either direction, represents an estimate of the particular bit associated to v. More precisely, it contains an estimate of its *sign* and, possibly, some estimate of its *reliability*. To be concrete, consider a discrete case in which the received alphabet, the alphabet of estimates provided to the decoder from the channel, is $\mathcal{O} := \{-q_o, -(q_o-1), \ldots, -1, 0, 1, \ldots, (q_o-1), q_o\}$ and where the message alphabet is $\mathcal{M} := \{-q, -(q-1), \ldots, -1, 0, 1, \ldots, (q-1), q\}$ where we assume $q \geq q_o$. The sign of the message indicates whether the transmitted bit is assumed to be -1 or $+1$, and the absolute value of the message is a measure of the reliability of this estimate. In particular, the value 0 represents an erasure. In the continuous case we may assume that $\mathcal{O} = \mathcal{M} = \mathbb{R}$. Again, the sign of the message indicates whether the transmitted bit is assumed to be -1 or $+1$, and the absolute value of the message is a measure of the reliability of this estimate. Of course, it is not necessary that 0 be included in either alphabet, nor is it necessary that the messages actually have this symmetric form, but they should be representable as such.

3.1. Restriction to the all-one codeword. For all algorithms of interest in the current context computation is performed at the nodes based on information passed along the edges. Thus, for every occurring variable degree d we have a map $\Psi^d_{\text{left}} : \mathcal{O} \times \mathcal{M}^d \to \mathcal{M}^d$ to represent the computation done at a variable node of degree d. Similarly, for every occurring check node of degree d we have a map $\Psi^d_{\text{right}} : \mathcal{O} \times \mathcal{M}^d \to \mathcal{M}^d$ to represent the computation done at a check node of degree d. (In some cases, i.e., turbo codes, outgoing messages to variable nodes of degree one have no effect on subsequent messages and are therefore usually not actually computed.) Although it is not necessary, one will usually have symmetry under permutation of the input messages so that if $\Psi^d_{\text{left}}(u_0, u_1, ..., u_d) = (w_1, ..., w_d)$ then $\Psi^d_{\text{left}}(u_0, u_{\pi(1)}, ..., u_{\pi(d)}) = (w_{\pi(1)}, ..., w_{\pi(d)})$ for any permutation π on d letters. In the case of LDPC codes we also have such symmetry for Ψ_{right} but for turbo codes, because of trellis termination, we do not.[4]

Under the message-passing paradigm the message maps have the further property that the i-th outgoing message does not depend on the i-th incoming message. In this case, and under the permutation symmetry assumption, we have functions $M^d_{\text{left}} : \mathcal{O} \times \mathcal{M}^{d-1} \to \mathcal{M}$ such that

$$\Psi^d_{\text{left}}(u_0, u_1, ..., u_d) =$$
$$(M^d_{\text{left}}(u_0, u_2, ..., u_d), M^d_{\text{left}}(u_0, u_1, u_3, ..., u_d), ..., M^d_{\text{left}}(u_0, u_1, ..., u_{d-1})).$$

---

[4] Asymptotically, however, some permutation symmetry is present in the turbo code case but the permutations must be restricted to, e.g., the information (x) bits.

Similarly, under permutation symmetry at the check nodes, we shall have a map $M_{\text{right}}^d : \mathcal{O} \times \mathcal{M}^d \to \mathcal{M}$.

Our subsequent analysis and notation will be greatly simplified by assuming the following symmetry conditions on the channel and the decoding algorithm.

- **Channel symmetry:** $p(Y = q | X = 1) = p(Y = -q | X = -1)$.
- **Check node symmetry:** If $\Psi_{\text{right}}^d(u_1, ..., u_d) = (w_1, ..., w_d)$ then $\Psi_{\text{right}}^d(b_1 u_1, ..., b_d u_d) = (b_1 w_1, ..., b_d w_d)$ for any ± 1 sequence $(b_1, ..., b_d)$ satisfying the constraint associated to the check node.
- **Variable node symmetry:**

$$\Psi_{\text{left}}^d(-u_0, -u_1, ..., -u_d) = -\Psi_{\text{left}}^d(u_0, u_1, ..., u_d).$$

We claim that under these conditions the (bit or block) error probability is independent of the transmitted codeword. For a proof in the message-passing case see [30]. The reader may recognize the channel symmetry condition as the condition that the channel be *output-symmetric*. Note that *linear* codes are known to be capable of achieving capacity for binary-input memoryless output-symmetric channels.[5] This is reassuring, since our search for low complexity coding schemes which can approach capacity takes place in the class of linear codes.

The great simplification which accrues from the symmetry assumption is that one need only determine the probability of error for a single codeword, the all-one codeword.[6] In this case, messages are 'correct' when their sign is positive and 'incorrect' when their sign is negative. When we consider the distribution of the messages we invoke the all-one codeword assumption so that the probability of an incorrect message is just the probability mass of the distribution supported on negative values. In other words, we are really interested in tracking the distribution of messages *relative to the transmitted codeword*. Under the symmetry assumptions this takes a particularly simple and appealing form.

In the sequel we will assume that the symmetry conditions are fulfilled and that the all-one codeword was transmitted. We will now present some examples of channels and decoders satisfying the above conditions

EXAMPLE 1 (BSC, LDPC code, Gallager A). *The channel provides only hard information* $\mathcal{O} = \{-1, +1\}$ *and the messages are also binary* $\mathcal{M} = \{-1, +1\}$. *The decoding is message passing with* $M_{\text{left}}^d(u_0, u_1, ..., u_{d-1})$ $= -u_0$ *if* $u_1 = u_2 = ... = u_{d-1} = -u_0$, *and, otherwise,* $M_{\text{left}}^d(u_0, u_1, ..., u_{d-1})$ $= u_0$. *The check node rule is given by* $M_{\text{right}}^d(u_1, ..., u_{d-1}) = u_1 u_2 ... u_{d-1}$. □

For this decoder thresholds and optimal codes can be analytically determined in many cases, see [2].

---

[5] The key for the proof of this statement lies in the observation that in this case an input distribution of $[1/2, 1/2]$ is optimal.

[6] This should be compared to the concept of *geometrically uniform* codes in the setting of maximum likelihood decoding.

EXAMPLE 2 (BEC, LDPC codes, Belief Propagation). *The channel either correctly relays the transmitted bit or this bit is erased. These events happen with probability $(1-\delta)$ and δ, respectively. Thus, assuming that the output of the channel is represented as log-likelihood ratios we have $\mathcal{O} = \{-\infty, 0, \infty\}$. Because of the special structure of this channel we have $\mathcal{M} = \mathcal{O}$, where the messages are again represented as log-likelihood ratios. The decoding is message passing with $M_{left}^d(u_0, u_1, ..., u_{d-1}) = \sum_{i=0}^{d-1} u_i$. Note that this rule is well defined since, by construction, the sum can never contain $+\infty$ and $-\infty$ together. The check node rule is given by*

$$M_{right}^d(u_1, ..., u_{d-1}) = \prod_{i=1}^{d-1} u_i \,,$$

where $0 \cdot \infty = 0$.   □

EXAMPLE 3 (BSC, LDPC codes, Belief Propagation). *The channel provides only hard information but associated to these values is the reliability magnitude given by $\log((1-\epsilon)/\epsilon)$ where ϵ is the cross-over probability of the BSC. Thus, we have $\mathcal{O} = \{-\log((1-\epsilon)/\epsilon), \log((1-\epsilon)/\epsilon)\}$. Messages, like the received values, are log-likelihoods, hence $\mathcal{M} = \mathbb{R}$. The decoding is message passing with $M_{left}^d(u_0, u_1, ..., u_{d-1}) = \sum_{i=0}^{d-1} u_i$. The check node rule is given by $M_{right}^d(u_1, ..., u_{d-1}) = 2 \tanh^{-1}(\prod_{i=1}^{d-1} \tanh(u_i/2))$. The decoding algorithm described in this example can be applied in general provided the received alphabet is mapped into the associated log-likelihood representation.*
□

EXAMPLE 4 (AWGNC, Turbo Codes, Belief Propagation). *We assume the channel provides the log-likelihood estimate of the bit based on the observation $y = x + z$ where $x \in \{\pm 1\}$ is the transmitted bit and z is zero mean Gaussian with variance σ^2. The log-likelihood estimate is given by $2\sigma^{-2}y$. Messages are log-likelihoods so $\mathcal{O} = \mathcal{M} = \mathbb{R}$. The decoding is message passing with $M_{left}^d(u_0, u_1, ..., u_{d-1}) = \sum_{i=0}^{d-1} u_i$. The check node rule is given by APP decoding of the associated constituent code, which corresponds to performing the BCJR algorithm on the trellis and outputting the extrinsic information.*   □

The final example is a flipping type algorithm. Various flipping algorithms have been considered in the literature. Some, such as list-based flipping, do not directly fit into our general description. This is because, in these algorithms, the decision on whether to flip a bit or not is based on a global criterion. Nevertheless, such an algorithm can be mimicked to any degree of accuracy by properly chosen local algorithms so that the analysis can still be applied to such algorithms by taking limits. We present here the flipping algorithm in a somewhat unusual form.

EXAMPLE 5 (BSC, LDPC codes, Flipping). *We assume a (d_v, d_c)-regular LDPC code and transmission over a BSC. Hence, the channel provides only hard information $\mathcal{O} = \{-1, +1\}$ and the messages are also binary*

$\mathcal{M} = \{-1, +1\}$. *We are given a threshold t and*

$$\Psi_{left}^{d_v}(u_0, u_1, ..., u_{d_v}) = (u_0, u_0, ..., u_0)$$

if $|\{u_i : u_i = u_0\}| > t$ and

$$\Psi_{left}^{d_v}(u_0, u_1, ..., u_{d_v}) = (-u_0, -u_0, ..., -u_0)$$

otherwise. For the check nodes we have

$$\Psi_{right}^{d_c}(u_1, ..., u_{d_c}) = (u_1 p, u_2 p, ..., u_{d_v} p)$$

where $p = \prod_{i=1}^{d_c} u_i$. □

Note that the check node message map satisfies the message-passing paradigm but the variable node message map does not. In general t will depend on the iteration number and on the degree of the node. To approximate list based flipping algorithms we might also allow t to depend on some exogenous random variable.

4. Decoding: Localization and concentration. As we have seen, the coding systems we consider can be represented as graphs on which the decoders directly operate. More precisely, decoding consists of *local* computations at the nodes in the graph together with exchange of information along the edges of the graph. Time is measured discretely so that if decoding proceeds through ℓ time steps then any random variable associated to a node v in the graph can influence computations at only those nodes whose distance in the graph from the associated node is ℓ, i.e., an influenced node must be able to reach to v in ℓ steps along the graph. Because of this, and because of the boundedness of the degree of the graph the range of influence of local properties (either of the graph or of received data) remains bounded independent of the size of the graph given a bounded number of iterations.

In this paper we focus mainly on message-passing algorithms although, as pointed out previously, there are decoding algorithms of interest, e.g., the flipping algorithm, which do not fit directly into this framework. Our main reason for this restriction is that the class of message-passing decoders contains the locally-optimal decoder, namely belief-propagation, as well as many other very good and low-complexity decoders. Further, the class of message-passing decoders can be analyzed in a unified manner whereas other decoders, like the flipping algorithm, require quite distinct methods for their analysis. Nevertheless, it is worth pointing out that the concentration theorem, which we will discuss in this section, in its most general form applies to *all* types of local algorithms and not only to message-passing algorithms.

What is the role of the concentration theorem in the context of iterative coding systems? Consider, e.g., the bit error probability for a particular code and decoding algorithm. Clearly, this is a quantity of great practical significance. Choosing a code from an ensemble amounts to choosing one of the possible random instances for the graph. E.g., in the case of turbo codes a considerable amount of attention has been paid to the issue of choosing a good interleaver. For short lengths different randomly chosen instances of the interleaver may exhibit significantly different behavior. It has been observed, however, that as the codes get longer the variation among interleavers becomes less significant (although it is easy to find bad interleavers). This is a direct consequence of the concentration theorem which asserts that, as n increases, certain quantities, such as the average decoded bit error probability, *concentrate*: In the probabilistic sense, different instances of the graph and different channel realizations will behave similarly with respect to average quantities. It should be understood that the rate at which this concentration occurs is not well predicted by the current theory: The theory predicts exponential convergence in n but the actual exponent might differ significantly from the bounds which one derives. To make an analogy, the concentration theorem plays a similar role for iterative coding systems that the asymptotic equipartition theorem plays in much of information theory.

The method of analysis has extremely broad application since it requires very few assumptions. The mathematical technique which leads to the concentration theorem is now virtually standard among probabilistic techniques used to analyze combinatorial problems. In the next section we shall present the technique as applied to graph coloring. For an extended introduction to the iterative decoding application we refer the reader to [32]. The migration of the technique into coding theory was initiated in the work of Luby *et. al.* [22]. They analyzed LDPC codes in an essentially combinatorial setting: transmission over the binary symmetric channel (BSC) or the binary erasure channel (BEC) with hard decision message-passing decoding. In [30] the basic technique, as applied to LDPC codes, was significantly generalized to encompass essentially any message-passing decoding algorithm, including belief propagation, and any memoryless channel.

The general concentration theorem itself consists of two distinct parts. The first part shows convergence of the expected performance to the asymptotic value, the second part shows concentration of the performance around the mean as a function of length. This concentration is further separated into concentration over input noise and concentration over the random ensemble of graphs. In practice it is the concentration over the random ensemble of graphs that is of most interest. Ideally one would like to find the graph that offers the best performance averaged over the channel noise statistics. Simulation performance curves convey precisely this information.

The key asymptotic behavior required for the first part is that, as n increases, the graphs become more and more locally tree-like, i.e., loop-free. Although it is readily apparent that this holds in the case of LDPC codes it is not so apparent for the case of turbo codes. In fact, as initially defined, the graphs associated to turbo codes do not exhibit this tree-like asymptotic. Nevertheless, the theory applies to this case because dependencies among variables at the check nodes, i.e., in the trellis, decay with 'distance' along the trellis. In the case of windowed decoding this is precise: edges sufficiently well separate participate in non-overlapping disjoint windows and, assuming incoming messages are independent, their outgoing messages will be independent. Since we fix the number of iterations and let n grow, the expected fraction of nodes whose decoding neighborhood is not tree-like decays as n^{-1}. Hence, the expected bit error rate converges to that for a random tree.

The key asymptotic behavior required for the second part is that graphs which are strongly similar, i.e., have most edges in common, will perform similarly on the same input. This can be seen most easily as follows. Consider a LDPC code and its associated graph. Imagine altering the graph by swapping the connections of two edges, i.e., replace edges $(v_1, c_1), (v_2, c_2)$ with $(v_1, c_2), (v_2, c_1)$. Now compare results when decoding a particular input. Since decoding proceeds locally and we consider only a finite number of iterations, the effect of the swap is bounded in the graph, limited only to those nodes and variables that are sufficiently close to v_1, c_1, v_2, and c_2. The larger the graph, the smaller will be the fraction of the affected portion. If we look at, e.g., the number of errors in the decoding, then the difference between the two graphs is bounded by a constant independent of n. In the following section we describe the mathematical technique which exploits this property.

4.1. Doob martingales and Azuma's inequality.

Although we are most interested in applying the mathematical techniques described in this section to iterative decoding, they are most conveniently presented via their application to graph coloring. It was in this application, due to Shamir and Spencer [36], that the method first made its impact in combinatorics.

Consider a random graph on n vertices in which each possible edge, i.e., vertex pair, is independently admitted with probability p. Consider the problem of coloring the vertices of a randomly constructed such graph so that no two vertices connected by an edge have the same color. The number of colors required is called the *chromatic number* of the graph H and is denoted by $\chi(H)$. What Shamir and Spencer showed, in effect, is that the chromatic number $\chi(H)$, viewed as a random variable, is tightly concentrated around its expected value $\mathbb{E}[\chi(H)]$. Perhaps most striking, this result was obtained without the determination of $\mathbb{E}[\chi(H)]$.

The argument used is now known as a *vertex exposure* martingale argument. Let H denote a graph constructed randomly as above. Pick an arbitrary ordering of the n vertices. Imagine that the graph H is unknown but that vertices will be *revealed* one at a time. Each time a vertex is revealed all edges connected to previously revealed vertices are revealed, i.e., when vertex i is revealed the presence or absence of the edge (j, i) for $j < i$ is revealed. Let $I_{j,i}$ denote a random variable which indicates whether the edge (j, i) is present or not. Let $X_0(H)$ denote the expected chromatic number of H when nothing has been revealed, i.e., $X_0(H) = \mathbb{E}[\chi(H)]$. Let $X_i(H)$ denote the expected chromatic number of H conditioned on the information obtained after revealing the first i vertices of H, i.e., $X_i(H) = \mathbb{E}[\chi(H)|\{I_{j,k}\}_{1 \leq j < k \leq i}]$. Thus, in particular, $X_n(H) = \chi(H)$. Now, recall that H is a random variable and therefore $X_i := X_i(H)$ is a random variable as well (with the randomness residing in the set of random variables $\{I_{j,k}\}_{1 \leq j < k \leq i}$). Note that

$$\mathbb{E}[X_{i+1}|\{I_{j,k}\}_{1 \leq j < k \leq i}]$$
$$= \mathbb{E}[\mathbb{E}[\chi(H)|\{I_{j,k}\}_{1 \leq j < k \leq i+1}]|\{I_{j,k}\}_{1 \leq j < k \leq i}]$$
$$= \mathbb{E}[\chi(H)|\{I_{j,k}\}_{1 \leq j < k \leq i}]$$
$$= X_i$$

so that the sequence $\mathbb{E}[\chi(H)] = X_0, X_1, ..., X_n = \chi(H)$ forms a martingale. More precisely, a martingale constructed this way, as sequence of conditional expectations where the information conditioned on is increasing, is usually called a *Doob's Martingale* [27, p. 90]. If we take the graph H and modify the edges involving the i-th vertex in an arbitrary way to obtain a graph G, then $|\chi(H) - \chi(G)| \leq 1$: given a coloring of H we need at most one more color to color G and vice versa. It follows that $|X_i(H) - X_{i-1}(H)| \leq 1$ since $X_{i-1}(H)$ is an average of all possible values for $X_i(H)$, hence, $|X_i - X_{i-1}| \leq 1$.

In the case of a random walk, $X_i = \sum_{j=1}^{i} Z_j$ where the Z_j are independent random variables with finite expectation, we can use the Chernoff bound to give an exponential bound on the probability that X_i will deviate from its mean by more than a fraction ϵ. Azuma's inequality provides a similar bound for the case of dependent random variables assuming that the sequence of these random variables forms a martingale.

THEOREM 4.1 (Azuma's Inequality). *Let $X_0, X_1, ...$ be a martingale sequence such that for each $k \geq 1$,*

$$|X_k - X_{k-1}| \leq \alpha_k,$$

where the constant α_k may depend on k. Then, for all $l \geq 1$ and any $\lambda > 0$

$$Pr\{|X_l - X_0| \geq \lambda\} \leq 2e^{-\frac{\lambda^2}{2\sum_{k=1}^{l} \alpha_k^2}}.$$

The sequence $X_0, ..., X_n$ constructed above satisfies the conditions required by Azuma's inequality with $\alpha_k = 1$. Since $X_0(H) = \mathbb{E}[\chi(H)]$ and $X_n(H) = \chi(H)$, we obtain

$$Pr[|\chi(H) - \mathbb{E}[\chi(H)]| > \lambda\sqrt{n-1}] \leq 2e^{-\lambda^2/2}.$$

This is Shamir and Spencer's result [36]. For further applications see [27, 1].

4.1.1. Application to iterative decoding. The application of the technique described above to iterative decoding involves some additional complexities but follows essentially the same lines of argument. For both LDPC codes and turbo codes, once the nodes are fixed, the ensemble of graphs is in one-to-one correspondence with the set of permutations on m letters for some m. A Doobs martingale is formed: conceptually, we reveal the permutation one element at a time. In the case of LDPC codes this corresponds to revealing one edge in the graph. In the case of turbo codes this corresponds to revealing a pair of edges in the graph. As the permutation is revealed we look at the expected number of decoding errors. The martingale may be extended by revealing received values after the graph is fully revealed. In this way one can prove concentration not only of average performance but also concentration of performance on individual inputs.

When revealing the graph, the bound required for Azuma's inequality is derived using the swapping argument indicated earlier. Different possibilities for the revealed edges are compared by identifying the continuations of the revelation process through swapping. In this way the impact of the revealed edge can be shown to be bounded. When revealing the received values the argument is easier. Since we consider a finite number of iterations, the effect of any received value is restricted to a finite portion of the graph independent of n.

Let $Z_\ell[n]$ be the random variable denoting the fraction of incorrect messages passed in the ℓ-th iteration of a randomly chosen code and set of received values for a code of length n. For either LDPC codes or turbo codes one obtains a theorem of the following form.

THEOREM 4.2. *For any ℓ there exists a constant β such that the probability that $|Z_\ell[n] - \mathbb{E}[Z_\ell[n]]| > \epsilon$ is less than $e^{-\beta\epsilon^2 n}$ Furthermore, $Z_\ell[\infty] := \lim_{n\to\infty} \mathbb{E}[Z_\ell[n]]$ exists and there exists a constant γ such that $|\mathbb{E}[Z_\ell[n]] - Z_\ell[\infty]| < \frac{\gamma}{n}$.*

5. The support tree of message-passing decoders. In anticipation of the analysis of message-passing algorithms, we shall describe in some detail the dependency structure of messages in the graphs of interest under the message-passing paradigm. The extension to non-message-passing is straightforward and we shall not consider it in detail.

Consider a message passed along an edge during a message-passing decoding. Depending on the iteration number this message will, in general, be a function of some subset of the received values and some sub-graph of the graph defining the code. The subset of received values on which the message depends are those associated to variable nodes of the sub-graph. Thus, we can view the message simply as a function of the sub-graph. Note that the sub-graph depends only on the iteration number and not on the particular message-passing algorithm used. (In the case of turbo codes one should bear in mind windowed decoding of width $W = 2w + 1$.) Since the message traverses the edge in a particular direction the sub-graph depends on this direction. If the iteration number is ℓ then the sub-graph is called the ℓ-directed neighborhood of the (directed) edge.

By proceeding backwards in time, one can 'unvolve' the directed neighborhood of the edge in terms of iteration number. Described in more detail below, the unvolved graph can be conveniently pictured as a tree and we refer to such a tree as a support tree. In general the support tree is not actually a tree because certain elements of the tree, i.e., nodes or edges, may be replicated.

If there are no repetitions in the support tree, then we say that the corresponding ℓ-directed neighborhood of the edge is a *tree*. In all cases of interest, as n increases the probability that the directed neighborhood is a tree approaches one at a rate of $O(1/n)$. The underlying reason for this property is the boundedness of the degrees of the nodes in the graph. To see this: imagine picking a node and revealing its neighbors up to distance ℓ in order of increasing distance, then, at any point there will be at most a constant number of edge placements which will create a loop. The total number of possible edge placements grows linearly in n and the total number of edges to be revealed is bounded by a constant independent of n. Since the connections are chosen randomly it follows that the probability of creating a loop decays like $1/n$.

5.1. Support trees of LDPC codes.

For simplicity we consider a (d_v, d_c)-regular LDPC code. Consider a randomly chosen edge $e = (v, c)$ from the graph and consider the message passed along e from v to c in the ℓ-th iteration under message-passing decoding. This message is a function of the received value at v and the messages arriving at v in the $(\ell - 1)$-th iteration along its *other* edges. These other edges are connected to some collection of constraint nodes. For each such constraint node and connecting edge the message sent to v along that edge in the $(\ell - 1)$-th iteration is a function of the messages sent to this constraint node along its *other* edges. Continuing in this recursive fashion we can determine the support tree of the original message under consideration by unvolving the graph, tracing back the dependency of the message on previous messages. Fig. 3 gives a pictorial representation of the support tree for a $(3, 6)$ LDPC code over one and a half iterations.[7] Given the appropriate labelling of the nodes, this picture represents the sub-graph of the original graph on which the message passed from v to c depends.

If we write the directed neighborhood in the form of a support tree rooted at e, as in Figure 3, then, under message passing, information flows up the tree from the leaves to produce the message carried along e. Messages are passed up the tree: a message from a node is a function only of messages which come from below and, in the case of variable nodes, the received value

---

[7]Although depicted as a tree, some of the nodes and edges in the unvolved picture may be identical to other nodes and edges.

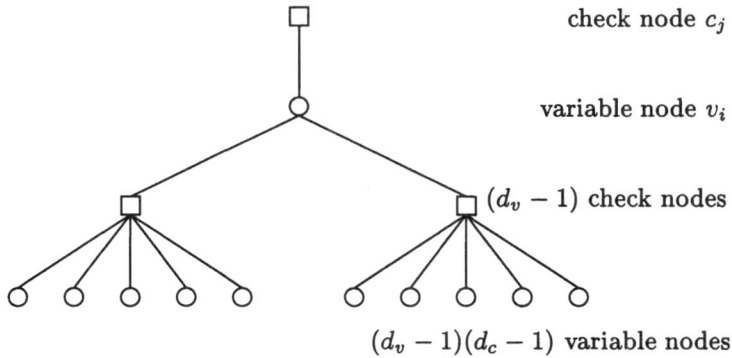

check node c_j

variable node v_i

$(d_v - 1)$ check nodes

$(d_v - 1)(d_c - 1)$ variable nodes

FIG. 3. *The directed neighborhood of depth 2 of the edge* $e = (v_i, c_j)$ *for an LDPC code.*

associated to the node.

In general, for fixed n the directed neighborhoods of an edge will contain repetitions (nodes and/or edges). As pointed out above, for any fixed ℓ and, assuming a randomly chosen edge e, the probability that some repetition does occur tends to zero as n increases at the rate of n^{-1}. Hence, asymptotically the behavior of the decoder after ℓ iterations is equivalent to the behavior of the messages on a true tree.

In the case of regular LDPC codes the support trees of different edges look the same. In the case of irregular LDPC codes the support trees vary because of the different possible degrees of the nodes. In this case one considers the distribution on trees obtained by picking an edge from the graph uniformly at random, and, in effect, averages over this distribution. How can this averaging be accomplished? Consider choosing uniformly at random an edge $e = (v, c)$ and developing its directed neighborhoods for increasing ℓ. The degree of the variable node v is distributed according to λ, i.e., the node v has $i - 1$ children with probability λ_i. Each of the check nodes connected to the variable node v has a degree which is distributed according to ρ., i.e., each such check node has $i - 1$ children with probability ρ_i, and so on.

5.2. Support trees of turbo codes under windowed decoding.
A standard technique for the implementation of Viterbi decoders is to limit the trace-back to a fixed window size [41, p. 258]. Similarly, the Bahl algorithm can be modified to work within a fixed window size and it is well-known that for a large enough window size the resulting loss in performance is negligible. Hence, to decode the i-th systematic bit we apply a window of length $W = 2w + 1$ symmetrically around the i-th bit. We locally construct the trellis of length W and initialize the Bahl algorithm by assigning the end states a uniform probability. We now run the forward, backward and combining iteration as for the standard Bahl algorithm and determine the

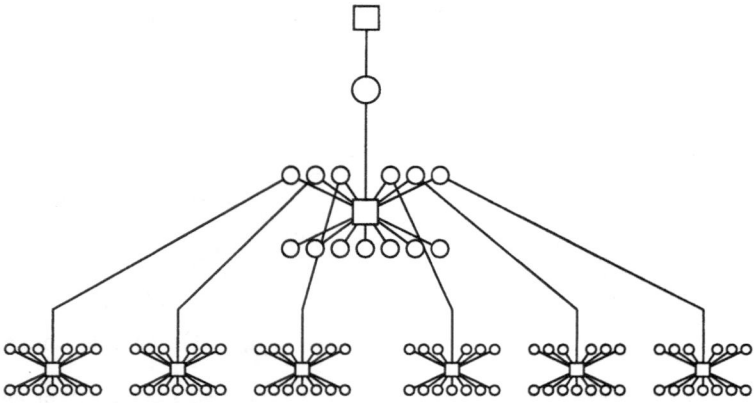

FIG. 4. *The support tree of a extrinsic information message for a turbo code. Note continuation of systematic bits only.*

extrinsic information for the i-th bit. We proceed in the same way for every bit. The important thing to notice is that the extrinsic information output for bit i is a function only of the inputs for bits $i - w$ to $i + 1 + w$, assuming a linear ordering of the systematic bits along the trellis, and the parity-check bits connected to the associated trellis sections.

Fig. 4 shows the resulting graphical representation for the support tree of a message from a particular information bit variable node to its neighboring check node when $w = 3$. Note that the value of each bit is now only a function of a finite number of other bits where the number is determined only by the window length and the number of iterations but not by the length of the code. This notion of the support tree for turbo codes and windowed decoding was introduced by Wiberg in his thesis [43] and it is a crucial ingredient for the subsequent analysis (see also [17]).

In the support tree representation given above, different check nodes will come from decoder 1 or decoder 2 depending on the depth of the check node. What is important here is that, as n tends to infinity, the trellis *segments* indicated by check nodes in the tree become disjoint with high probability. Thus, asymptotically, there will be no dependencies in the tree other than those indicated by the edges.

The importance of using windowed decoding for the analysis is that it allows one to assert that asymptotically the support tree really is a tree, i.e., there are no repetitions with high probability. In standard turbo decoding dependencies extend along the entire trellis so, strictly speaking, the same assertion cannot be made. Nevertheless, it is a fact that dependency decays along the trellis (for a proof see [31].) Thus, the standard algorithm will *behave* as if the trellises at different levels were independent when the length of the code is sufficiently large. Based on the fact indicated above it can be proved that the results of decoding using a window of size w converge as w tends to infinity to the standard turbo decoding except that each appearance of the trellis can be assumed to have *independent* realizations of the received values, and, for any fixed number of iterations, the performance of standard turbo decoding is indistinguishable from this in the infinite code length limit.

Therefore, in the analysis of the asymptotic performance of turbo decoding one may assume that the window sizes are in fact infinite and that there are no repetitions in the support tree.

6. Density evolution. In this section we consider the evolution of message distributions as the messages are passed up a support tree. As we have indicated, this amounts to determining the distribution of messages when decoding an infinitely long code. Only for certain message-passing algorithms do we presently know efficient algorithms to determine these distributions. Fortunately, the cases of greatest interest are among them.

6.1. Density evolution for LDPC codes. Let us imagine how we can determine the distribution of messages passed by a message-passing decoder on an infinitely large (d_v, d_c)-regular LDPC code.

Let $M_{\text{left}}^{d_v} : \mathcal{O} \times \mathcal{M}^{d_v - 1} \to \mathcal{M}$ denote the left (variable node) message map and let $M_{\text{right}}^{d_c} : \mathcal{M}^{d_c - 1} \to \mathcal{M}$ denote the right (check node) message map where, we recall, $\mathcal{O}$ denotes the alphabet of the received values and $\mathcal{M}$ denotes the message alphabet. For simplicity we shall assume that the maps are independent of the iteration number and that the initial messages sent out by the variables nodes are equal to their received values.

The received values given to the decoder have distribution P_0, hence this is the distribution of the messages initially sent out from the variable nodes to the check nodes. Now, choose an edge at random and consider the message sent along it from its neighboring check node c. That message is $M_{\text{right}}^{d_c}(u_1, ..., u_{d_c - 1})$ where $(u_1, ..., u_{d_c - 1})$ are the messages which came in along the other edges connected to c. The probability of a repetition among $(u_1, ..., u_{d_c - 1})$ is zero, so the messages u_i are i.i.d. with distribution P_0. Let R_1 denote the distribution of $M_{\text{right}}^{d_c}(u_1, ..., u_{d_c - 1})$. Now consider the distribution of messages sent in the next round from variable nodes to check nodes. Again, pick an edge at random and let v be its incident variable node. Then the message sent along this edge is given by $M_{\text{left}}^{d_v}(u_0, u_1, ..., u_{d_v - 1})$ where $(u_1, ..., u_{d_v - 1})$ are the messages which came in along the other edges connected to v and u_0 is the received value for v. Let us consider the check nodes $c_1, ..., c_{d_v - 1}$ from which the messages $(u_1, ..., u_{d_v - 1})$ originate. With probability one they are disjoint, i.e., there are no repetitions. Let us consider for each of them their other $d_c - 1$ variable node neighbors. With probability one there are no repetitions among all of these neighbors. Thus, $(u_1, ..., u_{d_v - 1})$ are i.i.d. with distribution R_1. Let P_1 denote the distribution of $M_{\text{left}}^{d_v}(u_0, u_1, ..., u_{d_v - 1})$ where $(u_1, ..., u_{d_v - 1})$ are i.i.d. with distribution R_1 and u_0 is independent with distribution P_0. More generally, let us recursively define distributions

$P_\ell, R_\ell, \ell = 1, 2, \dots$ by setting R_ℓ to be the distribution of $M_{\text{right}}^{d_v}(u_1, \dots, u_{d_c-1})$ where $u_1, \dots, u_{d_c-1}$ are i.i.d. with distribution $P_{\ell-1}$, and by setting P_ℓ to be the distribution of $M_{\text{left}}^{d_v}(u_0, u_1, \dots, u_{d_v-1})$ where $u_1, \dots, u_{d_v-1}$ are i.i.d. with distribution R_ℓ, and u_0 is independent with distribution P_0. Then P_ℓ is the distribution of messages sent from variable nodes to check nodes in iteration ℓ and R_ℓ is the distribution of messages sent from check nodes to variable nodes in iteration ℓ.

Now, fix the number of decoding iterations to be L and consider ensembles of finite graphs. For each n there will be distributions for the messages passed in the graph, call them $R_\ell^{(n)}$ and $P_\ell^{(n)}$. Not surprisingly the distributions $R_\ell^{(n)}, P_\ell^{(n)}$ converge (in an appropriate sense) to R_ℓ, P_ℓ for $\ell < L$ as n tends to infinity. The concentration results from the preceding section further imply that if one picks at random a code from the ensemble and a set of received values distributed according to P_0 then the distribution of the messages in the particular decoding of that particular received sequence will be 'close to'[8] R_ℓ, P_ℓ for $\ell < L$ with high probability as n tends to infinity.

The sequence of distributions R_ℓ, P_ℓ and their determination is collectively referred to as *density evolution*. The generalization to irregular LDPC codes is straightforward. In this case, when we pick an edge at random the distribution of the degree of its neighboring constraint node c is given by ρ. Thus, if $R_\ell[k]$ denotes the return message distribution assuming c has degree k, then R_ℓ is given by $R_\ell = \sum_k \rho_k R_\ell[k]$. Similarly, when we pick an edge at random its neighboring variable node v has degree distributed according to λ and, if $P_\ell[k]$ denote the return message distribution assuming c has degree k, then P_ℓ is given by $P_\ell = \sum_k \lambda_k P_\ell[k]$.

EXAMPLE 6 (BSC, Gallager A). *Recall Gallager's decoding algorithm A described in Example 1. For this decoder the message density can be represented by a single real variable x, namely the probability of error. Under the all-one codeword assumption we identify x with the probability of sending a minus-one. Let ϵ denote the cross-over probability of the BSC. If x is the probability of a minus-one message entering a variable node of degree d, then the probability that a minus-one is passed up the tree from that node is given by $\epsilon(1 - (1 - x)^{d-1}) + (1 - \epsilon)x^{d-1}$. If x is the probability of a minus-one message entering a check node of degree d, then the probability that a minus-one is passed up the tree from that node is given by $\frac{1}{2}(1 - (1 - 2x)^{d-1})$. Let x_l denote the probability of a minus-one being passed up from a variable node to a check node in the ℓ-th iteration where $x_0 = \epsilon$. Then, given an LDPC code with degree sequence pair (λ, ρ), we obtain for $\ell \geq 1$*

$$x_\ell = \epsilon(1 - \lambda(\frac{1 + \rho(1 - 2x_{\ell-1})}{2})) + (1 - \epsilon)\lambda(\frac{1 - \rho(1 - 2x_{\ell-1})}{2}).$$

---

[8]The appropriate metric will depend on, e.g., the message alphabets.

6.2. Density evolution for LDPC belief propagation. In this section we consider the important special case of density evolution for LDPC codes and belief propagation decoding. We start with the simplest example, namely belief propagation for the BEC.

EXAMPLE 7 (BEC, Belief Propagation Decoding). *Consider the belief propagation decoder for a BEC described in Example 2. In this case the message density can be represented by a single real variable x, the probability of erasure. Let δ denote the probability of erasure in the channel. If x is the probability of an erasure for messages entering a variable node of degree d, then the probability that an erasure is passed up the tree from that node is given by δx^{d-1}. If x is the probability of an erasure for messages entering a check node of degree d, then the probability that an erasure is passed up the tree from that node is given by $1 - (1-x)^{d-1}$. Let x_l denote the probability of an erasure being passed up from a variable node to a check node in the ℓ-th iteration where $x_0 = \delta$. Then given a LDPC code with degree sequence (λ, ρ) we obtain $x_\ell = \delta\lambda(1 - \rho(1 - x_{\ell-1}))$ for $\ell \geq 1$.*

The above example is uncharacteristically simple due to the special nature of the channel. At any point in the decoding algorithm there are only two possible states for a given bit: it can either be known (with infinite confidence) or its log-likelihood ratio is equal to 0. It is for that reason that the progress of the system can be characterized by a single real variable, namely the fraction of erasure messages.

The general scenario is more complicated. Recall that for belief propagation we may represent input and output messages as log-likelihood ratios,

$$\log \frac{p(y|\mathsf{x} = 1)}{p(y|\mathsf{x} = -1)} \, ,$$

where y represents all the observations, including the received value in the case of variable nodes, conveyed to the node at that time. In general, message distributions are probability distributions on the two-point compactification of $\mathbb{R}$, which we denote by $\bar{\mathbb{R}}$ (we require $\pm\infty$.) To simplify the presentation we shall generally assume that distributions are smooth and can therefore be represented by their densities.[9]

At a check node, i.e., internal to the computation at check nodes, it is actually more convenient, see [13, p. 46], to represent the input and output messages as a tuple (s, r) where $s \in \mathbb{F}_2 := \{\pm 1\}$ indicates the associated hard decision and where $r \in \mathbb{R}^+$ is equal to

$$\Big| \log |p(\mathsf{x} = 1|y) - p(\mathsf{x} = -1|y)| \Big| \, .$$

Here $\mathbb{F}_2$ denotes the two-element group with elements $\{\pm 1\}$ and group

---

[9]A couple of exceptions will be made for certain discrete measures. In particular, we shall require the distribution Δ_∞, the 'delta function at infinity.' We shall also require the distribution Δ_0 the 'delta function at zero,' corresponding to an erasure.

operation equal to real multiplication. Evidently, there is a straightforward transformation from one representation to the other.

Assume now that the received message from the channel and the incoming messages at a variable node are given in log-likelihood form. The outgoing message along an edge e is then simply the sum of the received message plus all incoming messages, excluding the message incoming along edge e. Therefore, assuming independence (tree assumption), the *density* of the outgoing message is the convolution of the densities of the messages participating in this sum.

An equivalent statement holds for the check nodes. Assume that the incoming messages to a check node are all in the form (s, r). The outgoing message (again in (s, r) representation) along an edge e is then simply the sum of all incoming messages, excluding the message incoming along edge e (here "addition" is performed component-wise, i.e., $(s_1, r_1) + (s_2, r_2) = (s_1 * s_2, r_1 + r_2)$, where the first component is in $\mathbb{F}_2 = \{\pm 1\}$ and "addition" corresponds to real multiplication, and the second component is an element of $\mathbb{R}$ with regular addition.) Thus, as in the previous case, the density of the outgoing message, over $\mathbb{F}_2 \times \bar{\mathbb{R}}^+$, can be written as the convolution of the corresponding densities of the incoming messages. Here, $\bar{\mathbb{R}}^+$ refers to $[0, \infty]$ the compactification of $\mathbb{R}^+$. To complete the description we need only specify the change of variables necessary to convert a density from one representation to the other and to incorporate the degree sequences.

To obtain the distribution of messages from check nodes back to variable nodes, we must perform a change of measure. To this end we define the operator γ which takes the space of distributions on $\bar{\mathbb{R}}^+$ into itself. For densities on $\mathbb{R}^+$ we define γ as

$$(6.1) \qquad \forall \, y \geq 0: \qquad \gamma(f)(y) := f(\ln \coth y/2) \operatorname{csch}(y).$$

The operator γ represents a change of variables $y \to \ln \coth y/2$, i.e., $f(y) \, \mathrm{d}y = \gamma(f)(y) \, \mathrm{d}y$, since $\frac{\mathrm{d}}{\mathrm{d}y} \ln \coth y/2 = -\operatorname{csch}(y)$.

For any distribution f on $\bar{\mathbb{R}}$ let f^- denote $1_{[-\infty, 0]} f$ and let f^+ denote $1_{[0, \infty]} f$, where 1_A denotes the characteristic function of A so that $f = f^+ + f^-$.[10]

Given a distribution f over $\bar{\mathbb{R}}$ of log-likelihoods l, we can represent it as a distribution over $\mathbb{F}_2 \times \bar{\mathbb{R}}^+$ by representing l as the pair (s, r). We therefore define Γ, the change of variable operator, by

$$\Gamma(f)(1, r) = \gamma(f^+)(r) \text{ and } \Gamma(f)(-1, r) = \gamma(\mathcal{R}f^-)(r).$$

where $\mathcal{R}f^-$ denotes the reflection of f about 0, i.e., $\mathcal{R}f^-(x) = f^-(-x)$.

Now, let g be a distribution over $\mathbb{F}_2 \times \bar{\mathbb{R}}^+$. Let g^+ and g^- be defined by $g^+(r) = \delta_{\{1\}}(s)g(s, r)$ and $g^-(r) = \delta_{\{-1\}}(s)g(s, r)$. Then the inverse of

---

[10]In the general case, where a point mass at 0 is allowed, the mass at 0 should be split equally between f^- and f^+.

Γ is given by

$$\Gamma^{-1}(g)(l) = \gamma(g^+)(l) + \mathcal{R}\gamma(g^-)(l).$$

We have now explicitly constructed the change of measure operator for both directions.

Let P_0 be the distribution of the received log-likelihoods and P_ℓ denote the distribution of log-likelihoods sent from the variable nodes to the check nodes in the ℓ-th iteration. Let R_ℓ denote the distribution of log-likelihoods sent from the check nodes to the variable nodes in the ℓ-th iteration. We may initialize with $R_0 = \Delta_0$.

The distribution of messages passed from a variable node of degree d_v to a check node in the ℓ-th iteration is given by the convolution $P_\ell[d_v] := P_0 \otimes R_{\ell-1}^{\otimes d_v - 1}$, where, here and in the sequel, $\otimes$ denotes convolution and $R^{\otimes m}$ denotes R convolved with itself m times.

Suppose that we have a graph with left and right edge degree distributions given by $\lambda(x) = \sum_{i=1}^{d_\ell} \lambda_i x^{i-1}$, and $\rho(x) = \sum_{i=1}^{d_r} \rho_i x^{i-1}$, respectively. We follow [22] and use the following convention: for a polynomial $q(x) = \sum_i q_i x^{i-1}$ with non-negative real coefficients q_i and $q(1) = 1$ we denote by $q(f)$ the distribution $\sum_i q_i f^{\otimes(i-1)}$.

THEOREM 6.1 (Density Evolution). *Let P_0 denote the initial message distribution, under the assumption that the all-one codeword was transmitted, of a low-density parity-check code specified by edge degree distributions $\sum_i \lambda_i x^{i-1}$ and $\sum_i \rho_i x^{i-1}$. If R_ℓ denotes the density of the messages passed from the check nodes to the variable nodes at round ℓ of the belief-propagation, with $R_0 := \Delta_0$, then we have*

$$R_\ell = \Gamma^{-1}\rho(\Gamma(P_0 \otimes \lambda(R_{\ell-1}))).$$

Note that the above indicated computations consist of two main components: convolutions and change of measures. It should therefore not be surprising that density evolution can efficiently be computed by means of Fourier transform based methods.

6.3. Density evolution for turbo code belief propagation. Although one can consider density evolution for other decoders of turbo codes, we will describe the process only for the belief propagation decoder as applied to standard parallel concatenated codes.

We consider the distribution of messages as they are passed up the support tree as described in Section 5.2. One can view the window w as fixed but it is appropriate to consider the limit as $w \to \infty$ since the resulting distributions converge in this limit and we thereby remove the ancillary variable w.

Imagine a very long trellis and suppose that the systematic variables have distribution P_ℓ and the parity-check variables have distribution P_0. In the forward-backward decoding algorithm one computes one-sided conditional densities for states in the trellis. Usually denoted by α and β, these vectors are recursively computed from the two terminal ends of the trellis and represent conditional distributions of the states in the trellis conditioned on the trellis section to the 'left' and the trellis section to the 'right' respectively. In our current view, the inputs to the trellis are random variables and hence so are the α and β vectors. Under very general hypotheses one can show that the distributions of the α and β vectors converge to a unique steady state distribution (depending on P_ℓ and P_0) regardless of the trellis termination. (In the case of periodic puncturing of the parity bits one has convergence to a periodic sequence of distributions.) These steady state distributions thereby induce steady state distributions for the extrinsic information which will be returned from the trellis. (In the case of periodic puncturing of the parity bits one has a periodic sequence of extrinsic distributions but if the interleaver ignores this additional structure then the distributions can be mixed into a single one for purposes of the asymptotic analysis.) Thus, in the limit of infinite length, one has a well-defined map Θ which takes pairs of input distributions, one for the systematic bits and one for the parity bits, into the output distribution of the extrinsic information. Let R_ℓ denote the distribution of extrinsic information returned to the systematic variable nodes at the ℓ-th (half) iteration. Then $R_\ell = \Theta(P_{\ell-1}, P_0)$.

In principle the steady state distributions of α and β can be directly computed. Unfortunately, the computational complexity of such an approach renders practically infeasible. For memory four, for example, the vector α is fifteen (16-1) dimensional. Even just representing an arbitrary probability distribution in fifteen dimensional space with sufficient accuracy seems hopeless. Of course, the distributions which arise are not arbitrary but there seems to be no obvious way to exploit the constraints which exist. Thus, in practice one estimates the distribution R_ℓ by using Monte-Carlo simulation of a very long trellis. Certain structural conditions on the form of R_ℓ contribute to make such an approach highly practical (see Section 8.1.)

The update of distributions at the variable nodes is identical to that for LDPC codes. In the case of parallel concatenated codes it is particularly simple. Assuming a log-likelihood representation for the messages, the distribution P_ℓ returned from the systematic check nodes is simply given by $P_\ell = R_\ell \otimes P_0$. Thus, one has

$$P_\ell = P_0 \otimes \Theta(P_{\ell-1}, P_0).$$

7. Monotonicity and thresholds. Assume we are given a class of channels fulfilling the required symmetry condition and that this class is parameterized by α. This parameter may be real valued as is the case for, e.g., the cross-over probability ϵ for the BSC and the standard deviation σ for the AWGNC, or it may take values in a different domain. For a fixed parameter α we can use the above algorithm to determine if, for a given code, the fraction of incorrect messages tends to zero with an increasing number of (loop free) iterations.

In many cases the parameter α actually reflects a natural ordering of the channels – the capacity decreases with increasing parameter α. It is therefore natural to ask in such cases whether convergence to zero of the probability of error for a parameter α' automatically implies convergence to zero of the probability of error for every parameter α such that $\alpha \leq \alpha'$. More generally, we might want to define a partial ordering of channels with respect to a given code and decoder. In the case of the belief propagation decoder quite general results are possible.

Let a channel W be represented by its transition probability $p_W(y|x)$. We say that a channel W' is *physically degraded* with respect to W if $p_{W'}(y'|x) = p_Q(y'|y)p_W(y|x)$ for some auxiliary channel Q, see [6].

Very often a family of channels is ordered by physical degradation, i.e., if $\alpha \leq \alpha'$, then the channel with parameter α' is physically degraded with respect to the channel with parameter α. Examples of families of channels ordered by physical degradation in this way include the BSC, the AWGNC, the Laplace channel, the erasure channel, and many others.

Our most general result on the question of monotonicity of performance is the following.

THEOREM 7.1 (Monotonicity for Physically Degraded Channels). *Let W and W' be two given memoryless binary input channels that fulfill the required channel symmetry conditions. Assume that W' is physically degraded with respect to W. For a given code and a belief propagation decoder, let p be the expected fraction of incorrect messages passed at the ℓ-th decoding step assuming tree-like neighborhoods and transmission over channel W, and let p' denote the equivalent quantity for transmission over channel W'. Then $p \leq p'$.*

For a proof see [30]. The main idea of the proof is that belief propagation is optimal, i.e., maximum likelihood, in graphs that are trees. As an alternative one could, in principle, degrade the observations before performing maximum likelihood decoding. This can clearly not be any better than performing maximum likelihood decoding on the original data and it is equivalent to decoding the output from the weaker channel.

The arguments used to prove monotonicity of performance under belief propagation do not carry over to other – non maximum-likelihood – message-passing decoders. Nevertheless, for many natural decoders of interest monotonicity can be proved directly. Virtually all message-passing decoders of interest can be interpreted as approximations to belief propagation and it is not surprising that many of them inherit the monotonicity property. We do not know of any decoder of interest for which monotonicity on any family of physically degraded channel has been disproved. There are several examples of decoders of interest, however, for which such monotonicity has not been proved. For various examples we refer the reader to [30].

Given an ordered family of channels, an ensemble of codes, and a decoder for which monotonicity holds one can then define a *threshold* as the maximum channel parameter σ^* such that the probability of error under density evolution converges to zero if $\sigma < \sigma^*$. In some degenerate cases this threshold can actually be zero. For example, for an LDPC code ensemble with a non-zero proportion of degree one variable nodes the threshold is zero since in this case these degree one variables prevent the error probability from reaching zero (see Section 8.3). In most cases of interest, however, a non-trivial threshold will exist demarking the noise limit beyond which the iterative decoding system can not be expected to perform well regardless of the length of the code and below which iterative decoding is expected to work well. In practice one observes that codes approach their asymptotic performance from below. Finite size effects only degrade performance. Also, the orderings induced by the asymptotic performance tend to hold over a wide range of lengths.

8. Analysis of density evolution for belief propagation. In the preceding section we stated a general result on the monotonicity of the asymptotic performance of the belief propagation decoder. There are many other interesting properties associated to density evolution of belief propagation. In this section we present the most important results in this direction.

8.1. Consistency. In the following, we call a distribution f on $\bar{\mathbb{R}}$ *consistent* if it satisfies $f(x) = f(-x)e^x$ for all $x \in \bar{\mathbb{R}}$. Let us first consider a few examples of received distributions. Note that in each case P_0 is consistent.

EXAMPLE 8 (BEC). *For the BEC the initial message distribution is* $\delta\Delta_0 + (1 - \delta)\Delta_\infty$, *where* Δ_x *denotes a delta function at position* x. $\qquad\Box$

EXAMPLE 9 (BSC). *For the BSC the initial message distribution is* $P_0(y) := \epsilon\Delta_{-\log\frac{1-\epsilon}{\epsilon}} + (1 - \epsilon)\Delta_{\log\frac{1-\epsilon}{\epsilon}}$. $\qquad\Box$

EXAMPLE 10 (AWGNC). *Here, the initial message distribution is* $P_0(y) := \sqrt{\frac{\sigma^2}{8\pi}}e^{-\frac{(y-\frac{2}{\sigma^2})^2\sigma^2}{8}}$. *The consistency condition is then easily verified:*

$$P_0(y) = \sqrt{\frac{\sigma^2}{8\pi}}e^{-\frac{(y-\frac{2}{\sigma^2})^2\sigma^2}{8}} = \sqrt{\frac{\sigma^2}{8\pi}}e^{-\frac{(-y-\frac{2}{\sigma^2})^2\sigma^2}{8}}e^y = P_0(-y)e^y.$$

$\Box$

In [29] the following proposition was stated and proved.

PROPOSITION 8.1. *Suppose we are given a binary-input, output-symmetric memoryless channel. Then P_0, the initial message distribution in log-likelihood ratio form under the all-one word assumption, is consistent.*

It was next shown in [29] that the set of consistent distributions is invariant under LDPC density evolution, i.e., each distribution R_ℓ and P_ℓ

which arise from density evolution is consistent. It turns out that the consistency condition is much more general than this.

PROPOSITION 8.2. *Suppose we are given a binary-input, output-symmetric memoryless channel and let x be a bit from a binary linear code. Then P, the density of the conditional log-likelihood of x assuming the all-one codeword was transmitted and the entire codeword observed, is consistent.*

For a proof of this proposition we refer the reader to [31]. This result implies Proposition 8.1 and it also implies that the distributions of extrinsic information which arise in asymptotic turbo decoding are consistent. Consistency is an important property both for theoretical and for practical reasons. As explained in more detail below, the consistency condition can be used to improve the accuracy of some numerical algorithms and it is also helpful in giving a compact description of the *stability condition*, see Section 8.4.

Let $f(x)$ be a consistent distribution. We define the error probability operator as[11]

$$\text{Pr}_{err}(f) := \int_{-\infty}^{0} f(x) \, dx .$$

An easy but helpful consequence of the consistency condition is noted in the following

COROLLARY 8.1. *If $f(x)$ is a consistent distribution then $\text{Pr}_{err}(f) = 0$ if and only if $f = \Delta_{\infty}$.*

Hence, requiring the probability of error to converge to zero is equivalent to requiring the message density to converge to a delta function at infinity.

8.1.1. Application of consistency to density evolution for turbo codes. One important application of the consistency condition is to the numerically accurate estimation of the threshold for turbo codes. For turbo codes, the message distributions induce distributions on the state probability vectors (usually denoted α and β) and the outgoing messages (extrinsic information) are functionals of the state probability vectors. It is not known (and seems unlikely) whether the distribution of the extrinsic information can be computed without first computing the distribution of the state probability vectors. The state probability vectors are, in general, vectors in $\mathbb{R}^{2^m - 1}$. Thus, even for quite small m, e.g., $m = 3$, representations of distributions of state probability vectors are prohibitively complex. Thus, direct computation of density evolution appears to be practically intractable except in the simplest cases.

Nevertheless, it is possible to estimate the distributions of density evolution via simulation of the decoding process. We simulate decoding on a

---

[11]If f has a point mass at 0 then exactly half of that mass is included in $\text{Pr}_{err}(f)$.

very long trellis in which all variables are independently sampled from their known distributions, determining the outgoing extrinsic information for each variable. These outgoing messages thus provide an empirical sample of the true message distribution. Since the decoding process, in this sense, is ergodic, the empirical distribution converges weakly to the true message distribution. In actual computations the distributions are quantized and hence discrete, thus, with sufficiently long simulation, an arbitrarily accurate estimate of the message distribution can be obtained with arbitrarily high probability.

A substantial savings in simulation time can be had by exploiting the consistency of the true message distribution. In simulation we need only determine the distribution of the *magnitude* of the extrinsic log-likelihoods. For a given log-likelihood magnitude x then we have $P(x) = \frac{1}{1+e^{-x}}(P(x) + P(-x))$ and $P(-x) = \frac{1}{1+e^{x}}(P(x) + P(-x))$. Since $P(x) + P(-x)$ is exponentially larger than $P(-x)$ this provides a very significant acceleration of the convergence of the empirical distribution to the true message distribution with regard to estimates of rare error events.

8.2. Fixed points. Recall from Example 7 that for the BEC and a belief propagation decoder the state of the system at any iteration is described by the remaining fraction of erasures. Denote the remaining fraction of erasures by x and let $h(x)$ be the remaining fraction of erasures after a further iteration. Then the threshold is given by the maximum fraction x_0 such that

$$(8.1) \qquad\qquad \forall 0 < x < x_0 : \quad h(x) < x.$$

Recall further from Example 1 that a similar statement is true for the BSC under Gallager's decoding algorithm A. In this case x signifies the remaining number of errors.

In these cases, an alternative description of the threshold can be given in terms of fixed points. The threshold is given by the maximum number x_0 such that the equation

$$(8.2) \qquad\qquad h(x) = x$$

has no solutions for $x \in (0, x_0)$.

Assume now we are given a general discrete memoryless channel which fulfills the channel symmetry conditions and we employ a belief propagation decoder. The state of the system is now described by the message density. Let f denote such a density and let $h(f)$ denote the corresponding density after a further iteration. Clearly, there is no characterization of the threshold which corresponds to (8.1) since there is, a priori, no given linear ordering of densities. The alternative formulation in (8.2) involving fixed points looks more promising. Unfortunately, the non-existence of fixed-points for iterated function systems in more than one dimension is in general not enough to guarantee convergence. So it is quite surprising, as it turns out, that, for message distributions of belief propagation decoders, fixed points sufficiently characterize the convergence of the sequences.

Let $P_\ell(x)$ denote the distribution of the messages at the ℓ-th iteration assuming, as usual, that the all-one word was transmitted. Suppose that we were to decide on the transmitted bit according to the sign of the message. In this case the conditional error probability is equal to $\mathrm{Pr}_{\mathrm{err}}(P_\ell)$. But, because of the symmetry conditions, this is equal to the error probability even without conditioning on the transmitted codeword. Since the sign of the message is equal to the MAP estimate, this error probability is clearly a non-increasing function of ℓ.

This monotonicity property is the key to the fixed point theorem below. It is not hard to see that there is actually a whole family of such monotonicity conditions.

THEOREM 8.1 (General Monotonicity Law). Let P_ℓ be the message distribution at the ℓ-th decoding step and let g be a consistent distribution on $\bar{\mathbb{R}}$. Then

$$\mathrm{Pr}_{\mathrm{err}}(P_\ell \otimes g)$$

is a non-increasing function of ℓ.

THEOREM 8.2 (Fixed Point Theorem). *Let f be a consistent distribution and assume that f has support over the entire real axis. Let $P_{\ell_1}(x)$ and $P_{\ell_2}(x)$ denote the message distributions at the ℓ_1-th and ℓ_2-th iteration respectively. If*

$$(8.3) \qquad \mathrm{Pr}_{\mathrm{err}}(P_{\ell_1} \otimes f) = \mathrm{Pr}_{\mathrm{err}}(P_{\ell_2} \otimes f)$$

then $P_{\ell_1}(x) = P_\ell(x)$ for $\ell \geq \min\{\ell_1, \ell_2\}$ i.e., $P_{\ell_1}(x) = P_{\ell_2}(x)$ is a fixed point of density evolution.

We note that the above theorem has some very strong implications. Although the message distributions for a belief propagation decoder live nominally on $\mathbb{R}^{\mathbb{R}}$, the above theorem implies that the possible message distributions can be ordered by one real parameter (the projection). Hence, roughly speaking, we are dealing with a one-dimensional manifold embedded in an infinite dimensional space. This is very much analogous to the simple one-dimensional case we encounter in the case of an erasure channel, or Gallager's decoding algorithm A.

8.3. Stability. Consider transmitting over the BEC with erasure probability δ using LDPC codes. Recall from example 7 that the expected fraction of erasure messages at the ℓ-th iteration is given by

$$(8.4) \qquad x_\ell = \delta\lambda(1 - \rho(1 - x_{\ell-1})),$$

where $x_0 = \delta$. As pointed out earlier, the threshold δ^* is the single most important parameter describing the performance of an iterative coding system. We recall that δ^* is the supremum of all values of δ such that x_ℓ tends to zero as ℓ tends to infinity. In general, it is not an easy task to determine the threshold analytically but we can give an analytical upper bound on δ^* by considering the behavior of (8.4) for very small values of x_ℓ. Hence, let $h(x) := \delta\lambda(1 - \rho(1 - x))$. Then $h(x) = \delta\lambda'(0)\rho'(1)x + O(x^2)$. Therefore, to first order in x, the fraction of erasure messages will evolve from x to $\delta\lambda'(0)\rho'(1)x$. Clearly, if we want the fraction of erasure messages to tend to zero then we need $\lambda'(0)\rho'(1) < 1/\delta$. From this we can deduce the bound $\delta^* < \frac{1}{\lambda'(0)\rho'(1)}$. Vice versa, if $\lambda'(0)\rho'(1) < 1/\delta$ then there exists an $\delta > 0$ such that the values of the recursion tend to zero if the recursion is initialized with a value which does not exceed δ. The condition $\lambda'(0)\rho'(1) < 1/\delta$, first discussed in [37], can be seen as a *stability condition* of the fixed point $x = 0$.

Such a stability condition can be given in a much more general setting and it plays an important role in the theory of iterative coding systems. As we have seen it leads to an upper bound on the threshold of an interative coding system. In the case of, e.g., *cycle codes* and the BSC this upper bound is tight, i.e., the stability condition determines the threshold of cycle codes for the BSC exactly [8]. The stability condition can also be interpreted in an alternative way. Assume we are trying to construct degree sequence pairs which achieve capacity on a given channel. In this case the desired threshold is determined by the capacity formula. E.g., for the case of the BEC we have $\delta^* = 1 - r$, where r is the rate of the code. Now we can write the stability condition as $\lambda'(0) \leq \frac{1}{(1-r)\rho'(1)}$, i.e., the stability condition imposes an upper bound on the fraction of edges which connect to degree two nodes. It has been shown in the case of the BEC that capacity achieving sequences must fulfill the above inequality with equality and we conjecture that this is true in general. We will now describe some important instances of the stability condition.

8.4. Stability condition for LDPC codes and belief propagation. Consider now a transmission over a general binary-input, memoryless, output-symmetric channel and assume that we use a belief propagation decoder. Observe that if $P_\ell = \Delta_\infty$ for some $\ell \geq 0$ then $P_{\ell+i} = \Delta_\infty$ for any $i \geq 0$, i.e., Δ_∞ is a fixed point of density evolution. Since we desire that the error probability associated with density evolution converge to zero, it is clear that the fixed point described above should, in some sense, be an attractor (recall Corollary 8.1). To analyze local convergence to this fixed point we shall consider a linearization of density evolution about the fixed point.

Consider a density $\epsilon P + (1-\epsilon)\Delta_\infty$ where P is any (consistent) density, $P \neq \Delta_\infty$. After a complete iteration of density evolution this density will evolve to

$$\lambda'(0)\rho'(1)\epsilon P \otimes P_0 + (1 - \lambda'(0)\rho'(1)\epsilon)\Delta_\infty + O(\epsilon^2),$$

where $\lambda'(x)$ and $\rho'(x)$ denote the derivatives of $\lambda(x)$ and $\rho(x)$, respectively.

In other words, the linearization of density evolution at the fixed point Δ_∞ is given by $P \to \lambda'(0)\rho'(1)P \otimes P_0$.

One would expect, therefore, that a necessary condition for convergence to zero of the probability of error be that the probability of error associated with

$$(\lambda'(0)\rho'(1))^n P \otimes P_0^{\otimes n}$$

be decaying to zero as n grows. Under very general conditions (see Lemma 8.1) the limit

(8.5) $$r := -\lim_{n\to\infty} \frac{1}{n}\log \mathrm{Pr}_{\mathrm{err}}(P_0^{\otimes n})$$

is well defined. In this case, since $P \neq \Delta_\infty$, the quantity

$$(\lambda'(0)\rho'(1))^n \mathrm{Pr}_{\mathrm{err}}(P \otimes P_0^{\otimes n})$$

converges to zero if and only if $(\lambda'(0)\rho'(1))^n \mathrm{Pr}_{\mathrm{err}}(P_0^{\otimes n})$ converges to zero. Although the above argument is only heuristic, the assertion is correct. This is summarized in the following.

THEOREM 8.3. *If* $\lambda'(0)\rho'(1) > e^r$, *then the probability of error of density evolution is strictly bounded away from 0. Conversely, if* $\lambda'(0)\rho'(1) < e^r$, *then there exists* $\epsilon > 0$ *such that if density evolution is initialized with a consistent message distribution* P *satisfying* $\mathrm{Pr}_{\mathrm{err}}(P) < \epsilon$, *then the probability of error will converge to zero.*

We note that in all cases of interest the exponent r can be computed using moment generating functions as stated in the following.

LEMMA 8.1. *Let* $g(s)$ *be the moment generating function corresponding to the distribution* $P_0(x)$, *i.e.,* $g(s) = E_{P_0}[e^{sX}]$, *and assume that* $g(s) < \infty$ *for all* s *in some neighborhood of zero. Then* $r = -\log(\inf_{s<0} g(s))$. *Further, if* P_0 *is consistent then* $r = -\log\left(2\int_0^\infty P_0^+(x)\,e^{-x/2}\,dx\right)$.

EXAMPLE 11 (BEC). *For the BEC (see Example 2) we have*

$$e^{-r} = 2\int_0^\infty [\frac{\delta}{2}\Delta_0 + (1-\delta)\Delta_\infty]\,e^{-x/2}\,dx = \delta\,.$$

Therefore, the stability condition reads

$$\lambda'(0)\rho'(1) < \frac{1}{\delta},$$

as noted above.   □

EXAMPLE 12 (BSC). *For the BSC (see Example 9) we have*

$$e^{-r} = 2\int_0^\infty (1-\epsilon)\Delta_{\log\frac{1-\epsilon}{\epsilon}}\,e^{-x/2}\,dx = 2\sqrt{\epsilon(1-\epsilon)}\,.$$

It follows that the stability condition for the BSC is given by

$$\lambda'(0)\rho'(1) < \frac{1}{2\sqrt{\epsilon(1-\epsilon)}}.$$

□

EXAMPLE 13 (AWGC). *For the AWGNC (see Example 10) we have*

$$e^{-r} = 2\int_0^\infty \sqrt{\frac{\sigma^2}{8\pi}}\, e^{-\frac{(x-\frac{2}{\sigma^2})\sigma^2}{8}}\, e^{-x/2}\, dx = e^{-\frac{1}{2\sigma^2}}.$$

Thus, the stability condition reduces to

$$\lambda'(0)\rho'(1) < e^{\frac{1}{2\sigma^2}}.$$

□

8.5. Examples of other instances of the stability condition.
The stability condition is not limited to LDPC codes and belief propagation decoders. Consider, e.g., the case of LDPC codes decoded with Gallager's decoding algorithm A and transmission over a binary symmetric channel with cross-over probability ϵ, as discussed in Example 1. In this case it was shown in [2] that the appropriate stability condition reads

$$\frac{1 - \lambda'(0)\rho'(1)}{\lambda'(1)\rho'(1) - \lambda'(0)\rho'(0)} > \epsilon.$$

We note that for some codes e.g., the $(4,8)$, $(5,10)$ and the $(4,6)$ regular codes, this condition is tight, i.e., it determines the threshold exactly.

Consider next the special case of a parallel concatenated turbo code with the simple component code $[1, \frac{1}{1+D}]$, no puncturing and transmission over the BEC with erasure probability δ. In this case it was shown in [31] that the erasure probability x_ℓ evolves as

$$x_\ell = \frac{x_{\ell-1}\delta^2(2 - 2\delta + x_{\ell-1}\delta)}{(1 - \delta(1 - x_{\ell-1}))^2}.$$

The stability condition is easily found to be $\delta < \frac{1}{2}$. Again, the stability is tight, i.e., the threshold δ^* can be shown to be equal to $\frac{1}{2}$. The general form of the stability condition for turbo codes has not been explored in any depth so far and this promises to be a fruitful direction for future research.

9. Acknowledgment. We would like to thank David Proietti and the anonymous reviewer for pointing out to us some typos in a previous version of this paper.

REFERENCES

[1] N. ALON, J. SPENCER AND P. ERDÖS, *The Probabilistic Method*, John Wiley & Sons, Inc., New York, 1992.

[2] L. BAZZI, T. RICHARDSON, AND R. URBANKE, *Exact thresholds and optimal codes for the binary symmetric channel and Gallager's decoding algorithm A.* submitted IEEE IT.

[3] S. BENEDETTO AND G. MONTORSI, *Unveiling turbo codes: Some results on parallel concatenated coding schemes*, IEEE Trans. Inform. Theory, **42** (1996), pp. 409–428.

[4] C. BERROU, A. GLAVIEUX, AND P. THITIMAJSHIMA, *Near Shannon limit error-correcting coding and decoding*, in Proceedings of ICC'93, Geneva, Switzerland, May 1993, pp. 1064–1070.

[5] S.-Y. CHUNG. personal communication.

[6] T.M. COVER AND J.A. THOMAS, *Elements of Information Theory*, Wiley, New York, 1991.

[7] M.C. DAVEY AND D.J.C. MACKAY, *Low density parity check codes over GF(q)*, IEEE Communications Letters, **2** (1998).

[8] L. DECREUSEFOND AND G. ZÉMOR, *On the error-correcting capabilities of cycle codes of graphs*, Combinatorics, Probability and Computing (1997), pp. 27–38.

[9] D. DIVSALAR, *A simple tight bound on error probability of block codes with application to turbo codes*. TMO Progress Report 42-139.

[10] T.M. DUMAN AND M. SALEHI, *Performance bounds for turbo-coded modulation systems*, IEEE Transactions on Communications, **47** (1999), pp. 511–521.

[11] D. FORNEY, *Codes on graphs: Generalized state realizations*. submitted to IEEE Trans. Inform. Theory.

[12] B. FREY AND F. KSCHISCHANG, *Probability propagation and iterative decoding*, in Allerton Conf. on Communication, Control and Computing, 1996.

[13] R.G. GALLAGER, *Low-Density Parity-Check Codes*, M.I.T. Press, Cambridge, Massachusetts, 1963.

[14] J. GARCIA-FRIAS AND J.D. VILLASENOR, *Combining hidden Markov source models and parallel concatenated codes*, IEEE Communications Letters, **1** (1997), pp. 111–113.

[15] J. GARCIA-FRIAS AND J.D. VILLASENOR, *Exploiting binary Markov channels with unknown parameters in Turbo decoding*, in Proc. Globecom'98, Sydney, Australia, Nov. 1998, pp. 3244–3249.

[16] ———, *Turbo decoders for Markov channels*, IEEE Commun. Lett., **2** (1998), pp. 257–259.

[17] E.A. GELBLUM, A.R. CALDERBANK, AND J. BOUTROS, *Understanding serially concatenated codes from a support tree approach*, in Proceedings of the International Symposium on Turbo Codes and Related Topics, Brest, France, Sept. 1997, pp. 271–274.

[18] E.K. HALL AND S.G. WILSON, *Design and analysis of turbo codes on Rayleigh fading channels*, IEEE Journal of Selected Areas in Communications, **16** (1998), pp. 160–174.

[19] P. HOEHER, J. LODGE, R. YOUNG AND J. HAGENAUER, *Separable map "filters" for the decoding of product and concatenated codes*, in Proceedings of ICC'93, Geneva, Switzerland, May 1993, pp. 1740–1745.

[20] F. KSCHISCHANG AND B. FREY, *Iterative decoding of compound codes by probability propagation in graphical models*, IEEE Journal on Selected Areas in Communications (1998), pp. 219–230.

[21] F. KSCHISCHANG, B. FREY, AND H.-A. LOELIGER, *Factor graphs and the sum-product algorithm*. submitted to IEEE Trans. Inform. Theory.

[22] M. LUBY, M. MITZENMACHER, A. SHOKROLLAHI, AND D. SPIELMAN, *Analysis of low density codes and improved designs using irregular graphs*, in Proceedings of the 30th Annual ACM Symposium on Theory of Computing, 1998, pp. 249–258.

[23] M. LUBY, M. MITZENMACHER, A. SHOKROLLAHI, D. SPIELMAN, AND V. STEMANN, *Practical loss-resilient codes*, in Proceedings of the 29th annual ACM Symposium on Theory of Computing, 1997, pp. 150–159.

[24] D.J.C. MACKAY AND R.M. NEAL, *Good codes based on very sparse matrices*, in Cryptography and Coding. 5th IMA Conference, C. Boyd, ed., no. 1025 in Lecture Notes in Computer Science, Springer, Berlin, 1995, pp. 100–111.

[25] I.D. MARSLAND AND P. MATHIOUPOULOS, *Multiple differential detection of parallel concatenated convolutional (turbo) codes in correlated fast rayleigh fading*, IEEE Journal of Selected Areas in Communications, 16 (1998), pp. 265–275.

[26] R. MCELIECE, E. RODEMICH, AND J.-F. CHENG, *The turbo decision algorithm*, in Proceedings of the 33rd Allerton Conference on Communication, Control, and Computing, Monticello, IL, 1995.

[27] R. MOTWANI AND P. RAGHAVAN, *Randomized Algorithms*, Cambridge University Press, Cambridge, 1995.

[28] J. PEARL, *Probabilistic reasoning in intelligent systems: networks of plausible inference*, Morgan Kaufmann Publishers, 1988.

[29] T. RICHARDSON, A. SHOKROLLAHI, AND R. URBANKE, *Design of provably good low-density parity check codes*. submitted IEEE IT.

[30] T. RICHARDSON AND R. URBANKE, *The capacity of low-density parity check codes under message-passing decoding*. submitted IEEE IT.

[31] ———, *The capacity of turbo codes and other concatenated codes under message-passing decoding*. in preparation.

[32] ———, *Concentrate!*, in Allerton Conf. on Communication, Control and Computing, 1999.

[33] P. ROBERTSON, *Illuminating the structure of code and decoder of parallel concatenated recursive systematic (turbo) codes*, in Proceedings of GLOBECOM'94, Nov. 1994, pp. 1298–1303.

[34] I. SASON AND S. SHAMAI (SHITZ), *Improved upper bounds on the decoding error probability of parallel and serial concatenated turbo codes via their ensemble distance spectrum*, in 1998 IEEE International Symposium on Information Theory, Boston, MA, Aug. 16–21 1998, p. 30.

[35] ———, *Bounds on the error probability of ML decoding for block and turbo-block codes*, Annales de Telecommuncations, 54 (1999), pp. 61–78.

[36] A. SHAMIR AND J. SPENCER, *Sharp concentration of the chromatic number on random graphs $G_{n,p}$*, Combinatorica, 7 (1987), pp. 121–129.

[37] A. SHOKROLLAHI, *New sequences of linear time erasure codes approaching the channel capacity*, in Proceedings of AAECC-13, Lecture Notes in Computer Science 1719, 1999, pp. 65–76.

[38] M. SIPSER AND D. SPIELMAN, *Expander codes*, IEEE Trans. on Information Theory, 42 (1996).

[39] N. SOURLAS, *Spin-glass models as error-correcting codes*, Nature, 339 (1989), pp. 693–695.

[40] R.M. TANNER, *A recursive approach to low complexity codes*, IEEE Trans. Inform. Theory, 27 (1981), pp. 533–547.

[41] A. VITERBI AND J. OMURA, *Principles of Digital Communication and Coding*, McGraw-Hill, 1979.

[42] A. VITERBI, A. VITERBI, J. NICOLAS, AND N. SINDHUSHAYANA, *Perspective on interleaved concatenated codes with iterative soft-output decoding*, in Proceedings of the International Symposium on Turbo Codes and Related Topics, Brest, France, Sept. 1997, pp. 47–54.

[43] N. WIBERG, *Codes and Decoding on General Graphs*, PhD thesis, Linköping University, S-581 83, Linköping, Sweden, 1996.

[44] N. WIBERG, H.-A. LOELIGER, AND R. KÖTTER, *Codes and iterative decoding on general graphs*, European Transactions in Telecommuncations, **6** (1995), pp. 513–526.

[45] V. ZYABLOV AND M. PINSKER, *Estimation of the error-correction complexity of Gallager low-density codes*, Problemy Peredachi Informatsii, **11** (1975), pp. 23–26.

CONNECTIONS BETWEEN LINEAR SYSTEMS AND CONVOLUTIONAL CODES*

JOACHIM ROSENTHAL†

Abstract. The article reviews different definitions for a convolutional code which can be found in the literature. The algebraic differences between the definitions are worked out in detail. It is shown that bi-infinite support systems are dual to finite-support systems under Pontryagin duality. In this duality the dual of a controllable system is observable and vice versa. Uncontrollability can occur only if there are bi-infinite support trajectories in the behavior, so finite and half-infinite-support systems must be controllable. Unobservability can occur only if there are finite support trajectories in the behavior, so bi-infinite and half-infinite-support systems must be observable. It is shown that the different definitions for convolutional codes are equivalent if one restricts attention to controllable and observable codes.

Key words. Convolutional codes, linear time-invariant systems, behavioral system theory.

AMS(MOS) subject classifications. Primary 37B10, 93B25, 94B10.

1. Introduction. It is common knowledge that there is a close connection between linear systems over finite fields and convolutional codes. In the literature one finds however a multitude of definitions for convolutional codes, which can make it confusing for somebody who wants to enter this research field with a background in systems theory or symbolic dynamics. It is the purpose of this article to provide a survey of the different points of view about convolutional codes.

The article is structured as follow: In Section 2 we will review the way convolutional codes have often been defined in the coding literature [20, 21, 28, 35, 38].

Section 3 reviews a definition for convolutional codes that can be found in the literature on symbolic dynamics. From the symbolic dynamics point of view [24, 29, 32], a convolutional code is a linear irreducible shift space.

In Section 4 we will review the class of time-invariant, complete linear behaviors in the sense of Willems [50, 51, 52]. We will show how these behaviors relate to the definitions given in Section 2 and 3.

In Section 5 we will give a definition for convolutional codes in which it is required that the code words have finite support. Such a definition was considered by Fornasini and Valcher [48, 5] and by the author in collaboration with Schumacher, Weiner and York [42, 44, 49]. The study of behaviors with finite support has been done earlier in the context of automata the-

---

*The work was supported in part by NSF grant DMS-96-10389. This research has been carried out while the author was a guest professor at EPFL in Switzerland. The author would like to thank EPFL for its support and hospitality.

†Department of Mathematics, University of Notre Dame, Notre Dame, Indiana 46556-5683. *E-mail:* Rosenthal.1@nd.edu.

ory and we refer to Eilenberg's book [1]. We show in Section 5 how this module-theoretic definition relates to complete, linear and time-invariant behaviors by Pontryagin duality.

In Section 6 we will study different first-order representations connected with the different viewpoints. Finally, in Section 7 we compare the different definitions. We also show how cyclic redundancy check codes can naturally be viewed in the context of finite-support convolutional codes.

Throughout the paper we will emphasize the algebraic properties of the different definitions. We will also restrict ourselves to the concrete setting of convolutional codes defined over finite fields. It is however known that many of the concepts in this paper generalize to group codes [3, 12, 9] and multidimensional convolutional codes [4, 5, 16, 48, 49]. All of the definitions which we are going to give are quite similar, but there are some notable differences.

Since the paper draws from results from quite different research areas, one is faced with the problem that there is no uniform notation. In this paper we will adopt the convention used in systems theory in which vectors are regarded as column vectors. For the convenience of the reader, we conclude this section with a summary of some of the notation used in this paper:

$\mathbb{F}$	A fixed finite field;
$\mathbb{F}[z]$	The polynomial ring over $\mathbb{F}$;
$\mathbb{F}[z, z^{-1}]$	The Laurent polynomial ring over $\mathbb{F}$;
$\mathbb{F}(z)$	The field of rationals;
$\mathbb{F}[[z]]$	The ring of formal power series of the form $\sum_{i=0}^{\infty} a_i z^i$;
$\mathbb{F}((z))$	Field of formal Laurent series having the form $\sum_{i=d}^{\infty} a_i z^i$;
$\mathbb{F}[[z, z^{-1}]]$	The ring of formal power series of the form $\sum_{i=-\infty}^{\infty} a_i z^i$;
$\mathbb{Z}$	The integers;
$\mathbb{Z}_+$	The nonnegative integers;
$\mathbb{Z}_-$	The nonpositive integers.

Consider the ring of formal power series $\mathbb{F}[[z, z^{-1}]]$. We will identify the set $\mathbb{F}[[z, z^{-1}]]$ with the (two-sided) sequence space $\mathbb{F}^{\mathbb{Z}}$. We have natural embeddings:

$$\mathbb{F} \longrightarrow \mathbb{F}[z] \longrightarrow \mathbb{F}[z, z^{-1}] \longrightarrow \mathbb{F}(z) \longrightarrow \mathbb{F}((z)) \longrightarrow \mathbb{F}[[z, z^{-1}]].$$

With these embeddings we can view e.g. the set of rationals $\mathbb{F}(z)$ as a subset of the sequence space $\mathbb{F}^{\mathbb{Z}}$, and we will make use of such identifications throughout the paper.

The set of n-vectors with polynomial entries will be denoted by $\mathbb{F}^n[z]$. Similarly we define the sets $\mathbb{F}^n(z), \mathbb{F}^n((z))$ etc. All these sets are subsets of the two sided sequence space $(\mathbb{F}^n)^{\mathbb{Z}} = \mathbb{F}^n[[z, z^{-1}]]$. The definitions of convolutional codes which we will provide in the next sections will all be $\mathbb{F}$-linear subspaces of $(\mathbb{F}^n)^{\mathbb{Z}}$.

The idea of writing a survey on the different points of view about convolutional codes was suggested to the author by Paul Fuhrmann during a stimulating workshop on "Codes, Systems and Graphical Models" at the Institute for Mathematics and its Applications (IMA) in August 1999. A first draft of this paper was circulated in October 1999 to about a dozen people interested in these research issues. This generated an interesting 'Internet discussion' on these issues, in which the different opinions were exchanged by e-mail. Some of these ideas have been incorporated into the final version of the paper and the author would like to thank Dave Forney, Paul Fuhrmann, Heide Gluesing-Luerssen, Jan Willems and Sandro Zampieri for having provided valuable thoughts. The author wishes also to thank the IMA and its superb staff, who made the above mentioned workshop possible.

2. The linear algebra point of view. The theory of convolutional codes grew out and extended the theory of linear block codes into a new direction. Because of this reason we start the section with linear block codes and we introduce convolutional codes in a quite intuitive way.

An $[n, k]$ linear block code is by definition a linear subspace $\mathcal{C} \subset \mathbb{F}^n$ having dimension $\dim \mathcal{C} = k$. Let G be a $n \times k$ matrix with entries in $\mathbb{F}$. The linear map

$$\varphi : \mathbb{F}^k \longrightarrow \mathbb{F}^n, \ m \longmapsto c = Gm$$

is called an *encoding map* for the code $\mathcal{C}$ if $\mathrm{im}\,(\varphi) = \mathcal{C}$. If this is the case then we say G is a *generator matrix* or an *encoder* for the block code $\mathcal{C}$.

Assume that a sequence of message blocks $m_0, \dots, m_t \subset \mathbb{F}^k$ should be encoded into a corresponding sequence of code words $c_i = Gm_i \in \mathbb{F}^n$, $i = 0, \dots, t$. By introducing the polynomial vectors $m(z) = \sum_{i=0}^{t} m_i z^i \in \mathbb{F}^k[z]$ and $c(z) = \sum_{i=0}^{t} c_i z^i \in \mathbb{F}^n[z]$ it is possible to describe the encoding procedure through the module homomorphism:[1]

(2.1) $$\varphi : \mathbb{F}^k[z] \longrightarrow \mathbb{F}^n[z], \ m(z) \longmapsto c(z) = Gm(z).$$

The original idea of a convolutional code goes back to the paper of Elias [2], where it was suggested to use a polynomial matrix $G(z)$ in the encoding procedure (2.1).

Polynomial encoders $G(z)$ are physically easily implemented through a feedforward linear sequential circuit. Massey and Sain [34, 45] showed that there is a close connection between linear systems and convolutional codes. Massey and Sain viewed the polynomial encoder $G(z)$ as a transfer function. More generally it is possible to realize a transfer function $G(z)$ with rational entries by (see e.g. [20, 21]) a linear sequential circuit whose elements include feedback components. If one allows rational entries in

---

[1]Throughout the paper we use the symbol φ to denote an encoding map. The context will make it clear what the domain and the range of this map is in each situation.

the encoding matrix then it seems natural to extend the possible message sequences to the set of rational vectors $m(z) \in \mathbb{F}^k(z)$ and to process this sequence by a 'rational encoder' resulting again in a rational code vector $c(z) \in \mathbb{F}^n(z)$. With this we have a first definition of a convolutional code as it can be found e.g. in the Handbook of Coding Theory [35, Definition 2.4]:

DEFINITION A. A $\mathbb{F}(z)$-linear subspace $\mathcal{C}$ of $\mathbb{F}^n(z)$ is called a convolutional code.

If $G(z)$ is a $n \times k$ matrix with entries in $\mathbb{F}(z)$ whose columns form a basis for $\mathcal{C}$, then we call $G(z)$ a generator matrix or an encoder for the convolutional code $\mathcal{C}$. $G(z)$ describes the encoding map:

$$\varphi : \ \mathbb{F}^k(z) \longrightarrow \mathbb{F}^n(z), \ m(z) \longmapsto c(z) = G(z)m(z).$$

The field of rationals $\mathbb{F}(z)$ viewed as a subset of the sequence space $\mathbb{F}^{\mathbb{Z}} = \mathbb{F}[[z, z^{-1}]]$ consists precisely of those sequences whose support is finite on the negative sequence space $\mathbb{F}^{\mathbb{Z}-}$ and whose elements form an ultimately periodic sequence on the positive sequence space $\mathbb{F}^{\mathbb{Z}+}$. It therefore seems that one equally well could restrict the possible message words $m(z) \in \mathbb{F}^k(z)$ to sequences whose coordinates consists of Laurent polynomials only, in other words to sequences of the form $m(z) \in \mathbb{F}^k[z, z^{-1}]$.

Alternatively one could allow message words $m(z)$ whose coordinates are not ultimately periodic and possibly not of finite support on the negative sequence space $\mathbb{F}^{\mathbb{Z}-}$. This would suggest that one should take as possible message words the whole sequence space $\left(\mathbb{F}^k\right)^{\mathbb{Z}} = \mathbb{F}^k[[z, z^{-1}]]$. The problem with this approach is that the multiplication of an element in $\mathbb{F}[[z, z^{-1}]]$ with an element in $\mathbb{F}(z)$ is in general not well defined. If one restricts however the message sequences to the field of formal Laurent series then the multiplication is well defined. This leads to the following definition which goes back to the work of Forney [7]. The definition has been adopted in the book by Piret [38] and the book by Johannesson and Zigangirov [21], and it appears as Definition 2.3 in the Handbook of Coding Theory [35]:

DEFINITION A'. A $\mathbb{F}((z))$-linear subspace $\mathcal{C}$ of $\mathbb{F}^n((z))$ which has a basis of rational vectors in $\mathbb{F}^n(z)$ is called a convolutional code.

The requirement that $\mathcal{C}$ has a basis with rational entries guarantees that $\mathcal{C}$ has also a basis with only polynomial entries. $\mathcal{C}$ can therefore be represented by a $n \times k$ generator matrix $G(z)$ whose entries consist only of rationals or even polynomials. The encoding map with respect to $G(z)$ is given through:

$$(2.2) \qquad \varphi : \ \mathbb{F}^k((z)) \longrightarrow \mathbb{F}^n((z)), \ m(z) \longmapsto c(z) = G(z)m(z).$$

If $G(z)$ is a polynomial matrix, then finitely many components of $m(z)$ influence only finitely many components of $c(z)$, and the encoding procedure may be physically implemented by a simple feedforward linear shift register.

If $G(z)$ contains rational entries, then it is in general the case that a finite (polynomial) message vector is encoded into an infinite (rational) code vector of the form $c(z) = \sum_{i=s}^{\infty} c_i z^i$. This might cause some difficulties in the decoder. For the encoding process, $G(z)$ can be physically realized by linear shift registers, in general with feedback (see e.g. [20, 21]).

From a systems theory point of view, it is classical [23] to view the encoding map (2.2) as an input-output linear system. This was the point of view taken by Massey and Sain [34, 45] and thereafter in most of the coding literature. However unlike in systems theory, the important object in coding theory is the code $\mathcal{C} = \operatorname{im}(\varphi)$. As a result one calls encoders φ which generate the same image $\operatorname{im}(\varphi)$ equivalent; we will say more about this in a moment. In Sections 3 and 4 we will view (2.2) as an image representation of a time-invariant behavior in the sense of Willems [50, 51], which we believe captures the coding situation in a more natural way.

Assume that $G(z)$ and $\tilde{G}(z)$ are two $n \times k$ rational encoding matrices defining the same code $\mathcal{C}$ with respect to either Definition A or A'. In this case we say that $G(z)$ and $\tilde{G}(z)$ are equivalent encoders. The following lemma is a simple result of linear algebra:

LEMMA 2.1. *Two $n \times k$ rational encoders $G(z)$ and $\tilde{G}(z)$ are equivalent with respect to either Definition A or A' if and only if there is a $k \times k$ invertible rational matrix $R(z)$ such that $\tilde{G}(z) = G(z)R(z)$.*

It follows from this lemma that Definition A and Definition A' are completely equivalent with respect to equivalence of encoders.

From an algebraic point of view we can identify a convolutional code in the sense of Definition A or Definition A' through an equivalence class of rational matrices. The following theorem singles out a set of very desirable encoders inside each equivalence class.

THEOREM 2.2. *Let $G(z)$ be a $n \times k$ rational encoding matrix of rank k defining a code $\mathcal{C}$. Then there is a $k \times k$ invertible rational matrix $R(z)$ such that $\tilde{G}(z) = G(z)R(z)$ has the properties:*
(i) *$\tilde{G}(z)$ is a polynomial matrix.*
(ii) *$\tilde{G}(z)$ is right prime.*
(iii) *$\tilde{G}(z)$ is column reduced with column degrees $\{e_1, \ldots, e_k\}$.*
Furthermore, every polynomial encoding matrix of $\mathcal{C}$ which is right prime and column-reduced has (unordered) column degrees $\{e_1, \ldots, e_k\}$. Thus these indices are invariants of the convolutional code.

The essence of Theorem 2.2 was proved by Forney [6, Theorem 3]. In [8] Forney related the indices appearing in (iii) to the controllability and observability indices of a controllable and observable system. Paper [8] had an immense impact in the linear systems theory literature. We will follow here the suggestion of McEliece [35] and call these indices the *Forney indices* of the convolutional code, despite the fact that Theorem 2.2 can be traced back to the last century, when Kronecker, Hermite and in particular Dedekind and Weber studied matrices over the rationals and more general function fields. In Sections 4 and 5 we will make a distinction between the Forney indices as defined above and the Kronecker indices of a submodule of $\mathbb{F}^n[z]$.

In the coding literature [21, 38], an encoder satisfying conditions (i), (ii) and (iii) of Theorem 2.2 is called a *minimal basic encoder*.

So far we have used encoding matrices to describe a convolutional code. As is customary in linear algebra, one often describes a linear subspace as the kernel of a matrix. This leads to the notion of a *parity-check matrix*. The following theorem is well known (see e.g. [38]).

THEOREM 2.3. *Let $\mathcal{C} \subset \mathbb{F}^n((z))$ be a rank-k convolutional code in the sense of Definition A'. Then there exists an $r \times n$ matrix $H(z)$ such that the code is equivalently described as the kernel of $H(z)$:*

$$\mathcal{C} = \{ \ c(z) \in \mathbb{F}^n((z)) \ \mid \ H(z)c(z) = 0 \ \}.$$

Moreover, it is possible to choose $H(z)$ in such a way that:
(i) $H(z)$ *is a polynomial matrix.*
(ii) $H(z)$ *is left prime.*
(iii) $H(z)$ *is row-reduced having row degrees $\{f_1, \ldots, f_r\}$.*
Furthermore, every polynomial parity check matrix of $\mathcal{C}$ which is left prime and row reduced will have (unordered) row degrees $\{f_1, \ldots, f_r\}$. Thus these indices are invariants of the convolutional code.

Properties (i)–(iii) essentially follow from the fact that the transpose $H^t(z)$ is a generator matrix for the dual (orthogonal) code $\mathcal{C}^\perp$.

The set of indices $\{e_1, \ldots, e_k\}$ and $\{f_1, \ldots, f_r\}$ differ in general, their sum is however always the same, and is called the *degree* of the convolutional code. One says that a rank-k code $\mathcal{C} \subset \mathbb{F}^n((z))$ has *transmission rate* k/n, *controller memory* $m := \max\{e_1, \ldots, e_k\}$ and *observer memory* $n := \max\{f_1, \ldots, f_r\}$.

Another important code parameter is the *free distance*. The free distance of a code measures the smallest distance between any two different code words, and is formally defined as:

$$(2.3) \qquad d_{\text{free}}(\mathcal{C}) := \min_{\substack{u,v \in \mathcal{C} \\ u \neq v}} \sum_{t \in \mathbb{Z}} d_H(u_t, v_t),$$

where $d_H(\ ,\ )$ denotes the usual Hamming distance on $\mathbb{F}^n$.

3. The symbolic dynamics point of view. In this section we present a definition of convolutional codes as it can be found in the symbolic dynamics literature [24, 29, 32]. Convolutional codes in this framework are exactly the linear, compact, irreducible and shift-invariant subsets of $\mathbb{F}^n[[z, z^{-1}]]$. In order to make this precise, we will have to develop some basic notions from symbolic dynamics.

In the sequel we will work with the finite alphabet $\mathcal{A} := \mathbb{F}^n$. A *block* over the alphabet $\mathcal{A}$ is a finite sequence $\beta = x_1 x_2 \ldots x_k$ consisting of k

elements $x_i \in \mathcal{A}$. If $w = w(z) = \sum_i w_i z^i \in \mathbb{F}^n[[z, z^{-1}]]$ is a sequence, one says that the block β occurs in w if there is some integer j such that $\beta = w_j w_{j+1} \ldots w_{k+j-1}$. If $X \subset \mathbb{F}^n[[z, z^{-1}]]$ is any subset, we denote by $\mathcal{B}(X)$ the set of blocks which occur in some element of X.

The fundamental objects in symbolic dynamics are the *shift spaces*. For this let $\mathcal{F}$ be a set of blocks, possibly infinite.

DEFINITION 3.1. The subset $X \subset \mathbb{F}^n[[z, z^{-1}]]$ consisting of all sequences $w(z)$ which do not contain any of the (forbidden) blocks of $\mathcal{F}$ is called a *shift space*.

The left-shift operator is the $\mathbb{F}$-linear map

$$(3.1) \qquad \sigma: \mathbb{F}[[z, z^{-1}]] \longrightarrow \mathbb{F}[[z, z^{-1}]], \quad w(z) \longmapsto z^{-1}w(z).$$

Let I_n be the $n \times n$ identity matrix. The shift map σ extends to the shift map

$$\sigma I_n: \mathbb{F}^n[[z, z^{-1}]] \longrightarrow \mathbb{F}^n[[z, z^{-1}]].$$

One says that $X \subset \mathbb{F}^n[[z, z^{-1}]]$ is a *shift-invariant set* if $(\sigma I_n)(X) \subset X$. Clearly shift spaces are shift-invariant subsets of $\mathbb{F}^n[[z, z^{-1}]]$.

It is possible to characterize shift spaces in a topological manner. For this we will introduce a metric on $\mathbb{F}^n[[z, z^{-1}]]$:

DEFINITION 3.2. If $v(z) = \sum_i v_i z^i$ and $w(z) = \sum_i w_i z^i$ are both elements of $\mathbb{F}^n[[z, z^{-1}]]$ we define their distance through:

$$(3.2) \qquad d(v(z), w(z)) := \sum_{i \in \mathbb{Z}} 2^{-|i|} d_H(v_i, w_i).$$

In this metric two elements $v(z), w(z)$ are 'close' if they coincide over a 'large block around zero'. One readily verifies that $d(\ ,\ )$ indeed satisfies all the properties of a metric and therefore induces a topology on $\mathbb{F}^n[[z, z^{-1}]]$. Using this topology we can characterize shift spaces:

THEOREM 3.3. *A subset of* $\mathbb{F}^n[[z, z^{-1}]]$ *is a shift space if and only if it is shift-invariant and compact.*

Proof. The metric introduced in Definition 3.2 is equivalent to the metric described in [29, Example 6.1.10]. The induced topologies are therefore the same. The result follows therefore from [29, Theorem 6.1.21]. $\qquad\square$

The topological space $\mathbb{F}^n[[z, z^{-1}]]$ is a typical example of a linearly compact vector space, a notion introduced by S. Lefschetz. There is a large theory on linearly compact vector spaces, and several of the results which we are going to derive are valid in this broader context. We refer the interested reader to [25, §10] for more details.

A further important concept is irreducibility which will turn out to be equivalent to the concept of controllability in our concrete setting.

DEFINITION 3.4. A shift space $X \subset \mathbb{F}^n[[z, z^{-1}]]$ is called *irreducible* if for every ordered pair of blocks β, γ of $\mathcal{B}(X)$ there is a block μ such that the concatenated block $\beta\mu\gamma$ is in $\mathcal{B}(X)$.

We are now prepared to give the symbolic dynamics definition for a convolutional code and to work out the basic properties for these codes.

DEFINITION B. A linear, compact, irreducible and shift-invariant subset of $\mathbb{F}^n[[z, z^{-1}]]$ is called a convolutional code.

This is an abstract definition and it is not immediately clear how one should encode messages with such convolutional codes. The following will make this clear.

Let $G(z)$ be a $n \times k$ matrix with entries in the ring of Laurent polynomials $\mathbb{F}[z, z^{-1}]$. Consider the encoding map:

$$(3.3) \qquad \varphi : \ \mathbb{F}^k[[z, z^{-1}]] \longrightarrow \mathbb{F}^n[[z, z^{-1}]], \ m(z) \longmapsto c(z) = G(\sigma)m(z).$$

In terms of polynomials the map φ is simply described through $m(z) \longmapsto c(z) = G(z^{-1})m(z)$.

Recall that a continuous map is called closed if the image of a closed set is closed. Using the fact that $\mathbb{F}^n[[z, z^{-1}]]$ is compact, one (easily) proves the following result:

LEMMA 3.5. *The encoding map* (3.3) *is* $\mathbb{F}$-*linear, continuous and closed.*

Clearly $\mathrm{im}\,(\varphi)$ is also shift-invariant, and one shows [29] that the image of an irreducible set under φ is irreducible again.

In summary we have shown that $\mathrm{im}\,(\varphi)$ describes a convolutional code in the sense of Definition B. Actually the converse is true as well:

THEOREM 3.6. $\mathcal{C} \subset \mathbb{F}^n[[z, z^{-1}]]$ *is a convolutional code in the sense of Definition B if and only if there exists a Laurent polynomial matrix* $G(z)$ *such that* $\mathcal{C} = \mathrm{im}\,(\varphi)$, *where* φ *is the map in* (3.3).

A proof of this theorem will be given in the next section after Theorem 4.8.

The question now arises how Definition B relates to Definition A and Definition A'. The following theorem will provide a partial answer to this question.

THEOREM 3.7. *Assume that* $\mathcal{C} \subset \mathbb{F}^n[[z, z^{-1}]]$ *is a nonzero convolutional code in the sense of Definition A or Definition A'. Then* $\mathcal{C}$ *is not closed, but the closure of* $\bar{\mathcal{C}}$ *of* $\mathcal{C}$ *is a convolutional code in the sense of Definition B.*

Proof. Let $G(z)$ be a minimal basic encoder of $\mathcal{C}$ and let $w(z) \in \mathbb{F}^n[z]$ be the first column of $G(z)$. Note that $w(z) \in \mathcal{C}$ and that there is at least one entry of $w(z)$ which does not contain the factor $(z - 1)$. Let $\phi_N(z) := \sum_{i=-N}^{N} z^i \in \mathbb{F}[z, z^{-1}]$ and consider the sequence of code words $w^N(z) := \phi_N(z)w(z)$. For each $N > 0$ one has that $w^N(z) \in \mathcal{C}$. However

$\lim_{N \to \infty} w^N(z)$ is in $\mathbb{F}^n[[z, z^{-1}]] \setminus \mathbb{F}^n((z)) \subset \mathbb{F}^n[[z, z^{-1}]] \setminus \mathcal{C}$. This shows that $\mathcal{C}$ is not a closed set inside $\mathbb{F}^n[[z, z^{-1}]]$. The closure $\bar{\mathcal{C}}$ is obtained by extending the input space $F^k((z))$ to all of $F^k[[z, z^{-1}]]$. The image of $F^k[[z, z^{-1}]]$ under the encoding map (3.3) is closed by Lemma 3.5, hence the closure is a code in the sense of Definition B. $\qquad \square$

Actually one can show that there is a bijective correspondence between the convolutional codes in the sense of Definition A (respectively Definition A′) and the convolutional codes in the sense of Definition B, as we will show in Theorem 7.1 and Theorem 7.2. It is also worthwhile to remark that already in 1983 Staiger published a paper [47] where he studied the closure of convolutional codes generated by a polynomial generator matrix.

In analogy to Lemma 2.1, one has:

LEMMA 3.8. *Two $n \times k$ encoding matrices $G(z)$ and $\tilde{G}(z)$ defined over the Laurent polynomial ring $\mathbb{F}[z, z^{-1}]$ are equivalent with respect to Definition B if and only if there is a $k \times k$ invertible rational matrix $R(z)$ such that $\tilde{G}(z) = G(z)R(z)$.*

We leave the proof again as an exercise for the reader. We remark that rational transformations of the form $R(z)$ are needed to describe the equivalence, even though it is in general not possible to use a rational encoder $G(z)$ in the encoding procedure (3.3). This is simply due to the fact that in general the multiplication of an element of $\mathbb{F}(z)$ with an element of $\mathbb{F}[[z, z^{-1}]]$ is not defined. The following example should make this clear. (Compare also with Remark 4.4.)

EXAMPLE 3.9. Consider $f(z) = \frac{1}{1-z} = \sum_{i=0}^{\infty} z^i \in \mathbb{F}(z)$ and $g(z) = \sum_{i=-\infty}^{\infty} z^i \in \mathbb{F}[[z, z^{-1}]]$. Trying to multiply the two power series $f(z), g(z)$ would result in a power series in which each coefficient would be infinite.

In the same way as at the end of Section 2 we define the transmission rate, the degree, the memory and the free distance of a convolutional code $\mathcal{C}$ in the sense of Definition B.

4. Linear time-invariant behaviors. In this section we will take the point of view that a convolutional code is a linear time-invariant behavior in the sense of Willems [50, 51, 52]. Of course behavioral system theory is quite general, allowing all kinds of time axes and signal spaces. In order to relate the behavioral concepts to the previous points of view, we will restrict our study to linear behaviors in $(\mathbb{F}^n)^{\mathbb{Z}} = \mathbb{F}^n[[z, z^{-1}]]$ and $(\mathbb{F}^n)^{\mathbb{Z}+} = \mathbb{F}^n[[z]]$.

Let σ be the shift operator defined in (3.1). One says that a subset $\mathcal{B} \subset \mathbb{F}^n[[z, z^{-1}]]$ is *time-invariant* if $(\sigma I_n)(\mathcal{B}) \subset \mathcal{B}$. The concept therefore coincides with the symbolic dynamics concept of shift-invariance.

In addition to linearity and time-invariance, there is a third important concept usually required of a time-invariant behavior:

DEFINITION 4.1. A behavior $\mathcal{B} \subset \mathbb{F}^n[[z, z^{-1}]]$ is said to be *complete* if $w \in \mathbb{F}^n[[z, z^{-1}]]$ belongs to $\mathcal{B}$ whenever $w|_J$ belongs to $\mathcal{B}|_J$ for every finite subinterval $J \subset \mathbb{Z}$.

The definition simply says that $\mathcal{B}$ is complete if membership can be decided on the basis of finite windows. Completeness is an important well behavedness property for linear time-invariant behaviors, as Willems [50, p. 567] emphasized with the remark:

> As such, it can be said that the study of non-complete systems does not fall within the competence of system theorists and could be left to cosmologists or theologians.

In Definition 3.2 we introduced a metric on the vector space $\mathbb{F}^n[[z, z^{-1}]]$. We remark that with respect to this metric a subset $\mathcal{B} \subset \mathbb{F}^n[[z, z^{-1}]]$ is complete if and only if every Cauchy sequence converges inside $\mathcal{B}$. In other words, the completeness notion of Definition 4.1 coincides with the usual topological notion of completeness.

The following result is known for linearly compact vector spaces, a proof can be found in [50]:

LEMMA 4.2. *A linear subset $\mathcal{B} \subset \mathbb{F}^n[[z, z^{-1}]]$ is complete if and only if it is closed and hence compact.*

With these preliminaries we can define a convolutional code as follows:

DEFINITION C. A linear, time-invariant and complete subset $\mathcal{B} \subset \mathbb{F}^n[[z, z^{-1}]]$ is called a convolutional code.

It is immediate from Lemma 4.2 that the convolutional codes defined in Definition B are complete and that Definition C is more general than Definition B, since no irreducibility is required. It also follows from Theorem 3.7 and Lemma 4.2 that the convolutional codes defined in Definition A and Definition A' are in general not complete.

Before we elaborate on these differences we would like also to treat the situation when the time axis is $\mathbb{Z}_+$ since traditionally a large part of linear systems theory has been concerned with systems defined on the positive time axis. We first define the left-shift operator acting on $(\mathbb{F}^n)^{\mathbb{Z}_+} = \mathbb{F}^n[[z]]$ through:

$$(4.1) \qquad \sigma : \mathbb{F}[[z]] \longrightarrow \mathbb{F}[[z]], \quad w(z) \longmapsto z^{-1}(w(z) - w(0)).$$

We have used the same symbol as in (3.1) since the context will always make it clear if we work over $\mathbb{Z}$ or $\mathbb{Z}_+$. In analogy to (3.1) σ extends to the shift map $\sigma I_n : \mathbb{F}^n[[z]] \longrightarrow \mathbb{F}^n[[z]]$, and one says a subset $X \subset \mathbb{F}^n[[z]]$ is time-invariant if $(\sigma I_n)(X) \subset X$. Notice however that the map of (4.1), unlike that of (3.1), is not invertible.

With this we have:

DEFINITION C'. A linear, time-invariant and complete subset $\mathcal{B} \subset \mathbb{F}^n[[z]]$ is called a convolutional code.

The following fundamental theorem was proved by Willems [50, Theorem 5].

THEOREM 4.3. *A subset $B \subset \mathbb{F}^n[[z, z^{-1}]]$ (respectively a subset $B \subset \mathbb{F}^n[[z]]$) is linear, time-invariant and complete if and only if there is a $r \times n$ matrix $P(z)$ having entries in $\mathbb{F}[z]$ such that*

$$(4.2) \qquad\qquad B = \{ w(z) \mid P(\sigma)w(z) = 0 \}.$$

By Lemma 3.5 the linear map $\psi : \mathbb{F}^n[[z, z^{-1}]] \longrightarrow \mathbb{F}^n[[z, z^{-1}]]$, $w(z) \longmapsto P(\sigma)w(z)$ is continuous and its kernel is therefore a complete set. It is therefore immediate that the behavior defined in (4.2) is linear, time-invariant and complete. The harder part of Theorem 4.3 is the converse statement.

Equation (4.2) is often referred to as a kernel (or AR) representation of a behavioral system. We will denote a behavior having the form (4.2) by $\ker P(\sigma)$. By contrast, the encoding map φ defined in (3.3) describes an image (or MA) representation of the behavior $\mathrm{im}\,(\varphi) = \mathrm{im}\,G(\sigma)$.

The most general representation is an ARMA representation. For this let $P(z)$ and $G(z)$ be matrices of size $r \times n$ and $r \times k$ respectively, having entries in the Laurent polynomial ring $\mathbb{F}[z, z^{-1}]$. Then

$$(4.3) \qquad B = \Big\{ w(z) \in \mathbb{F}^n[[z, z^{-1}]] \mid \exists m(z) \in \mathbb{F}^k[[z, z^{-1}]] : \\ P(\sigma)w(z) = G(\sigma)m(z) \Big\}$$

is called an ARMA model. One immediately verifies that the set B is linear and time-invariant. It is a direct consequence of Lemma 3.5 that B is also closed and hence complete. Theorem 4.3 therefore states that it is possible to eliminate the so called 'latent variable' $m(z)$ and describe the behavior B by a simpler kernel representation of the form (4.2). It follows in particular that the code $\mathrm{im}\,(\varphi) = \mathrm{im}\,G(\sigma)$ defined in (3.3) has an equivalent kernel representation of the form (4.2) but that in general the converse is not true.

REMARK 4.4. As we explained in Section 2 it is quite common to use rational encoders for convolutional codes. In the ARMA model (4.3) we required that the entries of $P(z)$ and $G(z)$ be from the Laurent polynomial ring. If $P(z)$ and $G(z)$ were rational matrices, then the behavior $B \subset \mathbb{F}^n[[z, z^{-1}]]$ appearing in (4.3) might not be well defined, as we showed in Example 3.9. On the other hand if one restricts the behavior to the positive time axis $\mathbb{Z}_+$, i.e. if one assumes that $B \subset \mathbb{F}^n[[z]]$, then the set (4.3) is defined even if $P(z)$ and $G(z)$ are rational encoders. This is certainly one reason why much classical system theory focused on shift spaces $B \subset \mathbb{F}^n[[z]]$ or $B \subset \mathbb{F}^n((z))$.

In the sequel we will concentrate on representations of the form (4.2). Again the question arises, when are two kernel representations equivalent?

LEMMA 4.5. *Two $r \times n$ matrices $P(z)$ and $\tilde{P}(z)$ defined over the Laurent polynomial ring $\mathbb{F}[z, z^{-1}]$ describe the same behavior $\ker P(\sigma) = \ker \tilde{P}(\sigma) \subset \mathbb{F}^n[[z, z^{-1}]]$ if and only if there is a $r \times r$ matrix $U(z)$, unimodular over $\mathbb{F}[z, z^{-1}]$, such that $\tilde{P}(z) = U(z)P(z)$.*

Proof. [52, Proposition III.3]. $\square$

Similarly, if $P(z)$ and $\tilde{P}(z)$ are defined over $\mathbb{F}[z]$, then these matrices define the same behavior $\ker P(\sigma) = \ker \tilde{P}(\sigma) \subset \mathbb{F}^n[[z]]$ if and only if there is a matrix $U(z)$, unimodular over $\mathbb{F}[z]$, such that $\tilde{P}(z) = U(z)P(z)$.

The major difference between Definition B and Definition C seems to be that Definition C does not require irreducibility. This last concept corresponds to the term controllability (see [10]) in systems theory. We first start with some notation taken from [42]:

For a sequence $w = \sum_{-\infty}^{\infty} w_i z^i \in \mathbb{F}^n[[z, z^{-1}]]$, we use the symbol w^+ to denote the 'right half' $\sum_0^{\infty} w_i z^i$ and the symbol w^- to denote the 'left half' $\sum_{-\infty}^0 w_i z^i$.

DEFINITION 4.6.   A behavior $\mathcal{B}$ defined on $\mathbb{Z}$ is said to be *controllable* if there is some integer ℓ such that for every w and w' in $\mathcal{B}$ and every integer j there exists a $w'' \in \mathcal{B}$ such that $(z^j w'')^- = (z^j w)^-$ and $(z^{j+\ell} w'')^+ = (z^{j+\ell} w')^+$.

REMARK 4.7.   Loeliger and Mittelholzer [30] speak of *strongly controllable* if a behavior satisfies the conditions of Definition 4.6. 'Weakly controllable' in contrast requires an integer ℓ which may depend on the trajectories w and w'. The notions are equivalent in our concrete setting.

We leave it as an exercise for the reader to show that irreducibility as introduced in Definition 3.4 is equivalent to controllability for linear, time-invariant and complete behaviors $\mathcal{B} \subset \mathbb{F}^n[[z, z^{-1}]]$. The next theorem gives equivalent conditions for a behavior to be controllable.

THEOREM 4.8.   *(cf. [51, Prop. 4.3]) Let $P(z)$ be a $r \times n$ matrix of rank r defined over $\mathbb{F}[z, z^{-1}]$. The following conditions are equivalent:*

(i)   *The behavior $\mathcal{B} = \ker P(\sigma) = \{w(z) \in \mathbb{F}^n[[z, z^{-1}]] \mid P(\sigma)w(z) = 0\}$ is controllable.*

(ii) $P(z)$ *is left prime over $\mathbb{F}[z, z^{-1}]$.*

(iii)   *The behavior $\mathcal{B}$ has an image representation. This means there exists an $n \times k$ matrix $G(z)$ defined over $\mathbb{F}[z, z^{-1}]$ such that*

$$\mathcal{B} = \{w(z) \in \mathbb{F}^n[[z, z^{-1}]] \mid \exists m(z) \in \mathbb{F}^k[[z, z^{-1}]] : w(z) = G(\sigma)m(z)\}.$$

Combining the theorem with the facts that completeness corresponds to compactness and irreducibility corresponds to controllability gives a proof of Theorem 3.6.

We conclude the section by defining some parameters of a linear, time-invariant and complete behavior. For simplicity we will do this in an algebraic manner. We will first treat behaviors $\mathcal{B} \subset \mathbb{F}^n[[z]]$, i.e. behaviors in the sense of Definition C'. In Remark 4.10 we will explain how the definitions have to be adjusted for behaviors defined on the time axis $\mathbb{Z}$.

Assume that $P(z)$ is a $r \times n$ polynomial matrix of rank r defining the behavior $\mathcal{B} = \ker P(\sigma)$. There exists a matrix $U(z)$, unimodular over

$\mathbb{F}[z]$, such that $\tilde{P}(z) = U(z)P(z)$ is row-reduced with ordered row degrees $\nu_1 \geq \ldots \geq \nu_r$. The indices $\nu = (\nu_1, \ldots, \nu_r)$ are invariants of the row module of $P(z)$ (and hence also invariants of the behavior $\mathcal{B}$), and are sometimes referred to as the *Kronecker indices* or *observability indices* of $\mathcal{B}$. The invariant $\delta := \sum_{i=1}^{r} \nu_i$ is called the *McMillan degree* of the behavior $\mathcal{B}$. If we think of $\mathcal{B}$ as a convolutional code in the sense of Definition C' then we say that $\mathcal{B}$ has transmission rate $\frac{n-r}{n}$. Finally, the free distance of the code is defined as in (2.3).

REMARK 4.9. The Kronecker indices ν are in general different from the minimal row indices (in the sense of Forney [8]) of the $\mathbb{F}(z)$-vector space generated by the rows of $P(z)$. They coincide with the minimal row indices if and only if $P(z)$ is left prime.

REMARK 4.10. If $\mathcal{B} \subset \mathbb{F}^n[[z, z^{-1}]]$ is a linear, time-invariant and complete behavior, then we can define parameters like the Kronecker indices and the McMillan degree in the following way: Assume $P(z)$ has the property that $\mathcal{B} = \ker P(\sigma)$. There exists a matrix $U(z)$, unimodular over $\mathbb{F}[z, z^{-1}]$, such that $\tilde{P}(z) = U(z)P(z)$ is row-reduced and $P(0)$ has full row rank r. One shows again that the row degrees of $\tilde{P}(z)$ are invariants of the behavior. The McMillan degree, the transmission rate and the free distance are then defined in the same way as for behaviors $\mathcal{B} \subset \mathbb{F}^n[[z]]$.

5. The module point of view. Fornasini and Valcher [5, 48] and the present author in joint work with Schumacher, Weiner and York [42, 44, 49] proposed a module-theoretic approach to convolutional codes. The module point of view simplifies the algebraic treatment of convolutional codes to a large degree, and this simplification is probably almost necessary if one wants to study convolutional codes in a multidimensional setting [5, 48, 49].

From a systems theoretic point of view, the module-theoretic approach studies linear time-invariant systems whose states start at zero and return to zero in finite time. Such dynamical systems have been studied by Hinrichsen and Prätzel-Wolters [18, 19], who recognized these systems as convenient objects for the study of systems equivalence.

In our development we will again deal with the time axes $\mathbb{Z}$ and $\mathbb{Z}_+$ in a parallel manner.

DEFINITION D. A submodule $\mathcal{C}$ of $\mathbb{F}^n[z, z^{-1}]$ is called a convolutional code.

We like the module-theoretic language. If one prefers to define everything in terms of trajectories then one could equivalently define $\mathcal{C}$ as $\mathbb{F}$-linear, time-invariant subset of $\mathbb{F}^n[[z, z^{-1}]]$ whose elements have finite support.

The analogous definition for codes supported on the positive time axis $\mathbb{Z}_+$ is:

DEFINITION D'. A submodule $\mathcal{C}$ of $\mathbb{F}^n[z]$ is called a convolutional code.

Since both the rings $\mathbb{F}[z, z^{-1}]$ and $\mathbb{F}[z]$ are principal ideal domains (PID), a convolutional code $\mathcal{C}$ has always a well-defined rank k, and there is a full-rank matrix $G(z)$ of rank k such that $\mathcal{C} = \text{colsp}_{\mathbb{F}[z, z^{-1}]} G(z)$ (respectively $\mathcal{C} = \text{colsp}_{\mathbb{F}[z]} G(z)$ if $\mathcal{C}$ is defined as in Definition D'). We will call $G(z)$ an encoder of $\mathcal{C}$, and the map

$$(5.1) \qquad \varphi : \ \mathbb{F}^k[z, z^{-1}] \longrightarrow \mathbb{F}^n[z, z^{-1}], \ m(z) \longmapsto c(z) = G(z)m(z)$$

an encoding map.

REMARK 5.1. In contrast to the situation of Section 3, it is possible to define a convolutional code in the sense of Definition D (respectively Definition D') using a rational encoder. For this, assume that $G(z)$ is an $n \times k$ matrix with entries in $\mathbb{F}(z)$. Then

$$\mathcal{C} = \left\{ \ c(z) \in \mathbb{F}^n[z, z^{-1}] \ | \ \exists m(z) \in \mathbb{F}^k[z, z^{-1}] : \ c(z) = G(z)m(z) \ \right\}$$

defines a submodule of $\mathbb{F}^n[z, z^{-1}]$. Note that the map (5.1) involving a rational encoding matrix $G(z)$ has to be 'input-restricted' in this case.

In analogy to Lemma 3.8 we have:

LEMMA 5.2. *Two $n \times k$ matrices $G(z)$ and $\tilde{G}(z)$ defined over the Laurent polynomial ring $\mathbb{F}[z, z^{-1}]$ (respectively over the polynomial ring $\mathbb{F}[z]$) generate the same code $\mathcal{C} \subset \mathbb{F}^n[z, z^{-1}]$ (respectively $\mathcal{C} \subset \mathbb{F}^n[z]$) if and only if there is a $k \times k$ matrix $U(z)$, unimodular over $\mathbb{F}[z, z^{-1}]$ (respectively over $\mathbb{F}[z]$), such that $\tilde{G}(z) = G(z)U(z)$.*

As we already mentioned earlier convolutional codes in the sense of Definitions D and D' are linear and time-invariant. The following theorem answers any question about controllability (i.e. irreducibility) and completeness.

THEOREM 5.3. *A nonzero convolutional code with either Definition D or D' is controllable and incomplete.*

Sketch of Proof. The proof of the completeness part of the Theorem is analogous to the proof of Theorem 3.7. In order to show controllability, let $G(z)$ be an encoding matrix for a code $\mathcal{C} \subset \mathbb{F}^n[z]$ and consider two code words $w(z) = G(z)(a_0 + a_1 + \cdots + a_s z^s)$ and $w'(z) = G(z)(b_0 + b_1 + \cdots + b_s z^s)$. The codeword $w''(z)$ required by Definition 4.6 can be constructed in the form

$$G(z)(a_0 + a_1 + \cdots + a_j z^j + b_{j+\ell} z^{j+\ell} + \cdots + \cdots + b_s z^s). \qquad \square$$

Submodules of $\mathbb{F}^n[z, z^{-1}]$ (respectively of $\mathbb{F}^n[z]$) form the Pontryagin dual of linear, time-invariant and complete behaviors in $\mathbb{F}^n[[z, z^{-1}]]$ (respectively $\mathbb{F}^n[[z]]$). In the following we follow [42] and explain this in a very explicit way when the time axis is $\mathbb{Z}$. Of course everything can be done *mutatis mutandis* when the time axis is $\mathbb{Z}_+$.

Consider the bilinear form:

$$(\, ,\,): \quad \mathbb{F}^n[[z, z^{-1}]] \times \mathbb{F}^n[z, z^{-1}] \quad \longrightarrow \quad \mathbb{F}$$

(5.2)

$$(w, v) \quad \mapsto \quad \sum_{i=-\infty}^{\infty} \langle w_i, v_i \rangle,$$

where $\langle\, ,\, \rangle$ represents the standard dot product on $\mathbb{F}^n$. One shows that $(\, ,\,)$ is well defined and nondegenerate, in particular because there are only finitely many nonzero terms in the sum. For any subset $\mathcal{C}$ of $\mathbb{F}^n[z, z^{-1}]$ one defines the annihilator

(5.3) $$\mathcal{C}^\perp = \{ w \in \mathbb{F}^n[[z, z^{-1}]] \mid (w, v) = 0, \forall v \in \mathcal{C} \}$$

and the annihilator of a subset $\mathcal{B}$ of $\mathbb{F}^n[[z, z^{-1}]]$ is

(5.4) $$\mathcal{B}^\perp = \{ v \in \mathbb{F}^n[z, z^{-1}] \mid (w, v) = 0, \forall w \in \mathcal{B} \}.$$

The relation between these two annihilator operations is given by:

THEOREM 5.4. *If $\mathcal{C} \subseteq \mathbb{F}^n[z, z^{-1}]$ is a convolutional code with generator matrix $G(z)$, then $\mathcal{C}^\perp$ is a linear, left-shift-invariant and complete behavior with kernel representation $P(z) = G^t(z)$. Conversely, if $\mathcal{B} \subseteq \mathbb{F}^n[[z, z^{-1}]]$ is a linear, left-shift-invariant and complete behavior with kernel representation $P(z)$, then $\mathcal{B}^\perp$ is a convolutional code with generator matrix $G(z) = P^t(z)$.*

REMARK 5.5. An elementary proof of Theorem 5.4 in the case of the positive time axis $\mathbb{Z}_+$ is given in [42].

REMARK 5.6. Theorem 5.4 is a special instance of a broad duality theory between solution spaces of difference equations on the one hand and modules on the other, for which probably the most comprehensive reference is Oberst [37]. In this article Oberst [37, p. 22] works with a bilinear form which is different from (5.2). This bilinear form induces however the same duality as shown in [16]. Extensions of duality results to group codes were derived by Forney and Trott in [12].

For finite support convolutional codes in the sense of Definition D or Definition D′ the crucial issue is *observability*. In the literature there have been several definitions of observability [4, 11, 5, 9, 30, 42] and it is not entirely clear how these definitions relate to each other.

In the sequel we will follow [4, 42].

DEFINITION 5.7. (cf. [4, Prop. 2.1]) A code $\mathcal{C}$ is *observable* if there exists an integer N such that, whenever the supports of v and v' are separated by a distance of at least N and $v + v' \in \mathcal{C}$, then also $v \in \mathcal{C}$ and $v' \in \mathcal{C}$.

With this we have the 'Pontryagin dual statement' of Theorem 4.8:

THEOREM 5.8. *(cf. [42, Prop. 2.10]) Let $G(z)$ be a $n \times k$ matrix of rank k defined over $\mathbb{F}[z, z^{-1}]$. The following conditions are equivalent:*

(i) *The convolutional code* $C = \mathrm{colsp}_{\mathbb{F}[z,z^{-1}]} G(z)$ *is observable.*

(ii) $G(z)$ *is right prime over* $\mathbb{F}[z, z^{-1}]$.

(iii) *The code* C *has a kernel representation. This means there exists an* $r \times n$ *'parity-check matrix'* $H(z)$ *defined over* $\mathbb{F}[z, z^{-1}]$ *such that*

$$C = \{ \, v(z) \in \mathbb{F}^n[z, z^{-1}] \mid H(z)v(z) = 0 \, \}.$$

REMARK 5.9. The concept of observability is clearly connected to the coding concept of non-catastrophicity. Indeed an encoder is non-catastrophic if and only if the code generated by this encoder is observable. In the context of Definition A (respectively Definition A') every code has a catastrophic as well as a non-catastrophic encoder. In the module setting of Definition D every encoder of an observable code is non-catastrophic and every encoder of an non-observable code is catastrophic. If one defines a convolutional code by Definition D then one could talk of a 'non-catastrophic convolutional code'. The term observable seems however much more appropriate.

As at the end of Section 4, we now define the code parameters. We do it only for codes given by Definition D' and leave it to the reader to adapt the definitions to codes given by Definition D.

Assume that $G(z)$ is an $n \times k$ polynomial matrix of rank k defining the code $C = \mathrm{colsp}_{\mathbb{F}[z]} G(z)$. There exists a unimodular matrix $U(z)$ such that $\tilde{G}(z) = G(z)U(z)$ is column-reduced with ordered column degrees $\kappa_1 \geq \ldots \geq \kappa_k$. The indices $\kappa = (\kappa_1, \ldots, \kappa_k)$ are invariants of the code C, which we call the *Kronecker indices* or *controllability indices* of C. The invariant $\delta := \sum_{i=1}^{r} \kappa_i$ is called the *degree* of the code C. The free distance of the code is defined as in (2.3). Finally we say that C has transmission rate $\frac{k}{n}$.

6. First-order representations. In this section we provide an overview of the different first-order representations (realizations) associated with the convolutional codes and encoding maps which we have defined.

We start with the encoding map (2.2). As is customary in most of the coding literature, we view the map (2.2) as an input-output operator from the message space to the code space. The existence of associated state spaces and realizations can be shown on an abstract level. Kalman [22, 23] first showed how the encoding map (2.2) can be 'factored' resulting in a realization of the encoding matrix φ. Fuhrmamnn [13] refined the realization procedure in an elegant way. (Compare also [15, 17].)

In the sequel we will simply assume that a realization algorithm exists. We summarize the main results in the following two theorems:

THEOREM 6.1. *Let* $T(z)$ *be a* $p \times m$ *proper transfer function of McMillan degree* δ. *Then there exist matrices* (A, B, C, D) *of size* $\delta \times \delta$, $\delta \times m$, $p \times \delta$ *and* $p \times m$ *respectively such that*

$$(6.1) \qquad T(z) = C(zI - A)^{-1}B + D.$$

The minimality conditions are that (A, B) forms a controllable pair and (A, C) forms an observable pair. Finally (6.1) is unique in the sense that if $T(z) = \tilde{C}(zI - \tilde{A})^{-1}\tilde{B} + \tilde{D}$ with $(\tilde{A}, \tilde{B})$ controllable and $(\tilde{A}, \tilde{C})$ observable, then there is a unique invertible matrix S such that

$$(6.2) \qquad (\tilde{A}, \tilde{B}, \tilde{C}, \tilde{D}) = (SAS^{-1}, SB, CS^{-1}, D).$$

Consider the encoding map (2.2) with generator matrix $G(z)$. Let $m(z) = \sum_{i=s}^{t} m_i z^i \in \mathbb{F}^k((z))$ and $c(z) = \sum_{i=s}^{t} c_i z^i \in \mathbb{F}^n((z))$ be the sequence of message and code symbols respectively. Then one has:

THEOREM 6.2. *Assume that $G(z)$ has the property that $\operatorname{rank} G(0) = k$. Then $G(z^{-1})$ is a proper transfer function, and by Theorem 6.1 there exist matrices (A, B, C, D) of appropriate sizes such that $G(z^{-1}) = C(zI - A)^{-1}B + D$. The dynamics of (2.2) are then equivalently described by:*

$$(6.3) \qquad \begin{aligned} x_{t+1} &= Ax_t + Bm_t, \\ c_t &= Cx_t + Dm_t. \end{aligned}$$

The realization (6.3) is useful if one wants to describe the dynamics of the encoder $G(z)$. It is however less useful if one is interested in the construction of codes having certain properties. The problem is that every code $\mathcal{C}$ has many equivalent encoders whose realizations appear to be completely different.

EXAMPLE 6.3. The encoders

$$G(z) = \begin{pmatrix} \dfrac{1-z}{z-4} \\ \dfrac{1+z}{z-4} \end{pmatrix} \qquad \text{and} \qquad \tilde{G}(z) = \begin{pmatrix} \dfrac{1-z}{(z-2)(z+3)} \\ \dfrac{1+z}{(z-2)(z+3)} \end{pmatrix}$$

are equivalent since they define the same code in the sense of Definition A. The transfer functions $G(z^{-1})$ and $\tilde{G}(z^{-1})$ are however very different from a systems theory point of view. Indeed, they have different McMillan degrees, and over the reals the first is stable whereas the second is not. The state space descriptions are therefor very different for these encoders.

This example should make it clear that for the purpose of constructing good convolutional codes, representation (6.3) is not very useful.

We are now coming to the realization theory of the behaviors and codes of Section 4 and 5. We will continue with our algebraic approach. The results are stated for the positive time axis $\mathbb{Z}_+$, but they hold mutatis mutandis for the time axis $\mathbb{Z}$.

THEOREM 6.4 (Existence). *Let $P(z)$ be an $r \times n$ matrix of rank r describing a behavior $\mathcal{B}$ of the form (4.2) with McMillan degree δ. Let*

$k = n - r$. *Then there exist (constant) matrices G, F of size $\delta \times (\delta + k)$ and a matrix H of size $n \times (\delta + k)$ such that $\mathcal{B}$ is equivalently described by:*

$$
(6.4) \quad
\begin{aligned}
\mathcal{B} = \Big\{ w(z) \in \mathbb{F}^n\,[[z]] \mid \exists \zeta(z) \in \mathbb{F}^{\delta+k}\,[[z]] : \\
(\sigma G - F)\zeta(z) = 0, w(z) = H\zeta(z) \Big\}.
\end{aligned}
$$

Moreover the following minimality conditions will be satisfied:
(i) *G has full row rank;*
(ii) $\begin{bmatrix} G \\ H \end{bmatrix}$ *has full column rank;*
(iii) $\begin{bmatrix} zG-F \\ H \end{bmatrix}$ *is right prime.*

For a proof, see [26, Thm. 4.3] or [27, 41]. Equation (6.4) describes the behavior locally in terms of a time window of length 1. The computation of the matrices G, F, H from a kernel description is not difficult. It can even be done 'by inspection', i.e., just by rearranging the data [41]. The next result describes the extent to which minimal first-order realizations are unique. A proof is given in [26, Thm. 4.34].

THEOREM 6.5 (Uniqueness). *The matrices (G, F, H) are unique in the following way: If $(\tilde{G}, \tilde{F}, \tilde{H})$ is a second triple of matrices describing the behavior $\mathcal{B}$ through (6.4) and if the minimality conditions (i), (ii) and (iii) are satisfied, then there exist unique invertible matrices S and T such that*

$$
(6.5) \qquad (\tilde{G}, \tilde{F}, \tilde{G}) = (SGT^{-1}, SFT^{-1}, HT^{-1}).
$$

The relation to the traditional state-space theory is as follows: Assume that $P(z)$ can be partitioned into $P(z) = (Y(z)\, U(z))$ with $U(z)$ a square $r \times r$ matrix and $\deg \det U(z) = \delta$, the McMillan degree of the behavior $\mathcal{B}$. Assume that (G, F, H) provides a realization for $\mathcal{B}$ through (6.4). Then one shows that the pencil $\begin{bmatrix} zG-F \\ H \end{bmatrix}$ is equivalent to the pencil:

$$
(6.6) \qquad
\begin{bmatrix}
zI_\delta - A & B \\
0 & I_k \\
C & D
\end{bmatrix}.
$$

The minimality condition (iii) simply translates into the condition that (A, C) forms an observable pair, showing that the behavior $\mathcal{B}$ is observable. One also verifies that the matrices (A, B, C, D) form a realization of the proper transfer function $U(z)^{-1}Y(z)$ and that this is a minimal realization if and only if (A, B) forms a controllable pair. Finally (A, B) is controllable if and only if the behavior $\mathcal{B}$ is controllable.

The Pontryagin dual statements of Theorem 6.4 and 6.5 are (see [42]):
THEOREM 6.6 (Existence). *Let $G(z)$ be an $n \times k$ polynomial matrix generating a rate $\frac{k}{n}$ convolutional code $\mathcal{C} \subseteq \mathbb{F}^n\,[z]$ of degree δ. Then there*

exist $(\delta + n - k) \times \delta$ matrices K, L and a $(\delta + n - k) \times n$ matrix M (all defined over $\mathbb{F}$) such that the code C is described by

$$(6.7) \quad C = \{v(z) \in \mathbb{F}^n[z] \mid \exists x(z) \in \mathbb{F}^\delta[z] : \ zKx(z) + Lx(z) + Mv(z) = 0\}.$$

Moreover the following minimality conditions will be satisfied:
(i) *K has full column rank;*
(ii) *$[K \ M]$ has full row rank;*
(iii) *$[zK + L \mid M]$ is left prime.*

Equation (6.7) describes the behavior again locally in terms of a time window of length 1.

THEOREM 6.7 (Uniqueness). *The matrices (K, L, M) are unique in the following way: If $(\tilde{K}, \tilde{L}, \tilde{M})$ is a second triple of matrices describing the code C through (6.7) and if the minimality conditions (i), (ii) and (iii) are satisfied, then there exist unique invertible matrices T and S such that*

$$(6.8) \qquad (\tilde{K}, \tilde{L}, \tilde{M}) = (TKS^{-1}, TLS^{-1}, TM).$$

If $G(z)$ can be partitioned into $G(z) = \begin{bmatrix} Y(z) \\ U(z) \end{bmatrix}$ with $U(z)$ a square $k \times k$ matrix and $\deg \det U(z) = \delta$, the degree of the code C, then the pencil $[zK + L \mid M]$ is equivalent to the pencil:

$$(6.9) \qquad \begin{bmatrix} zI_\delta - A & 0_{\delta \times (n-k)} & -B \\ -C & I_{n-k} & -D \end{bmatrix}.$$

The minimality condition (iii) then translates into the condition that (A, B) forms a controllable pair, showing that the code C is controllable. One also verifies that the matrices (A, B, C, D) form a realization of the proper transfer function $Y(z)U(z)^{-1}$, that this is a minimal realization if and only if (A, C) forms an observable pair, and that this is the case if and only if the code C is observable. Finally, the Kronecker indices of C coincide with the controllability indices of the pair (A, B) [44].

The systems-theoretic meaning of the representation (6.9) is as follows (see [44]). Partition the code vector $v(z)$ into:

$$v(z) = \begin{bmatrix} y(z) \\ u(z) \end{bmatrix} \in \mathbb{F}^n[z]$$

and consider the equation:

$$(6.10) \qquad \begin{bmatrix} zI_\delta - A & 0_{\delta \times (n-k)} & -B \\ -C & I_{n-k} & -D \end{bmatrix} \begin{bmatrix} x(z) \\ y(z) \\ u(z) \end{bmatrix} = 0.$$

Let

$$x(z) = x_0 z^\gamma + x_1 z^{\gamma-1} + \ldots + x_\gamma; \quad x_t \in \mathbb{F}^\delta, t = 0, \ldots, \gamma,$$
$$u(z) = u_0 z^\gamma + u_1 z^{\gamma-1} + \ldots + u_\gamma; \quad u_t \in \mathbb{F}^k, t = 0, \ldots, \gamma,$$
$$y(z) = y_0 z^\gamma + y_1 z^{\gamma-1} + \ldots + y_\gamma; \quad y_t \in \mathbb{F}^{n-k}, t = 0, \ldots, \gamma.$$

Then (6.10) is satisfied if and only if

(6.11)
$$x_{t+1} = Ax_t + Bu_t,$$
$$y_t = Cx_t + Du_t,$$
$$v_t = \begin{pmatrix} y_t \\ u_t \end{pmatrix}, \quad x_0 = 0, \ x_{\gamma+1} = 0,$$

is satisfied. Note that the state-space representation (6.11) is different from the representation (6.3). Equation (6.11) describes the dynamics of the *systematic* and *rational* encoder

$$G(z)U^{-1}(z) = \begin{bmatrix} Y(z)U(z)^{-1} \\ I_k \end{bmatrix}.$$

The encoding map $u(z) \mapsto y(z) = G(z)U^{-1}(z)u(z)$ is input-restricted, i.e. $u(z)$ must be in the column module of $U(z)$ in order to make sure that $y(z)$ and $x(z)$ have finite support. In terms of systems theory, this simply means that the state should start at zero and return to zero in finite time. Linear systems satisfying these requirements have been studied by Hinrichsen and Prätzel-Wolters [18, 19].

7. Differences and similarities among the definitions. After having reviewed these different definitions for convolutional codes, we would like to make some comparison.

The definitions of Section 2 and Section 3 viewed convolutional codes as linear, time-invariant, controllable and observable behaviors, not necessarily complete. Definition C and Definition C′ were more general in the sense that non-controllable behaviors were accepted as codes. Definition D and Definition D′ were more general in the sense that non-observable codes were allowed.

In the following subsection we show that all definitions are equivalent for all practical purposes if one restricts oneself to controllable and observable codes.

7.1. Controllable and observable codes. Consider a linear, time-invariant, complete behavior $\mathcal{B} \subset \mathbb{F}^n[[z, z^{-1}]]$, i.e. a convolutional code in the sense of Definition C. Let

$$\mathcal{C} := \mathcal{B} \cap \mathbb{F}^n((z)).$$

Then one has

THEOREM 7.1. *$\mathcal{C}$ is a convolutional code in the sense of Definition A', and its completion $\bar{\mathcal{C}}$ is the largest controllable sub-behavior of $\mathcal{B}$. Moreover, one has a bijective correspondence between controllable behaviors $\mathcal{B} \subset \mathbb{F}^n[[z, z^{-1}]]$ and convolutional codes $\mathcal{C} \subset \mathbb{F}^n((z))$ in the sense of Definition A'.*

Sketch of Proof. Let $\mathcal{B} = \ker P(\sigma) = \{w(z) \in \mathbb{F}^n[[z, z^{-1}]] \mid P(\sigma)w(z) = 0\}$. If $\mathcal{B}$ is not controllable, then $P(z)$ is not left prime and one has a factorization $P(z) = V(z)\tilde{P}(z)$, where $\tilde{P}(z)$ is left prime and describes the controllable sub-behavior $\ker \tilde{P}(\sigma) \subset \mathcal{B}$. Since $\ker V(\sigma)$ is an autonomous behavior it follows that

$$\mathcal{C} = \mathcal{B} \cap \mathbb{F}^n((z)) = \ker P(\sigma) \cap \mathbb{F}^n((z)) = \ker \tilde{P}(\sigma) \cap \mathbb{F}^n((z)).$$

It follows (compare with Theorem 3.7) that the completion $\bar{\mathcal{C}} = \ker \tilde{P}(\sigma)$. □

Consider now a convolutional code $\mathcal{C} \subset \mathbb{F}^n((z))$ in the sense of Definition A'. Define:

$$\check{\mathcal{C}} := \mathcal{C} \cap \mathbb{F}^n[z, z^{-1}]$$
$$\check{\check{\mathcal{C}}} := \mathcal{C} \cap \mathbb{F}^n[z].$$

Conversely if $\mathcal{C} \subset \mathbb{F}^n[z]$ is a convolutional code in the sense of Definition D', then define:

$$\hat{\mathcal{C}} := \mathrm{span}_{\mathbb{F}[z, z^{-1}]}\{v(z) \mid v(z) \in \mathcal{C}\}.$$
$$\hat{\hat{\mathcal{C}}} := \mathrm{span}_{\mathbb{F}((z))}\{v(z) \mid v(z) \in \mathcal{C}\}.$$

By definition it is clear that $\hat{\mathcal{C}} \subset \hat{\hat{\mathcal{C}}}$ are convolutional codes in the sense of Definition D and Definition A' respectively.

THEOREM 7.2. *Assume that $\mathcal{C} \subset \mathbb{F}^n((z))$ is a convolutional code in the sense of Definition A'. Then $\check{\check{\mathcal{C}}} \subset \mathbb{F}^n[z]$ is an observable code in the sense of Definition A'. Moreover the operations $\hat{\ }$ and $\check{\ }$ induce a bijective correspondence between the observable codes $\mathcal{C} \subset \mathbb{F}^n[z]$ and convolutional codes $\mathcal{C} \subset \mathbb{F}^n((z))$ in the sense of Definition A'.*

Theorem 7.2 is essentially the Pontryagin dual statement of Theorem 7.1; we leave it to the reader to work out the details. Theorem 7.1 and 7.2 together show that there is a bijection between controllable and observable codes in the sense of one definition and another definition. For controllable and observable codes the code parameters like the rate k/n, the degree δ and the Forney (Kronecker) indices are all the same. Moreover the free distance is in every case the same as well. For all practical purposes one can therefore say that the frameworks are completely equivalent, if one is only interested in controllable and observable codes.

The advantage of Definition D (respectively Definition D$'$) over the other definitions lies in the fact that non-observable codes become naturally part of the theory. It also seems that for construction purposes the relation between quasi-cyclic codes and convolutional codes [33, 46] is best described in a module-theoretic framework.

Definition C (respectively Definition C$'$) allows one to introduce non-controllable codes in a natural way.

A Laurent series setting as in Definition A$'$ seems to be most natural if one is interested in the description of the encoder and/or syndrome former. Extensions of the Laurent series framework to multidimensional convolutional codes is however much less natural than the polynomial framework, which is why the theory of multidimensional convolutional codes has mainly been developed in a module-theoretic framework [5, 48, 49].

7.2. Duality. In (5.2) we introduced a bilinear form which induced a bijection between behaviors $\mathcal{B} \subset \mathbb{F}^n[[z, z^{-1}]]$ and modules $\mathcal{C} \subset \mathbb{F}^n[z, z^{-1}]$. This duality is a special instance of Pontryagin duality, and generalizes to group codes [12] and multidimensional systems [37].

In this subsection we show that the bilinear form (5.2) can also be used to obtain a duality between modules and modules (both in $\mathbb{F}^n[z, z^{-1}]$) or between behaviors and behaviors (both in $\mathbb{F}^n[[z, z^{-1}]]$).

For this let $\mathcal{C} \subset \mathbb{F}^n[z, z^{-1}]$ be a submodule. Define:

$$(7.1) \qquad \mathcal{C}^{\vdash} := \mathcal{C}^{\perp} \cap \mathbb{F}^n[z, z^{-1}].$$

One immediately verifies that $\mathcal{C}^{\vdash}$ is a submodule of $\mathbb{F}^n[z, z^{-1}]$, which necessarily is observable. One always has $\mathcal{C} \subset (\mathcal{C}^{\vdash})^{\vdash}$.

One can do something similar for behaviors. For this let $\mathcal{B} \subset \mathbb{F}^n[[z, z^{-1}]]$ be a behavior. Define:

$$(7.2) \qquad \mathcal{B}^{\vdash} := \left(\mathcal{B} \cap \mathbb{F}^n[z, z^{-1}]\right)^{\perp} = \overline{\mathcal{B}^{\perp}}.$$

Then it is immediate that $\mathcal{B}^{\vdash}$ is a controllable behavior, $(\mathcal{B}^{\vdash})^{\vdash} \subset \mathcal{B}$ and $(\mathcal{B}^{\vdash})^{\vdash}$ describes the controllable sub-behavior of $\mathcal{B}$.

It is also possible to adapt (5.2) for a duality of subspaces $\mathcal{C} \subset \mathbb{F}^n((z))$. For such a subspace we define:

$$(7.3) \qquad \mathcal{C}^{\vdash} := \left(\mathcal{C} \cap \mathbb{F}^n[z, z^{-1}]\right)^{\perp} \cap \mathbb{F}^n((z)).$$

The duality (7.1) does not in general correspond to the linear algebra dual of the $R = \mathbb{F}[z, z^{-1}]$ module $\mathcal{C} \subset R^n$ since there is some 'time reversal' involved. The same is true for the duality (7.3), which does not correspond to the linear algebra dual of the $\mathbb{F}((z))$ vector space $\mathcal{C}$ without time reversal.

If one works however with the 'time-reversed' bilinear form:

$$(7.4) \qquad \begin{aligned} [\,,\,]: \quad & \mathbb{F}^n[[z, z^{-1}]] \times \mathbb{F}^n[z, z^{-1}] \;\longrightarrow\; \mathbb{F} \\ & (w(z), v(z)) \;\mapsto\; \sum_{i=-\infty}^{\infty} \langle w_i, v_{-i} \rangle \end{aligned}$$

then the definitions (7.1) and (7.3) do correspond to the module dual (and the linear algebra dual respectively), used widely in the coding literature [38]. In this case one has: If $G(z)$ is a generator matrix of $C^{\vdash}$ then $H(z) := G^t(z)$ is a parity check matrix of $(C^{\vdash})^{\vdash}$.

In the Laurent-series context it is also possible to induce the duality (7.3) directly through the time-reversed bilinear form defined on the set $\mathbb{F}^n((z)) \times \mathbb{F}^n((z))$:

(7.5)
$$[\,,\,]: \quad \mathbb{F}^n((z)) \times \mathbb{F}^n((z)) \quad \longrightarrow \quad \mathbb{F}$$
$$(w(z), v(z)) \quad \mapsto \quad \sum_{i=-\infty}^{\infty} \langle w_i, v_{-i} \rangle.$$

Note that the sum appearing in (7.5) is always well defined. This bilinear form has been widely used in functional analysis and in systems theory [14].

7.3. Convolutional codes as subsets of $\mathbb{F}[[z, z^{-1}]]$, a case study. In this subsection we illustrate the differences of the definitions in the peculiar case $n = 1$.

If one works with Definition A or Definition B then there exist only the two trivial codes having the 1×1 generator matrix (1) and (0) as subsets of $\mathbb{F}[[z, z^{-1}]]$.

The situation of Definition C is already more interesting. For each polynomial $p(z)$ one has the associated 'autonomous behavior':

(7.6)
$$\mathcal{B} = \{\ w(z)\ |\ p(\sigma)w(z) = 0\ \}.$$

Autonomous behaviors are the extreme case of uncontrollable behaviors. If $\deg p(z) = \delta$, then $\mathcal{B}$ is a finite-dimensional $\mathbb{F}$-vector space of dimension δ. For coding purposes $\mathcal{B}$ is not useful at all. Indeed, the code allows only δ symbols to be chosen freely, say the symbols $w_0, w_1, \ldots, w_{\delta-1}$. With this the codeword $w(z) = \sum_{i=-\infty}^{\infty} w_i z^i \in \mathcal{B}$ is determined, and the transmission of $w(z)$ requires infinite symbols in the past and infinite symbols in the future. In other words, the code has transmission rate 0. The distance of the code is however very good, namely $d_{\text{free}}(\mathcal{B}) = \infty$. If $\mathcal{B}$ is defined on the positive time axis, i.e. $\mathcal{B} \subset \mathbb{F}[[z]]$ then the situation is only slightly better. Indeed in this situation, one sends first δ message words and then an infinite set of 'check symbols'. As these remarks make clear, a code of the form (7.6) is not very useful.

The most interesting situation happens in the setup of Definition D and Definition D'. In this situation the codes are exactly the ideals $< g(z) > \subset \mathbb{F}[z, z^{-1}]$ (respectively $< g(z) > \subset \mathbb{F}[z]$). We now show that ideals of the form $< g(z) >$ are of interest in the coding context.

EXAMPLE 7.3. Let $\mathbb{F} = \mathbb{F}_2 = \{0, 1\}$. Consider the ideal generated by $g(z) = (z + 1)$. $< g(z) > \subset \mathbb{F}[z, z^{-1}]$ consists in this case of the even-weight sequences, namely the set of all sequences with a finite and even number of ones. This code is controllable but not observable.

Ideals of the form $< g(z) >$ are the extreme case of non-observable behaviors. In principle this makes it impossible for the receiver to decode a message. However with some additional 'side-information' decoding can still be performed, as we now explain.

One of the most often used codes in practice is probably the *cyclic redundancy check code* (CRC code). These codes are the main tool to ensure error-free transmissions over the Internet. They can be defined in the following way: Let $g(z) \in \mathbb{F}[z]$ be a polynomial. Then the encoding map is simply defined as:

$$(7.7) \qquad \varphi : \mathbb{F}[z] \longrightarrow \mathbb{F}[z], \quad m(z) \longmapsto c(z) = g(z)m(z).$$

The code is then the ideal $< g(z) > = \mathrm{im}\,(\varphi)$. The distance of this code is 2, since there exists an integer N such that $(z^N - 1) \in < g(z) >$. As we already mentioned the code is not observable. Assume now that the sender gives some additional side information indicating the start and the end of a message. This can be either done by saying: "I will send in a moment 1 Mb", or it can be done by adding some 'stop signal' at the end of the transmission. Once the receiver knows that the transmission is over, he applies long division to compute

$$c(z) = \tilde{m}(z)g(z) + r(z), \quad \deg r(z) < \delta.$$

If $r(z) = 0$ the receiver accepts the message $\tilde{m}(z)$ as the transmitted message $m(z)$. Otherwise he will ask for retransmission.

The code performs best over a channel (like the Internet) which has the property that the whole message is transmitted correctly with probability p and with probability $1-p$ whole blocks of the message are corrupted during transmission. One immediately sees that the probability that a corrupted message $\tilde{m}(z)$ is accepted is $q^{-\delta}$, where $q = |\mathbb{F}|$ is the field size.

One might argue that the code $< g(z) > = \mathrm{im}\,(\varphi)$ is simply a cyclic block code, but this is not quite the case. Note that the protocol does not specify any length of the code word and in each transmission a different message length can be chosen. In particular the code can be even used if the message length is longer than N, where N is the smallest integer such that $(z^N - 1) \in < g(z) >$.

EXAMPLE 7.4. Let $\mathbb{F} = \mathbb{F}_2 = \{0, 1\}$ and let $g(z) = z^{20} + 1$. Assume transmission is done on a channel with very low error probability where once in a while a burst error might happen destroying a whole sequence of bits. Assume that the sender uses a stop signal where he repeats the 4 bits 0011 for 100 times. Under these assumptions the receiver can be reasonably sure once a transmission has been complete. The probability of failure to detect a burst error is in this case 2^{-20} which is less than 10^{-6}. Note that $g(z)$ is a very poor generator for a cyclic code of any block length.

REMARK 7.5.   CRC codes are in practice often implemented in a slightly different way than we described it above (see e.g. [36]). The sender typically performs long division on $z^\delta m(z)$ and computes

$$z^\delta m(z) = f(z)g(z) + r(z), \quad \deg r(z) < \delta.$$

He then transmits the code word $c(z) := z^\delta m(z) - r(z) \in <g(z)>$. Clearly the schemes are equivalent. The advantage of the latter is that the message sequence $m(z)$ is transmitted in 'plain text', allowing processing of the data immediately.

7.4. Some geometric remarks. One motivation for the author to take a module-theoretic approach to convolutional coding theory has come from algebraic-geometric considerations. As is explained in [31, 39, 40], a submodule of rank k and degree δ in $\mathbb{F}^n[z]$ describes a quotient sheaf of rank k and degree δ over the projective line $\mathbb{P}^1$. The set of all such quotient sheaves having rank k and degree at most δ has the structure of a smooth projective variety denoted by $X_{k,n}^\delta$. This variety has been of central interest in the recent algebraic geometry literature. In the context of coding theory, it has actually been used to predict the existence of maximum-distance-separable (MDS) convolutional codes [43].

The set of convolutional codes in the sense of Definition A or A' or B having rate $\frac{k}{n}$ and degree at most δ form all proper Zariski open subsets of $X_{k,n}^\delta$. The points in the closure of these Zariski open sets are exactly the non-observable codes if the rate is $\frac{1}{n}$. These geometric considerations suggest that non-observable convolutional codes should be incorporated into a complete theory of convolutional codes. The following example will help to clarify these issues:

EXAMPLE 7.6.   Let $\delta = 2$, $k = 1$ and $n = 2$, i.e., consider $X_{1,2}^2$. Any code of degree at most 2 then has an encoder of the form:

$$G(z) = \begin{pmatrix} g_1(z) \\ g_2(z) \end{pmatrix} = \begin{pmatrix} a_0 + a_1 z + a_2 z^2 \\ b_0 + b_1 z + b_2 z^2 \end{pmatrix}$$

We can identify the encoder through the point $(a_0, a_1, a_2, b_0, b_1, b_2) \in \mathbb{P}^5$. The variety $X_{1,2}^2$ is in this example exactly the projective space $\mathbb{P}^5$. For codes in the sense of Definition A or A' or B, $G(z)$ must be taken as a basic minimal encoder in order to have a unique parameterization. This requires that $g_1(z)$ and $g_2(z)$ are coprime polynomials. The set of coprime polynomials $g_1(z), g_2(z)$ viewed as a subset of $\mathbb{P}^5$ forms a Zariski open subset $U \subset \mathbb{P}^5$ described by the resultant condition

$$\det \begin{pmatrix} a_0 & 0 & b_0 & 0 \\ a_1 & a_0 & b_1 & b_0 \\ a_2 & a_1 & b_2 & b_1 \\ 0 & a_2 & 0 & b_2 \end{pmatrix} \neq 0.$$

For codes in the sense of Definition D, we require that a_0 and b_0 are not simultaneously zero in order to have a unique parameterization. Definition D leads to a larger Zariski open set V, i.e. $U \subset V \subset \mathbb{P}^5$. Only with Definition D' does one obtain the whole variety $X_{1,2}^2 = \mathbb{P}^5$.

In the general situation $X_{k,n}^\delta$ naturally contains the non-observable codes as well. If $k = 1$, then $X_{1,n}^\delta = \mathbb{P}^{n(\delta+1)-1}$, and the codes in the sense of Definition D' having rate $\frac{1}{n}$ and degree at most δ are exactly parameterized by $X_{1,n}^\delta$.

8. Conclusion. The paper surveys a number of different definitions of convolutional codes. All definitions have in common that a convolutional code is a subset $\mathcal{C} \subset \mathbb{F}^n[[z, z^{-1}]]$ which is both linear and time-invariant. The definitions differ in requirements such as controllability, observability, completeness and restriction to finite support.

If one requires that a code be both controllable and observable, then the restriction to any finite time window will result in equivalent definitions. Actually Loeliger and Mittelholzer [30] define a convolutional code locally in terms of one trellis section and they require in their definition that a code is controllable and observable. Algebraically such a trellis section is simply described through the generalized first order description (6.4) or (6.7).

If one wants to have a theory which allows one to work with rational encoders, then it will be necessary that the code has finite support on the negative time axis $\mathbb{Z}_-$ (or alternatively on the positive time axis $\mathbb{Z}_+$). This is one reason why a large part of the coding literature works with the field of formal Laurent series.

If one wants in addition to have a theory which can accommodate non-observable codes (and such a theory seems to have some value) then it is best to work in a module-theoretic setting.

REFERENCES

[1] S. EILENBERG. *Automata, languages, and machines. Vol. A.* Academic Press, New York, 1974. Pure and Applied Mathematics, vol. **59**.
[2] P. ELIAS. Coding for noisy channels. *IRE Conv. Rec.*, 4:37–46, 1955.
[3] F. FAGNANI AND S. ZAMPIERI. Minimal syndrome formers for group codes. *IEEE Trans. Inform. Theory*, 45(1):3–31, 1999.
[4] E. FORNASINI AND M.E. VALCHER. Observability and extendability of finite support nD behaviors. In *Proc. of the 34th IEEE Conference on Decision and Control*, pp. 3277–3282, New Orleans, Louisiana, 1995.
[5] E. FORNASINI AND M.E. VALCHER. Multidimensional systems with finite support behaviors: Signal structure, generation, and detection. *SIAM J. Control Optim.*, 36(2):760–779, 1998.
[6] G.D. FORNEY. Convolutional codes I: Algebraic structure. *IEEE Trans. Inform. Theory*, IT-16(5):720–738, 1970.
[7] G.D. FORNEY. Structural analysis of convolutional codes via dual codes. *IEEE Trans. Inform. Theory*, IT-19(5):512–518, 1973.
[8] G.D. FORNEY. Minimal bases of rational vector spaces, with applications to multivariable linear systems. *SIAM J. Control*, 13(3):493–520, 1975.

[9] G.D. FORNEY. Group codes and behaviors. In G. Picci and D.S. Gilliam, editors, *Dynamical Systems, Control, Coding, Computer Vision: New Trends, Interfaces, and Interplay*, pp. 301–320. Birkäuser, Boston-Basel-Berlin, 1999.

[10] G.D. FORNEY, B. MARCUS, N.T. SINDHUSHAYANA, AND M. TROTT. A multilingual dictionary: System theory, coding theory, symbolic dynamics and automata theory. In *Different Aspects of Coding Theory*, Proceedings of Symposia in Applied Mathematics number 50, pp. 109–138. American Mathematical Society, 1995.

[11] G.D. FORNEY AND M.D. TROTT. Controllability, observability, and duality in behavioral group systems. In *Proc. of the 34th IEEE Conference on Decision and Control*, pp. 3259–3264, New Orleans, Louisiana, 1995.

[12] G.D. FORNEY AND M.D. TROTT. The dynamics of group codes: Dual abelian group codes and systems. Preprint, August 1997; submitted to IEEE Trans. Inform. Theory, January 2000.

[13] P.A. FUHRMANN. Algebraic system theory: An analyst's point of view. *J. Franklin Inst.*, 301:521–540, 1976.

[14] P.A. FUHRMANN. Duality in polynomial models with some applications to geometric control theory. *IEEE Trans. Automat. Control*, 26(1):284–295, 1981.

[15] P.A. FUHRMANN. *A Polynomial Approach to Linear Algebra.* Universitext. Springer-Verlag, New York, 1996.

[16] H. GLUESING-LUERSSEN, J. ROSENTHAL, AND P.A. WEINER. Duality between multidimensional convolutional codes and systems. E-print math.OC/9905046, May 1999.

[17] M.L.J. HAUTUS AND M. HEYMANN. Linear feedback—an algebraic approach. *SIAM J. Control*, 16:83–105, 1978.

[18] D. HINRICHSEN AND D. PRÄTZEL-WOLTERS. Solution modules and system equivalence. *Internat. J. Control*, 32:777–802, 1980.

[19] D. HINRICHSEN AND D. PRÄTZEL-WOLTERS. Generalized Hermite matrices and complete invariants of strict system equivalence. *SIAM J. Control Optim.*, 21:289–305, 1983.

[20] R. JOHANNESSON AND Z. WAN. A linear algebra approach to minimal convolutional encoders. *IEEE Trans. Inform. Theory*, IT-39(4):1219–1233, 1993.

[21] R. JOHANNESSON AND K. SH. ZIGANGIROV. *Fundamentals of Convolutional Coding.* IEEE Press, New York, 1999.

[22] R. E. KALMAN. Algebraic structure of linear dynamical systems. I. The module of $\sum$. *Proc. Nat. Acad. Sci. U.S.A.*, 54:1503–1508, 1965.

[23] R.E. KALMAN, P.L. FALB, AND M.A. ARBIB. *Topics in Mathematical System Theory.* McGraw-Hill, New York, 1969.

[24] B. KITCHENS. Symbolic dynamics and convolutional codes. Preprint, February 2000.

[25] G. KÖTHE. *Topological Vector Spaces I.* Springer Verlag, 1969.

[26] M. KUIJPER. *First-Order Representations of Linear Systems.* Birkhäuser, Boston, 1994.

[27] M. KUIJPER AND J.M. SCHUMACHER. Realization of autoregressive equations in pencil and descriptor form. *SIAM J. Control Optim.*, 28(5):1162–1189, 1990.

[28] S. LIN AND D.J. COSTELLO. *Error Control Coding: Fundamentals and Applications.* Prentice-Hall, Englewood Cliffs, NJ, 1983.

[29] D. LIND AND B. MARCUS. *An Introduction to Symbolic Dynamics and Coding.* Cambridge University Press, 1995.

[30] H.A. LOELIGER AND T. MITTELHOLZER. Convolutional codes over groups. *IEEE Trans. Inform. Theory*, 42(6):1660–1686, 1996.

[31] V. LOMADZE. Finite-dimensional time-invariant linear dynamical systems: Algebraic theory. *Acta Appl. Math*, 19:149–201, 1990.

[32] B. MARCUS. Symbolic dynamics and connections to coding theory, automata theory and system theory. In *Different aspects of coding theory (San Francisco, CA, 1995)*, vol. 50 of *Proc. Sympos. Appl. Math.*, pp. 95–108. Amer. Math. Soc., Providence, RI, 1995.

[33] J.L. MASSEY, D.J. COSTELLO, AND J. JUSTESEN. Polynomial weights and code constructions. *IEEE Trans. Inform. Theory*, IT-19(1):101–110, 1973.

[34] J.L. MASSEY AND M.K. SAIN. Codes, automata, and continuous systems: Explicit interconnections. *IEEE Trans. Automat. Contr.*, AC-12(6):644–650, 1967.

[35] R.J. McELIECE. The algebraic theory of convolutional codes. In V. Pless and W.C. Huffman, editors, *Handbook of Coding Theory*, volume 1, pages 1065–1138. Elsevier Science Publishers, Amsterdam, The Netherlands, 1998.

[36] H. NUSSBAUMER. *Computer Communication Systems*, vol. 1. John Wiley & Sons, Chichester, New York, 1990. Translated by John C.C. Nelson.

[37] U. OBERST. Multidimensional constant linear systems. *Acta Appl. Math*, 20:1–175, 1990.

[38] PH. PIRET. *Convolutional Codes, an Algebraic Approach*. MIT Press, Cambridge, MA, 1988.

[39] M.S. RAVI AND J. ROSENTHAL. A smooth compactification of the space of transfer functions with fixed McMillan degree. *Acta Appl. Math*, 34:329–352, 1994.

[40] M.S. RAVI AND J. ROSENTHAL. A general realization theory for higher order linear differential equations. *Systems & Control Letters*, 25(5):351–360, 1995.

[41] J. ROSENTHAL AND J.M. SCHUMACHER. Realization by inspection. *IEEE Trans. Automat. Contr.*, AC-42(9):1257–1263, 1997.

[42] J. ROSENTHAL, J.M. SCHUMACHER, AND E.V. YORK. On behaviors and convolutional codes. *IEEE Trans. Inform. Theory*, 42(6, Part I):1881–1891, 1996.

[43] J. ROSENTHAL AND R. SMARANDACHE. Maximum distance separable convolutional codes. *Appl. Algebra Engrg. Comm. Comput.*, 10(1):15–32, 1999.

[44] J. ROSENTHAL AND E.V. YORK. BCH convolutional codes. *IEEE Trans. Inform. Theory*, 45(6):1833–1844, 1999.

[45] M.K. SAIN AND J.L. MASSEY. Invertibility of linear time-invariant dynamical systems. *IEEE Trans. Automat. Contr.*, AC-14:141–149, 1969.

[46] R. SMARANDACHE, H. GLUESING-LUERSSEN, AND J. ROSENTHAL. Constructions of MDS-convolutional codes. Submitted to IEEE Trans. Inform. Theory, August 1999.

[47] L. STAIGER. Subspaces of $GF(q)^\omega$ and convolutional codes. *Information and Control*, 59:148–183, 1983.

[48] M.E. VALCHER AND E. FORNASINI. On 2D finite support convolutional codes: An algebraic approach. *Multidim. Sys. and Sign. Proc.*, 5:231–243, 1994.

[49] P. WEINER. *Multidimensional Convolutional Codes*. PhD thesis, University of Notre Dame, 1998. Available at http://www.nd.edu/~rosen/preprints.html.

[50] J.C. WILLEMS. From time series to linear system. Part I: Finite dimensional linear time invariant systems. *Automatica*, 22:561–580, 1986.

[51] J.C. WILLEMS. Models for dynamics. In U. Kirchgraber and H.O. Walther, editors, *Dynamics Reported*, volume 2, pp. 171–269. John Wiley & Sons Ltd, 1989.

[52] J.C. WILLEMS. Paradigms and puzzles in the theory of dynamical systems. *IEEE Trans. Automat. Control*, AC-36(3):259–294, 1991.

MULTI-DIMENSIONAL SYMBOLIC DYNAMICAL SYSTEMS

KLAUS SCHMIDT[*]

Abstract. The purpose of this note is to point out some of the phenomena which arise in the transition from classical shifts of finite type $X \subset A^{\mathbb{Z}}$ to *multi-dimensional* shifts of finite type $X \subset A^{\mathbb{Z}^d}$, $d \geq 2$, where A is a finite alphabet. We discuss rigidity properties of certain multi-dimensional shifts, such as the appearance of an unexpected intrinsic algebraic structure or the scarcity of isomorphisms and invariant measures. The final section concentrates on group shifts with finite or uncountable alphabets, and with the symbolic representation of such shifts in the latter case.

Key words. Multi-dimensional symbolic dynamics, Tiling systems, symbolic representation of $\mathbb{Z}^d$-actions.

AMS(MOS) subject classifications. Primary: 37B15, 37B50; Secondary: 37A15, 37A60.

1. Shifts of finite type. Let $d \geq 1$, A a finite set (the *alphabet*), and let $A^{\mathbb{Z}^d}$ be the set of all maps $x \colon \mathbb{Z}^d \longrightarrow A$. For every nonempty subset $F \subset \mathbb{Z}^d$, the map

$$\pi_F \colon A^{\mathbb{Z}^d} \longrightarrow A^F$$

is the projection which restricts each $x \in A^{\mathbb{Z}^d}$ to F. For every $\mathbf{n} \in \mathbb{Z}^d$ we define a homeomorphism $\sigma^{\mathbf{n}}$ of the compact space $A^{\mathbb{Z}^d}$ by

$$(1.1) \qquad (\sigma^{\mathbf{n}} x)_{\mathbf{m}} = x_{\mathbf{n}+\mathbf{m}}$$

for every $x = (x_{\mathbf{m}}) \in A^{\mathbb{Z}^d}$. The map $\sigma \colon \mathbf{n} \mapsto \sigma^{\mathbf{n}}$ is the *shift-action* of $\mathbb{Z}^d$ on $A^{\mathbb{Z}^d}$, and a subset $X \subset A^{\mathbb{Z}^d}$ is *shift-invariant* if $\sigma^{\mathbf{n}}(X) = X$ for all $\mathbf{n} \in \mathbb{Z}^d$. A closed, shift-invariant set $X \subset A^{\mathbb{Z}^d}$ is a *shift of finite type* (*SFT*) if there exist a finite set $F \subset \mathbb{Z}^d$ and a subset $P \subset A^F$ such that

$$(1.2) \qquad X = X(F, P) = \{x \in A^{\mathbb{Z}^d} : \pi_F \circ \sigma^{\mathbf{n}}(x) \in P \text{ for every } \mathbf{n} \in \mathbb{Z}^d\}.$$

A closed shift-invariant subset $X \subset A^{\mathbb{Z}^d}$ is a *SFT* if and only if there exists a finite set $F \subset \mathbb{Z}^d$ such that

$$(1.3) \qquad X = \{x \in A^{\mathbb{Z}^d} : \pi_F \circ \sigma^{\mathbf{n}}(x) \in \pi_F(X) \text{ for every } \mathbf{n} \in \mathbb{Z}^d\}.$$

An immediate consequence of this characterization of *SFT*'s is that the notion *SFT* is an invariant of topological conjugacy. For background and details we refer to [21]–[25].

---

[*]Mathematics Institute, University of Vienna, and Erwin Schrödinger Institute for Mathematical Physics, Boltzmanngasse 9, A-1090 Vienna, Austria. Email: klaus.schmidt@univie.ac.at.

If $X \subset A^{\mathbb{Z}^d}$ is a *SFT* we may change the alphabet A and assume that

$$F = \{0,1\}^d \text{ or } F = \{0\} \cup \bigcup_{i=1}^{d} \{e^{(i)}\},$$

where $e^{(i)}$ is the i-th basis vector in $\mathbb{Z}^d$.

Let $X \subset A^{\mathbb{Z}^d}$ be a *SFT*. A point $x \in X$ is *periodic* if its orbit under σ is finite. In contrast to the case where $d = 1$, a higher-dimensional *SFT* X may not contain any periodic points (we give an example below). This potential absence of periodic points is associated with certain **undecidability problems** (cf. e.g. [1], [9], [19] and [32]):

(1) It is algorithmically undecidable if $X(F, P) \neq \varnothing$ for given (F, P);
(2) It is algorithmically undecidable whether an allowed[1] partial configuration can be extended to a point $x \in X(F, P)$.

In dealing with concrete *SFT*'s undecidability is not really a problem, but it indicates the difficulty of making general statements about higher-dimensional *SFT*'s. There have been several attempts to define more restrictive classes of *SFT*'s with the hope of a systematic approach within such a class (cf. e.g. [16]–[17], the algebraic systems considered in [9], or certain specification properties — such as in [13] — which guarantee 'sufficient similarity' to full shifts).

2. Some examples.

EXAMPLE 1 (Chessboards). *Let $n \geq 2$ and $A = \{0, \ldots, n - 1\}$. We interpret A as a set of colours and consider the* SFT $X = X^{(n)} \subset A^{\mathbb{Z}^2}$ *consisting of all configurations in which adjacent lattice points must have different colours.*

For $n = 2$, $X^{(2)}$ consists of two points. For $n \geq 3$, $X^{(n)}$ is uncountable.

There is a big difference between $n = 3$ and $n \geq 4$: for $n = 3$ there exist frozen *configurations in $X^{(3)}$, which cannot be altered in only finitely many places. These points are the periodic extensions of*

0	1	2	0	1	2		0	2	1	0	2	1
2	0	1	2	0	1		1	0	2	1	0	2
1	2	0	1	2	0		2	1	0	2	1	0
0	1	2	0	1	2		0	2	1	0	2	1
2	0	1	2	0	1		1	0	2	1	0	2
1	2	0	1	2	0		2	1	0	2	1	0

EXAMPLE 2 (Wang tilings). *Let T be a finite nonempty set of distinct, closed 1×1 squares (tiles) with coloured edges such that no horizontal edge has the same colour as a vertical edge: such a set T is called a*

---

[1] If $X = X(F, P)$ is a *SFT* and $\varnothing \neq E \subset \mathbb{Z}^d$, then an element $x \in A^E$ is an *allowed partial configuration* if $\pi_{(F+\mathbf{n}) \cap E}(x)$ coincides (in the obvious sense) with an element of $\pi_{F \cap (E-\mathbf{n})}(P)$ whenever $F \cap (E - \mathbf{n}) \neq \varnothing$.

collection of Wang tiles. For each $\tau \in T$ we denote by $r(\tau), t(\tau), l(\tau), b(\tau)$ the colours of the right, top, left and bottom edges of τ, and we write $C(T) = \{r(\tau), t(\tau), l(\tau), b(\tau) : \tau \in T\}$ for the set of colours occurring on the tiles in T. A Wang tiling w by T is a covering of $\mathbb{R}^2$ by translates of copies of elements of T such that

(i) *every corner of every tile in w lies in $\mathbb{Z}^2 \subset \mathbb{R}^2$,*

(ii) *two tiles of w are only allowed to touch along edges of the same colour, i.e. $r(\tau) = l(\tau')$ whenever τ, τ' are horizontally adjacent tiles with τ to the left of τ', and $t(\tau) = b(\tau')$ if τ, τ' are vertically adjacent with τ' above τ.*

We identify each such tiling w with the point

$$w = (w_{\mathbf{n}}) \in T^{\mathbb{Z}^2},$$

where $w_{\mathbf{n}}$ is the unique element of T whose translate covers the square $\mathbf{n} + [0,1]^2 \subset \mathbb{R}^2$, $\mathbf{n} \in \mathbb{Z}^2$. The set $W_T \subset T^{\mathbb{Z}^2}$ of all Wang tilings by T is obviously a SFT, and is called the Wang shift of T.

Here is an explicit example of a two-dimensional Wang shift: let T_D be the set of Wang tiles

with the colours H, h, V, v on the solid horizontal, broken horizontal, solid vertical and broken vertical edges. The following picture shows a partial Wang tiling of $\mathbb{R}^2$ by T_D and explains the name 'domino tiling' for such a tiling: two tiles meeting along an edge coloured h or v form a single vertical or horizontal 'domino'.

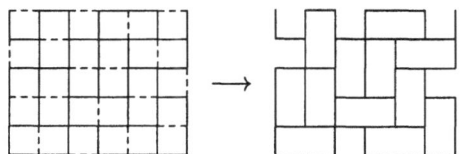

The Wang shift $W_D \subset T_D^{\mathbb{Z}^2}$ of T_D is called the domino (or dimer) shift, and is one of the few higher dimensional SFT's for which the dynamics is understood to some extent (cf. e.g. [2], [3], [7]). The shift-action σ_{W_D} of $\mathbb{Z}^2$ on W_D is topologically mixing, and its topological entropy $h(\sigma_{W_D})$ was computed by Kastelleyn in [7]:

$$h(\sigma_{W_D}) = \frac{1}{4} \int_0^1 \int_0^1 (4 - 2\cos 2\pi s - 2\cos 2\pi t) \, ds \, dt.$$

The domino-tilings again have frozen configurations which look like 'brick walls'.

EXAMPLE 3 (A shift of finite type without periodic points). *Consider the following set T' of six polygonal tiles, introduced by Robinson in [19],* *each of which which should be thought of as a 1×1 square with various bumps and dents.*

We denote by T the set of all tiles which are obtained by allowing horizontal and vertical reflections as well as rotations of elements in T' by multiples of $\frac{\pi}{2}$. Again we consider the set $W_T \subset T^{\mathbb{Z}^2}$ consisting of all tilings of $\mathbb{R}^2$ by translates of elements of T aligned to the integer lattice (as much as their bumps and dents allow). The set W_T is obviously a SFT, and W_T is uncountable and has no periodic points. If we allow each (or even only one) of these tiles to occur in two different colours with no restriction on adjacency of colours then we obtain a SFT with positive entropy, but still without periodic points.

The paper [19] also contains an explicit set T of Wang tiles for which the extension problem is undecidable.

3. Wang tiles and shifts of finite type.

THEOREM 3.1. *Every SFT can be represented (in many different ways) as a Wang tiling.*

Proof. Assume that $F = \{0,1\}^2 \subset \mathbb{Z}^2$. We set $T = \pi_F(X(F,P))$ and consider each

$$\tau = \begin{array}{|cc|} \hline x_{(0,1)} & x_{(1,1)} \\ x_{(0,0)} & x_{(1,0)} \\ \hline \end{array} \in T$$

as a unit square with the 'colours' $\left[\, x_{(0,0)} \;\; x_{(1,0)} \,\right]$ and $\left[\, x_{(0,1)} \;\; x_{(1,1)} \,\right]$ along its bottom and top horizontal edges, and $\left[\begin{smallmatrix} x_{(0,1)} \\ x_{(0,0)} \end{smallmatrix}\right]$ and $\left[\begin{smallmatrix} x_{(1,1)} \\ x_{(1,0)} \end{smallmatrix}\right]$ along its left and right vertical edges. With this interpretation we obtain a one-to-one correspondence between the points $x = (x_{\mathbf{n}}) \in X$ and the Wang tilings $w = (w_{\mathbf{n}}) = (\pi_F \circ \sigma^{\mathbf{n}}(x)) \in T^{\mathbb{Z}^2}$. $\square$

This correspondence allows us to regard each *SFT* as a Wang shift and vice versa. However, the correspondence is a bijection only up to topological conjugacy: if we start with a *SFT* $X \subset A^{\mathbb{Z}^2}$ with $F = \{0,1\}^2$, view it as the Wang shift $W_T \subset T^{\mathbb{Z}^2}$ with $T = \pi_F(X)$, and then interpret W_T as a *SFT* as above, we do not end up with X, but with the 2-block representation of X.

DEFINITION 3.1. *Let A be a finite set and $X \subset A^{\mathbb{Z}^2}$ a SFT, T a set of Wang tiles and W_T the associated Wang shift. We say that W_T represents X if W_T is topologically conjugate to X. Two Wang shifts W_T and $W_{T'}$ are equivalent if they are topologically conjugate as SFT's.*

Since any given infinite *SFT* X has many different representations by Wang shifts one may ask whether these different representations of X have anything in common. The answer to this question turns out to be related to a measure of the 'complexity' of the *SFT* X. For this we need to introduce the *tiling group* associated with a Wang shift.

Let T be a collection of Wang tiles and $W_T \subset T^{\mathbb{Z}^2}$ the Wang shift of T. Following Conway, Lagarias and Thurston ([4], [29]) we write

$$\Gamma(T) = \langle \mathcal{C}(T) | \mathsf{t}(\tau)\mathsf{l}(\tau) = \mathsf{r}(\tau)\mathsf{b}(\tau), \ \tau \in T \rangle$$

for the free group generated by the colours occurring on the edges of elements in T, together with the relations $\mathsf{t}(\tau)\mathsf{l}(\tau) = \mathsf{r}(\tau)\mathsf{b}(\tau)$, $\tau \in T$. The countable, discrete group $\Gamma(T)$ is called the *tiling group* of T (or of the Wang shift W_T). From the definition of $\Gamma(T)$ it is clear that the map $\theta: \Gamma(T) \to \mathbb{Z}^2$, given by

$$\theta(\mathsf{b}(\tau)) = \theta(\mathsf{t}(\tau)) = (1,0),$$
$$\theta(\mathsf{l}(\tau)) = \theta(\mathsf{r}(\tau)) = (0,1),$$

for every $\tau \in T$, is a group homomorphism whose kernel is denoted by

$$\Gamma_0(T) = \ker(\theta).$$

Suppose that $E \subset \mathbb{R}^2$ is a bounded set, and that $w \in T^{\mathbb{R}^2 \smallsetminus E}$ is a Wang-tiling of $\mathbb{R}^2 \smallsetminus E$. When can we complete w to a Wang-tiling of $\mathbb{R}^2$ (possibly after enlarging E by a finite amount)? After a finite enlargement we may assume that E is the empty rectangle in the left picture of Figure 1 (the tiles covering the rest of $\mathbb{R}^2 \smallsetminus E$ are not shown). If we add a tile legally

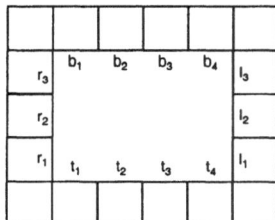 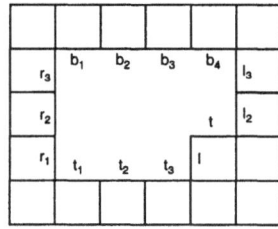

FIG. 1.

(as in the right picture), then the words in $\Gamma(T)$ obtained by reading off the colours along the edges of the two holes coincide because of the tiling relations:

(3.1)
$$r_1^{-1}r_2^{-1}r_3^{-1}b_1^{-1}b_2^{-1}b_3^{-1}b_4^{-1}l_3l_2l_1t_4t_3t_2t_1$$
$$= r_1^{-1}r_2^{-1}r_3^{-1}b_1^{-1}b_2^{-1}b_3^{-1}b_4^{-1}l_3l_2tlt_3t_2t_1$$

In particular, if the hole can be closed, then the word must be the identity.

If $X \subset A^{\mathbb{Z}^2}$ is a *SFT* and W_T a Wang representation of X then the tiling group $\Gamma(T)$ gives an obstruction to the *weak closing* of bounded holes (i.e. the closing of holes *after finite enlargement*) for points $x \in A^{\mathbb{Z}^2} \smallsetminus E$, where $E \subset \mathbb{Z}^2$ is a finite set. However, different Wang-representations of X may give different answers.

    EXAMPLE 4. *Let X be the 3-coloured chessboard, and let T be the set of Wang tiles*

with the colours

$$h_0 = \; \text{——} \;, \quad h_1 = \; \text{- - -} \;, \quad h_2 = \; \cdots \;,$$
$$v_0 = \; | \;, \quad v_1 = \; \vdots \;, \quad v_2 = \; \vdots$$

on the horizontal and vertical edges. Then W_T represents X. The tiling group $\Gamma(T)$ is of the form

$$\Gamma(T'_C) = \{h_i, v_i, \; i = 0, 1, 2 \,|\, v_1 h_0 = v_2 h_0 = h_1 v_0 = h_2 v_0,$$
$$v_2 h_1 = v_0 h_1 = h_2 v_1 = h_0 v_1, \quad v_0 h_2 = v_1 h_2 = h_0 v_2 = h_1 v_2\}.$$

Since $h_0 = h_1 = h_2$, $v_0 = v_1 = v_2$ and $h_0 v_0 = v_0 h_0$, $\Gamma(T) \cong \mathbb{Z}^2$, and every hole appears closable.

    With a different representation of X as a Wang shift we obtain more information. Let T' be the set of Wang tiles

1 0	1 2	2 0	1 0	2 0	2 1	0 1	0 2	2 1
0 1	0 1	0 1	0 2	0 2	0 2	1 0	1 0	1 0

0 1	2 0	2 1	0 1	0 2	1 2	0 2	1 0	1 2
1 2	1 2	1 2	2 0	2 0	2 0	2 1	2 1	2 1

with the colours $h_{ij} = [\,i\;j\,]$ on the horizontal and $v_i^j = \left[\begin{smallmatrix} j \\ i \end{smallmatrix}\right]$ on the vertical edges, where $i, j \in \{0, 1, 2\}$ and $i \neq j$. Then $W_{T'}$ represents X.

    There exists a group homomorphism $\phi \colon \Gamma(T') \longrightarrow \mathbb{Z}$ with

$$\phi(h_{01}) = \phi(h_{12}) = \phi(h_{20}) = \phi(v_0^1) = \phi(v_1^2) = \phi(v_2^0) = 1,$$
$$\phi(h_{10}) = \phi(h_{21}) = \phi(h_{02}) = \phi(v_1^0) = \phi(v_2^1) = \phi(v_0^2) = -1.$$

This homomorphism detects that the hole with the edge

$$\begin{matrix} 1 & 2 & 1 \\ 2 & ? & 0 \\ 0 & 1 & 2 \end{matrix}$$

cannot be closed, no matter how it is extended on the outside, and how much it is enlarged initially.

    This example raises the alarming possibility that more and more complicated Wang-representations of a *SFT* X will give more and more combinatorial information about X. Remarkably, this is not the case.

THEOREM 3.2. *For many topologically mixing $\mathbb{Z}^2$-SFT's there exists a Wang-representation W_T of X which contains all the combinatorial information obtainable from* all *possible Wang-representations of X.*

For examples we refer to [25] and [5]. In order to make this statement comprehensible one has to express it in terms of the *continuous cohomology* of X.

4. Wang tiles and cohomology. Let $X \subset A^{\mathbb{Z}^2}$ be a *SFT* and G a discrete group with identity element 1_G. A map $c\colon \mathbb{Z}^2 \times X \longrightarrow G$ is a *cocycle* for the shift-action σ of $\mathbb{Z}^2$ on X if $c(\mathbf{n}, \cdot)\colon X \longrightarrow G$ is continuous for every $\mathbf{n} \in \mathbb{Z}^2$ and

$$c(\mathbf{m} + \mathbf{n}, x) = c(\mathbf{m}, \sigma^{\mathbf{n}} x) c(\mathbf{n}, x)$$

for all $x \in X$ and $\mathbf{m}, \mathbf{n} \in \mathbb{Z}^2$. One can interpret this equation as *path-independence*.

A cocycle $c\colon \mathbb{Z}^2 \times X \longrightarrow G$ is a *homomorphism* if $c(\mathbf{n}, \cdot)$ is constant for every $\mathbf{n} \in \mathbb{Z}^2$, and c is a *coboundary* if there exists a continuous map $b\colon X \longrightarrow G$ such that

$$c(\mathbf{n}, x) = b(\sigma^{\mathbf{n}} x)^{-1} b(x)$$

for all $x \in X$ and $\mathbf{n} \in \mathbb{Z}^2$. Two cocycles $c, c'\colon \mathbb{Z}^2 \times X \longrightarrow G$ are *cohomologous* with continuous *transfer function* $b\colon X \longrightarrow G$, if

$$c(\mathbf{n}, x) = b(\sigma^{\mathbf{n}} x)^{-1} c'(\mathbf{n}, x) b(x)$$

for all $\mathbf{n} \in \mathbb{Z}^2$ and $x \in X$.

For every Wang representation W_T of X we define a *tiling cocycle* $c_T\colon \mathbb{Z}^2 \times W_T \longrightarrow \Gamma(T)$ (and hence a cocycle $c'_T\colon \mathbb{Z}^2 \times X \longrightarrow \Gamma(T)$) by setting

$$c_T((1,0), w) = \mathsf{b}(w_0), \quad c_T((0,1), w) = \mathsf{l}(w_0)$$

for every Wang tiling $w \in W_T \subset T^{\mathbb{Z}^2}$, and by using the cocycle equation to extend c_T to a map $\mathbb{Z}^2 \times W_T \longrightarrow \Gamma(T)$ (the relations $\mathsf{t}(\tau)\mathsf{l}(\tau) = \mathsf{r}(\tau)\mathsf{b}(\tau)$, $\tau \in T$, in the tiling group are precisely what is needed to allow such an extension). Conversely, if G is a discrete group and $c\colon \mathbb{Z}^2 \times X \longrightarrow G$ a cocycle, then Theorem 4.2 in [25] shows that there exists a Wang representation w_T of X and a group homomorphism $\eta\colon \Gamma(T) \longrightarrow G$ such that

(4.1) $$c = \eta \circ c_T.$$

In order to establish a link between cocycles and the 'closing of holes' discussed in the last section we return for a moment to the Wang tiles in Figure 1 and assume that the partial configuration $w_{\mathbf{n}}$, $\mathbf{n} \in \mathbb{Z}^2 \smallsetminus E$, shown there extends to an element $w \in W_T$ with $\mathsf{l}(w_0) = r_1$ and $\mathsf{b}(w_0) = t_1$

(i.e. the tile w_0 occupies the bottom left hand corner of the 'hole' E in Figure 1). Then

$$c_T((4,3),w) = l_3l_2l_1t_4t_3t_2t_3 = l_3l_2tlt_3t_2t_1 = b_4b_3b_2b_1r_3r_2r_1,$$

depending on the route chosen from $\mathbf{0}$ to $(4,3)$, which is equivalent to (3.1).

If $W_{T'}$ is another Wang representation of X, then there exists a topological conjugacy $\phi\colon W_T \longrightarrow W_{T'}$, and the coordinates $w_{\mathbf{n}}$, $\mathbf{n} \in \mathbb{Z}^2 \smallsetminus E$, determine the coordinates $\phi(w)_{\mathbf{m}}$, $\mathbf{m} \in \mathbb{Z}^2 \smallsetminus E'$, for some finite set $E' \subset \mathbb{Z}^2$ which we may again assume to be a rectangle. If the tiling cocycle $c_{T'}$ of $W_{T'}$ is a homomorphic image of w_T in the sense of (4.1), then $w_{T'}$ cannot lead to any new obstructions (other than those already exhibited by c_T). A slightly more refined version of the same argument shows that $w_{T'}$ will not lead to any new obstructions even if it is only cohomologous to a homomorphic image of c_T. This observation is the motivation for the following definition.

DEFINITION 4.1. *A cocycle* $c^*\colon \mathbb{Z}^2 \times X \longrightarrow G^*$ *with values in a discrete group* G^* *is* fundamental *if the following is true: for every discrete group* G *and every cocycle* $c\colon \mathbb{Z}^2 \times X \longrightarrow G$ *there exists a group homomorphism* $\theta\colon G^* \longrightarrow G$ *such that* c *is cohomologous to the cocycle* $\theta \circ c^*\colon \mathbb{Z}^2 \times X \longrightarrow G$.

In this terminology we can state a more precise (but still rather vague) form of Theorem 3.2 (cf. [25]).

THEOREM 4.1. *In certain examples of topologically mixing* $\mathbb{Z}^2$-SFT*'s there exists an explicitly computable Wang representation* W_T *of* X *whose tiling cocycle* $c'_T\colon \mathbb{Z}^2 \times X \longrightarrow \Gamma(T)$ *is fundamental.*

For a list of examples (which includes the chessboards in Example 4 and the domino-tilings in Example 2) we refer to [5] and [24]–[25].

Although one can make analogous definitions for classical (one-dimensional) *SFT*'s, they never have fundamental cocycles. The existence of fundamental cocycles is a *rigidity phenomenon* specific to multi-dimensional *SFT*'s.

5. Group shifts and their symbolic representations. In this section we leave the general setting of multi-dimensional shifts of finite type with all its inherent problems and restrict our attention to *SFT*'s with a group structure. This class of *SFT*'s is of interest in coding theory and allows much more detailed statements about conjugacy and dynamical properties than arbitrary *SFT*'s.

Let $d \geq 1$, and let X be a compact abelian group with normalized Haar measure λ_X. A $\mathbb{Z}^d$-action $\alpha\colon \mathbf{n} \mapsto \alpha^{\mathbf{n}}$ by continuous automorphisms of X is called an *algebraic* $\mathbb{Z}^d$-*action* on X. An algebraic $\mathbb{Z}^d$-action α on X is *expansive* if there exists an open neighbourhood $\mathcal{U}$ of the identity 0_X in X with $\bigcap_{\mathbf{n}\in\mathbb{Z}^d} \alpha^{-\mathbf{n}}(\mathcal{U}) = \{0_X\}$.

Suppose that α is an algebraic $\mathbb{Z}^d$-action on a compact abelian group X. An α-invariant probability measure μ on the Borel field $\mathcal{B}_X$ of X is *ergodic* if

$$\mu\left(\bigcup_{\mathbf{n}\in\mathbb{Z}^d} \alpha^{-\mathbf{n}}(B)\right) \in \{0,1\}$$

for every $B \in \mathcal{B}_X$, and *mixing* if

$$\lim_{\mathbf{n}\to\infty} \mu(B \cap \alpha^{-\mathbf{n}}(B')) = \mu(B)\mu(B')$$

for all $B, B' \in \mathcal{B}_X$. The action α is *ergodic* or *mixing* if λ_X is ergodic or mixing.

Let α_1, α_2 be algebraic $\mathbb{Z}^d$-actions on compact abelian groups X_1 and X_2, respectively. A Borel bijection $\phi\colon X_1 \longrightarrow X_2$ is a *measurable conjugacy* of α_1 and α_2 if

$$\lambda_{X_1}\phi^{-1} = \lambda_{X_2}$$

and

(5.1) $$\phi \circ \alpha_1^{\mathbf{n}}(x) = \alpha_2^{\mathbf{n}} \circ \phi(x)$$

for every $\mathbf{n} \in \mathbb{Z}^d$ and λ_{X_1}-a.e. $x \in X_1$.

A continuous group isomorphism $\phi\colon X_1 \longrightarrow X_2$ is an *algebraic conjugacy* of α_1 and α_2 if it satisfies (5.1) for every $\mathbf{n} \in \mathbb{Z}^d$ and $x \in X_1$.

The actions α_1, α_2 are *measurably* (resp. *algebraically*) *conjugate* if there exists a measurable (resp. algebraic) conjugacy between them.

Finally we call a map $\phi\colon X_1 \longrightarrow X_2$ *affine* if there exist a continuous group isomorphism $\psi\colon X_1 \longrightarrow X_2$ and an element $x' \in X_2$ such that

$$\phi(x) = \psi(x) + x'$$

for every $x \in X_1$.

Here we are interested in algebraic $\mathbb{Z}^d$-actions of a particularly simple form. Let A be a compact abelian group, and let $\Omega_A^{(d)} = A^{\mathbb{Z}^d}$ be the compact abelian group consisting of all maps $\omega\colon \mathbb{Z}^d \longrightarrow A$, furnished with the product topology and coordinate-wise addition. We write every $\omega \in \Omega_A^{(d)}$ as $\omega = (\omega_{\mathbf{n}})$ with $\omega_{\mathbf{n}} \in A$ for every $\mathbf{n} \in \mathbb{Z}^d$ and define the *shift-action* σ of $\mathbb{Z}^d$ on $\Omega_A^{(d)}$ by (1.1). Clearly, σ is an algebraic $\mathbb{Z}^d$-action on $\Omega_A^{(d)}$. A *group shift* is the restriction of the shift-action σ to a closed, shift-invariant subgroup $X \subset \Omega_A^{(d)}$.

Throughout the following discussion we shall assume that the 'alphabet' A is either finite or $A = \mathbb{T}$. In the former case every group shift $X \subset \Omega_A^{(d)}$ is automatically a d-dimensional shift of finite type (cf. [9]–[10] and [23]). In our earlier discussion of *SFT*'s we were interested in topological conjugacy invariants. Here we are interested in the connection between measurable and algebraic conjugacy.

EXAMPLE 5. *The shift automorphisms*

$$(\sigma x)_n = x_{n+1}$$

on the compact abelian groups

$$X = (\mathbb{Z}/4\mathbb{Z})^{\mathbb{Z}},$$
$$Y = ((\mathbb{Z}/2\mathbb{Z}) \times (\mathbb{Z}/2\mathbb{Z}))^{\mathbb{Z}}.$$

are measurably (even topologically) conjugate, but the groups X and Y are not algebraically isomorphic.

EXAMPLE 6. *For every nonempty finite set $E \subset \mathbb{Z}^d$ we denote by $X_E \subset \bar{X} = (\mathbb{Z}/2\mathbb{Z})^{\mathbb{Z}^d}$ the closed shift-invariant subgroup consisting of all $x \in \bar{X}$ whose coordinates sum to 0 in every translate of E in $\mathbb{Z}^d$. If E has at least two points then X_E is uncountable and the restriction σ_E of σ to X_E is an expansive algebraic $\mathbb{Z}^d$-action.*

For $d = 2$ and the subset

$$E = \{(0,0),(1,0),(0,1)\} \subset \mathbb{Z}^2,$$

the $\mathbb{Z}^2$-action σ_E on X_E is called Ledrappier's example: σ_E is mixing and expansive, but not mixing of order 3 (for every $n \geq 0$, $x_{(0,0)} + x_{(2^n,0)} + x_{(0,2^n)} = 0$).

In this example, 3-mixing breaks down in a particularly regular way: if we call a finite subset $S \subset \mathbb{Z}^d$ mixing for a group shift X if

$$(5.2) \qquad \lim_{k \to \infty} \lambda_X \left(\bigcap_{\mathbf{m} \in S} \sigma^{-k\mathbf{m}} B_{\mathbf{m}} \right) = \prod_{\mathbf{m} \in S} \lambda_X(B_{\mathbf{m}})$$

for all Borel sets $B_{\mathbf{m}}$, $\mathbf{m} \in S$, and nonmixing *otherwise, then the last paragraph shows that $S = \{(0,0),(1,0),(0,1)\}$ is nonmixing for Ledrappier's example.*

We also consider the subsets

$$E_1 = \{(0,0),(1,0),(2,0),(1,1),(0,2)\},$$
$$E_2 = \{(0,0),(2,0),(0,1),(1,1),(0,2)\},$$
$$E_3 = \{(0,0),(1,0),(2,0),(0,1),(1,1),(0,2)\}.$$

of $\mathbb{Z}^2$. The shift-actions $\sigma_i = \sigma_{E_i}$ of $\mathbb{Z}^2$ on $X_i = X_{E_i}$ are again mixing, but the set $S = \{(0,0),(1,0),(0,1)\}$ is nonmixing for each of these actions (cf. [11]).

For every $\mathbf{n} \in \mathbb{Z}^2$, the automorphisms $\sigma_i^{\mathbf{n}}$ are measurably conjugate. However, as was shown in [12], these three $\mathbb{Z}^2$-actions are not even measurably conjugate.

In general, if $X \subset A^{\mathbb{Z}^2}$ is a group shift with finite alphabet A, then X has nonmixing sets if and only if it does not have completely positive entropy *or, equivalently, if and only if it is not measurably conjugate to a full shift* $Y = B^{\mathbb{Z}^2}$, *where B is a finite set (cf. [15], [11] and [20]). However, even if X has completely positive entropy, it need not be topologically conjugate to a full shift.*

For $d = 1$, algebraic conjugacy of group shifts $X \subset A^{\mathbb{Z}^d}$ to full shifts $Y = B^{\mathbb{Z}^d}$, where A and B are finite abelian groups, is a matter of considerable interest in coding theory (cf. e.g. [18]), and results for $d > 1$ are just beginning to emerge.

Example 6 is based on a special case of another rigidity phenomenon specific to $\mathbb{Z}^d$-actions with $d > 1$. We call an algebraic $\mathbb{Z}^d$-action α on a compact abelian group *irreducible* if every closed, α-invariant subgroup $Y \subsetneq X$ is finite. The following statement is proved in [8] and [12].

THEOREM 5.1. *Let $d > 1$, and let α_1 and α_2 be mixing algebraic $\mathbb{Z}^d$-actions on compact abelian groups X_1 and X_2, respectively. If α_1 is irreducible, and if $\phi\colon X_1 \longrightarrow X_2$ is a measurable conjugacy of α_1 and α_2, then α_2 is irreducible and ϕ is λ_{X_1}-a.e. equal to an affine map. Hence measurable conjugacy of α_1 and α_2 implies algebraic conjugacy.*

Irreducibility of algebraic $\mathbb{Z}^d$-actions with $d > 1$ implies that these actions have zero entropy (as $\mathbb{Z}^d$-actions). For actions with positive entropy one cannot expect this kind of isomorphism rigidity, since positive entropy implies the existence of nontrivial Bernoulli factors (cf. [23]). However, it is sometimes still be possible to apply Theorem 5.1 to prove measurable nonconjugacy of actions with positive entropy.

EXAMPLE 7 (Conjugacy of $\mathbb{Z}^2$-actions with positive entropy). *We modify Example 6 by setting $\bar{Y} = (\mathbb{Z}/4\mathbb{Z})^{\mathbb{Z}^d}$ and consider, for every nonempty finite set $E \subset \mathbb{Z}^d$ the closed shift-invariant subgroup $Y_E \subset \bar{Y}$ consisting of all $y \in \bar{Y}$ whose coordinates sum to 0 (mod 2) in every translate of E in $\mathbb{Z}^d$. The group Y_E is always uncountable, and the restriction τ_E of the shift-action σ to Y_E is an expansive algebraic $\mathbb{Z}^d$-action with entropy $\log 2$. As in Example 6 we set $d = 2$ and consider the the subsets $E, E_1, E_2, E_3 \subset \mathbb{Z}^2$ defined there. Theorem 6.5 in [15] implies that the Pinsker algebra $\pi(\tau_{E_i})$ of τ_{E_i} is the sigma-algebra $\mathcal{B}_{Y_{E_i}/Z_{E_i}}$ of Z_{E_i}-invariant Borel sets in Y_{E_i}, where*

$$Z_{E_i} = \{x = (x_\mathbf{n}) \in Y_{E_i} : x_\mathbf{n} = 0 \,(\mathrm{mod}\, 2) \text{ for every } \mathbf{n} \in \mathbb{Z}^2\}.$$

Then the $\mathbb{Z}^2$-action τ'_{E_i} induced by τ_{E_i} on Y_{E_i}/Z_{E_i} is algebraically conjugate to the shift-action σ_{E_i} on the group X_{E_i} in Example 6.

Since any measurable conjugacy of τ_{E_i} and τ_{E_j} would map $\pi(\tau_{E_i})$ to $\pi(\tau_{E_j})$ and induce a conjugacy of τ'_{E_i} and τ'_{E_j} and hence of σ_{E_i} and σ_{E_j}, Example 6 implies that τ_i and τ_j are measurably nonconjugate for $1 \leq i < j \leq 3$.

EXAMPLE 8 (Group shifts with uncountable alphabet). *We write $\mathfrak{R}_d = \mathbb{Z}[u_1^{\pm 1}, \ldots, u_d^{\pm 1}]$ for the ring of Laurent polynomials with integral coefficients in the commuting variables $u_1, \ldots, u_d$, and represent every $f \in$*

$\mathfrak{R}_d$ as $f = \sum_{\mathbf{m} \in \mathbb{Z}^d} f_{\mathbf{m}} u^{\mathbf{m}}$ with $u^{\mathbf{m}} = u_1^{m_1} \cdots u_d^{m_d}$ and $f_{\mathbf{m}} \in \mathbb{Z}$ for every $\mathbf{m} = (m_1, \ldots, m_d) \in \mathbb{Z}^d$.

Let σ be the shift-action (1.1) of $\mathbb{Z}^d$ on $\Omega^{(d)} = \mathbb{T}^{\mathbb{Z}^d}$. For every nonzero $f \in \mathfrak{R}_d$ and $x \in X$ we set

$$(5.3) \qquad f(\sigma)(x) = \sum_{\mathbf{n} \in \mathbb{Z}^d} f_{\mathbf{n}} \sigma^{\mathbf{n}} x$$

and note that $f(\sigma) \colon \Omega^{(d)} \longrightarrow \Omega^{(d)}$ is a continuous surjective group homomorphism. For every ideal $I \subset \mathfrak{R}_d$ we set

$$(5.4) \qquad X_I = \bigcap_{f \in I} \ker(f(\sigma))$$

and denote by σ_I the restriction of σ to X_I. If $\{g_1, \ldots, g_L\}$ is a set of generators of I (such a finite set of generators always exists, since $\mathfrak{R}_d$ is Noetherian), then

$$X_I = \bigcap_{j=1}^{m} \ker(g_j(\sigma)).$$

The dynamical properties of group shifts of the form X_I are described in [22], [15] and [23]. In the special case where the ideal I is principal, i.e. where $I = (f) = f\mathfrak{R}_d$ for some $f \in \mathfrak{R}_d$, the entropy of $\sigma_{(f)}$ is given by

$$h(\sigma_{(f)}) = \begin{cases} \int_0^1 \cdots \int_0^1 \log |f(e^{2\pi i t_1}, \ldots, e^{2\pi i t_d})| \, dt_1 \cdots dt_d & \text{if } f \neq 0, \\ \infty & \text{otherwise.} \end{cases}$$

Furthermore, $\sigma_{(f)}$ is expansive if and only if

$(5.5) \quad f(\mathbf{z}) \neq 0$ for all $\mathbf{z} = (z_1, \ldots, z_d) \in \mathbb{C}^d$ with $|z_1| = \cdots = |z_d| = 1$.

If $\sigma_{(f)}$ is expansive then it is automatically mixing and Bernoulli (in particular, it has finite and positive entropy).

Although the group shifts $\sigma_{(f)}$, $f \in \mathfrak{R}_d$, in Example 8 are of finite type in the sense that they are determined by restrictions in a finite 'window' of coordinates (consisting of those $\mathbf{n} \in \mathbb{Z}^d$ with $f_{\mathbf{n}} \neq 0$), their uncountable alphabets put them outside the customary framework of symbolic dynamics. In view of this (and for a variety of other reasons) it seems desirable to find 'symbolic' representations of such systems, analogous to the representation of hyperbolic toral automorphisms as SFT's by means of Markov partitions.

Following [6] we consider the Banach space $\ell^\infty(\mathbb{Z}^d, \mathbb{R})$ and write

$$\ell^\infty(\mathbb{Z}^d, \mathbb{Z}) \subset \ell^\infty(\mathbb{Z}^d, \mathbb{R})$$

for the subgroup of bounded integer-valued functions. Consider the surjective map $\eta\colon \ell^\infty(\mathbb{Z}^d, \mathbb{R}) \longrightarrow \mathbb{T}^{\mathbb{Z}^d}$ given by

$$\eta(v)_{\mathbf{n}} = v_{\mathbf{n}} \pmod 1$$

for every $v = (v_{\mathbf{n}}) \in \ell^\infty(\mathbb{Z}^d, \mathbb{R})$ and $\mathbf{n} \in \mathbb{Z}^d$. Let $\bar\sigma$ be the shift-action of $\mathbb{Z}^d$ on $\ell^\infty(\mathbb{Z}^d, \mathbb{R})$, defined as in (1.1), and set, for every $h = \sum_{\mathbf{n} \in \mathbb{Z}^d} h_{\mathbf{n}} u^{\mathbf{n}} \in \mathfrak{R}_d$ and $v \in \ell^\infty(\mathbb{Z}^d, \mathbb{Z})$,

$$h(\bar\sigma)(v) = \sum_{\mathbf{n} \in \mathbb{Z}^d} h_{\mathbf{n}} \bar\sigma^{\mathbf{n}} v.$$

The expansiveness of $\sigma_{(f)}$ can be expressed in terms of the kernel of $f(\bar\sigma)$: $\sigma_{(f)}$ is expansive if and only if $\ker(f(\bar\sigma)) = \{0\} \subset \ell^\infty(\mathbb{Z}^d, \mathbb{R})$.

According to Lemma 4.5 in [14] there exists a unique element $w^\Delta \in \ell^\infty(\mathbb{Z}^d, \mathbb{R})$ with the property that

$$f(\bar\sigma)(w^\Delta)_{\mathbf{n}} = \begin{cases} 1 & \text{if } \mathbf{n} = \mathbf{0}, \\ 0 & \text{otherwise.} \end{cases}$$

The point w^Δ also has the property that there exist constants $c_1 > 0, 0 < c_2 < 1$ with

$$|w_{\mathbf{n}}^\Delta| \le c_1 c_2^{\|\mathbf{n}\|}$$

for every $\mathbf{n} = (n_1, \ldots, n_d) \in \mathbb{Z}^d$, where $\|\mathbf{n}\| = \max_{i=1,\ldots,d} |n_i|$. From the properties of w^Δ it is clear that

$$\bar\xi(v) = \sum_{\mathbf{n} \in \mathbb{Z}^d} v_{\mathbf{n}} \bar\sigma^{-\mathbf{n}} w^\Delta$$

is a well-defined element of $\ell^\infty(\mathbb{Z}^d, \mathbb{R})$ for every $v \in \ell^\infty(\mathbb{Z}^d, \mathbb{Z})$, and we set

$$\xi = \eta \circ \bar\xi \colon \ell^\infty(\mathbb{Z}^d, \mathbb{Z}) \longrightarrow X_{(f)}.$$

The map $\xi\colon \ell^\infty(\mathbb{Z}^d, \mathbb{Z}) \longrightarrow X_{(f)}$ is a surjective group homomorphism, and

$$\xi \circ \bar\sigma^{\mathbf{n}} = \sigma_{(f)}^{\mathbf{n}} \circ \xi \quad \text{for every } \mathbf{n} \in \mathbb{Z}^d,$$
$$\ker(\xi) = f(\bar\sigma)(\ell^\infty(\mathbb{Z}^d, \mathbb{Z})).$$

The point $x^\Delta = \xi(w^\Delta)$ is *homoclinic*:

(5.6) $$\lim_{\mathbf{n} \to \infty} \sigma_{(f)}^{\mathbf{n}}(x^\Delta) = 0.$$

Furthermore, x^Δ is a *fundamental homoclinic point* in the sense that every homoclinic point of $\sigma_{(f)}$ (i.e. every $x \in X_{(f)}$ satisfying (5.6) with x replacing x^Δ) lies in the countable subgroup of $X_{(f)}$ generated by $\{\sigma_{(f)}^{\mathbf{n}} x^\Delta :$

$\mathbf{n} \in \mathbb{Z}^d\}$. It can be shown that an expansive algebraic $\mathbb{Z}^d$-action α on a compact abelian group X has a fundamental homoclinic point if and only if it is of the form $\alpha = \sigma_{(f)}$, $X = X_{(f)}$, for some $f \in \mathfrak{R}_d$ satisfying (5.5) (cf. [26]).

From the definition of ξ it is clear that its restriction to every bounded subset of $\ell^\infty(\mathbb{Z}^d, \mathbb{Z})$ is continuous in the weak*-topology. One can easily find bounded (and thus weak*-compact) subset $\mathcal{V} \subset \ell^\infty(\mathbb{Z}^d, \mathbb{Z})$ with $\xi(\mathcal{V}) = X_{(f)}$:

PROPOSITION 5.1. *For every* $h = \sum_{\mathbf{n} \in \mathbb{Z}^d} h_{\mathbf{n}} u^{\mathbf{n}} \in \mathfrak{R}_d$ *we set*

$$h^+ = \sum_{\mathbf{n} \in \mathbb{Z}^d} \max(0, h_{\mathbf{n}}) u^{\mathbf{n}}, \qquad h^- = -\sum_{\mathbf{n} \in \mathbb{Z}^d} \min(0, h_{\mathbf{n}}) u^{\mathbf{n}},$$

$$\|h^+\|_1' = \max(\|h^+\|_1 - 1, 0), \qquad \|h^-\|_1' = \max(\|h^-\|_1 - 1, 0),$$

$$\|h\|_1^* = \|h^+\|_1' + \|h^-\|_1'.$$

Then the set

$$\mathcal{V} = \{v \in \ell^\infty(\mathbb{Z}^d, \mathbb{Z}) : 0 \le v_{\mathbf{n}} \le \|f\|_1^* \text{ for every } \mathbf{n} \in \mathbb{Z}^d\}$$

satisfies that $\xi(\mathcal{V}) = X_{(f)}$.

The restriction of the homomorphism ξ to $\mathcal{V}$ is surjective, but generally not injective, and the key problem in constructing symbolic representations of the $\mathbb{Z}^d$-action $\sigma_{(f)}$ is to find closed, shift-invariant subsets $\mathcal{W} \subset \mathcal{V}$ with the following properties:

(a) $\mathcal{W}$ is a *SFT* or at least *sofic*, i.e. a topological factor of a *SFT*,
(b) $\xi(\mathcal{W}) = X_{(f)}$, and the restriction of ξ to a dense G_δ-set in $\mathcal{W}$ is injective.

Examples of such choices of $\mathcal{W} \subset \ell^\infty(\mathbb{Z}, \mathbb{Z})$ for appropriate polynomials $f \in \mathfrak{R}_1$ can be found in [26], [27], [28], [30] and [31]. Examples in higher dimensions (with $d \ge 2$) are much more difficult to find, and there are many unresolved problems in this area. We end this section with one of the few successful examples.

EXAMPLE 9 ([6]). *Let* $d = 2$ *and* $f = 3 - u_1 - u_2 \in \mathfrak{R}_2$. *Then* $\|f\|_1^* = 3$, *but the set*

$$(5.7) \qquad \mathcal{W} = \{v \in \ell^\infty(\mathbb{Z}^2, \mathbb{Z}) : 0 \le v_{\mathbf{n}} \le 2\} \subsetneq \mathcal{V}$$

also satisfies that $\xi(\mathcal{W}) = X_{(f)}$. *Furthermore, the restriction of* ξ *to* $\mathcal{W}$ *is almost injective in the sense of Condition (b).*

REFERENCES

[1] R. BERGER, *The undecidability of the Domino Problem*, Mem. Amer. Math. Soc. **66** (1966).

[2] R. BURTON AND R. PEMANTLE, *Local characteristics, entropy and limit theorems for spanning trees and domino tilings via transfer-impedances*, Ann. Probab. **21** (1993), 1329–1371.

[3] C. COHN, N. ELKIES AND J. PROPP, *Local statistics for random domino tilings of the Aztec diamond*, Duke Math. J. **85** (1996), 117–166.

[4] J.H. CONWAY AND J.C. LAGARIAS, *Tilings with polyominoes and combinatorial group theory*, J. Combin. Theory Ser. A **53** (1990), 183–208.

[5] M. EINSIEDLER, *Fundamental cocycles of tiling spaces*, Ergod. Th. & Dynam. Sys. (to appear).

[6] M. EINSIEDLER AND K. SCHMIDT, *Markov partitions and homoclinic points of algebraic $\mathbb{Z}^d$-actions*, Proc. Steklov Inst. Math. **216** (1997), 259–279.

[7] P.W. KASTELEYN, *The statistics of dimers on a lattice. I*, Phys. D **27** (1961), 1209–1225.

[8] A. KATOK, S. KATOK AND K. SCHMIDT, *Rigidity of measurable structure for algebraic actions of higher-rank abelian groups*, in preparation.

[9] B. KITCHENS AND K. SCHMIDT, *Periodic points, decidability and Markov subgroups*, in: Dynamical Systems, Proceeding of the Special Year, Lecture Notes in Mathematics, vol. 1342, Springer Verlag, Berlin-Heidelberg-New York, 1988, 440–454.

[10] B. KITCHENS AND K. SCHMIDT, *Automorphisms of compact groups*, Ergod. Th. & Dynam. Sys. **9** (1989), 691–735.

[11] B. Kitchens and K. Schmidt, *Mixing sets and relative entropies for higher dimensional Markov shifts*, Ergod. Th. & Dynam. Sys. **13** (1993), 705–735.

[12] B. KITCHENS AND K. SCHMIDT, *Isomorphism rigidity of irreducible algebraic $\mathbb{Z}^d$-actions*, Preprint (1999).

[13] S. Lightwood, *An aperiodic embedding theorem for square filling subshifts of finite type*, Preprint (1999).

[14] D. LIND AND K. SCHMIDT, *Homoclinic points of algebraic $\mathbb{Z}^d$-actions*, J. Amer. Math. Soc. **12** (1999), 953–980.

[15] D. LIND, K. SCHMIDT AND T. WARD, *Mahler measure and entropy for commuting automorphisms of compact groups*, Invent. Math. **101** (1990), 593–629.

[16] N.G. Markley and M.E. Paul, *Matrix subshifts for $\mathbb{Z}^n$ symbolic dynamics*, Proc. London Math. Soc. **43** (1981), 251–272.

[17] ———, *Maximal measures and entropy for $\mathbb{Z}^n$ subshifts of finite type*, Preprint (1979).

[18] R.J. McEliece, *The algebraic theory of convolutional codes*, in: Handbook of Coding Theory (2 vols.), North Holland, Amsterdam, 1998, 1065–1138.

[19] R.M. ROBINSON, *Undecidability and nonperiodicity for tilings of the plane*, Invent. Math. **12** (1971), 177–209.

[20] D.J. Rudolph and K. Schmidt, *Almost block independence and Bernoullicity of $\mathbb{Z}^d$-actions by automorphisms of compact groups*, Invent. Math. **120** (1995), 455–488.

[21] K. SCHMIDT, *Algebraic ideas in ergodic theory*, in: CBMS Lecture Notes, vol. 76, American Mathematical Society, Providence, R.I., 1990.

[22] K. SCHMIDT, *Automorphisms of compact abelian groups and affine varieties*, Proc. London Math. Soc. **61** (1990), 480–496.

[23] K. SCHMIDT, *Dynamical systems of algebraic origin*, Birkhäuser Verlag, Basel-Berlin-Boston, 1995.

[24] K. SCHMIDT, *The cohomology of higher-dimensional shifts of finite type*, Pacific J. Math. **170** (1995), 237–270.

[25] K. SCHMIDT, *Tilings, fundamental cocycles and fundamental groups of symbolic $\mathbb{Z}^d$-actions*, Ergod. Th. & Dynam. Sys. **18** (1998), 1473–1525.

[26] K. SCHMIDT, *Algebraic coding of expansive group automorphisms and two-sided beta-shifts*, Monatsh. Math. (to appear).

[27] N.A. SIDOROV AND A.M. VERSHIK, *Ergodic properties of Erdös measure, the entropy of the goldenshift, and related problems*, Monatsh. Math. **126** (1998), 215–261.

[28] N. SIDOROV AND A. VERSHIK, *Bijective arithmetic codings of the 2-torus, and binary quadratic forms*, to appear.

[29] W. THURSTON, *Conway's tiling groups*, Amer. Math. Monthly **97** (1990), 757–773.

[30] A. VERSHIK, *The fibadic expansion of real numbers and adic transformations*, Preprint, Mittag-Leffler Institute, 1991/92.

[31] A.M. VERSHIK, *Arithmetic isomorphism of hyperbolic toral automorphisms and sofic shifts*, Funktsional. Anal. i Prilozhen. **26** (1992), 22–27.

[32] H. WANG, *Proving theorems by pattern recognition II*, AT&T Bell Labs. Tech. J. **40** (1961), 1–41.

Part 2. Codes on graphs

LINEAR-CONGRUENCE CONSTRUCTIONS OF LOW-DENSITY PARITY-CHECK CODES*

J. BOND[†], S. HUI[‡], AND H. SCHMIDT[§]

Abstract. Low-Density Parity-Check codes (LDPCC) with Iterative Belief Propagation (Message Passing) decoding are attractive alternatives to Turbo codes. LDPCC previously discussed in the literature have involved matrices constructed using random techniques. In this paper, we discuss construction techniques for LDPCC involving multiple permutation matrices, each specified by a linear congruence. Construction options depend on the size of the parity-check matrix and the rate of the code. We relate desirable properties of the code to the parameters defining the linear congruences specifying the permutation matrices used to construct the code. For example, codes with few or no 4-cycles can be readily constructed. We summarize the construction options and describe selection processes for the parameters of the congruences. We then provide performance results for regular parity-check matrices constructed by random and the linear-congruence techniques for rate 1/2 transmit block-size 980 and rate 4/7 transmit block-size 847 codes. We introduce a symmetric channel model for decoding with the iterative belief propagation algorithm and describe its use as a heuristic for deciding whether a code is likely better or worse than most codes of the given rate and block size.

1. Introduction. Renewed interest in Low-Density Parity-Check Codes (LDPCC), originally discovered by Gallager [8], followed the work of David MacKay [11], who rediscovered the codes through his work on the Iterative Belief Propagation (or Message Passing) Decoding algorithm for codes with sparse parity-check matrices. Our work began in the summer of 1996 after hearing a presentation by MacKay, with the realization that LDPCC were promising candidates for use in a new communication system we were developing. We have focused on constructing the best possible codes of rates 1/2 or more and transmit block sizes of at most several thousand bits for our particular application.

LDPCC have been constructed by MacKay and others using random techniques [11, 12, 13, 14, 7]. In contrast, our focus has been on using systematic or algebraic techniques to construct codes. In [1, 2, 3], we describe codes for which each parity-check matrix contains a full rank circulant matrix in conjunction with the random construction of the remaining columns of the parity-check matrix. These codes performed as well as those described in the literature with both portions of the matrices constructed randomly. A LDPCC code is regular if each row and each column of the parity-check matrix contains a constant number of ones. The iterative be-

*The authors would like to thank the reviewer and Pascal O. Vontobel for useful comments.

[†]Science Applications International Corporation, 4015 Hancock Street, San Diego, CA 92110; bond_jw@nosc.mil.

[‡]Department of Mathematical Sciences, San Diego State University, San Diego, CA 92182; hui@saturn.sdsu.edu.

[§]Technology Service Corporation, 962 Wayne Avenue, Suite 800, Silver Spring, MD 20910; hschmidt@tscwo.com.

lief propagation appears to perform best when the columns of the regular matrices have exactly three ones. For this reason, most of our work has concentrated on codes with rates $\xi/(\xi + 3)$ for an integer ξ (for example, 1/2 and 4/7) and relatively small transmit block sizes (less than 1000).

In this paper, we describe techniques for the construction of regular LDPCC in terms of permutation matrices, each specified by a single linear congruence. In this manner, a given code is associated with a few algebraic equations and it becomes feasible to relate desirable code properties with the parameters of the linear congruences. At present, we do not have a complete theoretical analysis of this approach. We do not know the best way to choose the code parameters except to avoid 4-cycles (note that there are no 2-cycles since any two nodes can be connected by at most one edge). However, we have succeeded in constructing codes that perform at least as well as, if not better than, most codes constructed randomly, especially when performance is measured in terms of block error rates. This is done by choosing linear congruences to control the occurrences of cycles and thereby presumably low-weight codewords. We also give a condition that will guarantee the absence of cycles up to any desired length. We hope that when this approach is better understood, we can obtain better performance and better bounds for the minimum distance.

We also attempted to construct irregular codes for these small block sizes using two different approaches: the Richarson-Urbanke method and MacKay's super-Poisson method. The performance of the irregular codes we constructed did not match the performance of the regular codes. This indicates that constructing irregular LDPCC codes for small block sizes may be quite subtle and difficult. Since our focus in this paper is on linear-congruence codes, we did not pursue this further.

2. Linear-congruence constructions. Let $H = [R\ C]$, with C an invertible matrix, denote the parity-check matrix. In this paper, we introduce the method of constructing C and R in terms of permutation matrices each described by linear congruences and develop a theory for appropriately choosing the parameters of the linear congruences so that the matrices are regular and have no 4-cycles. We will also include the condition for avoiding cycles of any length without the technical details.

We first consider the construction of a permutation matrix using a linear congruence. Let a, b be integers. Let P be an $n \times n$ matrix with a one in the j-th column of the i-th row if and only if $j = ai + b \bmod n$. If a given column has two ones, then there exist i_1 and i_2 such that $ai_1 + b = ai_2 + b \bmod n$. It is easy to see that P is a permutation matrix if and only if a and n are relatively prime, and when this condition holds, the matrix P could also be described by the location of the one in each column, $i = \widehat{a}j + \widehat{b} \bmod n$, with $\widehat{a} = a^{-1}$ and $\widehat{b} = -a^{-1}b$. Since these representations are equivalent, we will use both without further notice but always with the i's referring to the rows and the j's to the columns. Observe that a circulant is a sum of matrices P described by congruences of the form $j = i + b \bmod n$.

Circulants are attractive for constructing the matrix C (see [4]). The best constructions to be discussed herein for regular rate $1/2$ and $4/7$ codes have R constructed in terms of permutation matrices described by general linear congruences and C a circulant. Codes constructed with carefully chosen circulants for C with R constructed randomly performed as well as codes with both R and C constructed randomly. Furthermore, it is easy to condition the primitive polynomial so that the C portion of the matrix has no 4-cycles, although it necessarily has some 6-cycles.

Codes constructed with both R and C circulants perform poorly because R and C commute. See Appendix A for details. However, by generalizing R to a sum of matrices associated with general linear congruences, R and C need no longer commute. Furthermore, since traditionally random numbers were sometimes generated by appropriate linear congruences, we hoped that the judicious choices of the parameters a and b will yield sufficiently random matrices. Indeed, we have found sufficient conditions on the parameters of the congruences that lead to candidate codes that should perform well. The nature of these conditions depends on n. The most important class of constructions for practical applications are those when n is composite with a factor greater than 3. We have also found a construction for the special case when n is a prime, but we will not present this rather special case.

2.1. Square matrices defined by linear congruences. In this section, we discuss the linear-congruence construction for square matrices. Before introducing the linear-congruence construction, we first observe that the linear-congruence construction can be considered a generalization of the circulant construction.

Let C be an $n \times n$ circulant matrix and let $b_1, \ldots, b_\xi$ be the positions of the nonzero entries of the first row of C. Then for $i = 0, 1, \ldots, n - 1$, the nonzero positions of the $i + 1$-st row of C are

$$i + b_1, \ i + b_2 \ \ldots, \ i + b_\xi \ \bmod n,$$

which can obviously be written as

$$a_1 i + b_1, \ a_2 i + b_2 \ \ldots, \ a_\xi i + b_\xi \ \bmod n,$$

where $a_1 = \cdots = a_\xi = 1$. The linear-congruence method uses the same technique with distinct a_i's. Of course, the a_i's must be chosen appropriately so the resulting matrices will have certain desired properties, such as regularity and no 4-cycles.

Unless stated otherwise, we assume that R is $n \times n$ and the required number of ones in each row and each column is ξ. We use linear congruences to construct $n \times n$ matrices with exactly ξ ones in each row and column. For each row, we need to find ξ positions for the ones so that each column of each matrix will also have ξ ones.

Let R denote the matrix to be constructed. Let a, b be positive integers less than n. Consider the sequence

$$r_i = (ai + b) \bmod n, \qquad i = 0, \dots, n-1.$$

For each i, let $R_{i,r_i} = 1$. Note that we identify 0 and n. If a, b are chosen properly, then $r_i \neq r_j$ for $i \neq j$ and there is exactly one 1 in each row and column. We iterate this procedure ξ times, with a different set of a, b for each iteration, to generate the matrix R. Of course, the a's and b's need to be chosen so that the ones are at different locations. We need the following theorems from number theory (see, for example, Hua [10] and Stewart [15]).

THEOREM 2.1. *Let $a_1, \dots, a_\xi$ and b be integers. The equation*

$$a_1 x_1 + \cdots + a_\xi x_\xi + b = 0 \bmod n$$

has a solution $(x_1, \dots, x_\xi)$ if and only if the greatest common divisor of $a_1, \dots, a_\xi, n$ divides b. If the equation is solvable, there are

$$n^{\xi-1} \gcd(a_1, \dots, a_\xi, n)$$

solutions that are not equivalent modulo n.

THEOREM 2.2. *For $j = 1, \dots, \xi$, let $a_{j1}, \dots, a_{j\xi}$ and b_j be integers and let*

$$L_j = a_{j1} x_1 + \cdots + a_{j\xi} x_\xi + b_j.$$

Let $L'_j = L_j$ for $j \neq k$ and

$$L'_k = c_1 L_1 + \cdots + c_k L_k + \cdots + c_\xi L_\xi.$$

If $\gcd(c_k, n) = 1$, then the system

$$L_j = 0 \bmod n, \qquad j = 1, \dots, \xi$$

is equivalent to

$$L'_j = 0 \bmod n, \qquad j = 1, \dots, \xi.$$

Let a_j, b_j be positive integers used in the j-th iteration in the construction of the matrix. To ensure that the ones for each iterations are in different locations, the equation

$$a_j i + b_j = a_m i + b_m \bmod n$$

cannot have solutions for $i = 0, \dots, n-1$ and $j \neq m$. The above equation is equivalent to

$$(2.1) \qquad (a_j - a_m)i + (b_j - b_m) = 0 \bmod n,$$

and using Theorem 2.1, we see that this equation is not solvable if and only if

$$(2.2) \qquad \gcd(a_j - a_m, n) \nmid b_j - b_m.$$

Observe that if n is prime, then $\gcd(a_j - a_m, n) = 1$ and it is not possible to construct R using this method directly. However, by using n prime and the b_j's zero, we can obtain an R with size $(n - 1) \times (n - 1)$. In general, it is easier to find $\{a_j, b_j\}_{j=1}^{\xi}$ if n has more factors.

2.2. General linear-congruence construction. We have seen in the previous section how to construct square regular linear-congruence matrices using sums of permutation matrices. Clearly, this method can only be used to construct square regular matrices, which can be used as the R or C part of the parity-check matrix as described above. To construct matrices that are not square, we need to use different methods. The methods impose certain restrictions on the size of the matrix and the number of entries in each row and column.

A large class of non-square regular parity-check matrices can be constructed from permutation matrices using:

1. matrix overlap and
2. matrix partitioning.

2.2.1. Matrix overlap. We first discuss the matrix overlap method. Suppose we wish to construct an $m \times n$, $m < n$, regular matrix with ξ ones in each column. The idea is to construct s permutation matrices of size $m \times m$ and place them as blocks in a row to form an $m \times sm$ matrix and then "fold" the long block matrix modulo m into an $m \times n$ matrix so that the ones from the different blocks do not collide. By simple counting of the ones, we have

$$sm = n\xi \qquad \text{or} \qquad s = \frac{n\xi}{m}.$$

Therefore, to use the overlap method, m must divide $n\xi$. It is clear that if there are no collisions of ones, the resulting matrix is regular.

For example, suppose we wish to generate a 363×484 regular matrix with three ones in each column. Then we construct $s = (484)(3)/363 = 4$ permutation matrices of size 363×363, say P_0, P_1, P_2, P_3. Form the 363×1452 matrix $[P_0 \ P_1 \ P_2 \ P_3]$, which we rewrite as $[Q_0 \ Q_1 \ Q_2]$, where each Q_j is 363×484. The final matrix is then $Q_0 + Q_1 + Q_2$. Of course, we must be careful that the nonzero entries are all in different positions. We will see how this can be arranged when the permutation matrices are constructed using linear congruences.

We next show that if m does not divide n, the above construction is equivalent to embedding each $m \times m$ matrix into an $m \times n$ matrix, cyclically shifting the s matrices by multiples of $\gcd(m, n)$, and then summing. For example, in our 363×484 example given above, we cyclically shift the matrices P_0, P_3, P_2, P_1 (note the order of the matrices), by $121k$, $k = 0, 1, 2, 3$, columns modulo 484 and add. More explicitly, let $P_k = [P_{k1} \; P_{k2} \; P_{k3}]$, where each P_{kj} is 363×121. Then the overlap matrix is formed by a row of four 363×121 matrices:

$$\left[\; P_{01} + P_{12} + P_{23}, \cdots, P_{11} + P_{22} + P_{33} \; \right].$$

Suppose m does not divide n. Let $u = \gcd(m, n)$ and let $n = n'u$ and $m = m'u$. Then $s = n\xi/m = n'\xi/m'$ and $m' \neq 1$. It follows that m' divides ξ. Since $n'm = m'n$, n' matrices of size $m \times m$ can be folded into an $m \times n$ matrix with m' ones in each column. Therefore, folding a block of s matrices of size $m \times m$ into an $m \times n$ matrix with ξ ones in each column is equivalent to breaking the s matrices into ξ/m' blocks of n' matrices, fold each n' block into an $m \times n$ matrix with m' ones in each column and then summing the ξ/m' resulting matrices.

We now consider each n' long block of $m \times m$ matrices separately. Each block is folded into an $m \times n$ matrix with m' ones in each column. Since m' and n' are relatively prime, there exists t, $1 \leq t \leq n'$, such that $m't = 1 \bmod n'$. Let $B = [P_0 \; P_1 \; \cdots \; P_{n'-1}]$ be the block of $m \times m$ matrices to be folded into a $m \times n$ matrix. Assume that P_0 starts at column 0. It is easy to see that for $i = 1, \ldots, n' - 1$, P_{it} starts at position iu modulo n. Therefore, the folding of B modulo n to form a $m \times n$ matrix is equivalent to cyclically shifting $P_0, P_t, \ldots, P_{(n'-1)t}$ by $0, u, \ldots, (n'-1)u$ columns modulo n, respectively, and then summing.

To ensure that the ones from the different permutation matrices occupy different positions, it is sufficient that the equation

$$a_p i + b_p = a_q i + b_q + ku \bmod n$$

does not have any solutions for $i = 0, \ldots, m-1$, $p \neq q$, and $k = 1, \ldots, s-1$. As in the square matrix case, the above equation will have no solution if

$$\gcd(a_p - a_q, n) \text{ does not divide } b_q - b_p + ku.$$

One way to guarantee the above nondivisibility condition is the following. Let $v > 1$ be a factor of u. Choose the $\{a_j, b_j\}$ so that the a_j's are relatively prime to m, v divides $a_p - a_q$ and v does not divide $b_q - b_p$ for all $p \neq q$. Then v divides $\gcd(a_p - a_q, n)$ and u but not $b_q - b_p$, and thus $\gcd(a_p - a_q, n)$ does not divide $b_q - b_p + ku$. For example, suppose we wish to construct a 363×484 regular matrix with three ones in each column. In this case,

$$\xi = 3, \; m = 363, \; n = 484, \; u = 121, \; m' = 3, \; n' = 4.$$

We can let $v = 11$, $a_1 = 23$, $a_2 = 34$, $a_3 = 67$, $a_4 = 89$, and $b_j = j - 1$ for $j = 1, \ldots, 4$.

2.2.2. Partition. We next describe the partition method. Again, let the dimension of the desired matrix be $m \times n$, $m < n$, and the number of ones in each column be ξ. To use the partition method, we need $m = ku$ and $n = qu$, where q and u are positive integers. With this assumption, we can partition the $m \times n$ matrix into a $\xi \times q$ block matrix, where each block is $u \times u$. One can then put a permutation in each block and obtain a regular matrix with ξ ones in each column. However, it is easier to control the number of short cycles if we use $[q/k]$ blocks of $m \times m$ linear-congruence matrices and then fill the remaining $u \times u$ blocks with either random or linear-congruence permutation matrices.

For example, to construct a 363×484 regular matrix with three ones in each column, we first construct a 363×363 regular matrix that has three ones in each column given by congruences. The remaining 363×121 matrix is constructed by stacking three 121×121 permutation matrices in a column. The permutation matrices can be constructed randomly or by linear congruences.

3. The avoidance of cycles. In this section, we give equations whose solutions give cycles in the bipartite graph corresponding to the matrix R. The matrix R generates a bipartite graph with the rows and columns as the two vertex or node sets. Row vertex i is connected by a single undirected edge to column vertex j if and only if $R_{ij} = 1$. A *cycle* is a path in the graph with the same starting and end points such that each undirected edge in the path appears only once. The number of edges in the path is called the length of the cycle and a cycle of length k is called an k-cycle. Clearly, a cycle in a bipartite graph must have even length and each vertex in a cycle must have an even number of edges. The number of edges ending at a vertex is called the *degree* of the vertex.

3.1. The square matrix case. We first treat the 4-cycle case with three ones in each column in detail, which is the case of main concern in practice and will also serve as an introduction to the more complicated general case.

3.1.1. The 4-cycle case. To avoid 4-cycles, two distinct rows of R can have at most one 1 at the same position. To have a 4-cycle, two different rows, say i and j, must have ones in two different columns and so the following system of equations must be solvable with $i \neq j \bmod n$:

$$\text{(3.1)} \qquad \begin{aligned} a_p i + b_p &= a_r j + b_r \bmod n \\ a_q j + b_q &= a_s i + b_s \bmod n, \end{aligned}$$

where $\{p, q, r, s\} \subset \{1, \dots, \xi\}$ with $p \neq r$, $q \neq s$, $p \neq s$, and $q \neq r$, since each permutation matrix can only have one 1 in each row and column. Equation (3.1) can be rewritten as

$$\text{(3.2)} \qquad \begin{bmatrix} a_p & -a_r \\ -a_s & a_q \end{bmatrix} \begin{bmatrix} i \\ j \end{bmatrix} = \begin{bmatrix} b_r - b_p \\ b_s - b_q \end{bmatrix} \bmod n.$$

Pre-multiply both sides of equation (3.2) by

$$\begin{bmatrix} a_q & a_r \\ a_s & a_p \end{bmatrix}$$

to obtain

$$(3.3) \quad (a_p a_q - a_r a_s) \begin{bmatrix} i \\ j \end{bmatrix} = \begin{bmatrix} a_p(b_r - b_p) + a_r(b_s - b_q) \\ a_s(b_r - b_p) + a_q(b_s - b_q) \end{bmatrix} \bmod n.$$

Using Theorem 2.1, we see that the matrix has no 4-cycles if one of the following two conditions holds:

$$(3.4) \qquad \begin{aligned} &\gcd(a_p a_q - a_r a_s, n) \nmid a_p(b_r - b_p) + a_r(b_s - b_q) \\ &\gcd(a_p a_q - a_r a_s, n) \nmid a_s(b_r - b_p) + a_q(b_s - b_q). \end{aligned}$$

Since $\gcd(a_p a_q - a_r a_s, n)$ need not be 1, the systems given by equations (3.2) and (3.3) may not be equivalent. However, if $\{a_p, a_q, a_r, a_s\}$ are relatively prime to n, then the following is true.

THEOREM 3.1. *If* $\{a_p, a_q, a_r, a_s\}$ *are relatively prime to n, the following are equivalent:*

(i) *equation (3.2) has no solutions;*
(ii) $\gcd(a_p a_q - a_r a_s, n) \nmid a_p(b_r - b_p) + a_r(b_s - b_q);$
(iii) $\gcd(a_p a_q - a_r a_s, n) \nmid a_s(b_r - b_p) + a_p(b_s - b_q).$

Proof. We have seen that (ii) or (iii) implies (i). We show that (i) implies (iii); the proof that (i) implies (ii) is similar and will be omitted.
Let

$$\begin{aligned} L_1 &= a_p i - a_r j + b_p - b_r \\ L_2 &= -a_s i + a_q j + b_q - b_s \end{aligned}.$$

Then equation (3.2) is the same as

$$\begin{cases} L_1 = 0 \\ L_2 = 0 \end{cases} \bmod n.$$

Since $\gcd(a_p, n) = 1$, a_p^{-1} exists in $\mathbb{Z}_n$. Let

$$(3.5) \qquad L_2' = a_s L_1 + a_p L_2$$
$$(3.6) \qquad = (a_p a_q - a_r a_s)j + a_p(b_q - b_s) + a_s(b_p - b_r).$$

By Theorem 2.2, the system

$$\begin{cases} L_1 = 0 \\ L_2' = 0 \end{cases} \bmod n$$

is equivalent to the system given by equation (3.2). If

$$\gcd(a_p a_q - a_r a_s, n) \mid a_p(b_q - b_s) + a_s(b_p - b_r),$$

then $L'_2 = 0 \bmod n$ can be solved for j by Theorem 2.1. Replace j in L_1 by any solution of $L'_2 = 0 \bmod n$, say $\hat{j}$, and call this formula $\hat{L}_1$. The equation $\hat{L}_1 = 0 \bmod n$ can be solved for i since $\gcd(a_p, n) = 1$. Let $\hat{i}$ be a solution. Then it is clear that $(\hat{i}, \hat{j})$ is a solution of equation (3.2). The proof of (i) implies (iii) is complete. $\square$

We give an example of how the condition (3.4) can be used to find matrices with no 4-cycles.

EXAMPLE 1. *Suppose n is divisible by an odd prime $v > 3$ and we wish to generate an $n \times n$ matrix with three ones in each column. Let t_1, t_2, t_3 be distinct integers modulo n and let d be an integer not divisible by v. For $j = 1, 2, 3$, let $a_j = t_j v + d$ and choose b_j such that v does not divide $b_k \pm b_\ell$ for $k \neq \ell$. In particular, we can take $b_1 = 0$, $b_2 = 1$, and $b_3 = 2$. We show that if the a_j's are relatively prime to n, then the matrix generated is regular and has no 4-cycles.*

In every collection $\{a_p, a_q, a_r, a_s\}$, at least two of them must be the same since there are only three a_j's and three b_j's. Taking into account the condition that $p \neq r$ and $p \neq s$, we can assume without loss of generality that $p = q$. Note that

$$v \,|\, \gcd(a_j - a_p, n)$$

and v does not divide $b_j - b_p$, equation (2.1) is not satisfied. By direct computation, it is easy to see that

$$v \,|\, a_p^2 - a_r a_s$$

and thus

$$v \,|\, \gcd(a_p^2 - a_r a_s, n).$$

To ensure that there are no 4-cycles, we need to check equation (3.4). First consider the case of $r = s$. Then

$$a_p(b_r - b_p) + a_r(b_s - b_p) = (a_p + a_r)(b_r - b_p)$$
$$= [v(t_p + t_r) + 2d](b_r - b_p).$$

Since the b_j's are chosen so that $b_r - b_p$ is not divisible by v for $r \neq s$,

$$\gcd(a_p^2 - a_r^2, n) \nmid a_p(b_r - b_p) + a_r(b_r - b_p).$$

We next consider the case of $p \neq q$. Then

$$a_p(b_r - b_p) + a_r(b_s - b_p) = a_p b_r + a_r b_s + (a_p - a_r)b_p$$
$$= (t_p b_r + t_r b_s)v + (b_r + b_s)d + (a_p - a_r)b_p$$

is not divisible by v if v does not divide $(b_r + b_s)d$. We conclude that there are no 4-cycles.

These results indicate that it is usually quite easy to construct regular sums of up to three permutation matrices that have no 4-cycles.

3.1.2. The general square case. We briefly describe the general case without giving all the technical details. A cycle of length $2s$ can be represented in the a natural way by

$$(3.7) \qquad r(i_1)c(j_1)r(i_2)c(j_2)\cdots r(i_s)c(j_s)r(i_1),$$

where $r(i)$ denotes the i-th row and $c(j)$ the j-th column. The above should be taken to mean that row i_1 is connected to column j_1, which is connected to row i_2, etc. Since the edges in a cycle cannot repeat, we must have

$$i_p \neq i_{p+1}, \ j_p \neq j_{p+1}, \text{ for } p = 1, \ldots, s,$$

where $i_{s+1} = i_1$ and $j_{s+1} = j_1$. If $i_p = i_q$ or $j_p \neq j_q$ for $p \neq q$, then the cycle contains another cycle with shorter length.

Using the representation given in (3.7), we have

$$R_{i_1,j_1} = R_{i_2,j_1} = R_{i_2,j_2} = \cdots = R_{i_s,j_s} = R_{i_1,j_s} = 1,$$

which we can express, using the relationship between rows and columns in our construction, by

$$(3.8) \qquad \begin{cases} j_p = a_{j_p} i_p + b_{j_p} \bmod n \\ j_p = a_{j'_p} i_{p+1} + b_{j'_p} \bmod n \end{cases}, \ p = 1, \ldots, s,$$

where $\{a_{j_p}, a_{j'_p} : p = 1, \ldots, s\} \subset \{a_1, \ldots, a_\xi\}$. Equating the two equations for j_p, we obtain

$$a_{j_p} i_p + b_{j_p} = a_{j'_p} i_{p+1} + b_{j'_p} \bmod n, \ p = 1, \ldots, s.$$

We can summarize the above system of equations by the following matrix equation:

$$(3.9) \qquad \begin{bmatrix} a_{j_1} & -a_{j'_1} & 0 & \cdots & 0 & 0 \\ 0 & a_{j_2} & -a_{j'_2} & \cdots & 0 & 0 \\ & & \vdots & & & \\ 0 & 0 & 0 & \cdots & a_{j_{s-1}} & -a_{j'_{s-1}} \\ -a_{j'_s} & 0 & 0 & \cdots & 0 & a_{j_s} \end{bmatrix} \begin{bmatrix} i_1 \\ i_2 \\ \vdots \\ i_{s-1} \\ i_s \end{bmatrix}$$

$$= \begin{bmatrix} d_{j_1} \\ d_{j_2} \\ \vdots \\ d_{j_{s-1}} \\ d_{j_s} \end{bmatrix} \bmod n,$$

where $d_j = b_{j'} - b_j$. Note that the coefficient matrix is upper triangular except for one nonzero entry at the $(s, 1)$ position. One must verify that equation (3.9), or any of its equivalents, has no solutions that satisfy the no-repeat condition of a cycle to guarantee that the matrix R has no $2s$-cycles. The equation can be transformed into an equivalent upper triangle form to make checking its solvability easier. We have the following theorem.

THEOREM 3.2. *Let*

$$\gamma = \gcd(\prod_{p=1}^{s} a_{j'_p} - \prod_{p=1}^{s} a_{j_p}, n).$$

The system given by equation (3.9) is solvable if and only if

$$\gamma \ divides \ d_{j_s} \prod_{q=1}^{s-1} a_{j_q} + a_{j'_s} \sum_{p=1}^{s-1} \left(\prod_{q=1}^{p-1} a_{j'_q} \right) \left(\prod_{q=p+1}^{s-1} a_{j_q} \right) d_{j_p}.$$

If the system is solvable, it has exactly γ solutions.

We briefly comment on the computational requirement. The pairs $\{j_1, j'_1\}, \ldots, \{j_s, j'_s\}$ each has $\xi(\xi-1)$ possible combinations from $\{1, \ldots, \xi\}$. Hence the total possible number of systems that need to be checked is $[\xi(\xi-1)]^s$. In fact, by permuting the indices, one can show that we only need to check the cases when $j_1 < j'_1$, which reduces the total number of systems to be checked to $[\xi(\xi-1)]^s/2$. For example, suppose $\xi = 3$ and we need to avoid 6-cycles, we only need to check 108 systems. In particular, the number of checks is independent of n.

3.2. Cycle avoidance for matrix overlap. We will transform the equations for the avoidance of cycles for the overlap construction to the same form as those for the square matrix case. We use the same approach and notation as in Section 3.1.2. Suppose the matrix to be constructed is $m \times n$ with m not a factor of n. Let $u = \gcd(m, n)$. Using the representation given in (3.7), we have

$$R_{i_1,j_1} = R_{i_2,j_1} = R_{i_2,j_2} = \cdots = R_{i_s,j_s} = R_{i_1,j_s} = 1,$$

which we can express, using the relationship between the rows and columns in our construction, by

$$(3.10) \qquad \begin{cases} j_p = a_{j_p} i_p + b_{j_p} + k_p u \bmod n \\ j_p = a_{j'_p} i_{p+1} + b_{j'_p} + k'_p u \bmod n \end{cases}, \ p = 1, \ldots, s.$$

These equations can be put in the form

$$(3.11) \qquad \begin{cases} j_p = a_{j_p} i_p + \widehat{b}_{j_p} \bmod n \\ j_p = a_{j'_p} i_{p+1} + \widehat{b}_{j'_p} \bmod n \end{cases}, \ p = 1, \ldots, s,$$

where $\widehat{b}_{j_p} = b_{j_p} + k_p u$, and $\widehat{b}_{j'_p} = b_{j'_p} + k'_p u$. Let $d_{j_p} = \widehat{b}_{j'_p} - \widehat{b}_{j_p}$. The above system can be put in the form given by equation (3.9) and the condition for the existence of 4-cycles is given by Theorem 3.1 and the condition for the general case is given by Theorem 3.2.

In the next example, we show how the above condition can be used to construct regular overlap matrices with no 4-cycles.

EXAMPLE 2. *Suppose the matrix to be constructed is $m \times n$ with $m \nmid n$ and ξ ones in each column. Thus we need $n\xi/m$ permutation matrices. Let $u = \gcd(m,n)$. As we have seen in Section 2.2.1, if $v > 1$ is a factor of u and $\{a_j, b_j\}_{j=1}^{n\xi/m}$ are chosen so that the a_j's are relatively prime to m, that v divides $a_p - a_q$ and does not divide $b_q - b_p$ for all $p \neq q$, then the matrix constructed will be regular with ξ ones in each column. We showed how*

$$a_j = t_j v + d$$

can be used to construct a regular matrix. We now use this form for the a_j's to simplify the problem of constructing a regular matrix with no 4-cycles.

By Theorem 3.1, the matrix will have no 4-cycles if

$$\gcd(a_p a_q - a_r a_s, n) \nmid a_p(\widehat{b}_r - \widehat{b}_p) + a_r(\widehat{b}_s - \widehat{b}_q)$$

for all indices $\{p,q,r,s\} \subset \{1,\dots,\xi\}$ such that $p \neq r$, $q \neq s$, $p \neq s$, and $q \neq r$. From $a_j = t_j v + d$ and $\widehat{b}_j = b_j + k_j u$, we obtain

$$a_p a_q - a_r a_s = [(t_p t_q - t_r t_s)v + (t_p + t_q - t_r - t_s)d]v.$$

and

$$a_p(\widehat{b}_r - \widehat{b}_p) + a_r(\widehat{b}_s - \widehat{b}_q) = [t_p(\widehat{b}_r - \widehat{b}_p) + t_r(\widehat{b}_s - \widehat{b}_q)]v + $$
$$d[b_r - b_p + b_s - b_q] + [k_r - k_p + k_s - k_q]du.$$

Therefore

$$v \mid \gcd(a_p a_q - a_r a_s, n)$$

and since $v \mid u$,

$$\gcd(a_p a_q - a_r a_s, n) \nmid a_p(\widehat{b}_r - \widehat{b}_p) + a_r(\widehat{b}_s - \widehat{b}_q)$$

if

(3.12) $$v \nmid d(b_r - b_p + b_s - b_q).$$

If v and d are relatively prime, then the matrix will have no 4-cycles if

$$v \nmid (b_r - b_p + b_s - b_q).$$

We now give an explicit construction of a 363×484 regular matrix with three ones in each column and no 4-cycles. In this case $m = 363$, $n = 484$, $u = \gcd(363, 484) = 121$, and $n' = n/u = 4$. We have seen in

Section 2.2.1 that if we let $v = 11$, $a_1 = 23$, $a_2 = 34$, $a_3 = 67$, $a_4 = 89$, and $b_j = j - 1$ for $j = 1, \ldots, 4$, then the overlap matrix is regular. It is easy to see that these b_j's will not satisfy equation (3.12) since

$$b_1 - b_2 + b_4 - b_3 = 0.$$

In fact, if $b_1 = 0$, $b_2 = 1$, and $b_3 = 2$, then no choice of b_4 will work since

$$b_2 - b_1 + b_2 - b_3 = 0.$$

A simple search shows that $b_1 = 0$, $b_2 = 1$, $b_3 = 3$, and $b_4 = 11$ work. The matrix constructed with these a_j's and b_j's is regular, has no 4-cycles and 24 6-cycles.

4. Simulation results. To show the potential of the linear-congruence method for constructing LDPCC, we compare the performances of the various construction methods with the best randomly constructed regular matrices. We generated rate 4/7 codes with block length 847 and rate 1/2 codes with block length 980 using the different methods described in the previous sections. For each rate, we plotted the bit error rate and the block error rate for the codes. For the block errors, we also include for comparison a curve that we derived from a simple model of the decoding process. We have found that this curve generated from the model gives an approximate upper bound for performance for the codes that we have considered. The details about this model is given in Appendix B.

The parity-check matrices for the codes are regular and have three ones in each column. For the rate 4/7 codes, there are 7 ones in each row and for the rate 1/2 codes, there are 6 ones in each row. For each rate, we use the best randomly constructed codes that we have found. For the rate 4/7 codes, we use parity-check matrices of the form $[LC\ C]$, where LC is a 363×484 linear-congruence matrix and C is a 363×363 circulant matrix. We give three different constructions for LC:

1. the left 363×363 submatrix is a linear-congruence matrix and the right 363×121 matrix is a stack of three random 121×121 permutation matrices;
2. the left 363×363 submatrix is a linear-congruence matrix and the right 363×121 matrix is a stack of three 121×121 linear-congruence permutation matrices;
3. the matrix is constructed using the overlap method.

It is clear from Figure 1 that the different methods of construction give essentially the same result. For this particular code rate and block size, we are confident that the linear-congruence method can produce codes that match the best randomly constructed regular parity-check codes.

For the rate 1/2 code, we use parity-check matrices of the form $[LC\ RC]$, where LC is a 490×490 linear-congruence matrix and RC has dimension 490×490 and is either a linear-congruence matrix or is a circulant matrix. The bit error rate and the block error rate curves are in

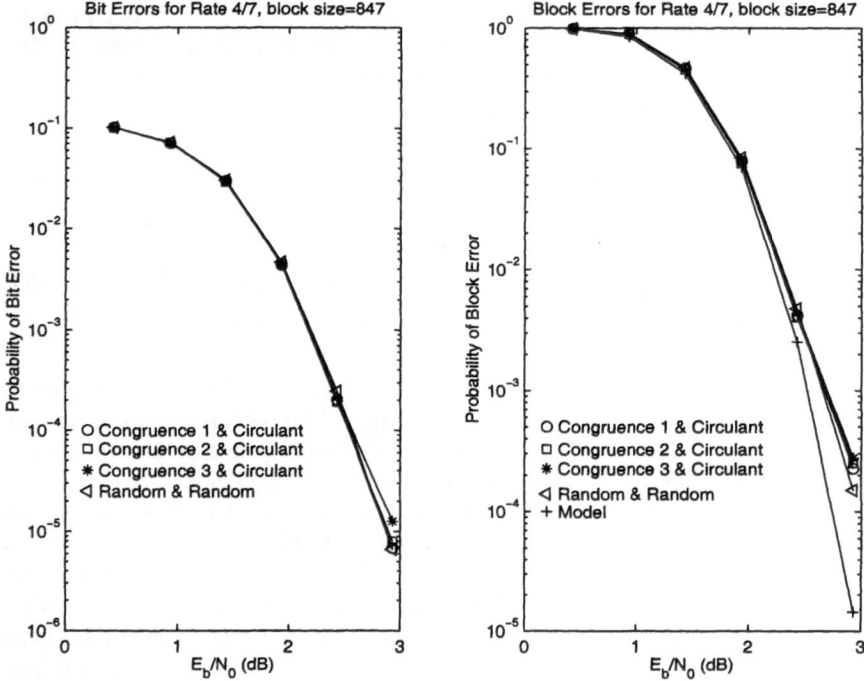

FIG. 1. *Performance comparison for the rate 4/7 codes.*

Figure 2. There is a little more difference in the performances for this code rate but the differences are small, within 0.2 dB of each other. For the bit error rate, there is no clear winner and for the block error rate, the random code is better by a very small margin.

APPENDIX

A. Parity-check matrix with two circulant matrices. In this section, we consider codes with parity-check matrices of the form

$$H = [C_2 \; C_1],$$

where C_1 and C_2 are circulant matrices. We show that the minimum distance of a code with parity-check matrix of this form is bounded above by 2ξ, where ξ is the number of ones in each column of C_1 and C_2. Hence, for small values of ξ, such as $h = 3$, we expect this class of code to have inferior performance. Simulations show that this is indeed the case. This class of codes should in general be avoided.

We assume that C_1 is invertible. Let C be a circulant matrix with first

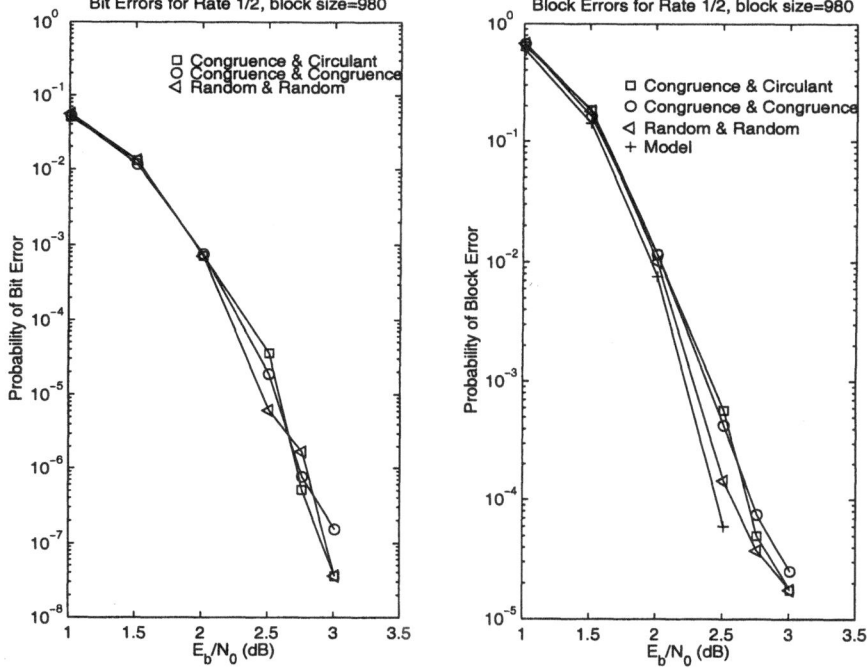

FIG. 2. *Performance comparison for the rate 1/2 Codes.*

row

$$\begin{bmatrix} c_0 & c_1 & \cdots & c_{n-1} \end{bmatrix}$$

It is known that C is invertible over $\mathbb{Z}_2$ if and only if the associated polynomial

$$c_0 + c_1 x + \cdots + c_{n-1} x^{n-1}$$

is primitive over $\mathbb{Z}_2$ provided that the polynomial is not a factor of $x^n - 1$. Therefore the requirement that C_1 be invertible is easy to fulfill.

Let B be the circulant matrix

$$B = \begin{bmatrix} 0 & 1 & 0 & \cdots & 0 \\ 0 & 0 & 1 & \cdots & 0 \\ & & \vdots & & \\ 0 & 0 & 0 & \cdots & 1 \\ 1 & 0 & 0 & \cdots & 0, \end{bmatrix}$$

where each row contains exactly one 1. Then a circulant matrix C with first row

$$\begin{bmatrix} c_0 & c_1 & \cdots & c_{n-1} \end{bmatrix}$$

can be expressed as

$$C = c_0 I + c_1 B + \cdots + c_{n-1} B^{n-1}.$$

Since it is obvious that powers of a fixed matrix commute, we conclude that the circulant matrices commute. This property of circulant matrices leads to an upper bound on the minimum distance of codes with parity-check matrix of the form $[C_2 \; C_1]$. Let C_1 and C_2 have ξ ones in each column. Let

$$\hat{G} = \begin{bmatrix} C_1 \\ C_2 \end{bmatrix}$$

Using the commutativity property, we have

$$\hat{G} = \begin{bmatrix} C_1 \\ C_2 \end{bmatrix} = \begin{bmatrix} I \\ C_2 C_1^{-1} \end{bmatrix} C_1 = \begin{bmatrix} I \\ C_1^{-1} C_2 \end{bmatrix} C_1 = G C_1,$$

where G is the generator matrix in systematic form. Since C_1 is invertible, $\hat{G}$ generates the same code as G. By assumption, C_1 and C_2 have ξ ones in each column, and it follows that the code has minimum distance no greater than 2ξ.

Since C_1 is invertible, the matrix $\hat{G}$ has full rank and thus there are at least n codewords of weight 2ξ.

B. A heuristic criterion for code selection. In this section, we present an experimental observation that a simple model gives a surprisingly good approximation of the block error rate of LDPCC with Belief Propagation decoding for a fairly broad range of E_b/N_0. In the range where the approximation is less accurate, the model seems to give an upper bound to performance. We model the decoding process by assuming that the decoding algorithm first uses hard decisions to decode each bit and then corrects a certain number of bit errors in a codeword up to a certain maximum, which we refer to as the *radius of convergence*. We next show how this maximum number in our model can be computed.

Let R_c be the radius of convergence and let p be the probability that a bit is in error if the bit is decoded using hard decisions. Note that p can be computed from E_b/N_0. Using p and the assumption that the decoding algorithm can correct up to R_c errors, the probability of block error is

$$P_{block} = 1 - \sum_{k=0}^{R_c} \binom{n}{k} p^k (1-p)^{n-k}.$$

To find R_c, we generate a family of codes at the desired rate and use simulations to estimate the probability of block error at an E_b/N_0, usually chosen to be at the center of the operating range of the code for this family. We use the best probability of block error from the family of codes and use that as P_{block}. The above equation relating R_c and P_{block} is then used to solve for R_c. Once R_c has been found, we assume that it is fixed for all E_b/N_0 and use the above equation to generate a curve using p from the different E_b/N_0.

We have found that the radius of convergence provides an excellent indicator of code performance and we have found that it can be used for deciding that a high-performance code has been constructed.

Note. The reviewer pointed out that one possible explanation for the observed phenomenon is that for large block lengths, the probability of error as a function of the number of input errors has a sharp transition exactly at the observed radius of convergence. For smaller block lengths, which is the case we are interested in, the transition is less sharp. The reviewer also pointed out that this heuristc fails if there are nodes with degree 2 in the graph of the parity-check matrix. The matrices we used in this paper do not have any nodes of degree 2.

REFERENCES

[1] J.W. BOND, S. HUI, H. SCHMIDT, Low Density Parity Check Codes based on Sparse Matrices with No Small Cycles, Institute of Mathematics Applications, Coding and Cryptography Conference, December 1997.

[2] J.W. BOND, S. HUI, H. SCHMIDT, Decoding Low-Density Generator Matrix Codes with the Aid of Comma-free Source Codes, Proceedings of 1988 Information Theory Workshop, Killarney, Ireland, 22-26 June 1998.

[3] J.W. BOND, S. HUI, H. SCHMIDT, Constructing low-density parity-check codes with circulant matrices, 1999 IEEE International Symposium on Information Theory, Metsovo, Greece, June 27-July 1, 1999.

[4] J.W. BOND, S. HUI, H. SCHMIDT, The Euclidean Algorithm and Primitive Polynomials over Finite Fields, to be published in a book associated with AAECC-13 on Applied Algebra, Fall 1999.

[5] J.W. BOND, S. HUI, H. SCHMIDT, Belief Propagation Decoding for Gaussian Channels, September 21, 1998.

[6] J.F. CHENG AND R.J. MCELIECE, Some High-Rate Near Capacity Codes for the Gaussian Channel, presented at the 34th Allerton Conference on Communication, Control, and Computing, 1996.

[7] M.C. DAVEY AND D.J.C. MACKAY, Monte Carlo Simulations of Infinite Low Density Parity-Check Codes over GF(q), March 30, 1998, Proceedings of 1988 Information Theory Workshop, Killarney, Ireland, 22-26 June 1998.

[8] R. GALLAGER, *Low-Density Parity-Check Codes*, MIT Press, Cambridge Mass, July 1963.

[9] G.H. HARDY, AND E.M. WRIGHT, *An Introduction to Number Theory*, Fourth Edition, Clarendon Press, Oxford, 1968.

[10] L.K. HUA, *Introduction to Number Theory*, Springer-Verlag, Berlin, 1982.

[11] D.J.C. MACKAY, Good Error-Correcting Codes Based on Very Sparse Matrices, IEEE Transactions on Information Theory, pp 399-431, Vol. 45, No. 2, March 1999.

[12] D.J.C. MACKAY AND M.C. DAVEY, Evaluation of Gallager Codes for Short Block Length and High Rate Applications, Preprint.

[13] T. RICHARDSON, A. SHOKROLLAHI AND R. URBANKE, Design of Provably Good Low-Density Parity-Check Codes, 1999, Preprint.

[14] T. RICHARDSON AND R. URBANKE, The Capacity of Low-Density Parity Check Codes under Message-Passing Decoding, Preprint.

[15] B.M. STEWART, *Theory of Numbers*, Second Edition, Macmillan, New York, 1964.

ON THE EFFECTIVE WEIGHTS OF PSEUDOCODEWORDS FOR CODES DEFINED ON GRAPHS WITH CYCLES

G. DAVID FORNEY, JR.[*], RALF KOETTER[†], FRANK R. KSCHISCHANG[‡], AND ALEX REZNIK[*§]

Abstract. The behavior of an iterative decoding algorithm for a code defined on a graph with cycles and a given decoding schedule is characterized by a cycle-free computation tree. The pseudocodewords of such a tree are the words that satisfy all tree constraints; pseudocodewords govern decoding performance. Wiberg [12] determined the effective weight of pseudocodewords for binary codewords on an AWGN channel. This paper extends Wiberg's formula for AWGN channels to nonbinary codes, develops similar results for BSC and BEC channels, and gives upper and lower bounds on the effective weight. The 16-state tail-biting trellis of the Golay code [2] is used for examples. Although in this case no pseudocodeword is found with effective weight less than the minimum Hamming weight of the Golay code on an AWGN channel, it is shown by example that the minimum effective pseudocodeword weight can be less than the minimum codeword weight.

Key words. Codes on graphs, iterative decoding, pseudocodewords, effective weights, tail-biting.

AMS(MOS) subject classifications. 94B99.

1. Introduction. The subject of codes defined on graphs was founded by Tanner [10], inspired by Gallager's low-density parity-check (LDPC) codes [4]. The thesis of Wiberg [12, 13], along with the practical successes of turbo codes and LDPC codes, has stimulated great current interest in this subject. For recent developments, see [1, 6, 7, 8].

By now it is well known that if C is a block code defined on a cycle-free graph G (*i.e.*, a tree), then the min-sum decoding algorithm is guaranteed to converge to the maximum-likelihood (ML) code sequence [12, 13].

The min-sum algorithm may also be applied to a graph with cycles, but its behavior then depends on the decoding schedule, and convergence is not guaranteed. Given a decoding schedule, there exists a cycle-free computation tree G' such that the behavior of the min-sum algorithm on G' with the given schedule is identical to that of the iterative algorithm on G [3, 11, 12, 13]. In general, a node in G has more than one representation in G'.

A codeword in C is a sequence **c** of node values in G that satisfies all the constraints of G. A pseudocodeword [3] is a sequence of node values in G' that satisfies all the constraints of G'. There exists a pseudocodeword

---

[*]Laboratory for Information and Decision Systems, M.I.T., Cambridge, MA 02139, USA.

[†]Coordinated Science Laboratory, University of Illinois, Urbana, IL 61801, USA.

[‡]Dept. of Electrical and Computer Engineering, University of Toronto, Toronto, Ont. M5S 3G4, Canada.

[§]InterDigital Communications Corporation, Melville, NY 11747, USA.

corresponding to every codeword $\mathbf{c} \in C$, obtained by assigning the values of the nodes of G to the corresponding nodes of G'. In general there will also exist pseudocodewords that do not correspond to valid codewords, because different values are assigned to nodes of G' that correspond to the same node of G.

The examples in this note will focus on pseudocodewords of tail-biting trellises, whose significance is particularly clear. A tail-biting trellis (TBT) corresponds to a graph G that consists of a single cycle. A computation tree G' is obtained by "unwrapping" G into a conventional trellis defined on an ordered time axis. A codeword $\mathbf{c}$ of G corresponds to a pseudocodeword on G' that repeats periodically with a period equal to one cycle of G. But there also exist periodic pseudocodewords on G' whose period is a multiple of the cycle length of G and which do not correspond to any valid codeword.

In Chapter 6 of [12], Wiberg developed a formula for the effective Hamming weight ("generalized weight") w_{eff} of a pseudocodeword ("tree configuration") for the case of binary codes and binary antipodal signaling on an additive white Gaussian noise (AWGN) channel, namely

$$(1.1) \qquad w_{\text{eff}} = \frac{\left(\sum_j n_j \right)^2}{\sum_j n_j^2},$$

where n_j is the number of nodes in G' corresponding to the jth node in G that have value equal to 1. If the pseudocodeword corresponds to a valid codeword $\mathbf{c}$, then $w_{\text{eff}} = w_{\text{H}}(\mathbf{c})$, the Hamming weight of $\mathbf{c}$. For a binary linear code, the probability of a decoding error on G' is governed by the minimum effective weight w_{eff}; therefore it is important that all pseudocodewords of G' have effective weight w_{eff} at least as great as the minimum Hamming weight d_{H} of C if performance is not to be degraded.

In this note we develop some extensions of Wiberg's result, as follows:

1. We extend Wiberg's formula to the nonbinary case;
2. We give lower and upper bounds on w_{eff};
3. We develop similar results for the binary symmetric channel (BSC) and binary erasure channel (BEC).

As examples, we compute the effective weights of certain pseudocodewords in the 16-state TBT of the binary (24, 12, 8) Golay code of [2]. For the AWGN channel, we have not found any examples of pseudocodewords with effective weight less than 8; however, neither have we been able to prove that 8 is the minimum effective weight for this case. For the binary symmetric channel, on the other hand, we exhibit a pseudocodeword with effective weight 6.

2. Effective weight on AWGN channels. Let C be a block code of length n defined on a graph G, and let G' be a computation tree corresponding to some schedule for min-sum decoding of C. Let $N_j, 1 \leq j \leq n$, be the number of occurrences of the jth node of G in G'.

Let $\mathbf{c} = \{c_j, 1 \leq j \leq n\}$ be a codeword of G, and let $\{p_{ji}, 1 \leq j \leq n, 1 \leq i \leq N_j\}$ be a pseudocodeword of G', where p_{ji} represents the value of the ith occurrence of the jth node, $1 \leq i \leq N_j$.

For each symbol x_m in the symbol alphabet A, let n_{jm} be the number of times that $p_{ji} = x_m$, and let $f_{jm} = n_{jm}/N_j$; i.e., f_{jm} is the frequency with which x_m appears in the N_j occurrences of the jth node. In the following, we think of the fractions f_{jm} as defining a random pseudocodeword $\mathbf{p} = \{p_j\}$ in which p_j takes on value $x_m \in A$ with probability f_{jm}, and we will take expectations over this distribution.

Note that $\mathbf{p}$ is non-random ($\mathbf{p} = \mathsf{E}[\mathbf{p}]$) if and only if $\mathbf{p}$ corresponds to a valid codeword $\mathbf{c}'$, for then and only then $f_{jm} = 1$ when $x_m = c_j'$ and $f_{jm} = 0$ otherwise.

Define the variance

$$(2.1) \qquad \sigma_p^2 = \sum_j \left(\mathsf{E}\left[p_j^2\right] - \mathsf{E}\left[p_j\right]^2 \right) = \mathsf{E}\left[\|\mathbf{p}\|^2\right] - \|\mathsf{E}\left[\mathbf{p}\right]\|^2.$$

We have the following obvious lemma:

LEMMA 2.1. *The variance σ_p^2 is greater than or equal to 0, with equality if and only if $\mathbf{p}$ is non-random; i.e., if and only if $\mathbf{p}$ corresponds to a valid codeword $\mathbf{c}'$.*

Let $\mathbf{c}$ be the input codeword to an AWGN channel whose output sequence is $\mathbf{r} = \mathbf{c} + \mathbf{n}$, where $\mathbf{n}$ is an i.i.d. Gaussian sequence with mean 0 and variance σ^2 per symbol. A maximum-likelihood (ML) decoder on G' chooses the pseudocodeword $\mathbf{p}$ that minimizes the squared distance

$$\|\mathbf{r} - \mathbf{p}\|^2 = \sum_j N_j \sum_m f_{jm}(r_j - x_m)^2.$$

For simplicity we will assume that G' is balanced [3, 11]— i.e., that $\frac{N_j}{N_i} \approx 1$ for a large number of iterations. For such balanced graphs we can normalize by N_j, which we will tacitly assume in the sequel. Then an ML decoder chooses the pseudocodeword $\mathbf{p}$ that minimizes the expected squared distance

$$\mathsf{E}\left[\|\mathbf{r} - \mathbf{p}\|^2\right] = \sum_j \sum_m \Pr[p_j = x_m](r_j - x_m)^2 = \sum_j \sum_m f_{jm}(r_j - x_m)^2.$$

The probability $\Pr(\mathbf{c} \to \mathbf{p})$ that an ML decoder will choose the pseudocodeword $\mathbf{p}$ over $\mathbf{c}$ is thus

$$\Pr(\mathbf{c} \to \mathbf{p}) = \Pr\left\{\mathsf{E}\left[\|\mathbf{r} - \mathbf{p}\|^2\right] \leq \|\mathbf{r} - \mathbf{c}\|^2\right\}.$$

(If G' is not balanced, then a similar result holds if we replace $\mathsf{E}[\|\mathbf{r} - \mathbf{p}\|^2]$ by the expectation of the weighted squared distance $N_j(r_j - x_m)^2$.)

Defining $\langle \mathbf{r}, \mathbf{c} \rangle = \sum_j r_j c_j$, we can write

$$\|\mathbf{r} - \mathbf{c}\|^2 = \|\mathbf{r}\|^2 - 2\langle \mathbf{r}, \mathbf{c} \rangle + \|\mathbf{c}\|^2;$$
$$\mathsf{E}\left[\|\mathbf{r} - \mathbf{p}\|^2\right] = \|\mathbf{r}\|^2 - 2\langle \mathbf{r}, \mathsf{E}[\mathbf{p}] \rangle + \mathsf{E}\left[\|\mathbf{p}\|^2\right],$$

where $\langle \mathbf{r}, \mathsf{E}[\mathbf{p}] \rangle = \sum_j r_j \mathsf{E}[p_j] = \sum_j r_j \sum_m f_{jm} x_m$. Thus

$$\Pr(\mathbf{c} \to \mathbf{p}) = \Pr\left\{ 2\langle \mathbf{r}, \mathbf{c} - \mathsf{E}[\mathbf{p}] \rangle \leq \|\mathbf{c}\|^2 - \mathsf{E}\left[\|\mathbf{p}\|^2\right] \right\}.$$

Define $\mathbf{d} = \mathbf{c} - \mathsf{E}[\mathbf{p}]$ and $D = \|\mathbf{c}\|^2 - \mathsf{E}\left[\|\mathbf{p}\|^2\right]$; then

$$\Pr(\mathbf{c} \to \mathbf{p}) = \Pr\left\{ 2\langle \mathbf{r}, \mathbf{d} \rangle \leq D \right\}.$$

Given c_j, the received symbol r_j is a Gaussian random variable (r.v.) with mean c_j and variance σ^2. Therefore $r_j d_j$ is a Gaussian r.v. with mean $c_j d_j$ and variance $\sigma^2 d_j^2$. The inner product $\langle \mathbf{r}, \mathbf{d} \rangle = \sum_j r_j d_j$ therefore has mean $\langle \mathbf{c}, \mathbf{d} \rangle = \sum_j c_j d_j$ and variance $\sigma^2 \sum_j d_j^2 = \sigma^2 \|\mathbf{d}\|^2$. The probability that $\langle \mathbf{r}, \mathbf{d} \rangle \leq D/2$ is thus the probability that a Gaussian r.v. with mean $\langle \mathbf{c}, \mathbf{d} \rangle - D/2$ and variance $\sigma^2 \|\mathbf{d}\|^2$ is less than zero, which is given by

$$Q\left(\frac{\langle \mathbf{c}, \mathbf{d} \rangle - D/2}{\sigma \|\mathbf{d}\|} \right),$$

where $Q(x) = \frac{1}{2\pi} \int_x^\infty \exp(-x^2/2) \, dx$ is the usual Q function.

Therefore if we define the effective squared Euclidean distance as

$$(2.2) \qquad d_{\text{eff}}^2(\mathbf{c}, \mathbf{p}) = \frac{(2\langle \mathbf{c}, \mathbf{d} \rangle - D)^2}{\|\mathbf{d}\|^2},$$

then we obtain the familiar expression

$$(2.3) \qquad \Pr(\mathbf{c} \to \mathbf{p}) = Q\left(\frac{d_{\text{eff}}(\mathbf{c}, \mathbf{p})}{2\sigma} \right).$$

If $\mathbf{p}$ corresponds to a codeword $\mathbf{c}'$, then $\mathsf{E}[\mathbf{p}] = \mathbf{c}'$ and $\mathsf{E}[\|\mathbf{p}\|^2] = \|\mathbf{c}'\|^2$, so

$$2\langle \mathbf{c}, \mathbf{d} \rangle - D = 2\langle \mathbf{c}, \mathbf{c} - \mathbf{c}' \rangle - \|\mathbf{c}\|^2 + \|\mathbf{c}'\|^2 = \|\mathbf{c} - \mathbf{c}'\|^2.$$

Also $\|\mathbf{d}\|^2 = \|\mathbf{c} - \mathbf{c}'\|^2$, so we have as usual

$$d_{\text{eff}}^2(\mathbf{c}, \mathbf{c}') = \|\mathbf{c} - \mathbf{c}'\|^2.$$

More generally, if $\sigma_p^2 = \mathsf{E}\left[\|\mathbf{p}\|^2\right] - \|\mathsf{E}[\mathbf{p}]\|^2$ is the variance of $\mathbf{p}$ defined in (2.1), then we have

$$2\langle \mathbf{c}, \mathbf{d} \rangle - D = 2\langle \mathbf{c}, \mathbf{c} - \mathsf{E}[\mathbf{p}] \rangle - \|\mathbf{c}\|^2 + \mathsf{E}\left[\|\mathbf{p}\|^2\right]$$
$$= \|\mathbf{c} - \mathsf{E}[\mathbf{p}]\|^2 + \mathsf{E}\left[\|\mathbf{p}\|^2\right] - \|\mathsf{E}[\mathbf{p}]\|^2$$
$$= \|\mathbf{d}\|^2 + \sigma_p^2.$$

This leads to our main theorem:

THEOREM 2.1. *Let* **c** *be a codeword and* **p** *a pseudocodeword in a balanced computation tree. Then the effective squared Euclidean distance between* **c** *and* **p** *is*

$$(2.4) \qquad d^2_{\text{eff}}(\mathbf{c}, \mathbf{p}) = \frac{\left(\|\mathbf{d}\|^2 + \sigma_p^2\right)^2}{\|\mathbf{d}\|^2},$$

where $\mathbf{d} = \mathbf{c} - \mathsf{E}[\mathbf{p}]$ *and* $\sigma_p^2 = \mathsf{E}[\|\mathbf{p}\|^2] - \|\mathsf{E}[\mathbf{p}]\|^2$. *If* **c** *is transmitted, then the probability that the received word* **r** *is closer to* **p** *than* **c** *is*

$$(2.5) \qquad Pr(\mathbf{c} \to \mathbf{p}) = Q\left(\frac{d_{\text{eff}}(\mathbf{c}, \mathbf{p})}{2\sigma}\right).$$

By Lemma 1, $\sigma_p^2 \geq 0$, with equality if and only if **p** corresponds to a codeword **c'**. Thus we obtain the following two lower bounds on $d^2_{\text{eff}}(\mathbf{c}, \mathbf{p})$:

COROLLARY 2.1. *The effective squared Euclidean distance* $d^2_{\text{eff}}(\mathbf{c}, \mathbf{p})$ *satisfies*

$$(2.6) \qquad d^2_{\text{eff}}(\mathbf{c}, \mathbf{p}) \geq \|\mathbf{d}\|^2 + \sigma_p^2 \geq \|\mathbf{d}\|^2 = \|\mathbf{c} - \mathsf{E}[\mathbf{p}]\|^2,$$

with equality in both cases if and only if **p** *corresponds to a codeword* **c'**.

This result shows that the variation σ_p^2 in non-codeword pseudocodewords **p** causes $d^2_{\text{eff}}(\mathbf{c}, \mathbf{p})$ to be greater than the squared distance $\|\mathbf{d}\|^2 = \|\mathbf{c} - \mathsf{E}[\mathbf{p}]\|^2$ between **c** and the average $\mathsf{E}[\mathbf{p}]$. Thus the more variation σ_p^2 in a pseudocodeword **p**, the less troublesome it is likely to be, for a given average $\mathsf{E}[\mathbf{p}]$.

3. Binary signaling on the AWGN channel. With binary antipodal signaling using the symbol alphabet $A = \{\pm 1\}$, we have $\|\mathbf{c}\|^2 = \mathsf{E}[\|\mathbf{p}\|^2]$, so $D = 0$. Define

$$d_j = 1 - \mathsf{E}[p_j] = 1 - \sum_m f_{jm} x_m = 2f_j,$$

where f_j is the fraction of p_{ji} equal to -1; *i.e.,* $f_j = (1 - \mathsf{E}[p_j])/2$. If we take $\mathbf{c} = \mathbf{1}$ (the all-zero codeword), then

$$d^2_{\text{eff}}(\mathbf{1}, \mathbf{p}) = 4w_{\text{eff}}(\mathbf{p}) = 4\frac{\left(\sum_j d_j\right)^2}{\sum_j d_j^2}.$$

If we further define $|\mathbf{f}| = \sum_j f_j$ and $\|\mathbf{f}\|^2 = \sum_j f_j^2$, then we have

$$\|\mathbf{d}\|^2 = \|\mathbf{1} - \mathsf{E}[\mathbf{p}]\|^2 = 4\|\mathbf{f}\|^2;$$
$$\sigma_p^2 = \mathsf{E}\left[\|\mathbf{p}\|^2\right] - \|\mathsf{E}[\mathbf{p}]\|^2 = 4(|\mathbf{f}| - \|\mathbf{f}\|^2);$$
$$d^2_{\text{eff}}(\mathbf{1}, \mathbf{p}) = \frac{\left(\|\mathbf{d}\|^2 + \sigma_p^2\right)^2}{\|\mathbf{d}\|^2} = 4\frac{|\mathbf{f}|^2}{\|\mathbf{f}\|^2}.$$

This yields Wiberg's formula (1.1) for $w_{\text{eff}}(\mathbf{p})$ in the balanced case:

COROLLARY 3.1.   *On an AWGN channel, the effective Hamming weight of a binary pseudocodeword* $\mathbf{p}$ *with frequency* f_j *of ones in the* jth *position is*

$$(3.1) \qquad w_{\text{eff}}(\mathbf{p}) = \frac{|\mathbf{f}|^2}{\|\mathbf{f}\|^2},$$

where $|\mathbf{f}| = \sum_j f_j$ *and* $\|\mathbf{f}\|^2 = \sum_j f_j^2$.

The quantity $|\mathbf{f}| = \sum_j f_j$ may be interpreted as the average Hamming weight of $\mathbf{p}$. Corollary 2.1 then has the following corollary:

COROLLARY 3.2.   *If* $\mathbf{p}$ *is binary, then its effective weight* $w_{\text{eff}}(\mathbf{p})$ *satisfies*

$$(3.2) \qquad w_{\text{eff}}(\mathbf{p}) \geq |\mathbf{f}| \geq \|\mathbf{f}\|^2,$$

with equality if and only if $\mathbf{p}$ *corresponds to a codeword* $\mathbf{c}'$.

In other words, $w_{\text{eff}}(\mathbf{p})$ is lowerbounded by the average Hamming weight $|\mathbf{f}|$, with strict inequality if $\mathbf{p}$ does not correspond to a codeword.

For example, three pseudocodewords of low effective weight in the Golay TBT have the following parameters:

Example 1. Suppose f_j equals $\frac{1}{2}$ in 8 places and 0 elsewhere. Then

$$\|\mathbf{f}\|^2 = 8 \times (1/4) = 2;$$
$$|\mathbf{f}| = 8 \times (1/2) = 4;$$
$$w_{\text{eff}}(\mathbf{p}) = \frac{|\mathbf{f}|^2}{\|\mathbf{f}\|^2} = \frac{4^2}{2} = 8.$$

Example 2. Suppose f_j equals $\frac{1}{2}$ in 8 places, 1 in 2 places, and 0 elsewhere. Then

$$\|\mathbf{f}\|^2 = 8 \times (1/4) + 2 \times 1 = 4;$$
$$|\mathbf{f}| = 8 \times (1/2) + 2 \times 1 = 6;$$
$$w_{\text{eff}}(\mathbf{p}) = \frac{|\mathbf{f}|^2}{\|\mathbf{f}\|^2} = \frac{6^2}{4} = 9.$$

Example 3. Suppose f_j equals $\frac{1}{3}$ in 6 places, $\frac{2}{3}$ in 2 places, 1 in 2 places, and 0 elsewhere. Then

$$\|\mathbf{f}\|^2 = 6 \times (1/9) + 2 \times (4/9) + 2 = 32/9;$$
$$|\mathbf{f}| = 6 \times (1/3) + 2 \times (2/3) + 2 = 16/3;$$
$$w_{\text{eff}}(\mathbf{p}) = \frac{|\mathbf{f}|^2}{\|\mathbf{f}\|^2} = 2^3 = 8.$$

Example 1 illustrates the general proposition that if all nonzero f_j are equal, then $w_{\text{eff}}(\mathbf{p})$ is equal to their number, $w_{\text{eff}}(\mathbf{p}) = |\operatorname{supp}(\mathbf{f})|$. (We will see that the support size $|\operatorname{supp}(\mathbf{f})|$ is the effective weight on a binary erasure channel.) More generally, we can show that $|\operatorname{supp}(\mathbf{f})|$ is an upper bound on $w_{\text{eff}}(\mathbf{p})$:

THEOREM 3.1. *If* $\mathbf{p}$ *is binary, then its effective weight* $w_{\text{eff}}(\mathbf{p})$ *satisfies*

$$(3.3) \qquad w_{\text{eff}}(\mathbf{p}) \leq |\operatorname{supp}(\mathbf{f})|,$$

with equality if and only if all nonzero f_j *are equal.*

Proof. Define $\mathbf{1}$ as the vector whose components are equal to one on the support of $\mathbf{f}$ and equal to zero elsewhere. Then $\|\mathbf{1}\|^2 = |\operatorname{supp}(\mathbf{f})|$, and $\langle \mathbf{1}, \mathbf{f} \rangle = |\mathbf{f}|$. Now by Schwarz's inequality,

$$\langle \mathbf{1}, \mathbf{f} \rangle^2 \leq \|\mathbf{1}\|^2 \|\mathbf{f}\|^2 = |\operatorname{supp}(\mathbf{f})| \cdot \|\mathbf{f}\|^2,$$

with equality if and only if $\mathbf{f} = \alpha \mathbf{1}$ for some α. The conclusion follows from

$$w_{\text{eff}}(\mathbf{p}) = \frac{|\mathbf{f}|^2}{\|\mathbf{f}\|^2} = \frac{\langle \mathbf{1}, \mathbf{f} \rangle^2}{\|\mathbf{f}\|^2} \leq |\operatorname{supp}(\mathbf{f})|,$$

with equality if and only if $\mathbf{f}$ is proportional to $\mathbf{1}$.   □

4. Binary symmetric channels. Now let us consider binary signaling on a binary symmetric channel (BSC) with crossover probability $\varepsilon < 1/2$. As in the previous section, a pseudocodeword $\mathbf{p}$ will be represented by a vector $\mathbf{f}$, where f_j is the fraction of pseudocodeword components equal to 1 in the jth position.

Suppose that the all-zero word is sent and that the received word is $\mathbf{e}$, where $e_j = 1$ if there is an error in the jth position and $e_j = 0$ otherwise. The Hamming distance between $\mathbf{e}$ and the all-zero word is $|\mathbf{e}| = \sum_j e_j$. Given a pseudocodeword $\mathbf{p}$ represented by $\mathbf{f}$, the average Hamming distance in the jth component is equal to f_j if $e_j = 0$ and $1 - f_j$ if $e_j = 1$, so the average Hamming distance $d_{\text{H}}(\mathbf{e}, \mathbf{p})$ is

$$(4.1) \qquad d_{\text{H}}(\mathbf{e}, \mathbf{p}) = \sum_j f_j(1 - e_j) + e_j(1 - f_j).$$

Thus the error event $\{ d_{\text{H}}(\mathbf{e}, \mathbf{p}) \leq |\mathbf{e}| \}$ is the event

$$(4.2) \qquad \left\{ \langle \mathbf{f}, \mathbf{1} - 2\mathbf{e} \rangle = \sum_j f_j(1 - 2e_j) = \sum_j f_j(-1)^{e_j} \leq 0 \right\}.$$

The probability of error is thus the probability that $\sum_j f_j(-1)^{e_j} \leq 0$. This is a sum of independent random variables v_j, where $v_j = f_j$ with probability ε and $v_j = -f_j$ with probability $1 - \varepsilon$.

We would like again to define the effective Hamming weight $w_{\text{eff}}(\mathbf{p})$ of $\mathbf{p}$ as a single parameter that has the same significance as the usual

Hamming weight $w_{\mathrm{H}}(\mathbf{c})$ of a codeword $\mathbf{c}$ for error probability, and that reduces to Hamming weight when $\mathbf{p}$ is actually a codeword. This cannot be done quite as neatly in the BSC case as in the Gaussian case, but a reasonable approach is as follows.

We ask for the minimum number of errors $|\mathbf{e}|$ that can cause a decoding error to $\mathbf{p}$. Clearly, given the total weight $|\mathbf{f}| = \sum_j f_j$ of $\mathbf{f}$, the worst case occurs when e errors occur in the e positions for which f_j is greatest. A decoding error may occur if the sum of the weights f_j in these e positions is equal to $|\mathbf{f}|/2$, and must occur if the sum exceeds $|\mathbf{f}|/2$. The effective weight in the former case will be taken as $w_{\mathrm{BSC}}(\mathbf{p}) = 2e$, and in the latter case as $w_{\mathrm{BSC}}(\mathbf{p}) = 2e - 1$.

To a first approximation, the decoding error probability to $\mathbf{p}$ will therefore be of the order of $K\varepsilon^{e(\mathbf{p})}$, where $e(\mathbf{p}) = \lceil w_{\mathrm{BSC}}(\mathbf{p})/2 \rceil$ is the minimum number of channel errors required to make a decoding error to $\mathbf{p}$.

The correspondence between effective weight and Hamming weight is not precise, because whereas with an ordinary codeword the multiplicity K is the number of ways that $e(\mathbf{c})$ errors can occur in $w_{\mathrm{H}}(\mathbf{c})$ positions, with pseudocodewords the number of possible combinations of $e(\mathbf{p})$ errors will in general be less.

The three examples given earlier illustrate these points and show that the effective weight for Gaussian channels and for BSCs are in general different.

Example 1. (cont.) If f_j equals $\frac{1}{2}$ in 8 places and 0 elsewhere, then $|\mathbf{f}| = 4$. There exist error patterns of weight $e = 4$ such that the sum of the e largest components of $\mathbf{f}$ is equal to $|\mathbf{f}|/2 = 2$, namely any error pattern with 4 errors in places where $f_j = \frac{1}{2}$. Thus the effective weight of such a pseudocodeword is $w_{\mathrm{BSC}}(\mathbf{p}) = 8$. In this case the number of error patterns of weight 4 that could cause a decoding error is $\frac{8!}{4!4!} = 70$, as in the usual case.

Example 2. (cont.) If f_j equals $\frac{1}{2}$ in 8 places, 1 in 2 places, and 0 elsewhere, then $|\mathbf{f}| = 6$. There exist error patterns of weight $e = 4$ such that the sum of the e largest components of $\mathbf{f}$ is equal to $|\mathbf{f}|/2 = 3$, namely error patterns with errors in the two places where $f_j = 1$ and in two other places where $f_j = \frac{1}{2}$. Thus the effective weight of such a pseudocodeword is $w_{\mathrm{BSC}}(\mathbf{p}) = 8$. The number of error patterns of weight 4 that could cause a decoding error is $\frac{8!}{6!2!} = 28$, compared to $\frac{8!}{4!4!} = 70$ in the usual case.

Example 3. (cont.) If f_j equals $\frac{1}{3}$ in 6 places, $\frac{2}{3}$ in 2 places, 1 in 2 places, and 0 elsewhere, then $|\mathbf{f}| = \frac{16}{3}$. There exist error patterns of weight $e = 3$ such that the sum of the e largest components of $\mathbf{f}$ is equal to $|\mathbf{f}|/2 = \frac{8}{3}$, namely error patterns with errors in the two places where $f_j = 1$ and in one other place where $f_j = \frac{2}{3}$. The effective weight of such a pseudocodeword is $w_{\mathrm{BSC}}(\mathbf{p}) = 6$. The number of error patterns of weight 3 that could cause a decoding error is 2, compared to $\frac{6!}{3!3!} = 20$ in the usual case.

From these examples one might conjecture that the effective Hamming weight of a binary pseudocodeword on a BSC is always less than or equal to its effective weight on an AWGN channel. We can construct a counterexample to such a conjecture as follows. Let d be an even integer greater than 4, let δ be a very small number such as $\delta = 0.001$, and let

$$N = d - 2 + \frac{1}{\delta},$$

where we assume that $1/\delta$ is an integer. Consider a set of nonzero weights f_j with one weight equal to 1 and N weights equal to δ. Then

$$|\mathbf{f}| = 2 + \delta(d - 2) \approx 2;$$
$$\|\mathbf{f}\|^2 = 1 + N\delta^2 \approx 1;$$
$$w_{\text{eff}}(\mathbf{p}) = \frac{|\mathbf{f}|^2}{\|\mathbf{f}\|^2} \approx 4;$$
$$e = 1 + \frac{d - 2}{2} = \frac{d}{2};$$
$$w_{\text{BSC}}(\mathbf{p}) = 2e = d > w_{\text{eff}}(\mathbf{p}),$$

where we note that on a BSC it takes one error in the position where $f_j = 1$ and $(d - 2)/2$ errors in positions where $f_j = \delta$ to accumulate a weight of $|\mathbf{f}|/2 = 1 + \delta(d - 2)/2$.

5. Binary erasure channels. Now let us consider binary signaling on a binary erasure channel (BEC) with erasure probability ε. Again, a pseudocodeword $\mathbf{p}$ will be represented by a vector $\mathbf{f}$, where f_j is the fraction of pseudocodeword symbols equal to 1 in the jth position.

Suppose that the all-zero word is sent and that $|S|$ erasures occur in a certain set S of coordinates. The remaining unerased symbols will all agree with the all-zero word. They will evidently also all agree with a pseudocodeword $\mathbf{p}$ represented by $\mathbf{f}$ if and only if $f_j = 0$ for all $j \notin S$.

Therefore we define the effective Hamming weight $w_{\text{BEC}}(\mathbf{p})$ of a pseudocodeword on a BEC as $|\operatorname{supp}(\mathbf{f})|$, the number of nonzero components of $\mathbf{f}$. Then:

(a) A decoding error to $\mathbf{p}$ may occur if and only if $|S| \geq w_{\text{BEC}}(\mathbf{p})$;

(b) If $\mathbf{p}$ is actually a codeword $\mathbf{c}$, then $w_{\text{BEC}}(\mathbf{p}) = w_{\text{H}}(\mathbf{c})$.

By Theorem 3.1, the effective weight $w_{\text{BEC}}(\mathbf{p})$ of a pseudocodeword $\mathbf{p}$ on a BEC is greater than or equal to its effective weight on an AWGN channel, with equality if and only if $\mathbf{p}$ is actually a codeword. Similarly, by the discussion in Section 4, $w_{\text{BEC}}(\mathbf{p}) \geq w_{\text{BSC}}(\mathbf{p})$, with equality if and only if $\mathbf{p}$ is actually a codeword.

For example, for Examples 1, 2 and 3, the effective weights on a BEC are 8, 10 and 10, respectively, compared to 8, 9 and 8 on an AWGN channel and 8, 8 and 6 on a BSC.

6. Examples with low pseudocodeword weights. In this section, we present a family of examples of binary tail-biting trellises for which the minimum effective pseudocodeword weight on an AWGN channel is strictly less than the minimum codeword weight.

For a first example, let C be the binary linear $(16, 6, 4)$ code generated by the following 6 generators:

$$
\begin{array}{cccc}
100 & 110000 & 110000 & 1 \\
010 & 001100 & 001100 & 1 \\
001 & 000011 & 000011 & 1 \\
\\
000 & 110000 & 001100 & 0 \\
000 & 001100 & 000011 & 0 \\
000 & 000011 & 110000 & 0 \\
\end{array}
$$

With a TBT constructed from these generators, the sum of the shifted generators shown below gives a low-weight three-cycle pseudocodeword:

```
000    000011   110000   0
                110000   1   100   110000
                         000   110000   001100   0
                                        001100   1   010   001100
                                                     000   001100   000011   0
001    000011                                                       000011   1
```
```
001    000000   000000   1   100   000000   000000   1   010   000000   000000   1
```

The resulting pseudocodeword has $f_j = \frac{1}{3}$ in 3 places, $f_j = 1$ in 1 place, and 0 elsewhere. Thus $|\mathbf{f}| = 2$, $||\mathbf{f}||^2 = \frac{4}{3}$, and the effective Hamming weight on an AWGN channel is $|\mathbf{f}|^2/||\mathbf{f}||^2 = 3$. Notice that since one position has weight $|\mathbf{f}|/2 = 1$, the effective weight on a BSC is only 2.

A generalization of this construction yields for every integer $a \geq 3$ a binary linear (n, k, d) code C with $d = 2\lceil a/2 \rceil$ and a TBT with an a-cycle pseudocodeword with effective weight

$$
w_{\text{eff}} = \frac{4a}{a+1} < d.
$$

The generator matrix has the form

$$
\begin{bmatrix}
I & B & B & 1 \\
O & B & C & 0
\end{bmatrix},
$$

where I is an $a \times a$ identity matrix, B is the matrix obtained from I by repeating every column b times where $b = \lceil a/2 \rceil$, 1 is a column of a ones, O is an $a \times a$ zero matrix, C is the cyclic shift of B to the right b times, and 0 is a column of a zeroes. It is straightforward to verify that the minimum nonzero codeword weight is $d = 2b$.

By a similar concatenation to that above, we obtain an a-cycle pseudocodeword with $f_j = \frac{1}{a}$ in a places, $f_j = 1$ in 1 place, and 0 elsewhere.

Thus $|\mathbf{f}| = 2$, $||\mathbf{f}||^2 = (a+1)/a$, and the effective Hamming weight on an AWGN channel is $|\mathbf{f}|^2/||\mathbf{f}||^2 = 4a/(a+1) < 4$. Notice that in general one position has weight $|\mathbf{f}|/2 = 1$, so the effective weight on a BSC is only 2. On the other hand, the effective weight on a BEC is $a+1 \geq d$.

For $a = 3, 4, 5, \ldots$, the minimum nonzero codeword weight is $4, 4, 6, \ldots$, while the pseudocodeword weight on an AWGN channel is $3, 3.2, 3.33, \ldots$, approaching a limit of 4.

We note that the TBT that produces this low-weight pseudocodeword is not in general minimal in the sense of [5]; however, it is linear, biproper and one-to-one.

7. Conclusions. We have determined the effective weight and distance of pseudocodewords on the AWGN channel, the BSC, and the BEC. In general pseudocodewords are least troublesome on a BEC.

For the Golay TBT, we have found pseudocodewords whose effective weight on the BSC is less than the minimum distance of the code, which indicates that ML decoding using this TBT will be distinctly suboptimal. For the AWGN channel, we have found no such pseudocodewords; moreover, simulations have shown that ML decoding using the Golay TBT is near-optimal [9]. However, as far as we know, there is no proof yet that the minimum nonzero pseudocodeword weight is 8.

For more general graphs, the concept of pseudocodeword may need some refinement. Just as in Viterbi decoding the influence of symbols far in the past eventually dies out, at least probabilistically, we expect that the influence of nodes far away from the root node in the computation tree will eventually die out. The concepts of pseudocodeword weight used in this paper do not have this property, which suggests that they need refinement.

Acknowledgments. We wish to acknowledge helpful discussions with S.M. Aji, B.J. Frey, G.B. Horn, H.-A. Loeliger, R.J. McEliece, A. Vardy, N. Wiberg and M. Xu.

REFERENCES

[1] S.M. AJI AND R.J. MCELIECE, *The generalized distributive law*, IEEE Trans. Inform. Theory, **46** (2000), 325–343.

[2] A.R. CALDERBANK, G.D. FORNEY, JR., AND A. VARDY, *Minimal tail-biting trellises: The Golay code and more*, IEEE Trans. Inform. Theory, **45** (1999), 1435–1455.

[3] B.J. FREY, R. KOETTER, AND A. VARDY, *Skewness and pseudocodewords in iterative decoding*, in Proc. 1998 IEEE Intl. Symp. Inform. Theory, Cambridge, MA, Aug. 1998, p. 148.

[4] R.G. GALLAGER, *Low-Density Parity-Check Codes*, MIT Press, Cambridge, MA, 1963.

[5] R. KOETTER AND A. VARDY, *Construction of minimal tail-biting trellises*, in Proc. 1998 Inform. Theory Workshop, Killarney, June 1998, 72–74.

[6] F.R. KSCHISCHANG AND B.J. FREY, *Iterative decoding of compound codes by probability propagation in graphical models*, IEEE J. Selected Areas Commun., **16** (1998), 219–230.

[7] F.R. Kschischang, B.J. Frey, and H.-A. Loeliger, *Factor graphs and the sum-product algorithm.* submitted to IEEE. Trans. Inform. Theory, July 1998.

[8] R.J. McEliece, D.J.C. MacKay, and J.-F. Cheng, *Turbo decoding as an instance of Pearl's 'belief propagation' algorithm*, IEEE J. Selected Areas Commun., **16** (1998), 140–152.

[9] A. Reznik, *Iterative decoding of codes defined on graphs*, Master's thesis, M. I. T., Cambridge, MA, May 1998.

[10] R.M. Tanner, *A recursive approach to low complexity codes*, IEEE Trans. Inform. Theory, **27** (1981), 533–547.

[11] Y. Weiss, *Correctness of local probability propagation in graphical models with loops*, Neural Comp., **12** (2000), 1–41.

[12] N. Wiberg, *Codes and decoding on general graphs*, PhD thesis, University of Linköping, Linköping, Sweden, 1996.

[13] N. Wiberg, H.-A. Loeliger, and R. Koetter, *Codes and iterative decoding on general graphs*, Euro. Trans. Telecomm., **6** (1995), 513–525.

EVALUATION OF GALLAGER CODES FOR SHORT BLOCK LENGTH AND HIGH RATE APPLICATIONS

DAVID J.C. MACKAY* AND MATTHEW C. DAVEY*

Abstract. Gallager codes with large block length and low rate (*e.g.*, $N \simeq 10,000$–$40,000$, $R \simeq 0.25$–0.5) have been shown to have record-breaking performance for low signal-to-noise applications. In this paper we study Gallager codes at the other end of the spectrum. We first explore the theoretical properties of binary Gallager codes with very high rates and observe that Gallager codes of any rate offer runlength-limiting properties at no additional cost.

We then report the empirical performance of high rate binary and non-binary Gallager codes on three channels: the binary input Gaussian channel, the binary symmetric channel, and the 16-ary symmetric channel.

We find that Gallager codes with rate $R = 8/9$ and block length $N = 1998$ bits outperform comparable BCH and Reed-Solomon codes (decoded by a hard input decoder) by more than a decibel on the Gaussian channel.

Key words. Error-correcting codes, Sum-product algorithm, Magnetic recording.

1. Introduction.

1.1. Definition of Gallager codes. A regular Gallager code [4] has a parity check matrix with uniform column weight j and uniform row weight k, both of which are very small compared to the blocklength. If the code has transmitted blocklength N and rate R then the parity check matrix **H** has N columns and M rows, where $M \geq N(1 - R)$. [Normally parity check matrices have $M = N(1 - R)$, but the matrices we construct may have a few redundant rows so that their rate could be a little higher than $1 - M/N$.]

In this paper we explore whether Gallager codes are useful for high rates ($R > 2/3$) and small block lengths ($N < 5000$).

1.2. High-rate codes. Reed-Solomon codes are the industry standard error-correcting codes for high rate, low block length applications such as magnetic disc drives and compact discs. They have good distance properties and they have an efficient bounded-distance decoder.

When we proposed evaluating Gallager codes with high rate and small block length for disc drive applications, a common response was 'why bother? You'll never beat Reed-Solomon codes.' But there are several reasons for checking the performance of Gallager codes.

1. Gallager codes with large block length N have good distance properties, with high probability [5, 10]. Given an optimal decoder, Gallager codes can get arbitrarily close to the Shannon limit of a wide variety of channels [10].

---

*Department of Physics, University of Cambridge, Cavendish Laboratory, Madingley Road, Cambridge, CB3 0HE, United Kingdom. `mackay|mcdavey@mrao.cam.ac.uk`.

2. There is a practical sum-product decoder for Gallager codes which works well for codes with block lengths of order $N = 10,000$ and rates of order $R = 1/4\text{--}2/3$ [11].

At rates of $R = 1/2$ and $R = 1/4$, regular binary Gallager codes decoded using this algorithm have near-Shannon limit performance. Irregular binary and non-binary Gallager codes with these rates perform better on the binary Gaussian channel than all known practical codes, including turbo codes [2, 14].

This decoder has three important features:

   (a) It is better than a bounded-distance decoder — it works well at noise levels significantly larger than the Gilbert noise level (that is, the noise level at which typical error events have weight greater than half the minimum distance of a code at the Gilbert bound).

   (b) It is a soft-input decoder, able to make use of likelihood information from the channel output. Such decoders can have considerable advantages over decoders that take hard inputs [5].

   (c) The decoder can be generalized to infer bursts if the channel is believed to be a bursty channel [15].

3. According to Berlekamp [1], one reason that high rate Reed-Solomon codes are used is that lower rate Reed-Solomon codes are more costly to encode and decode — the complexity increases with increasing redundancy.

In contrast, the encoding and decoding complexity for Gallager codes hardly depend on rate. Furthermore, Gallager codes of any desired rate and block length can easily be constructed.

If Reed-Solomon codes can be surpassed, the disc drive industry could benefit in various ways. A higher rate code with the same probability of error would allow a small increase in the storage capacity of a drive. Alternatively, a code that can cope with larger raw error rates would make the system more tolerant to tracking errors, and the disc could be spun faster, offering a higher data rate.

1.3. Outline of paper. In section 2, we explore four theoretical issues. First, we ask how good is the ensemble of random, high-rate, regular Gallager codes, if we do not constrain the overlap between the columns? We find that the expected distance properties of these codes are not good. Second, we ask what are the highest possible rates that regular Gallager codes could have if we do constrain the overlap between columns — these codes correspond to 'Steiner systems' — and could these codes have good distance properties? We prove that such codes, with $j = 3$, have bad distance, and we give a conjecture that, for larger j, the codes might be good. Third, we prove that a particular construction of Gallager codes in terms of permutation matrices leads to bad codes. Fourth, we show that Gallager codes can be constructed to have fortuitous runlength-limiting properties.

In section 3, we describe the empirical performance of high-rate binary regular Gallager codes with $j = 4$ and of a high-rate non-binary regular Gallager code with $j = 3$. We compare these Gallager codes with codes similar to those used in discdrives and show that their performance is good.

In section 4, we discuss difference-set cyclic codes, which are codes similar to Gallager codes, but having the special property that they satisfy many more than M low-weight parity constraints. They outperform equivalent Gallager codes by a significant margin [13, 7]. If we could find more codes like these, they could be very useful.

2. Theory of high rate Gallager codes.

2.1. Distance properties of random Gallager codes.

The expectation of the weight enumerator function of a random Gallager code with $M \times N$ parity check matrix can be computed for two ensembles.

Ensemble G: In Gallager's ensemble [5], a row weight k is selected, and a blocklength N. We find the weight enumerator function $A(w; 1)$ of the following submatrix with column weight 1 (illustrated for the case $k = 4$):

$$(1) \qquad \mathbf{H}^{(1)} = \begin{bmatrix} 1\,1\,1\,1\,0\,0\,0\,0\,0\,0\,0\,0\,0\,0\,0\,0 \\ 0\,0\,0\,0\,1\,1\,1\,1\,0\,0\,0\,0\,0\,0\,0\,0 \\ 0\,0\,0\,0\,0\,0\,0\,0\,1\,1\,1\,1\,0\,0\,0\,0 \\ 0\,0\,0\,0\,0\,0\,0\,0\,0\,0\,0\,0\,1\,1\,1\,1\ldots \end{bmatrix}.$$

As shown by Gallager [5], $A(w; 1)$ is given by convolving ($\star$) together N/k copies of the function

$$(2) \qquad a(w) = \left\{ \begin{array}{cc} \binom{k}{w} & w \text{ even} \\ 0 & w \text{ odd} \end{array} \right. :$$

$$(3) \qquad A(w; 1) = a(w) \star a(w) \star \ldots \star a(w).$$

We define an ensemble of random Gallager codes with column weight j by stacking j copies of $\mathbf{H}^{(1)}$ vertically above each other, each individual copy having its columns randomly permuted. We can then find the expected weight enumerator function $A(w; j)$ of the resulting (N, M, j, k) code with $M = \frac{j}{k} N$ using:

$$(4) \qquad \langle A(w; j) \rangle = A_{\mathrm{G}}(w; j) \equiv A(w; 1) \left[\frac{A(w; 1)}{\binom{N}{w}} \right]^{j-1}.$$

Ensemble M: An alternative ensemble of matrices that have column weight at most j, and arbitrary M and N, but do not have fixed row weight k, was used by [10]. Each column of the matrix $\mathbf{H}$ is created by flipping j not-necessarily-distinct entries. With high probability, any particular column has weight j, but it may, with smaller probability, have weight $j - 2$, etc. The expected weight enumerator function is

$$(5) \qquad \langle A(w; j) \rangle = A_M(w; j) \equiv \binom{N}{w} p_{00}^{(wj)}$$

where

$$(6) \qquad p_{00}^{(r)} = 2^{-M} \sum_{i=0}^{M} \binom{M}{i} \left(1 - \frac{2i}{M}\right)^r.$$

These ensembles are not the best ensembles for making good Gallager codes, but they are convenient for estimating weight enumerator functions and getting a feel for the dependence on block length and rate. Figure 1 shows the expected weight enumerator functions for a sequence of codes with block length 540 bits and rate increasing from 1/3 to 8/9.

It seems that for small block lengths and large rates such as $R = 8/9$, codes constructed by this random construction will almost certainly be bad codes, in that their distance will be nowhere near the Gilbert distance.

We therefore use constrained random constructions. The constraint used by Gallager [5] and MacKay and Neal [11], which we also use here, constrains the maximum overlap between any two columns in the matrix to be one. We will call this the overlap constraint.

2.2. Steiner systems and Gallager codes.

2.2.1. Existence of high rate codes. If we insist on the constraint that the overlap between any two columns in the parity check matrix should be at most one, then it is not possible to build Gallager codes with arbitrary values of (N, M, j); in particular, we cannot make the blocklength N arbitrarily large for fixed number of rows M. The blocklength N of such a code with column weight j and M rows is bounded above by the size of a Steiner system $S(M, j, 2)$.

A Steiner system $S(M, j, t)$ is a set $\mathcal{M}$ of M points, and a collection $\mathcal{N}$ of subsets of $\mathcal{M}$ of size j, called blocks, such that any subset of t points of $\mathcal{M}$ are in exactly one of the blocks. The size of the Steiner system, N, is defined to be the number of blocks $N = |\mathcal{N}|$. The special case $j = 3$, $t = 2$ is called a Steiner triple system.

The number of subsets of size t in $\mathcal{M}$ is $\binom{M}{t}$ and the number of subsets of size t in a block is $\binom{j}{t}$, so the size of an (M, j, t) Steiner system is

$$(7) \qquad N_S(M, j, t) = \binom{M}{t} \bigg/ \binom{j}{t}.$$

In the case of interest, $t = 2$, we obtain:

$$(8) \qquad N_S(M, j) = \frac{M(M - 1)}{j(j - 1)}.$$

Rate

1/3

2/3

8/9

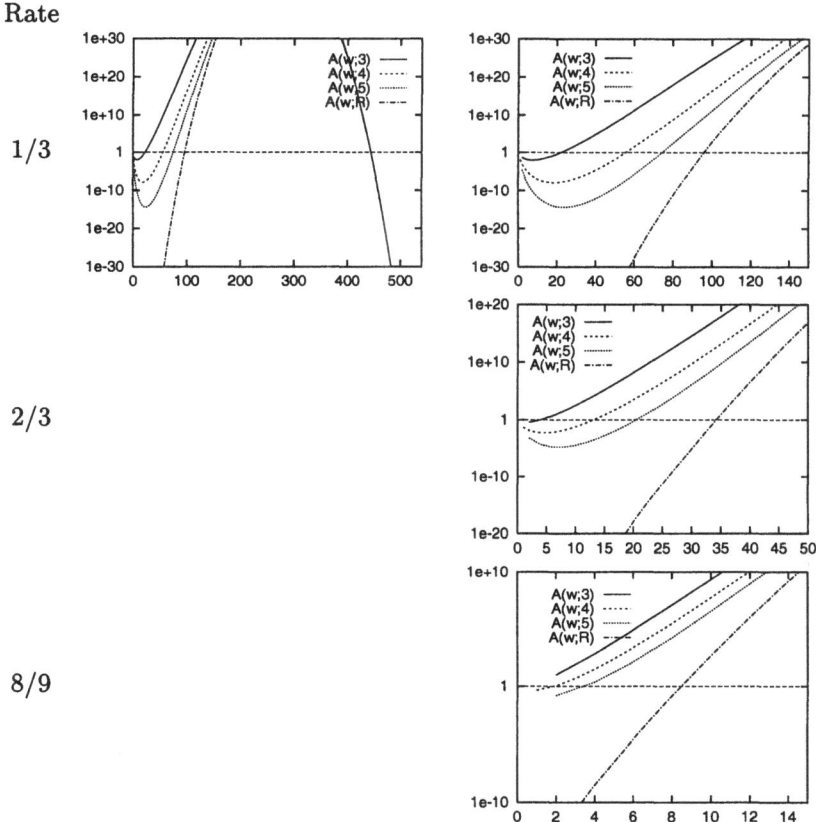

FIG. 1. *Expected weight enumerator functions computed using ensemble M. [The results for ensemble G are similar.] Block length is 540 bits in all cases. Top figure shows the expected weight enumerator function for codes with rate 1/3 having j = 3, 4 and 5. The lowest line shows the expected weight enumerator function of a random linear code with the same N and M. This line crosses the horizontal line A(w) = 1 at the Gilbert distance. The neighbouring figure shows detail from the first figure. Subsequent figures show the corresponding graphs for rates 2/3 and 8/9. It is evident that the typical distance of a Gallager code is becoming an increasingly small fraction of the Gilbert distance as the rate increases.*

The row weight of **H**, k, is

$$(9) \qquad k = j \frac{N_S(M,j)}{M} = \frac{(M-1)}{(j-1)} \; .$$

Any $(M, j, 2)$ Steiner system defines an (N, M, j) Gallager code with $N = N_S(M,j)$. If N exceeds $N_S(M,j)$, it is impossible to make a regular Gallager code with parameters N, M, j that satisfies the overlap constraint. So for any chosen M and j the overlap constraint implies a maximum possible rate, and for any rate R and column weight j, it implies a minimum possible blocklength.

These constraints are illustrated in Figure 2, which shows $N_S(M, j)$ as a function of M for various values of j. This figure also shows some actual values of (N, M, j) that have been constructed by the random constructions mentioned above. Fortunately, codes with column weight $t = 4$, blocklength $N \simeq 2000$ and rate $R \simeq 0.9$ are just buildable. Considerable computer time was spent searching for the highest rate codes that appear in later sections.

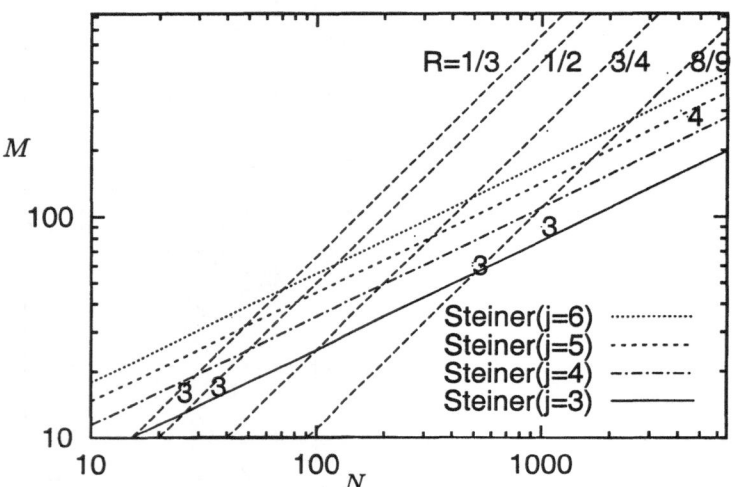

FIG. 2. *Parameters (N, M, j) that can be built without violating the constraint on column overlaps. Horizontal axis: N; vertical axis: M; the labels '3' and '4' show examples of parameters that have been built by random construction methods, including codes presented in this paper. The 45-degree lines are lines of constant rate $R = (N - M)/N$. The near-horizontal lines show the curves (N, M) defined by Steiner systems for various j (Equation (8)). Points (N, M, j) below the Steiner curve $S(j)$ are not buildable. For low rates such as 1/2 or 1/3, very small blocklengths are buildable but as the rate is increased, the smallest possible blocklength becomes quite large.*

2.2.2. Weakness of Steiner system codes with $j = 3$.

Having established that certain high rate, small blocklength codes can be constructed, we now ask whether we expect these codes to be good codes. Randomly chosen Gallager codes with large enough blocklength N have good distance properties [5, 10], but for small N, some of these properties deteriorate. In the case $j = 3$, there is bad news.

THEOREM 1. *Any Gallager code defined by a Steiner system with $j = 3$ has minimum distance less than or equal to 10.*

Proof. We define a (w, v) *near-codeword* of a code with parity check matrix $\mathbf{H}$ to be a vector $\mathbf{x}$ with weight w whose syndrome $\mathbf{z}(\mathbf{x}) \equiv \mathbf{Hx}$ has weight v.

We prove that the Gallager code derived from a Steiner triple system has words of weight 10 by counting how many $(5, 1)$ near-codewords it has. If a code with $M \times N$ parity check matrix $\mathbf{H}$ has more than M distinct $(w, 1)$ near-codewords, then its minimum distance is at most $2w$, because, by the pigeonhole principle, there must be at least two of them whose syndromes are identical; the sum of these two near-codewords must be a codeword of weight at most $2w$.

We can generate a $(5, 1)$ near-codeword using the Steiner system property as follows. First, pick a row of $\mathbf{H}$; we call this row a. (M choices.) Second, pick two columns n_1, n_2 satisfying $H_{an} = 1$. ($\binom{k}{2}$ choices.) These two columns define a $(2, 4)$ near-codeword. Call the rows in which the syndrome of this word is non-zero rows b_1, b_2, b_3 and b_4. Third, add two more columns n_3, n_4 to make either a $(4, 2)$ near-codeword or a weight 4 codeword. There are two choices for n_3, n_4, one of which is illustrated diagrammatically in Figure 3. Either, as shown in the figure, n_3 is the column in which points b_1 and b_3 appear and column n_4 contains b_2 and b_4; or n_3 contains b_1 and b_4 and n_4 contains b_2 and b_3. Call the new rows introduced by columns n_3 and n_4 rows c_1 and c_2. These rows might be the same as each other, in which case we have found a weight 4 codeword (n_1, n_2, n_3, n_4). Otherwise, rows c_1 and c_2 take us to a unique fifth column n_5 which contains points c_1 and c_2. Adding this column, we have a $(5, 1)$ near-codeword. The final row d is distinct from rows b_*-c_* but might be identical to row a.

	n_1	n_2	n_3	n_4	n_5
a	1	1			
b_1	1		1		
b_2	1			1	
b_3		1	1		
b_4		1		1	
c_1			1		1
c_2				1	1
d					1

FIG. 3. *Construction of a $(5, 1)$ near-codeword (or a weight 4 codeword) in a Steiner system code.*

We can create such $(5, 1)$ near-codewords in $M \times \binom{k}{2} \times 2$ ways. We will assume that none of these constructions generated a weight 4 codeword — if one did, then we already have the desired result that the minimum distance $d \leq 10$. Now, are all these

$$(10) \qquad 2M \binom{k}{2} = Mk(k-1) = M(M-1)(M-3)/4$$

$(5, 1)$ near-codewords distinct, or have we created duplicates? If rows a and d are different, then they are all distinct, because we can hang each subgraph defined by a $(5, 1)$ near-codeword from row d; the nearest neighbours of row d are rows c_1 and c_2; the next nearest neighbours are the rows b_*; and the furthest row in the subgraph from d is row a. Thus we can recover the $2M\binom{k}{2}$ choices that produced the word. If a and d are equal in one $(5, 1)$ near-codeword, however, then the above procedure will generate the same near-codeword in three ways (starting from (n_1, n_2), (n_1, n_5), and (n_2, n_5)). So the number of $(5, 1)$ near-codewords is at least

$$(11) \qquad C = M(M - 1)(M - 3)/(4 \times 3).$$

If M exceeds 7, then C exceeds M, so, by the pigeonhole principle, the code has minimum distance at most 10.   □

This negative result for binary Gallager codes with $j = 3$ gives a reason for concentrating on larger values of j when dealing with high rate binary codes.

2.2.3. Properties of high rate codes with small blocklength and $j \geq 4$. Do the codes derived from Steiner systems with $j \geq 4$ have better distance properties? We do not have a theorem, but using similar pigeonhole arguments to those used above, we conjecture that the best codes corresponding to Steiner systems have minimum distance satisfying the following scaling laws:

$$(12) \qquad j = 4 : d \gtrsim \log M$$

$$(13) \qquad j \geq 5 : d \gtrsim M^{\frac{j-4}{j-2}} \quad e.g., \quad \begin{cases} j = 5 : & d \gtrsim M^{1/3} \\ j = 6 : & d \gtrsim M^{1/2} \\ j = 8 : & d \gtrsim M^{2/3} \end{cases}.$$

2.3. Weakness of any Gallager codes built from commuting permutations. Some Steiner systems and other constructions of Gallager codes have the property that the parity check matrix contains a grid of non-overlapping permutation matrices. For example, the matrix

$$(14) \qquad \mathbf{H} = \begin{bmatrix} \mathbf{R}_{11} & \mathbf{R}_{12} & \mathbf{R}_{13} & \mathbf{R}_{14} \\ \mathbf{R}_{21} & \mathbf{R}_{22} & \mathbf{R}_{23} & \mathbf{R}_{24} \\ \mathbf{R}_{31} & \mathbf{R}_{32} & \mathbf{R}_{33} & \mathbf{R}_{34} \end{bmatrix},$$

where $\{\mathbf{R}_i\}$ are permutation matrices, defines a rate $1/4$ Gallager code with $j = 3$ and $k = 4$. If the permutations commute, that is, $\mathbf{R}_{ij}\mathbf{R}_{kl} = \mathbf{R}_{kl}\mathbf{R}_{ij}$ — which need not be the case for random constructions, but is the case for some algebraic constructions (John Fan, personal communication) — then the distance properties of the code are limited by the following theorem.

THEOREM 2. *If a parity check matrix of height M contains a submatrix of height M and width $(j+1)M/j$ containing $j(j+1)$ non-overlapping permutation matrices that all commute with each other, then the corresponding code has minimum distance less than or equal to $(j + 1)!$.* This result applies to Gallager codes of any rate, not just high rate codes. For example, a code with $j = 3$ built from commuting permutations has distance at most 24, and a similar code with $j = 4$ has distance at most 120.

Proof. We call the vertical and horizontal divisions of the matrix of size M/j 'blocks'. We will construct a codeword of a matrix $\mathbf{H}$ whose size is j row-blocks $\times (j+1)$ column-blocks, for example, if $j = 3$, the matrix in Equation (14). This matrix is in general a sub-matrix of the parity check matrix from which we started. In each of the column-blocks $1, 2, \ldots, (j+1)$ we will set $j!$ bits to 1 as follows. Define the operator d_h associated with column-block h ($h = 1 \ldots (j+1)$) to be the 'determinant' (modulo 2) obtained from the $j \times j$ matrix given by deleting column-block h from the matrix $\mathbf{H}$. For example, for the case $j = 3$,

$$(15) \quad \begin{aligned} d_2 = {} & \mathbf{R}_{11}\mathbf{R}_{23}\mathbf{R}_{34} + \mathbf{R}_{11}\mathbf{R}_{24}\mathbf{R}_{33} + \mathbf{R}_{13}\mathbf{R}_{21}\mathbf{R}_{34} \\ & + \mathbf{R}_{13}\mathbf{R}_{24}\mathbf{R}_{31} + \mathbf{R}_{14}\mathbf{R}_{21}\mathbf{R}_{33} + \mathbf{R}_{14}\mathbf{R}_{23}\mathbf{R}_{31} \, . \end{aligned}$$

Each of these operators has weight $j!$, that is, if we hit a weight-one vector of length M/j with d_h, we get a vector of weight at most $j!$. We can now make a codeword $\mathbf{w}$ starting from any weight-one vector of length M/j, $\mathbf{x}$, thus:

$$(16) \quad \mathbf{w} = (d_1\mathbf{x}, d_2\mathbf{x}, d_3\mathbf{x}, \ldots d_{j+1}\mathbf{x}).$$

Here, the commas correspond to the block boundaries. That this is a codeword can be seen by computing the syndrome in each row. In the top row-block, for example, the syndrome is:

$$(17) \quad \mathbf{R}_{11}d_1\mathbf{x} + \mathbf{R}_{12}d_2\mathbf{x} + \mathbf{R}_{13}d_3\mathbf{x} + \ldots + \mathbf{R}_{1(j+1)}d_{j+1}\mathbf{x}$$

which is equal to the product of $\mathbf{x}$ and the determinant of the square matrix:

$$(18) \quad \begin{vmatrix} \mathbf{R}_{11} & \mathbf{R}_{12} & \cdots & \mathbf{R}_{1(j+1)} \\ \mathbf{R}_{11} & \mathbf{R}_{12} & \cdots & \mathbf{R}_{1(j+1)} \\ \mathbf{R}_{21} & \mathbf{R}_{22} & \cdots & \mathbf{R}_{2(j+1)} \\ \vdots & \vdots & \ddots & \vdots \\ \mathbf{R}_{j1} & \mathbf{R}_{j2} & \cdots & \mathbf{R}_{j(j+1)} \end{vmatrix},$$

which is zero, since the top two rows are equal. Similarly, the syndrome in any row-block h is the product of a determinant that is equal to zero with $\mathbf{x}$. Thus $\mathbf{w}$ is a codeword, and the distance is at most the weight of $\mathbf{w}$, which is at most $(j+1)!$. □

2.4. Gallager codes are fortuitous runlength-limiting codes.
A potential benefit of Gallager codes is that they can be constructed to have a runlength-limiting property.

Optimal runlength-limiting codes for noiseless channels are nonlinear. But if we are using an error correcting code for a noisy channel, it would be nice if we could get the runlength-limiting property for free, as part of the error-correcting code. The standard procedure in discdrives is to use a small inner runlength-limiting code, for example, a nonlinear $(N, K) = (16, 15)$ code, and an outer code such as a Reed-Solomon code. This method has the disadvantage that the outer code cannot be given detailed likelihood information from the noisy channel; the errors introduced by the decoder of the inner code are complex.

2.4.1. Getting runlength constraints for free. If a Gallager code has row weight k and there are N/k rows in the parity check matrix like this (if $k = 5$):

$$
(19) \qquad
\begin{bmatrix}
1\,1\,1\,1\,1\,0\,0\,0\,0\,0\,0\,0\,0\,0\,0\,0\,0\,0\,0\,0 \\
0\,0\,0\,0\,0\,1\,1\,1\,1\,1\,0\,0\,0\,0\,0\,0\,0\,0\,0\,0 \\
0\,0\,0\,0\,0\,0\,0\,0\,0\,0\,1\,1\,1\,1\,1\,0\,0\,0\,0\,0 \\
0\,0\,0\,0\,0\,0\,0\,0\,0\,0\,0\,0\,0\,0\,0\,1\,1\,1\,1\,1\ldots
\end{bmatrix}
$$

or this (if $k = 4$):

$$
(20) \qquad
\begin{bmatrix}
1\,1\,1\,1\,0\,0\,0\,0\,0\,0\,0\,0\,0\,0\,0\,0 \\
0\,0\,0\,0\,1\,1\,1\,1\,0\,0\,0\,0\,0\,0\,0\,0 \\
0\,0\,0\,0\,0\,0\,0\,0\,1\,1\,1\,1\,0\,0\,0\,0 \\
0\,0\,0\,0\,0\,0\,0\,0\,0\,0\,0\,0\,1\,1\,1\,1\ldots
\end{bmatrix}
$$

then these constraints enforce local properties that we can use.
- If k is odd (as in (19)), then these constraints force each block of k successive transmitted bits to have even parity. Since k is odd, this means that there must be at least one 0 in every block of k bits. Thus a Gallager code with odd k is automatically a runlength-limiting code with maximum runlength of 1s equal to $2(k-1)$. There is no constraint on the maximum runlength of 0s.
- If k is even, then the original Gallager code is not necessarily a runlength-limiting code, but we can modify the code by adding a constant vector to all codewords in the code. For example, if $k = 4$, we could add the vector

$$
(21) \qquad \begin{bmatrix} 1\,0\,0\,0\,1\,0\,0\,0\,1\,0\,0\,0\cdots1\,0\,0\,0 \end{bmatrix}
$$

to all codewords, modifying the decoder appropriately. Now, the number of 1s in any block of k bits is odd, and so is the number of 0s. So there must be at least one 1 and one 0 in each of these blocks. So the maximum runlength for 1s is $2(k - 1)$, and the maximum runlength for 0s is $2(k - 1)$.

One could also construct a Gallager code, without impairing its error-correcting capabilities (at least if the channel is a memoryless channel), so that its top rows look like this:

$$(22) \qquad \begin{bmatrix} 1\,1\,1\,1\,1\,0\,0\,0\,0\,0\,0\,0\,0\,0\,0\,0 \\ 0\,0\,0\,0\,1\,1\,1\,1\,1\,0\,0\,0\,0\,0\,0\,0 \\ 0\,0\,0\,0\,0\,0\,0\,0\,1\,1\,1\,1\,1\,0\,0\,0\,0 \\ 0\,0\,0\,0\,0\,0\,0\,0\,0\,0\,0\,0\,1\,1\,1\,1\,1 \ldots \end{bmatrix},$$

in which case the maximum runlength would be $2k - 3$.

For practical purposes, Gallager codes have to have a column weight roughly equal to $j = 3$ or 4. A code with rate $R = (N - M)/N$ and column weight j has a row weight $k \simeq jN/M = j/(1 - R)$. So examples of the fortuitous runlength limits that can be obtained with Gallager codes are as follows:

R	$j = 3$		$j = 4$	
	$2(k - 1)$	$2k - 3$	$2(k - 1)$	$2k - 3$
0.25	6	5	9	8
0.50	10	9	14	13
0.75	22	21	30	29
0.90	58	57	78	77

If the rate is about 0.9 then these runlengths are not of much use — the maximum runlength used in current discdrives is about 15 — but perhaps as technology advances, this idea will become useful, especially if lower rate codes are used. The benefits could be substantial: not only would there be an increase of about 6% in the storage capacity if the inner code were removed, but the outer code could be provided with better likelihood information, which, as we will see below, can give a great improvement in performance for codes with appropriate decoders.

Almost-certainly runlength-limiting codes. Further methods for making high-rate runlength-limited transmissions without using a nonlinear inner code are described in [8]. The three key ideas are (a) make the outer code a coset code with the offset varying pseudorandomly from block to block; (b) space the parity bits of the outer code uniformly through the block and put aside a small number of source bits that can be set arbitrarily so that the parity bits take on the values required to satisfy the runlength constraints; and (c) flip any remaining bits that need to be changed, and rely on the outer code to correct them. By combining these methods, we can remove the need for any complicated inner code.

3. Empirical results.

3.1. Construction of Gallager codes. There are various methods for randomly constructing a parity check matrix with given j and k. When we make codes with large blocklengths these alternative methods generally give codes with equivalent performance. When the blocklengths are small, however, good codes become more difficult to find. We have implemented two construction methods. Both these methods attempt to constrain the maximum overlap between two columns in the matrix to be one. We find this constraint to be more important in codes with small blocklengths than it was with large blocklengths.

Permutation matrix method. This method is similar to Gallager's (see the appendix of his book), except that the largest possible sizes of random permutation matrix are used. This distinction is shown pictorially for rate 3/4 codes by the following figures, in which integers show the number of superposed permutation matrices in each square.

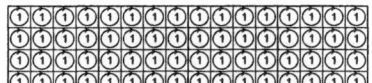

Small permutation matrices   Large permutation matrices

Left to right method. This is construction 1A from MacKay and Neal.

We have concentrated on regular constructions with columns weights $j = 3$ and $j = 4$. For high rate codes with small blocklengths, a column weight $j = 3$ gives weak codes, having small numbers of low-weight codewords. We therefore report results for column weight $j = 4$ only.

3.2. Binary Gallager codes, Gaussian channel.

3.2.1. Method. A sequence of noise levels was selected. At each noise level, a large number of block decodings was simulated. The decoding algorithm was run for up to 1000 iterations, halting earlier if the best guess of the decoder corresponded to a valid codeword. The outcome of each decoding was either a success (*i.e.*, the algorithm returns the transmitted codeword without any errors) or a failure. There are two possible types of failures.

Detected errors. The decoding algorithm failed to find a valid codeword. We could call these failures block erasures.

Undetected errors. The decoding algorithm halts in a valid codeword that differs from the transmitted codeword. This failure mode is expected to be very rare in codes that have good distance properties.

3.2.2. Errors observed. The codes with column weight $j = 4$ have never made undetected errors in these experiments. In the graphs, the 'undetected' error bars show empirical *upper bounds* on the probability of undetected error. We might conjecture that these codes have minimum distance similar to the Gilbert-Varshamov distance, and that undetected errors only occur when the maximum likelihood decoder would also make undetected errors. Using this conjecture, the probability of undetected error can be bounded above by the probability of error of the maximum likelihood decoder, which could probably be computed.

3.2.3. Decoding times. Figure 4 shows the cumulative distribution of decoding times for the code s2.94.594 at three noise levels. The decoding usually halts in fewer than ten iterations. Under good conditions three iterations usually suffice.

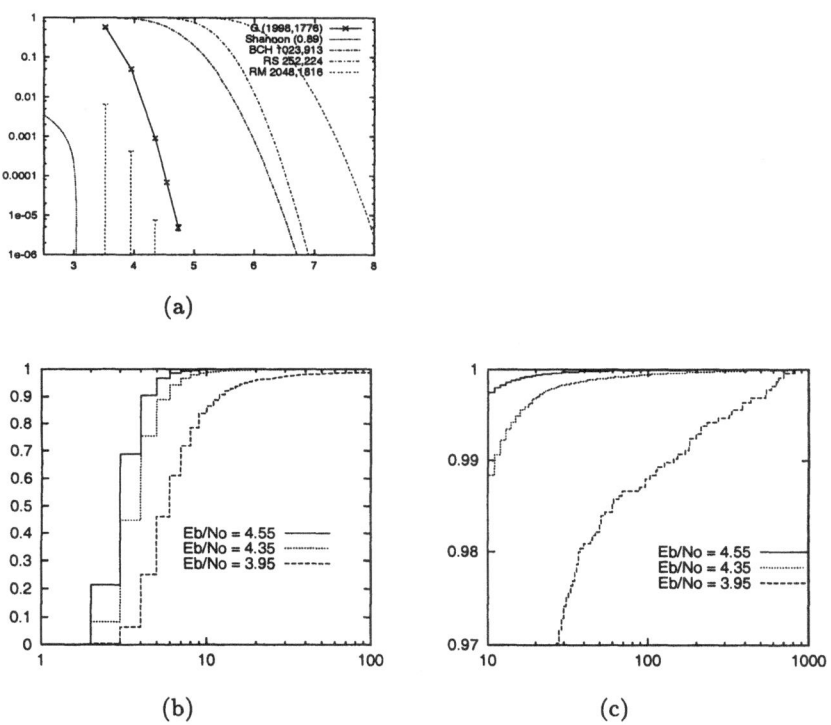

FIG. 4. *Regular Gallager code with rate $R = 8/9$ and $N = 1998$.*

(a) Dependence of block error rate on signal to noise ratio. Weight per column $t = 4$ and transmitted blocklength $N = 1998$. Vertical axis: block error rate. Horizontal axis: E_b/N_0 (decibels). Also shown are performance curves for Reed-Solomon, Reed-Muller and BCH codes with similar rate. These curves assume that the channel outputs are thresholded to give binary signals to the decoder. (That is, no soft decoders for the algebraic codes.)

(b) Decoding times, cumulative distribution.

(c) Detail from (b). The parity check matrix of this code, s2.94.594, can be found in the online archive [9].

The number of arithmetical operations per iteration is about four times the number of 1s in the parity check matrix. That makes 16 operations per iteration per transmitted bit, or 32000 operations per iteration if $N = 2000$.

3.3. Non-binary Gallager codes, q-ary symmetric channel. Gallager codes over $GF(q)$ were first reported in [3]; improved performance was gained at the expense of a decoding complexity that scaled as q^2. This complexity can be reduced using a Fourier transform of the probabilities [12].

In the following, we use the notation of [3]. Let $\mathcal{N}(m) := \{n : H_{mn} \neq 0\}$ be the set of noise symbols that participate in check m. The decoder needs to update quantities r^a_{mn}, the probability of check m being satisfied if symbol n of the message $\mathbf{x}$ is considered fixed at $a \in GF(q)$ and the other noise symbols have a separable distribution given by the probabilities $\{s^b_{mn'} : n' \in \mathcal{N}(m)\backslash n\}$. The new value of r^a_{mn} is:

$$(23) \qquad r^a_{mn} = \sum_{\mathbf{x}:x_n=a} \delta\left(\sum_{n' \in \mathcal{N}(m)} H_{mn'}x_{n'} = z_m\right) \prod_{j \in \mathcal{N}(m)\backslash n} s^{x_j}_{mj}.$$

This function of a is a convolution of the quantities s^a_{mj}, and so the summation can be replaced by a product of the Fourier transforms (taken over the additive group of $GF(q)$) of s^a_{mj} for $j \in \mathcal{N}(m)\backslash n$, followed by an inverse Fourier transform. The Fourier transform F of a function f over $GF(2)$ is given by $F^0 = f^0 + f^1, F^1 = f^0 - f^1$. Transforms over $GF(2^k)$ can be viewed as a sequence of binary transforms in each of k dimensions. Hence for $GF(4)$ we have

$$(24) \qquad\qquad\qquad F^0 = [f^0 + f^1] + [f^2 + f^3]$$
$$(25) \qquad\qquad\qquad F^1 = [f^0 - f^1] + [f^2 - f^3]$$
$$(26) \qquad\qquad\qquad F^2 = [f^0 + f^1] - [f^2 + f^3]$$
$$(27) \qquad\qquad\qquad F^3 = [f^0 - f^1] - [f^2 - f^3].$$

The inverse transform is the same, except that we also divide by 2^k.

With a slight abuse of notation, let $(S^0_{mj}, \ldots, S^{q-1}_{mj})$ represent the Fourier transform of the vector $(s^0_{mj}, \ldots, s^{q-1}_{mj})$, after permuting the components to take account of the matrix entry H_{mj}. Now r^a_{mn} is the ath coordinate of the inverse transform of

$$(28) \qquad \left(\left(\prod_{j \in \mathcal{N}(m)\backslash n} S^0_{mj}\right), \ldots, \left(\prod_{j \in \mathcal{N}(m)\backslash n} S^{q-1}_{mj}\right)\right).$$

The update of the quantities s is unchanged.

Each fast Fourier transform takes $q \log_2 q$ additions and q multiplications. Assuming a column weight $j = 3$ and taking $q = 16$, the total cost per iteration is 96 additions and 72 multiplications per bit. All these operations can be implemented in low precision arithmetic with a small loss in performance [12].

Figure 6 shows the performance of a column weight 3 rate 8/9 Gallager code over $GF(16)$ applied to 16-ary symmetric channel. The code is compared with two Reed-Solomon codes.

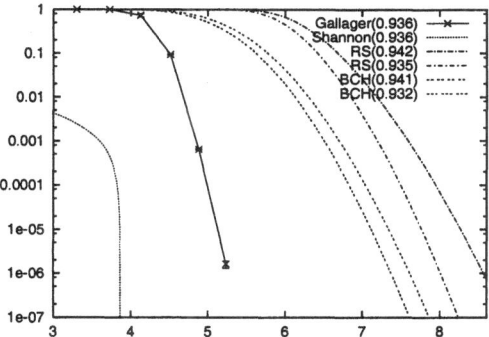

FIG. 5. *Regular Gallager code with blocklength 4376 and rate 0.936. Dependence of block error rate on signal to noise ratio. Weight per column $t = 4$. Also shown are performance curves for two nearby RS codes and two BCH codes with similar rate and blocklength 1023. These curves assume that the RS symbols are transmitted over the binary Gaussian channel and that the outputs are thresholded to give binary signals to the decoder. (That is, no soft decoders for the algebraic codes.) The parity check matrix of this code,* 4376.282.4.9598, *can be found in the online archive [9].*

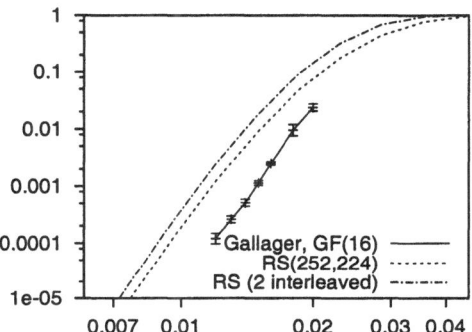

FIG. 6. *Gallager code over GF(16) applied to 16-ary symmetric channel. Weight per column $t = 3$. Vertical axis: block error rate. Horizontal axis: symbol error probability. The code is compared with two Reed-Solomon codes similar to those used in discdrives.*

4. Difference-set cyclic codes. The performance of Gallager codes can be enhanced by making a non-random code with *redundant sparse constraints* [13, 7]. There is a difference-set cyclic code, for example, that has $N = 273$, and $K = 191$, but the code satisfies not $M = 82$ but N, *i.e.*, *273*, low-weight constraints (Figure 7). It is impossible to make random Gallager codes that have anywhere near this much redundancy among their checks. The redundant checks allow the sum-product algorithm to work substantially better, as shown in Figure 7, in which a DSC code outperforms a comparable regular binary Gallager code by about 0.7 dB. The (73,45) DSC code has been implemented on a chip by Karplus and Krit [6] following a design of Tanner [13]. Product codes are another family of codes with redundant constraints. For example, the product with itself of a $(n, k) = (64, 57)$ Hamming code satisfying $m = 7$ constraints is a $(N, K) = (n^2, k^2) = (4096, 3249)$ code. The number of independent constraints is $M = 847$, but the sum-product decoder can make use of $2nm = 896$ equivalent constraints. Such codes have recently been named 'turbo product codes', but we think they should be called 'Tanner product codes', since they were first investigated by Tanner [13]. Product codes have the disadvantage, however, that their distance does not scale well with blocklength; the distance of a product code with blocklength n^2, built from two codes with distance d, is only d^2, so the ratio of distance to blocklength falls.

An open problem is to discover codes sharing the remarkable properties of the difference-set cyclic codes but with larger blocklengths and arbitrary rates. I call this task the Tanner challenge, in honour of Michael Tanner, who recognised the importance of such codes twenty years ago.

4.1. Notes on DSC codes.

Do the extra checks help? To confirm that the extra 191 redundant parity checks are responsible for the good performance of the DSC code, we tried decoding the code using only 82 of the parity checks. In case 82h, we took the first 82 rows of the cyclic parity check matrix; in case 82r we picked 82 random non-redundant rows from the matrix. This choice appears to make little difference. Either way, the performance is much worse than that of the code using the full $M = 273$ checks (Figure 7(b)). The random Gallager code with $j = 4$ performs slightly better than either of the crippled DSC codes.

Undetected errors. The (273,191) DSC code makes undetected errors. The frequency of these errors is shown in Figure 7(c).

Rescaling the log-probability messages. An *ad hoc* procedure found helpful by Tanner (personal communication) involves scaling down the log-probabilities by a 'fudge factor' f at the end of each iteration. We seize the message $a_{mn} \equiv \log \frac{q_{mn}^{(1)}}{q_{mn}^{(0)}}$ on its way from bit n to check m, and replace it by a_{mn}/f. Experimenting with a range of values of f, we find that values slightly greater than 1 reduce the error probability a little at large E_b/N_0. Figure 7(d) shows graphs for $f = 1$, $f = 1.25$, $f = 1.37$, $f = 1.50$, and $f = 2$. This fudge appears to reduce the frequency of detected errors and has little effect on the frequency of undetected errors, so that the error probability is dominated by undetected errors. We speculate that the fudged algorithm is indistinguishable from the optimal decoder for this DSC code, and the performance is only limited by the code's distance properties.

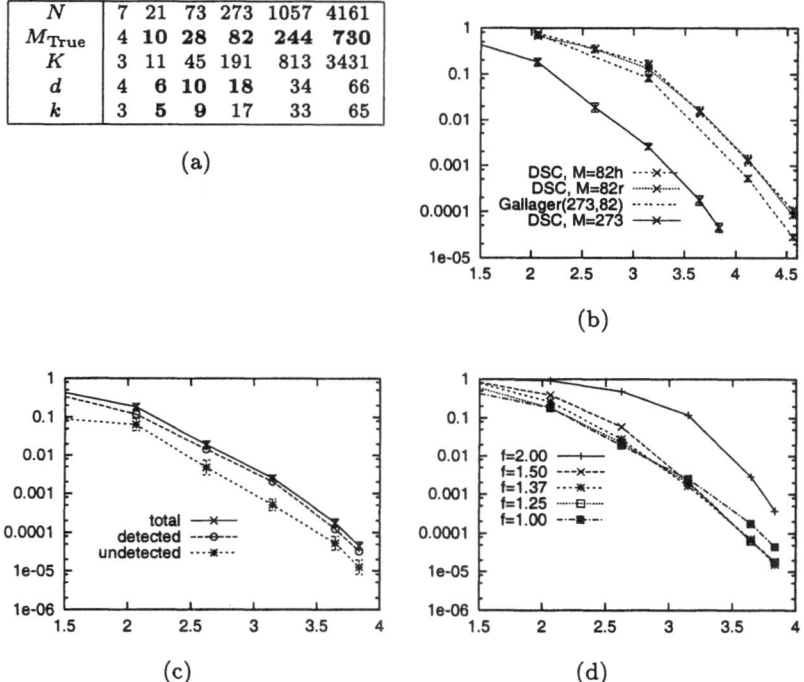

N	7	21	73	273	1057	4161
M_{True}	4	10	28	82	244	730
K	3	11	45	191	813	3431
d	4	6	10	18	34	66
k	3	5	9	17	33	65

(a)

(b)

(c)   (d)

FIG. 7. *Difference-set cyclic codes — low–density parity–check codes satisfying many redundant constraints — outperform equivalent Gallager codes.*

(a) The table shows the N, M_{True} (the number of independent rows in the parity check matrix, as opposed to the total number of rows M), K, distance d, and row weight k of some difference-set cyclic codes, highlighting the codes that have large d/N, small k, and large N/M. All DSC codes satisfy N constraints of weight k.

(b) In the comparison the Gallager code had $(j,k) = (4,13)$, and rate identical to the DSC codes. Vertical axis: block error probability; horizontal axis: $E_b/N_0/dB$.

(c) Decomposition of the DSC code's errors into detected and undetected errors.

(d) The error rate of the DSC code can be slightly reduced by using a 'fudge factor' of 1.25 or 1.37 during the sum-product decoding.

5. Discussion. Comparisons of Gallager codes with Reed-Solomon codes have been made before by Worthen and Stark [15]. They made a belief propagation decoder appropriate for bursty channels and achieved 3 dB performance gain over the Reed-Solomon code for rate 1/2 and block size of about 10000. Worthen and Stark attributed 2 dB of the gain to the use of soft decisions rather than hard decisions and 1 dB to code improvement. Using that reasoning, the gain of the shorter-blocklength Gallager codes over Reed-Solomon codes in our paper can be attributed entirely to using soft decisions. Our work makes a case for finding good short length codes that use soft decisions. Gallager codes over $GF(16)$ appear to be good candidates for this role.

The task of constructing Gallager codes with very short block lengths remains an interesting area for further research.

Acknowledgements. This work was supported by the Gatsby Foundation. DJCM thanks the researchers at the Sloane Center, Department of Physiology, University of California at San Francisco, for their generous hospitality, and Elywn Berlekamp, Michael Luby, Virginia de Sa, Bob McEliece, Emina Soljanin, Clifton Williamson, John Morris and Bernardo Rub for helpful discussions.

REFERENCES

[1] BERLEKAMP, E.R. (1968). *Algebraic Coding Theory*. New York: McGraw-Hill.

[2] DAVEY, M.C. AND MACKAY, D.J.C. (1998a). Low density parity check codes over GF(q). In *Proceedings of the 1998 IEEE Information Theory Workshop*, pp. 70–71. IEEE.

[3] DAVEY, M.C. AND MACKAY, D.J.C. (1998b). Low density parity check codes over GF(q). *IEEE Communications Letters* **2**(6):165–167.

[4] GALLAGER, R.G. (1962). Low density parity check codes. *IRE Trans. Info. Theory* **IT-8**:21–28.

[5] GALLAGER, R.G. (1963). *Low Density Parity Check Codes*. Number **21** in Research monograph series. Cambridge, Mass.: MIT Press.

[6] KARPLUS, K. AND KRIT, H. (1991). A semi–systolic decoder for the PDSC–73 error–correcting code. *Discrete Applied Mathematics* **33**:109–128.

[7] LUCAS, R., FOSSORIER, M., KOU, Y., AND LIN, S., (1999). Iterative decoding of one-step majority logic decodable codes based on belief propagation. Submitted.

[8] MACKAY, D.J.C. (1999a). Almost–certainly runlength–limiting codes. wol.ra.phy.cam.ac.uk/mackay.

[9] MACKAY, D.J.C. (1999b). Encyclopedia of sparse graph codes (hypertext archive). http://wol.ra.phy.cam.ac.uk/mackay/codes/data.html.

[10] MACKAY, D.J.C. (1999c). Good error correcting codes based on very sparse matrices. *IEEE Transactions on Information Theory* **45**(2):399–431.

[11] MACKAY, D.J.C. AND NEAL, R.M. (1996). Near Shannon limit performance of low density parity check codes. *Electronics Letters* **32**(18):1645–1646. Reprinted *Electronics Letters* **33**(6):457–458, March 1997.

[12] RICHARDSON, T. AND URBANKE, R. (1998). The capacity of low-density parity check codes under message-passing decoding. Submitted to IEEE Trans. on Information Theory.

[13] TANNER, R.M. (1981). A recursive approach to low complexity codes. *IEEE Transactions on Information Theory* **27**(5):533–547.

[14] URBANKE, R., RICHARDSON, T., AND SHOKROLLAHI, A. (1999). Design of provably good low density parity check codes. Submitted.

[15] WORTHEN, A. AND STARK, W. (1998). Low–density parity check codes for fading channels with memory. In *Proceedings of the 36th Allerton Conference on Communication, Control, and Computing, Sept. 1998*, pp. 117–125.

TWO SMALL GALLAGER CODES

DAVID J.C. MACKAY* AND MATTHEW C. DAVEY*

Abstract. We present a pair of Gallager codes with rate $R = 1/3$ and transmitted blocklength $N = 1920$ as candidates for the proposed international standard for cellular telephones.

A regular Gallager code [2] has a parity check matrix with uniform column weight j and uniform row weight k, both of which are very small compared to the blocklength. If the code has transmitted blocklength N and rate R then the parity check matrix $\mathbf{H}$ has N columns and M rows, where $M \geq N(1-R)$. [Normally parity check matrices have $M = N(1-R)$, but the matrices we construct may have a few redundant rows so that their rate could be a little higher than $1 - M/N$.]

It is easy to make good Gallager codes. We have found that for just about any blocklength N and rate R, a randomly chosen Gallager code with $j \simeq 3$ gives performance (in terms of word error probability on a Gaussian channel with signal to noise ratio E_b/N_0) that is within a fraction of a decibel of the best known codes [6]. Furthermore, with a little effort, irregular Gallager codes can be found which equal, or even exceed, the performance of what were the best known codes [1, 8].

We show in Figure 1 the performance of two rate $1/3$ codes with transmitted blocklength $N = 1920$ and source blocklength $K = 640$ that we constructed and tested within the space of one week. The first code is a nearly–regular code over GF(2) with column weights 3 and 2 and row weight 4 (Construction 2A from [6]). The second is an irregular code over GF(4) with a profile of column weights and row weights that was found by a brief Monte Carlo search. The irregular code was constructed according to the profile given in Figure 3, using the Poisson construction method described in [7]. The irregular code — which we expect could be further improved — is less than 0.2dB from the line showing the performance of the CCSDS turbo codes on the JPL code imperfectness web-page [3].

Gallager codes have advantages that these comparisons do not make evident. First, whereas turbo codes sometimes make undetected errors, Gallager codes are found to make only detected errors — all incorrectly decoded blocks are flagged by the decoder. Second, their decoding complexity is low — lower than that of turbo codes. The costs per bit per iteration are about 4 additions and 18 multiplies for the binary code, and 14 additions and 31 multiplies for the $GF(4)$ code. The expected number of iterations depends on the noise level, as we now describe.

---

*Department of Physics, University of Cambridge, Cavendish Laboratory, Madingley Road, Cambridge, CB3 0HE, United Kingdom; mackay|mcdavey@mrao.cam.ac.uk.

Bit error probability   Word error probability

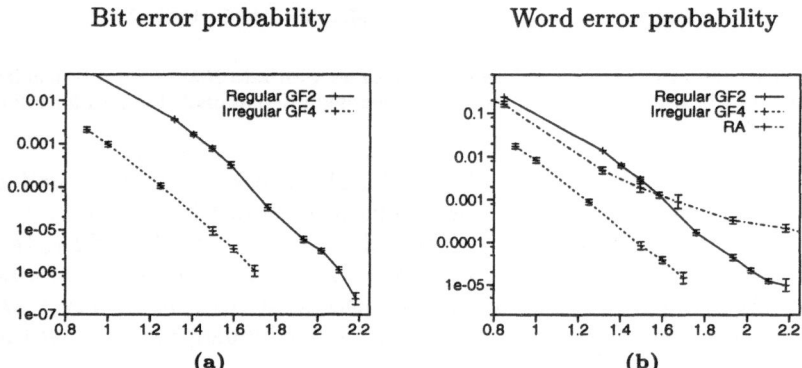

(a)   (b)

FIG. 1. *(a,b) Performance of Gallager codes with* $N = 1920$, $R = 1/3$, *as a function of* E_b/N_0. *In (b) we also show the performance of a repeat–accumulate code with* $N = 3000$.

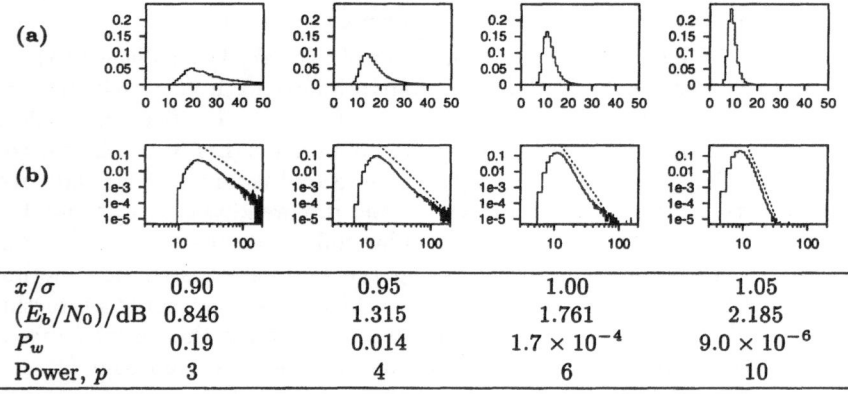

x/σ	0.90	0.95	1.00	1.05
$(E_b/N_0)/\mathrm{dB}$	0.846	1.315	1.761	2.185
P_w	0.19	0.014	1.7×10^{-4}	9.0×10^{-6}
Power, p	3	4	6	10

FIG. 2. *Histograms showing the frequency distribution of decoding times for the binary Gallager code from Figure 1: (a) linear plot; (b) log–log plot. The graphs show the number of iterations taken to reach a valid decoding; the value of* P_w *gives the frequency with which no valid decoding was reached after 1000 iterations. The power* p *which gives a good fit of the power law distribution* $P(\tau) \propto \tau^{-p}$ *(for large* τ*) is also shown.*

Decoding times follow power laws. We have previously noted [5, 7] that the probability distribution of the decoding time τ of Gallager codes and turbo–like codes appears to follow a power law, $P(\tau) \propto \tau^{-p}$, with the exponent p depending on the noise level and the code. Figure 2 shows the histograms of decoding times at four signal–to–noise ratios for the binary Gallager code shown in Figure 1.

The parity check matrix of the binary code, 1920.1280.3.303, can be found in the online archive [4]. We expect that these codes could be further improved by a careful optimization of their profiles.

Acknowledgements. This work was supported by the Gatsby Foundation. We thank Bob McEliece for helpful discussions.

Columns		Rows	
2	449	5	635
3	389	6	5
4	72		
5	1		
11	2		
13	4		
19	39		

FIG. 3. *The profile of the irregular code over GF(4), whose blocklength was* 960 *symbols.*

REFERENCES

[1] DAVEY, M.C. AND MACKAY, D.J.C. (1998). Low density parity check codes over GF(q). In *Proceedings of the 1998 IEEE Information Theory Workshop*, pp. 70–71. IEEE.

[2] GALLAGER, R.G. (1962). Low density parity check codes. *IRE Trans. Info. Theory*, **IT-8**: 21–28.

[3] JPL (1999). Code imperfectness.
http://www331.jpl.nasa.gov/public/imperfectness.html.

[4] MACKAY, D.J.C. (1999a). Encyclopedia of sparse graph codes (hypertext archive).
http://wol.ra.phy.cam.ac.uk/mackay/codes/data.html.

[5] MACKAY, D.J.C. (1999b). Gallager codes — recent results. In *Proceedings of International Symposium on Communication Theory and Applications, Ambleside, 1999*, ed. by M.D.B. Honary and P. Farrell. Research Studies Press.

[6] MACKAY, D.J.C. AND NEAL, R.M. (1996). Near Shannon limit performance of low density parity check codes. *Electronics Letters*, **32**(18):1645–1646. Reprinted *Electronics Letters*, **33**(6):457–458, March 1997.

[7] MACKAY, D.J.C., WILSON, S.T., AND DAVEY, M.C. (1999). Comparison of constructions of irregular Gallager codes. *IEEE Transactions on Communications*, **47** (10):1449–1454.

[8] URBANKE, R., RICHARDSON, T., AND SHOKROLLAHI, A. (1999). Design of provably good low density parity check codes. Submitted.

MILDLY NON-LINEAR CODES*

ALAN PARKS†

Abstract. We consider codes coming from systems of sparse linear equations. (These low-density parity check codes are examples of what are called Gallager codes.) We suggest how non-linear equations very close to the given linear equations might be used to improve decoding properties while retaining the same level of coding and decoding complexity.

1. Linear and single product predicates. We will write the coordinates of an element y of the set $\mathbb{Z}_2^n$ of n-bit vectors as $y[j]$ for $1 \leq j \leq n$. An *n-bit predicate* is a function f taking $\mathbb{Z}_2^n$ to the set $\{0,1\}$, which latter set is regarded as real because of wanting fourier coefficients. This describes both the setting where the equation $f(y) = 1$ defines a code and the case where this equation constitutes one check equation among many. In the former case, f is obviously the product of predicate functions for single check equations.

A *linear n-bit predicate* $f : \mathbb{Z}_2^n \to \{0,1\}$ comes about from a subset Q of the numbers $1, 2, \ldots, n$, where $f(y) = 1$ if and only if

$$\sum_{j \in Q} y[j] = 0 \mod 2.$$

To obtain a "mildly non-linear" perturbation of f, choose $S \subseteq Q$, and define $g(y) = 1$ if and only if

$$\prod_{j \in S} y[j] + \sum_{j \in Q} y[j] = 0.$$

We call g a *single product predicate*. If S is "small," then we think of g as "mildly" non-linear.

We are interested in comparing single product predicates to their linear counterparts. To facilitate the comparison, we will consider the following set-up, designed to keep the code rate the same for both types of predicates and to keep the coding and decoding complexity comparable.

We begin with a regular, low-density parity check matrix, of the kind considered in [1] and constructed in [2]. (Our ideas were motivated by the construction techniques of the latter paper.) We suppose that this matrix is $m \times n$ and that it has rank m (full row rank). Then the null space of the matrix forms a code in which the code constraints are n-bit linear predicates f_i for $1 \leq i \leq m$. The code rate is $(n - m)/m$.

*Much of this work was done under the auspices of Lawrence University's sabbatical program.

†Department of Mathematics, Lawrence University, Appleton, WI 54912. Email: parksa@lawrence.edu.

Recall that coding can be accomplished if we have a reduced row-echelon version of the parity check matrix. Given some particular elimination, let M be the set of pivot bit positions and let F be the set of free bit positions. Then, of course, $|M| = m$ and $|F| = n - m$, and the bit positions of F are the *information bits*. If bits $y[j]$ are given, for $j \in F$, then there are unique bits $y[i]$ for $i \in M$ such that y is a code vector.

We modify the code predicates as follows. Temporarily, we will need to identify the subset Q_i of $1, 2, \ldots, n$ such that $f_i(y) = 1$ if and only if $\sum_{i \in Q_i} y[i] = 0 \pmod 2$. For each f_i, choose a "small" subset S_i of $Q_i \backslash M$. Define

$$\pi_i(y) = \prod_{k \in S_i} y[k] \qquad \text{for} \qquad 1 \leq i \leq m$$

and define modified predicates f_i' where $f_i'(y) = 1$ if and only if

$$\pi(y) + \sum_{i \in Q_i} y[i] = 0 \qquad \text{mod 2, for} \qquad 1 \leq i \leq m.$$

Thus, the constraints f_i' are single product predicates formed by suffixing a product term to the mod 2 equation involved with f_i. Notice that no bit position from M is involved in the product part of the f_i'.

We can regard the product term π_i as a new variable $y[n + i]$, so that the modified code constraints are linear in $y[1], \ldots, y[n + m]$. The same elimination that brought the original system of constraints into row-echelon form will bring this modified system into row-echelon form and with the same pivots M. The information vectors are also the same in both codes, involving bit positions $y[j]$ with $j \in F$. Indeed, in the modified system, these values determine the product variables $y[n+i]$ for $1 \leq i \leq m$, and then the row-echelon form of the modified system determines the pivot variables $y[k]$ for $k \in M$. This proves that the rates of the linear code and the modified code are equal, and it shows that encoding in the modified code is comparable to that in the linear code. (In the modified code, the products must be computed, requiring a number of arithmetic operations proportional to m.)

We could also consider codes coming from non-homogeneous linear equations. If we restrict, as we did above, to check matrices with full row rank, and if we modify as we did above, then the modified codes still have the same rate and coding complexity as the linear codes they come from. We will stay with homogeneous codes to simplify the exposition somewhat.

In Sections 4 and 5 we describe theoretical and empirical comparisons made between linear and modified codes; in Sections 2 and 3, we build up the notation and results we need to describe those comparisons. The theoretical result is encouraging; there is improved probability of error correction when a modified code is used in place of its linear ancestor. The empirical results are somewhat weak, but since we have not attempted a definitive experiment, a thorough follow-up seems in order.

2. Bias and bias leverage. We have an n-bit code predicate f, and, as usual, we imagine a transmitted vector $y \in \mathbb{Z}_2^n$, such that $f(y) = 1$, incurring a noise vector z with (known or unknown) probability distribution P, resulting in a received vector $x = y + z$ (mod 2). Decoding algorithms employ the equation $f(x + z) = 1$ to make probabilistic statements about the bits in z. For a given bit position j, the *bias toward $z[j]$* is the ratio of the probability that $z[j] = 1$ with the probability that $z[j] = 0$. Given that x is received, this bias is equal to

$$\frac{\text{prob}(z[j] = 1 \mid x)}{\text{prob}(z[j] = 0 \mid x)} = \frac{\sum_{z,z[j]=1} P(z) \cdot f(x + z)}{\sum_{z,z[j]=0} P(z) \cdot f(x + z)} .$$

It will be convenient to define the *bias leverage $B_j(x)$* to be the ratio of the bias just given and the *apriori bias* — the bias on the noise in general. Thus,

$$(2.1) \qquad B_j(x) = \frac{\sum_{z,z[j]=1} P(z) \cdot f(x + z)}{\sum_{z,z[j]=0} P(z) \cdot f(x + z)} \cdot \frac{\sum_{z,z[j]=0} P(z)}{\sum_{z,z[j]=1} P(z)}$$

(N.B. In the notation $B_j(x)$, the predicate f and the distribution P are assumed.)

If the noise distribution, restricted to the event that x is received, were a product distribution on the bit positions, then the bias from the entire code predicate would tell us what $z[j]$ is, or the bias would identify an essential ambiguity in that bit. In practice, the biases from individual check equations are used to approximate the bias from the entire code predicate. For instance, the event that the bias leverage, from an individual check equation, is greater than 1 is used both to design decoders employing voting schemes (if a received bit gets enough "votes" it is flipped) and as a tool to estimate the effect of message passing algorithms that keep track of the estimated bias on each noise bit.

We show how to use the fourier coefficients of linear and single product predicates to calculate the leverage in cases of interest. For the predicates we will consider, very few of the fourier coefficients will be non-zero, and so the use of the coefficients is efficient numerically. Let f be an n-bit linear predicate, defined by the subset Q of $1, 2, \ldots, n$. We write y_Q to stand for the vector of bit positions from Q; in other words $y_Q : Q \to \mathbb{Z}_2$ where $y_Q[j] = y[j]$ for $j \in Q$. We also write $|y|$ for the *weight* of the vector y (the number of non-zero coordinates). This allows us to write f in the following form:

$$(2.2) \qquad f(y) = \frac{1}{2} \cdot \left[1 + (-1)^{|y_Q|} \right] .$$

The fourier coefficients of f are obvious from this form: they are 0 except for being $1/2$ at the zero vector and at the vector having ones in the positions of Q.

To obtain the fourier coefficients for a single product predicate, we will need the following notation. Assume that $S \subseteq Q$, and for $w : S \to \mathbb{Z}_2$ define $\overline{w} \in \mathbb{Z}_2^n$ where

$$\overline{w}[j] = \begin{cases} w[j] & \text{if } j \in S \\ 1 & \text{if } j \in Q - S \\ 0 & \text{if } j \notin Q \end{cases} .$$

We will also need the particular function $u : S \to \mathbb{Z}_2$, where $u[j] = 1$ for all $j \in S$. The straightforward proof of the following is left to the reader.

PROPOSITION 2.1. *Assume all the notation just given, and let f be the single product predicate defined by S and Q. Let $s = |S|$. Then, for $v \in \mathbb{Z}_2^n$, we have fourier coefficients*

$$f_v = \begin{cases} \frac{1}{2} & \text{if } v = 0 \\ cr2^{-s} \cdot (-1)^{s-1+|w|} & \text{if } v = \overline{w} \text{ where } w : S \to \mathbb{Z}_2 \text{ and } w \neq u \\ \frac{1}{2} - 2^{-s} & \text{if } v = \overline{u} \\ 0 & \text{otherwise} \end{cases} .$$

We will also employ the fourier coefficients of probability distributions on (noise) bit-vectors. Our distributions $P : \mathbb{Z}_2^n \to \mathbb{R}$ will always come about, by hypothesis or as an approximation, as product distributions over the bit positions. For position j, we will write $P_j(a)$ for the probability that bit j has value $a \in \mathbb{Z}_2$. For a product probability distribution P on $\mathbb{Z}_2^n$, and for a subset Q of $1, 2, \ldots, n$, define

$$\Pi(Q, P) = \prod_{j \in Q} (1 - 2 \cdot P_j(1)) .$$

The following is well known; we include a proof for the sake of completeness.

PROPOSITION 2.2. *Let P be a product probability distribution over the bit positions $1 \leq j \leq n$. Let Q be a subset of $1, 2, \ldots, n$, and let v be the n-bit vector having $v[j] = 1$ if and only if $j \in Q$. Then $2^{-n} \cdot \Pi(Q, P)$ is the fourier coefficient P_v of P with respect to v.*

Proof. We have

$$2^n \cdot P_v = \sum_z P(z) \cdot (-1)^{v \cdot z} .$$

We can view the sum as the expected value of

$$(-1)^{v \cdot z} = \prod_{j=1}^n (-1)^{v[j] \cdot z[j]} .$$

Since P is a product distribution over the bit positions, this product random variable has expected value equal to the product of the expected values of each of its factors $(-1)^{v[j] \cdot z[j]}$. This latter expected value is

$$P_j(0) \cdot (-1)^{v[j] \cdot 0} + P_j(1) \cdot (-1)^{v[j] \cdot 1} = P_j(0) + P_j(1) \cdot (-1)^{v[j]} .$$

If $v[j] = 0$, then the expected value of its factor is 1. If $v[j] = 1$, the expected value is $1 - 2P_j(1)$. □

In order to calculate the bias leverage in the situations of interest to us, we will need two convolution formulas.

PROPOSITION 2.3. *Let the set Q define the n-bit linear predicate f. Let the probability distribution P on $\mathbb{Z}_2^n$ be a product distribution. If $x \in \mathbb{Z}_2^n$, then*

$$\sum_{z \in \mathbb{Z}_2^n} P(z) \cdot f(z + x) = \frac{1}{2} \left[1 + (-1)^{|x_Q|} \cdot \Pi(Q, P) \right] .$$

Proof. Using the fourier series for P and for f, we have that

$$\sum_{z \in \mathbb{Z}_2^n} P(z) \cdot f(x + z) = 2^n \cdot \sum_{v \in \mathbb{Z}_2^n} P_v \cdot f_v \cdot (-1)^{v \cdot x} .$$

The result follows from the formula for f_v and from Proposition 2.2. □

A proof of the following more complicated result is given in the Appendix. For $x \in \mathbb{Z}_2^n$, the set S, and the distribution P, we define

$$\Pi_S(x, P) = \prod_{x_S[j]=1} P_j(0) \cdot \prod_{x_S[j]=0} (-P_j(1)) .$$

In each of the products on the right, we consider only $j \in S$.

PROPOSITION 2.4. *Let the sets $S \subseteq Q$ define the n-bit single product predicate f. Let P be a product probability distribution on $\mathbb{Z}_2^n$. If $x \in \mathbb{Z}_2^n$, then*

$$\sum_{z \in \mathbb{Z}_2^n} P(z) \cdot f(z+x) = \frac{1}{2} + (-1)^{|x_Q|} \cdot \Pi(Q \backslash S, P) \cdot \left[\frac{1}{2} \cdot \Pi(S, P) - \Pi_S(x, P) \right] .$$

We will also record the following triviality.

PROPOSITION 2.5. *Let f be an n-bit predicate, and let P be a probability distribution on $\mathbb{Z}_2^n$. If $x \in \mathbb{Z}_2^n$, then*

$$\sum_{z \in \mathbb{Z}_2^n} P(z) \cdot (1 - f(z + x)) = 1 - \sum_{z \in \mathbb{Z}_2^n} P(z) \cdot f(z + x) .$$

To connect these ideas, we find the bias leverage in the case of a linear predicate and in the case of a single product predicate. These formulas have relevance to the more interesting case of a predicate that is a product of linear or single product predicates.

PROPOSITION 2.6. *Let the set Q define the n-bit linear predicate f. Let the probability distribution P on $\mathbb{Z}_2^n$ be a product distribution, and assume that $P_j(1) < 1/2$ for each j. Let $x \in \mathbb{Z}_2^n$. If $j \notin Q$, then $B_j(x) = 1$. If $j \in Q$, then*

$$B_j(x) = \frac{1 - (-1)^{|x_Q|} \cdot \Pi(Q\backslash\{j\}, P)}{1 + (-1)^{|x_Q|} \cdot \Pi(Q\backslash\{j\}, P)} .$$

Furthermore, in this case we have $B_j(x) > 1$ if and only if $f(x) = 0$.

Proof. For $z \in \mathbb{Z}_2^n$, define $z' \in \mathbb{Z}_2^{n-1}$ by deleting the j-th bit from z. Then $P(z) = P_j(z[j]) \cdot P(z')$, where $P(z')$ evaluates the distribution on $\mathbb{Z}_2^{n-1}$ obtained from P in the obvious way.

If $j \notin Q$, then since bit j does not enter into the calculation of f, we view f as a function on $\mathbb{Z}_2^{n-1}$, and then $f(x + z) = f(x' + z')$ for all $x, z \in \mathbb{Z}_2^n$. It follows, for each $a \in \mathbb{Z}_2$, that

$$\sum_{z[j]=a} P(z) \cdot f(z + x) = P_j(a) \cdot \sum_{z' \in \mathbb{Z}_2^{n-1}} P(z') \cdot f(x' + z') .$$

Equation (2.2) then yields that $B_j(x) = 1$ in this case.

In the case $j \in Q$, the formula for B_j is Theorem 4.1 of [1]. In this case, $B_j(x) > 1$ if and only if

$$(-1)^{|x_Q|} \cdot \Pi(Q\backslash\{j\}, P) < 0 .$$

Since each $P_j(1) < 1/2$, this condition is that $|x_Q| \equiv 1 \mod 2$, in other words, $f(x) = 0$. □

Proposition 2.6 simply explains, in terms of leverage, how a linear constraint furnishes a check equation on a received vector. Next, we compute $B_j(x)$ in the case of a single product predicate when bit position j is not involved in the product, and we will observe this same property. The formula in Proposition 2.7 is a perturbed version of that of Proposition 2.6. When we use these formulas to study decoding, the perturbed version will show an apparent advantage over the leverage in the linear case.

In the following proposition there is a condition that the bit probabilities are "not too large"; this will cause no difficulty in practice.

PROPOSITION 2.7. *Let the sets $S \subseteq Q$ define the n-bit single product predicate f. Let P be a product probability distribution on $\mathbb{Z}_2^n$. Let p be the maximum of the $P_j(1)$, and assume both that $p < 1/2$ and that $2p < (1 - 2p)^{|S|}$. Let $x \in \mathbb{Z}_2^n$. If $j \notin Q$, then $B_j(x) = 1$. For $j \in Q - S$, let*

$$X_j(x) = (-1)^{|x_Q|} \cdot \Pi(Q\backslash S \cup \{j\}, P) \cdot \left[\frac{1}{2} \cdot \Pi(S, P) - \Pi_S(x, P)\right] .$$

Then

$$B_j(x) = \frac{1/2 - X_j(x)}{1/2 + X_j(x)} \cdot$$

Furthermore, we have $B_j(x) > 1$ if and only if $f(x) = 0$.

Proof. The case $j \notin Q$ proceeds exactly as in the proof of Proposition 2.6. Let $j \in Q - S$, and let $Q' = Q - \{j\}$. Let $a \in \mathbb{Z}_2$. If $z, x \in \mathbb{Z}_2^n$ and $z[j] = x[j] = a$, then $f(x + z) = 1$ if and only if

$$\prod_{k \in S} (x[k] + z[k]) + \sum_{j \in Q'} (x[k] + z[k]) = 0 \ .$$

Let g be the single product predicate arising from this equation. As before, define z' by deleting coordinate j. Put

$$C = \sum_{z[j]=a} P(z) \cdot f(x + z) = P_j(a) \cdot \sum_{z' \in \mathbb{Z}_2^{n-1}} P(z') \cdot g(x' + z') \ .$$

Proposition 2.4 then gives

$$\frac{C}{P_j(a)} = \frac{1}{2} + (-1)^{|x_{Q'}|} \cdot \Pi(Q' \backslash S, P) \cdot \left[\frac{1}{2} \cdot \Pi(S, P) - \Pi_S(x, P) \right] \ .$$

Since $a = x[j]$, we can replace the exponent $|x_{Q'}|$ by $a + x[j] + |x_{Q'}|$ without changing the value of the expression. Notice that $x[j] + |x_{Q'}| \equiv |x_Q|$ mod 2, and we have that

(2.3)
$$\frac{C}{P_j(a)} = \frac{1}{2} + (-1)^a \cdot X_j(x) \ .$$

We were assuming that $x[j] = a = z[j]$. If $x[j] \neq a = z[j]$, then $f(z + x) = 1$ if and only if

$$\prod_{k \in S} (x[k] + z[k]) + \sum_{j \in Q'} (x[k] + z[k]) = 1 \ .$$

Let g be the single product predicate arising from the left side of this equation equalling 0, and then $f(z + x) = 1$ if and only if $g(z' + x') = 0$. Let

$$C = \sum_{z[j]=a} P(z) \cdot f(x + z) = P_j(a) \cdot \sum_{z' \in \mathbb{Z}_2^{n-1}} P(z') \cdot (1 - g(x' + z')) \ .$$

Dividing by $P_j(a)$ and using Proposition 2.5 and Proposition 2.4 yields

$$\frac{C}{P_j(a)} = 1 - \frac{1}{2} - (-1)^{|x_{Q'}|} \cdot \Pi(Q' \backslash S, P) \cdot \left[\frac{1}{2} \cdot \Pi(S, P) - \Pi_S(x, P) \right] \ .$$

Since $x[j] \neq a$, we can replace the exponent $|x_{Q'}|$ by $a + x[j] + |x_{Q'}|$ at the expense of cancelling the minus sign that precedes this term. We obtain

$$\frac{C}{P_j(a)} = \frac{1}{2} + (-1)^a \cdot X_j(x) \ .$$

This is identical to (2.3); in other words, whether $x[j]$ is equal to a or not, equation (2.3) holds. Equation (2.1) now establishes the claimed formula for $B_j(x)$.

Continuing the case $j \in Q \backslash S$, we see that $B_j(x) > 1$ if and only if $X_j(x) < 0$. Recall the definition of p as the maximum $P_k(1)$.

There are two cases to consider. Suppose first that x_S is not all 1's; recall the definition of Π_S, and we claim that

$$\frac{1}{2} \cdot \Pi(S, P) - \Pi_S(x, P) > 0 \ .$$

Indeed, since there is $k \in S$ such that $x[k] = 0$, we estimate

$$\prod_{x_S[k]=0} (-P_k(1)) \leq p \quad \text{and} \quad \prod_{x_S[k]=1} P_k(0) \leq 1 \quad \text{and} \quad \Pi(S, P) \geq (1 - 2p)^{|S|} \ .$$

By hypothesis, the rightmost term is greater than $2p$. Thus,

$$\frac{1}{2} \cdot \Pi(S, P) - \Pi_S(x, P) > \frac{1}{2} \cdot 2p - p = 0 \ .$$

as claimed. Turning to the $\Pi(Q' \backslash S, P)$ factor in X, since $p < 1/2$, this factor is positive. We conclude that $X_j(x) < 0$ in this case if and only if $|x_Q| \equiv 1 \bmod 2$. Since x_S is not all 1's, the product term in f is 0; since $|x_Q| \equiv 1$, we see that $f(x) = 0$.

We have left to consider the case that x_S is all 1's. Compute

$$(2.4) \qquad \frac{1}{2} \cdot \Pi(S, P) - \Pi_S(x, P) = \frac{1}{2} \cdot \Pi(S, P) - \prod_{k \in S} P_k(0) \ .$$

Each factor $1 - 2 \cdot P_k(1)$ in $\Pi(S, P)$ is positive and no greater than the corresponding factor $P_k(0) = 1 - P_k(1)$ in the other product. It follows that the expression in (2.4) is negative. Thus, $X_j(x) < 0$ if and only if $|x_Q| \equiv 0 \bmod 2$. Since x_S is all 1's, this is that $f(x) = 0$. $\qquad \square$

The formula for $B_j(x)$ in Proposition 2.7 shows that it can be calculated numerically using a number of arithmetic operations proportional to $|Q|$. This is at the same level of complexity as the formula in the linear case in Proposition 2.6.

3. Hub bits. In most coding paradigms, the bias leverage (or some equivalent quantity) at a particular bit position is calculated for the set of constraints that involve that bit. We will say that a bit position is a *hub* if the constraints involving it have no other bit positions in common. Hub bits prevent short cycles from occurring in the dependence graph of a linear system, and so their desirability is fairly obvious. This idea goes back at least as far as [1] and has been used in many other papers.

We write $V(f)$ for the set of bit positions involved in $f : \mathbb{Z}_2^n \to \mathbb{R}$ (the set of bit positions in which f is not constant). Then, say for bit position 0, we have code predicates f_j for $1 \leq j \leq r$, such that $0 \in V(f_j)$ but otherwise the distinct $V(f_j)$ are disjoint. Let n be the total number of bit positions in the $V(f_j)$, and, for the moment, we will ignore any other bit positions there are (coming from constraints not involving bit position 0). The code predicate for bit position 0 is the product of the f_j.

For $x \in \mathbb{Z}_2^n$, we write $x = (x[0], x_1, \ldots, x_r)$ where $(x[0], x_j)$ hold the bit positions in $V(f_j)$. We regard f_j as a function of $(x[0], x_j)$. Because position 0 is a hub, there is an obvious correspondence between the set of $x \in \mathbb{Z}_2^n$ and tuples of the type just written.

Bringing in the probability distribution P, we write

$$P(x) = P_0(x[1]) \cdot P(x_1) \cdots P(x_r)$$

where $P(x_j)$ uses the correct probabilities for the bit positions held by x_j.

Proposition 2.6 and Proposition 2.7 used the convolution formulas in Proposition 2.3 and Proposition 2.4 to calculate the bias leverage for a single constraint equation. The notation we have just established allows us to write a formula for the numerator and denominator of the leverage in the current setting; each of these will be seen to be a product of convolutions of the type considered in Proposition 2.3 and Proposition 2.4, since, if each predicate f_j is linear or single product, then the predicate $f_j(a + x[0], z_j + x_j)$, where $a+x[0]$ is given, is either linear, single product, or the negation of one of those types. The proof of the following is obvious from the notation above.

PROPOSITION 3.1. *Given the notation above, let $a \in \mathbb{Z}_2$ and $x \in \mathbb{Z}_2^n$. Then*

$$\sum_{z \in \mathbb{Z}_2^n, z[0]=a} P(z) \cdot \prod_{j=1}^r f_j(z + x)$$

is the product of $P_0(a)$ and

$$\prod_{j=1}^r \left[\sum_{z_j} P(z_j) \cdot f_j(a + x[0], z_j + x_j) \right].$$

4. Comparing linear and single product constraints. Using Proposition 3.1, we compared the bias leverage at given hub bit positions for linear constraints to the leverage for single product constraints. Empirical results are summarized in Table 1. Each line of each table is parameterized by q (the number of bit positions in each linear or single product constraint), r (the number of constraints on the given hub bit), and the weight of the noise vector. The single product predicates had product terms with 2 factors (so $|S| = 2$ in the notation used above). In each line of the table, we randomly chose 1000 pairs (code vector, noise vector); each vector has $(q - 1) \cdot r + 1$ bit positions, of which $(q - 2) \cdot r + 1$ are free and determine r pivot bits. We set $P_j(1) = 0.01$ uniformly for the sake of experiment. Of course, this models the binary symmetric channel; the particular choice of $P_j(1)$ does not seem to matter much.

TABLE 1
Bias leverage.

q,r,weight	Non-lin%	Lin.%	Ave.non-lin.B	Ave.lin.B
6,3,0	1.00	1.00	1.7e+04	7.8e+03
6,3,1	1.00	1.00	1.7e+04	7.8e+03
10,5,1	1.00	1.00	3.0e+05	1.6e+05
6,3,2	1.00	1.00	3.5e+03	2.0e+01
10,5,2	1.00	1.00	4.5e+04	1.3e+03
6,3,2	0.56	0.29	3.7e+03	2.3e+03
10,5,2	1.00	1.00	4.4e+04	2.6e+04
14,7,2	1.00	1.00	3.3e+05	2.0e+05
6,3,3	0.55	0.29	3.7e+03	2.2e+03
10,5,3	1.00	1.00	4.5e+04	2.8e+04
14,7,3	1.00	1.00	3.2e+05	1.9e+05
6,3,3	0.61	0.75	2.2e+03	1.5e+01
10,5,3	0.63	0.49	1.6e+04	6.5e+02
14,7,3	1.00	1.00	7.3e+04	1.0e+04
6,3,4	0.55	0.72	2.1e+03	1.4e+01
10,5,4	0.65	0.51	2.2e+04	6.8e+02
14,7,4	1.00	1.00	8.9e+04	9.6e+03

When $z[j] = 1$, larger values of $B_j(x)$ are good, since they increase the (estimated) probability that we will conclude from the received vector that $z[j] = 1$. When $z[j] = 0$, we want small values of $B_j(x)$ to decrease the same estimated probability. In order to avoid distinguishing these two cases, we recorded the reciprocal of $B_j(x)$ in the case $z[j] = 0$. Thus, the larger value is the "winner" in each case, and in every case the single product predicates win. We also recorded the percentage of code, noise pairs for which the bias leverage correctly estimates whether $z[j] = 1$. (If $z[j]$ is, in fact 1, then the bias leverage should be greater than 1; if $z[j]$ is 0, it should be less than 1.) In all lines of the table except for one, the leverage for the single product predicates estimate $z[j]$ correctly at least as often as do the linear predicates.

We also compared the two types of predicates more directly in terms of decoding characteristics. We considered a standard decoding algorithm which flips bits in an input (received) vector according to "votes" by the constraints that estimate the likelihood of noise in those bits. We imagine a set F of code predicate constraints. Recall the notation $V(f)$ for the set of bit positions in which the function f is not constant.

Algorithm: Given the input n-bit vector x, produce the output n-bit vector x', as follows. For each k with $1 \le k \le n$, if it is the case that the bias leverage is greater than 1 for every code predicate $f \in F$ such that $k \in V(f)$, then set $x'[k] = 1 - x[k]$; otherwise, set $x'[k] = x[k]$.

A simple decoder may be obtained by iterating the Algorithm, stopping when the output vector is a code vector or when a failed state occurs at a maximum number of iterations.

One way to estimate the performance of this algorithm is to compute the expected change in the weight of the noise vector. Going back to the transmitted vector y and the noise vector z, we have received vector $x = y + z$ (mod 2). If input x to the Algorithm produces output x', set $z' = y + x'$, and we can regard z' as the updated noise vector, the Algorithm as replacing (x, z) by (x', z'). (Of course, z and z' are not observed.) The change in the weight of the noise vector is $|z| - |z'|$; if this quantity is negative, then the weight of z' is less than that of z, and x' is a more accurate approximation to the transmitted vector y. The expected change in weight may be computed over the set of pairs (x, z) of received and noise vectors, whose probability distribution comes about from the probability distribution P of the noise, along with the assumption that each code vector is equally likely.

We will approximate the expected change by limiting, as we did in tabulating the bias leverage, to a set of predicates involved in the decision whether to flip some particular hub bit position — say position 0. Invoking the notation used before, let f_j with $1 \le j \le r$ be the code predicates with $0 \in V(f_j)$. For a received vector x, only $(x[0], x_j)$ is necessary to compute $f_j(x[0], x_j)$ or to compute the bias leverage for this function. Let $g_j(x[0], x_j)$ be the predicate that tells whether the bias leverage toward bit position 0 and involving f_j is greater than 1. Proposition 2.6 and Proposition 2.7 show that the predicate $g_j(x[0], x_j)$, as a function of x_j, is linear, single product, or the negation of one of those types.

Suppose there are K vectors $(y[0], y_1, \ldots, y_r)$ such that $f_j(y[0], y_j) = 1$ for all $1 \le j \le r$. If P is the distribution of the noise, as usual, then the (received, noise) pair x, z has probability

$$\frac{1}{K} \cdot P(z) \cdot \prod_{j=1}^{r} f_j(x+z) = \frac{1}{K} \cdot P(z[0]) \cdot \prod_{j=1}^{r} [P(z_j) \cdot f_j(x[0]+z[0], x_j+z_j)] \, .$$

For a given x, the event that the Algorithm flips bit 0 is that $g_j(x[0], x_j)$ is equal to 1 for all $1 \leq j \leq r$. Under these circumstances, if $z[0] = 1$, then the pair (x, z) contributes a -1 to the change in the noise weight (bit 0 is correctly flipped by the Algorithm); whereas, if $z[0] = 0$, then the pair contributes a +1 to the change in noise weight (bit 0 is incorrectly flipped). In both cases the contribution is $(-1)^{z[0]}$. Thus, the expected change μ_0 in bit 0 is

$$\sum_{x,z}(-1)^{z[0]} \cdot \frac{1}{K} \cdot P_0(z[0]) \cdot \prod_{j=1}^{r} [P(z_j) \cdot f_j(x[0]+z[0], x_j+z_j) \cdot g_j(x[0], x_j)]$$

and if, for $a, b \in \mathbb{Z}_2$, we write

$$C(a, b) = \prod_{j=1}^{r} \left[\sum_{z_j, x_j} P(z_j) \cdot f_j(a+b, z_j+x_j) \cdot g_j(b, x_j) \right]$$

then we have

$$\mu_0 = \sum_{a,b \in \mathbb{Z}_2} (-1)^a \cdot \frac{1}{K} \cdot P_0(a) \cdot C(a, b) \ .$$

The factor $C(a, b)$ is the product of convolutions at the zero vector. If the code constraints are linear or single product, then, as we have remarked above, $f_j(a + b, z_j + x_j)$ as a function of $z_j + x_j$ is linear, single product, or the negation of one of these. Then Proposition 2.6 and Proposition 2.7 show that the predicate $g_j(b, x_j)$ as a function of x_j is of the same type. We know the fourier coefficients of these functions and of the distribution P, and this shows that μ_0 can be calculated efficiently.

We have graphed μ_0, under the set of assumptions we used in constructing the tables of leverages. Bit 0 is a hub bit, involved in r predicates, each of which involves position 0 and the same number $q - 1$ of other positions. Single product predicates always had two factors. We compared three different cases.

- **Case 1. Pure linear.**   Each predicate is linear.
- **Case 2. Linear part of single product.**   Each predicate is a single product predicate; position 0 is not in the product part of any of these predicates.
- **Case 3.   One product ignored.**   Each predicate is a single product predicate; position 0 is in the product part of at most one of these predicates, in which case this predicate is ignored in the voting part of the Algorithm.

We have graphed μ_0 as a function of $P_0(1)$ (the crossover probability) in these cases, for various choices of q, r. The definition of μ_0 shows that the more negative it is, the better. Cases 2 and 3, involving single product predicates, frequently post values of μ_0 less than that for Case 1, giving at least preliminary evidence that single product predicates may be capable of providing improved decoding. The μ_0 curve for Case 1 is a solid line, Case 2 a dotted line, Case 3 a dashed line. The units on the vertical axis are multiples of K (the number of code vectors).

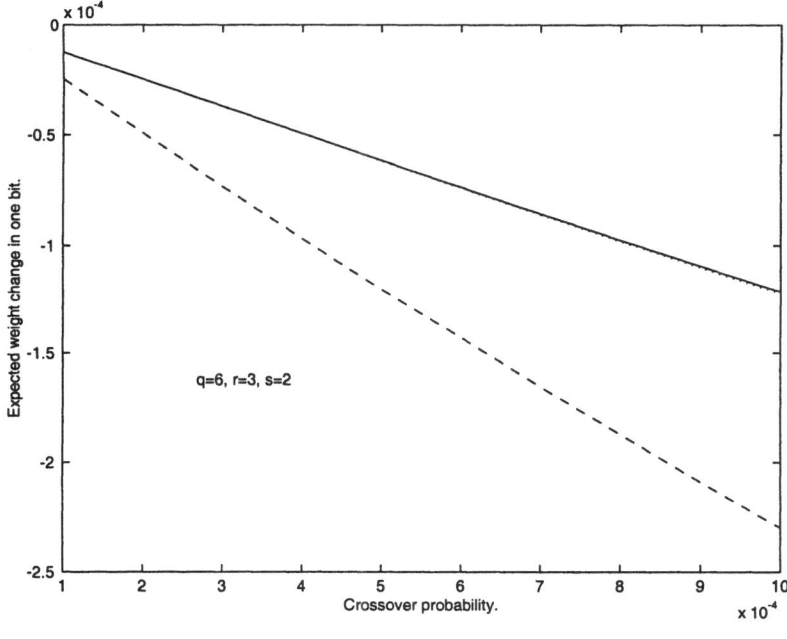

FIG. 1.

In all cases, Case 2 is no worse than Case 1, showing that code modification does not hinder decoding at bit positions not involved in the product terms used. Notice that the same crossover interval is considered in the second and third graphs, with differing q, r.

We have also performed random sampling to estimate the standard deviation of μ_0 in each case; apparently the deviations in Cases 2 and 3 are less than the deviations for strictly linear codes. This did not seem telling one way or the other.

5. Implementation experiments. Here is an explicit implementation of the foregoing ideas.

We ran computer tests on modified codes along with the linear code which, in each case, was used to construct the modified version. Our linear codes were constructed by a variation of the techniques discussed in [2]; to construct an $m \times n$ parity check matrix, we utilize parameters q, r, s where q is the number of bit positions in each linear constraint, r is the number of constraints involving each given bit position. If we count the number of ones in the matrix, we see that $m \cdot q = n \cdot r$. We want the rank of the

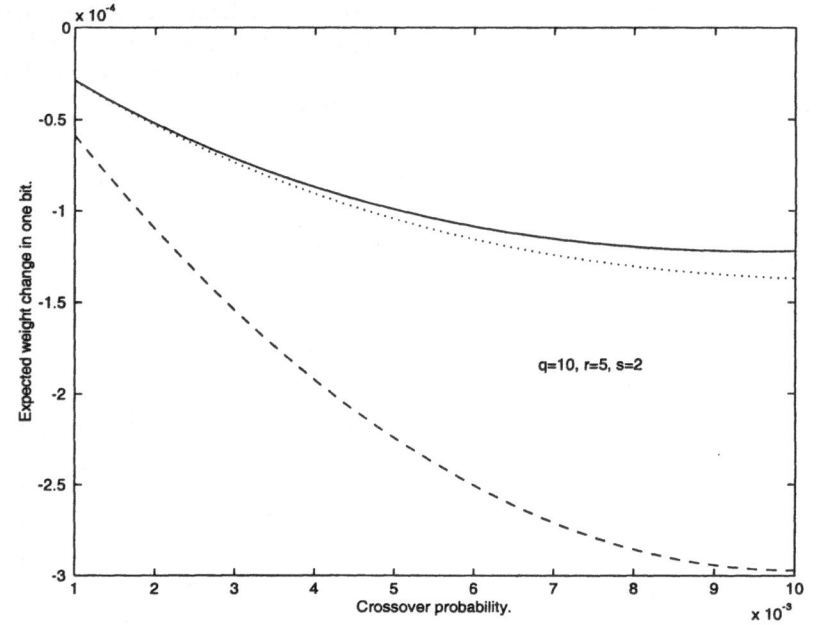

FIG. 2.

matrix to be m, and the code rate will then be $(n - m)/m = (q - r)/r$. The parameter s is the number of terms used in each product term of each single product predicate.

We formed random, square invertible blocks, making sure that each column of each block contained r ones and each row less than q ones. We placed the blocks in diagonal fashion in the parity check matrix. Suppose that a row of some particular block had t ones; then we chose $q - t$ random ones from among the variables not involved in any of the diagonal blocks to complete that row, in such a way that each column ends up with r ones. This construction is unsuccessful about half the time; when it succeeds, it constructs a regular Gallager code.

The columns in the diagonal blocks furnish the pivots of the resulting matrix. For each row, if we could choose s non-pivot columns having a one in that row, then those variables were used to form a product term in the modified code. Some equations, therefore, were not modified, and this suited our purposes as well, since we did not want to perturb the linear code over much. In this construction, the information vectors and block length are the same for both the linear and modified code, although the actual code vectors differ.

In testing the codes, because the modified code is non-linear, one cannot assume transmission of the zero vector; for each test, we chose between

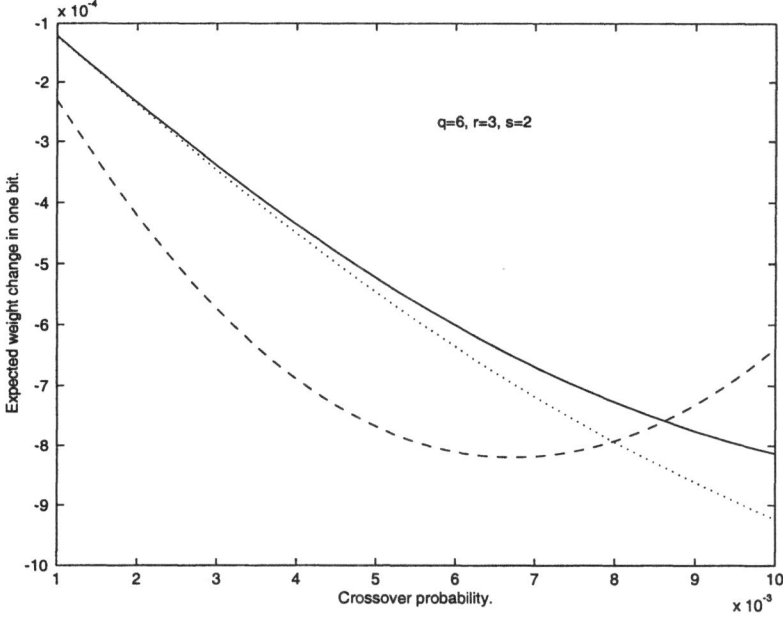

FIG. 3.

100 and 1000 pairs (information vector, noise vector). Each information vector produced a linear code vector and a modified code vector. The noise vector, whose length was equal to the block length, was then added.

We employed two familiar decoding schemes: an iterated version of the Algorithm and iterative re-estimation of the biases. The block lengths used were 200, 400, and 800. In the first decoding paradigm, in each loop we flipped those bits involved in a maximal number of equations having bias leverage greater than 1. With crossovers between 0.001 and 0.01, the linear and modified codes performed comparably. The standard deviation of the noise weight change per iteration was often but not always smaller for the modified codes, indicating slightly more reliability. With crossover increased above 0.02, the modified codes began to degrade, decoding roughly 85% of the test vectors, while the linear codes still decoded over 95% of the test vectors.

In the second decoding paradigm, the modified codes usually achieved quicker convergence of estimated probabilities at each bit to 1 or 0. We would like to run tests on larger block sizes, but our experiments suggest that modification of linear constraints to mildly non-linear constraints might lead to improved decoding. We chose a very particular kind of mildly non-linear constraint, and we wonder about optimal choices under more specialized hypotheses about the noise.

APPENDIX

Proof of Proposition 2.4. Taking the hypothesis and notation of that result, define V to be the set of all $w : S \to \mathbb{Z}_2$. Write

$$C = \sum_{z \in \mathbb{Z}_2^n} P(z) \cdot f(z + x) .$$

Let $s = |S|$. We use the fourier coefficients of P, given in Proposition 2.2, writing $h_j = 1 - 2 \cdot P_j(1)$, and the coefficients of f, given in Proposition 2.1, to compute

$$C = 2^n \cdot \sum_v P_v \cdot f_v \cdot (-1)^{v \cdot x}$$

$$= \sum_v \prod_{v[j]=1} h_j \cdot f_v$$

$$= \frac{1}{2} + \sum_{w \in V} \Pi(Q \backslash S, P) \cdot \prod_{w[j]=1} h_j \cdot \frac{1}{2^s} \cdot (-1)^{s-1+|w|} \cdot (-1)^{\overline{w} \cdot x}$$

$$\qquad + \Pi(Q, P) \cdot \frac{1}{2} \cdot (-1)^{\overline{u} \cdot x} .$$

The sum over V can be evaluated by parameterizing V in the following way. For $w \in V$, let a hold the bits of $w[j]$ with $x_S[j] = 1$ and b hold the bits of $w[j]$ with $x_S[j] = 0$. Thus, w corresponds to the pair (a, b). Let x' hold the bits $x[j]$ with $j \in Q \backslash S$. Then $|w| = |a| + |b|$ and $\overline{w} \cdot x \equiv |a| + |x'|$. We write the sum over V, collecting the a-factors and b-factors, thus.

$$C = \frac{1}{2} + \sum_{a,b} \Pi(Q \backslash S, P) \cdot \frac{1}{2^s} \cdot (-1)^{s-1+|x'|} \cdot \prod_{a[j]=1} h_j \cdot \prod_{b[j]=1} h_j \cdot (-1)^{|b|}$$

$$\qquad + \Pi(Q, P) \cdot \frac{1}{2} \cdot (-1)^{\overline{u} \cdot x} .$$

Since a, b run over sets of functions into $\mathbb{Z}_2$, the sums over these variables can be reversed with the product terms.

$$C = \frac{1}{2} + \Pi(Q \backslash S, P) \cdot \frac{1}{2^s} \cdot (-1)^{s-1+|x'|} \cdot \prod_{x_S[j]=1} (1 + h_j) \cdot \prod_{x_S[j]=0} (1 - h_j)$$

$$\qquad + \Pi(Q, P) \cdot \frac{1}{2} \cdot (-1)^{\overline{u} \cdot x} .$$

We have $1 + h_j = 2(1 - P_j(1)) = 2 P_j(0)$, and $1 - h_j = 2 P_j(1)$. There are s factors of 2 involved in all the h_j-factors. Thus,

$$C = \frac{1}{2} + \Pi(Q \backslash S, P) \cdot \frac{1}{2^s} \cdot (-1)^{s-1+|x'|} \cdot 2^s \cdot \prod_{x_S[j]=1} (1 - P_j(1)) \cdot \prod_{x_S[j]=0} P_j(1)$$

$$\qquad + \Pi(Q, P) \cdot \frac{1}{2} \cdot (-1)^{\overline{u} \cdot x} .$$

We have $\bar{u} \cdot x \equiv |x_Q| \bmod 2$. We write $1 - P_j(1)$ as $P_j(0)$, we cancel the 2^s terms, and we replace each $P_j(1)$ factor, where $x_S[j] = 0$, by $-P_j(1)$. This amounts to multiplication of the product by $(-1)^{s-|x_S|}$. We insert this factor next to the factor $(-1)^{s-1+|x'|}$. The result of this latter insertion is to change $(-1)^{s-1+|x'|}$ into $(-1)^{1+|x_Q|}$.

$$C = \frac{1}{2} + \Pi(Q \backslash S, P) \cdot (\text{-}1)^{1+|x_Q|} \cdot \prod_{x_S[j]=1} P_j(0) \cdot \prod_{x_S[j]=0} (-P_j(1))$$

$$+ \Pi(Q, P) \cdot \frac{1}{2} \cdot (\text{-}1)^{|x_Q|}$$

$$= \frac{1}{2} + \Pi(Q \backslash S, P) \cdot (\text{-}1)^{|x_Q|} \cdot \left[\frac{1}{2}\Pi(S, P) - \prod_{x_S[j]=1} P_j(0) \cdot \prod_{x_S[j]=0} (\text{-}P_j(1)) \right] \cdot$$

as claimed.   □

REFERENCES

[1] R.G. GALLAGER, *Low-Density Parity-Check Codes*, no. 21 in Research Monograph Series, Cambridge MA: MIT Press, 1963.

[2] D.J.C. MACKAY, *Good Error-Correcting Codes Based on Very Sparse Matrices*, IEEE Transactions on Information Theory, Vol. **45**, no. 2, March 1999.

CAPACITY-ACHIEVING SEQUENCES

Abstract. A capacity-achieving sequence of degree distributions for the erasure channel is, roughly speaking, a sequence of degree distributions such that graphs sampled uniformly at random satisfying those degree constraints lead to codes that perform arbitrarily close to the capacity of the erasure channel when decoded with a simple erasure decoder described in the paper. We will prove a necessary property called *flatness* for a sequence of degree distributions to be capacity-achieving, and will comment on possible applications to the design of capacity-achieving sequences on other communication channels.

Key words. Low-density parity-check codes, erasure channel.

1. Introduction. Low-density parity-check codes, discovered in the early 1960's by Gallager [2] have received a lot of attention lately. Advances in the theory of graphs, and connections to other fields like theoretical computer science have made it possible to rigorously analyze Gallager's original ideas, and to improve them in several directions [4–9, 11, 12].

In this paper, we will mostly focus on codes for the erasure channel. The Internet is perhaps the most appealing example of such a channel. When data is sent over the Internet, it is divided into packets. Each packet has an identifier which uniquely describes the entity it comes from and its location within that entity. Packets are then routed through the network from a sender to a recipient. Often, some packets do not arrive at their destination; in certain protocols like the TCP/IP the recipient requests in this case a retransmission of those packets, upon which the sender initiates the retransmission. These steps are iterated several times until the receiver has obtained the complete data. This protocol is excellent in certain cases, but is very poor in scenarios in which feedback channels do not exist (satellite links), or when one sender has to serve a large number of recipients (multicast).

Based on this motivation, the authors introduce in [6] a simple erasure recovery algorithm for low-density parity-check codes and derive a condition for that algorithm to finish successfully. Low-density parity-check codes are constructed from sparse bipartite graphs, and their combinatorial behavior is intricately related to that of the underlying graph. From a practical point of view, it is very appealing that the codes come with an efficient decoder. To make the decoder work well on the induced code, however, one has to design the graph in the right way. The situation is thus opposite to traditional coding, in which the codes are known and one wants to find efficient algorithms for them.

---

*Bell-Labs, Rm. 2C-381, 700 Mountain Ave, Murray Hill, NJ 07974, USA.

A priori, it would seem like an almost impossible task to analyze the behavior of decoding algorithms on a graph, let alone designing the graph in such a way that the algorithm performs well. One of the main and surprising results of [6] was that the *only* parameter that decides on the success of the algorithm is the distribution of nodes of various degrees on both sides of the graph. More precisely, the paper shows that, given a fixed distribution on the nodes of the graph, there is a threshold value p such that if the graph is sampled uniformly at random from the ensemble of graphs with that distribution, then the erasure recovery algorithm can recover from any fraction of erasures less than p, and will fail if the fraction of erasures is larger than p. The exact description of the conditions will be provided later in Section 2.3.

The question that arises is whether it is possible to create degree sequences such that the corresponding codes approach the capacity of the erasure channel when decoded by the simple erasure decoder. This question is answered in the affirmative in [6] by exhibiting such a sequence of *capacity-achieving* degree distributions. Meanwhile, there is a second such sequence [10], but these two sequences remain the only ones known. These sequences share certain properties, and we will prove later in Section 3 that this is no coincidence. We will prove an analytic condition, called the *flatness condition* which any capacity-achieving sequences of degree distributions have to satisfy. We hope that this condition will help to design methods for constructing capacity-achieving sequences for other types of channels, like the Binary Symmetric or the AWGN Channel. Thoughts about possible directions and some open problems are presented in the final section of the paper.

2. Low-density parity-check codes.

2.1. Code construction. In the following we will assume that the code-alphabet A is the binary field $\mathbb{F}_2$. Let G be a bipartite graph between n nodes on the right called *message nodes* and r nodes on the right called *constraint (or check) nodes*. The graph gives rise to a code in two different ways, see Figure 1: in the first version (which is Gallager's original version), the coordinates of a codeword are indexed by the message nodes $1, \ldots, n$ of G. A vector $(x_1, \ldots, x_n)$ is a valid codeword if and only if for each constraint node the sum (over $\mathbb{F}_2$) of the values of its adjacent message nodes is zero. Since each constraint node imposes one linear condition on the x_i, the rate of the code is at least $(n - r)/n$.

In the second version, the message nodes are indexed by the original message. The constraint nodes contain the redundant information: the value of each such node is equal to the sum (over $\mathbb{F}_2$) of the values of its adjacent message nodes. The block-length of this code is $n + r$, and its rate is $n/(n + r)$.

These two versions look quite similar, but differ fundamentally from a computational point of view. The encoding time of the second version is

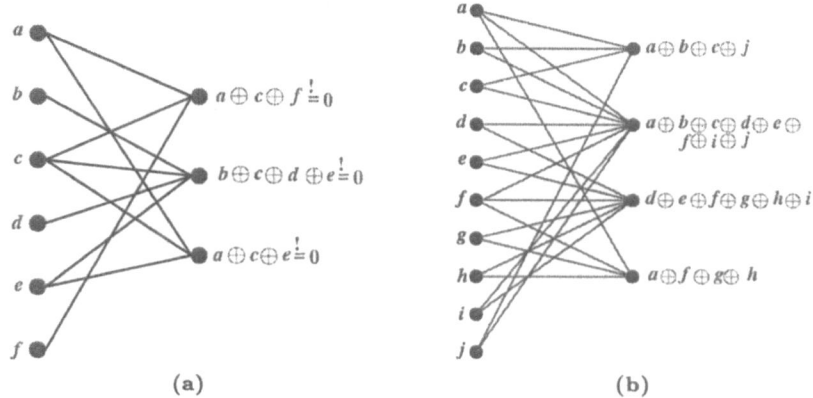

FIG. 1. *The two versions of low-density parity-check codes: (a) Original version, and (b) dual version.*

proportional to the number of edges in the graph G, while it is not clear how to encode the first version without solving systems of linear equations. (This needs to be done once for the graph; each encoding afterwards corresponds to a matrix/vector multiplication.) If the graph is sparse, the encoding time for the second version is essentially linear in the block-length, while that of the first version is essentially quadratic (after a pre-processing step).

While the second version is advantageous for the encoding, the first version is more suited to decoding. I don't want to go into further details on this issue, and will in the following only consider Gallager's original version of low-density parity-check codes. Readers are invited to consult [12, 6] to learn more about the second version.

2.2. Decoding on the erasure channel. It is not hard to see [1] that a linear code of minimum distance d is capable of correcting any pattern of $d - 1$ or less erasures. The algorithm is rather straight-forward, and uses a multiplication of a matrix of size $(n - k) \times (n - d + 1)$ with a vector of length $n - d + 1$, followed by solving a quadratic system of equations of size $d - 1$, where k is the dimension of the code. Further, Elias showed that random linear codes achieve capacity of the erasure channel with high probability. In typical applications that we have in mind, the running time of the recovery algorithm is majorized by the time $O(d^3)$ to solve the system of linear equations, and this is far too slow considering that typical values of d are in the $100,000$'s.

In contrast, the decoder that we use for the low-density parity-check codes is extremely fast and simple. It maintains a register for each of the message and constraint nodes. All of these registers are initially set to zero. In the first round of the decoding, the value of each received message node is added to the values of all of its adjacent constraint nodes, and then the message nodes and all the edges emanating from it are deleted. Once this *direct recovery step* is complete, the second *substitution recovery phase* kicks in. Here, one looks for a constraint node of degree one. Note that since the value of a constraint node in an intact codeword should be zero, a constraint node of degree one contains the value of its unique adjacent message node. This value is copied into the corresponding message node, that value is added to those of all its adjacent constraint nodes, and the message node together with all edges emanating from it are deleted from the graph. If there are no nodes left, or if there are no constraint nodes of degree one left, then the decoder stops. Note that the decoding time is proportional to the number of edges in the graph. If the graph is sparse, i.e., if the number of edges is linear in the number of nodes, then the decoder is linear time (at least on a RAM with unit cost measure). An example of a complete decoding is given in Figure 2.

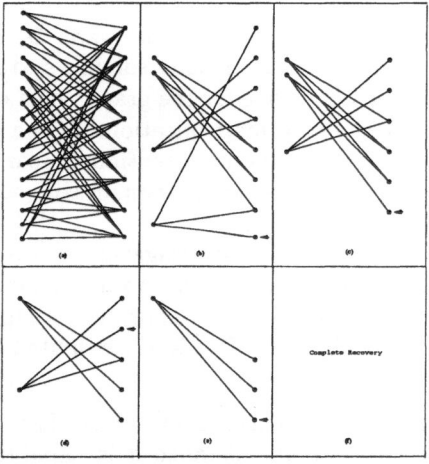

FIG. 2. *An example of complete decoding.*

The hope is that there is always enough supply of degree one constraint nodes so that the decoder finishes successfully. Whether or not this is the case depends on the original fraction of erasures and on the graph. Surprisingly, however, the only important parameter of the underlying graph is the distribution of nodes of various degrees. This analysis is the topic of the next section.

2.3. The analysis. To describe the conditions for successful decoding concisely, we need one further piece of notation. We call an edge in the graph G *of left (right) degree i* if it is connected to a message (constraint) node of degree i. Let λ_i and ρ_i denote the fraction of edges of left degree i

and right degree i, respectively. Further, we define the generating functions $\lambda(x) = \sum_i \lambda_i x^{i-1}$ and $\rho(x) = \sum_i \rho_i x^{i-1}$. The rather peculiar look of the exponent of x in these polynomials is an artifact of the particular *message passage decoding* that we are using. This is best explained by the analysis itself, which I will now describe in an informal way.

Let e be an edge between the message node m and the constraint node c. What is the probability that this edge is deleted at the ℓth round of the algorithm? This is the probability that the check node c is of degree one at the ℓth round, and, equivalently, it is the probability that the message node m is corrected at that round. To compute this probability, we unroll the graph in the neighborhood of the node m and consider the subgraph obtained by the neighborhood of depth ℓ of m. This is the subgraph of all the nodes in the graph except those that are connected to m via the edge e, for which there is a path of length at most 2ℓ connecting them to m. In the following we will assume that this graph is a tree. Suppose that the graph is sampled uniformly at random from the set of graphs which have an edge distribution according to the polynomials $\lambda(x)$ and $\rho(x)$. Let p_ℓ denote the probability that m is *not* corrected at round ℓ. Further, let δ denote the original fraction of erasures. Then, obviously $p_0 = \delta$. Further, because we have assumed that the neighborhood of m is a tree, at each level ℓ of the tree the message nodes are still erased with independent probability p_ℓ. (We assume that only the message nodes contribute to levels in the tree, so that the message nodes forming the leaves are at level 0 and the root m is at level ℓ.) From this, we can establish a recursion for p_ℓ. A message node at level $\ell + 1$ is not corrected if and only if it has not been received directly, and all the constraint nodes it is connected to have degree larger than 1. A constraint node has degree one if and only if all its descending message nodes at level ℓ have already been corrected. This happens with independent probability $1 - p_\ell$, and since the message node has j edges emanating from it with probability ρ_j, and $j - 1$ of them are descending message nodes in the tree, the probability that such a check node is of degree one is $\rho(1 - p_\ell)$. Hence, the probability that a message node at level $\ell + 1$ is connected only to descending constraint nodes of degree larger than 1 is $\lambda(1 - \rho(1 - p_\ell))$. That node is thus not corrected with probability $\delta\lambda(1 - \rho(1 - p_\ell))$, where the factor δ explains the probability that the node has not been received directly. Hence, this gives $p_{\ell+1} = \delta\lambda(1 - \rho(1 - p_\ell))$. Altogether, we obtain the condition

$$(2.1) \qquad \delta\lambda(1 - \rho(1 - p_\ell)) < p_\ell$$

for successful decoding. More precisely, this says that if neighborhoods of depth ℓ of message nodes are trees, and if $\delta\lambda(1 - \rho(1 - x)) < (1 - \epsilon)x$ for $x \in (0, \delta)$ and some $\epsilon > 0$, then after ℓ rounds of the algorithm the probability that a message node has not been corrected is at most $(1 - \epsilon)^\ell \delta$. For large random graphs the probability that the neighborhood of a message node is not a tree is small, and the argument shows that the decoding algorithm reduces the probability of undecoded message node below any constant. To show that the process finishes successfully, one needs expansion [6].

The above informal discussion can be made completely rigorous using proper martingale arguments [4, 9]. Summarizing, the condition for successful decoding after a δ-fraction of erasures is

$$(2.2) \qquad \delta\lambda(1 - \rho(1 - x)) < x \quad \text{for} \quad x \in (0, \delta).$$

It is easy to see that this condition is equivalent to

$$(2.3) \qquad \rho(1 - \delta\lambda(1 - x)) < x \quad x \in (0, 1).$$

To see this, first note that $\lambda(x)$ is invertible as a function on the interval $(0, 1)$, since it is a convex linear combination of monotonically increasing functions. Hence, (2.2) is equivalent to

$$\rho(1 - x) \geq 1 - \lambda^{-1}\left(\frac{x}{\delta}\right).$$

Now, set $x = \delta\lambda(1 - u)$, where now $u \in (0, 1)$.

2.4. Capacity-achieving sequences. The condition (2.2) is very handy if one wants to analyse the performance of random graphs with a *given* degree distribution. For instance, it turns out that the performance of regular graphs deteriorates as the degree of the message nodes increases [6]. In fact, the best performance is obtained if all message nodes have degree three. On the other hand, this condition does not give a clue on how to design good degree distributions λ and ρ. Our aim is to construct sequences that asymptotically achieve the capacity of the erasure channel. In other words, we want δ in (2.2) to be arbitrarily close to $1 - R$, where R is the rate of the code. To make this definition more rigorous, we call a sequence $(\lambda_n, \rho_n)_{n \geq 0}$ *capacity-achieving of rate R* if for all $\delta < 1 - R$ there exists $n_0 \in \mathbb{N}$ such that for all $n \geq n_0$ we have

$$(2.4) \qquad \rho_n(1 - \delta\lambda_n(1 - x)) > x \quad \forall x \in (0, 1).$$

Note that any δ satisfying the inequality above can be at most $1 - R$. This follows either from the fact that, information theoretically, no recovery is possible if the fraction of erasures is larger than $1 - R$. Equivalently, one can prove this by elementary integration [10].

It is surprising that such sequences do really exist. The first such sequence was discovered in [6]. To describe it, we first need to mention that, given λ and ρ, the average left and right degree of the graph is $1/\sum_i \lambda_i/i$ and $1/\sum_i \rho_i/i$, respectively. These quantities can be conveniently expressed as $1/\int_0^1 \lambda(x)dx$ and $1/\int_0^1 \rho(x)dx$. As a result, the rate of the code is at least $1 - \int_0^1 \rho(x)dx / \int_0^1 \lambda(x)dx$. It is a nice exercise to deduce from the equation (2.2) alone that δ is always less than or equal to $1 - R$, i.e., less than or equal to $\int_0^1 \rho(x)dx / \int_0^1 \lambda(x)dx$.

Fix a parameter D and let $\lambda_D(x) := \frac{1}{H(D)} \sum_{i=1}^{D} x^i/i$, where $H(D)$ is the harmonic sum $\sum_{i=1}^{D} 1/i$. Let $\rho_D(x) := e^{\mu(x-1)}$, where μ is the unique solution to the equation

$$\frac{1}{\mu}(1 - e^{-\mu}) = \frac{1-R}{H(D)}\left(1 - \frac{1}{D+1}\right).$$

Since $\int_0^1 \lambda_D(x)dx = \frac{1}{H(D)}(1 - 1/(D+1))$ and $\int_0^1 \rho_D(x) = (1 - e^{-\mu})/\mu$, the sequence $(\lambda_D(x), \rho_D(x))_{D \geq 1}$ gives rise to codes of rate at least R. Further, we have

$$\delta\lambda_D(1 - \rho_D(1-x)) = \delta\lambda_D(1 - e^{-\mu x})$$
$$\leq \frac{-\delta}{H(D)} \ln(e^{-\mu x})$$
$$= \frac{\delta\mu x}{H(D)}.$$

Hence, successful decoding is possible if the fraction of erasures is no more than $H(D)/\mu$. Note that this quantity equals $(1 - R)(1 - 1/(D+1))/(1 - e^{-\mu})$, and that this quantity is larger than $(1 - R)(1 - 1/D)$. Hence, we have that

$$(1 - R)(1 - 1/D)\lambda_D(1 - \rho_D(1 - x)) < x \quad \text{for} \quad x \in (0, (1 - R)(1 - 1/D)).$$

This shows that the sequence is indeed capacity-achieving. We have named these sequences the *Heavy-Tail/Poisson* sequences, or, more commercially oriented, *Tornado codes*.

In the meantime, I have obtained yet another capacity-achieving sequence whose left side is closely related to the power series expansion of $(1 - x)^{1/D}$, and which is right-regular, i.e., all nodes on the right have the same degree [10]. More precisely, the new sequence is defined as follows. For integers $a \geq 2$ and $n \geq 2$ let

$$\rho_a(x) := x^{a-1}, \qquad \lambda_{a,n}(x) := \frac{\sum_{k=1}^{n-1} \binom{\alpha}{k}(-1)^{k+1}x^k}{1 - n\binom{\alpha}{n}(-1)^{n+1}},$$

where $\alpha := 1/(a - 1)$. For the correct choice of the parameter n and other properties of these sequences we refer the reader to [10].

I would like to close this section with a few comments on the trade-off between proximity to the channel capacity and the running time of the decoder. For the Heavy-Tail/Poisson sequence the average degree of a message node was less than $H(D)$, and it could tolerate up to $(1 - R)(1 - 1/D)$ fraction of erasures. Hence, to get close to within $1 - \epsilon$ of the capacity $1 - R$, we needed codes of average degree $O(\log(1/\epsilon))$. This is shown to be essentially optimal in [10]. In other words, to get within $1 - \epsilon$ of the channel capacity, we need graphs of average degree $\Omega(\log(1/\epsilon))$. The same relation also holds for the right-regular sequences. Hence, these codes are essentially optimal for our simple decoders.

3. Flatness. The capacity-achieving sequences of the previous section share an interesting property: As n goes to infinity, the function $\delta_n \lambda_n(1 - \rho_n(1 - x)) - x$ and all its derivatives converge uniformly to zero on the interval $[0, 1]$, where δ_n is the supremum over all δ such that $\delta \lambda_n(1 - \rho_n(1 - x)) \leq x$ on $[0, 1]$. This can be verified by looking at the Taylor expansion of this function, and we leave it as an exercise to the reader. In terms of the decoding process, this means that as n grows large, the decoding process converges more slowly. In this section, we are going to prove that any capacity-achieving sequence will necessarily have this property, which we call *flatness*.

We will need several preliminary results before being able to prove the flatness condition. The following result gives equivalent formulations for a sequence of degree distributions to be capacity-achieving.

LEMMA 3.1. *Let $(\lambda_n(x), \rho_n(x))$ be a sequence of degree distributions of rate R. The following statements are equivalent:*

(a) *The sequence is capacity-achieving.*

(b) *The sequence (δ_n) converges towards $1 - R$ as n goes to infinity, where δ_n is the supremum over all δ for which (2.2) is satisfied.*

(c) *The sequence (δ_n) converges towards $1 - R$ as n goes to infinity, where δ_n is the supremum over all δ for which the following holds:*

$$(3.1) \qquad \delta \lambda_n(x) < 1 - \rho_n^{-1}(1 - x) \qquad for \; x \in (0, 1).$$

(d) *For any fixed $\xi < 1$, any $\varepsilon > 0$ and any $\tau > 0$ there is an n_0 such that for all $n \geq n_0$ and all $x \in (0, \xi]$ we have*

$$-\tau < (1 - R - \varepsilon)\lambda_n(x) - 1 + \rho_n^{-1}(1 - x) < 0.$$

Proof. Statements (b) and (c) are shown to be equivalent using a simple algebraic manipulation:

$$\rho_n(1 - \delta \lambda_n(x)) > 1 - x \iff \delta \lambda_n(x) < 1 - \rho_n^{-1}(1 - x).$$

(Note that $\rho_n(x)$ is invertible on $(0, 1)$, being a polynomial with nonnegative coefficients). We thus need to prove the equivalence of (a), (b) and (d).

(a) $\iff$ (b): This is very easy, but we will nevertheless carry out the argument. Suppose that $(\lambda_n(x), \rho_n(x))$ is capacity-achieving Then, for all $\varepsilon > 0$ there exists n_0 such that for all $n \geq n_0$ the condition (2.2) is satisfied with $\delta = 1 - R - \varepsilon$. Hence, for all $n \geq n_0$ we have $\delta_n \geq 1 - R - \varepsilon$. Noting that $\delta_n \leq 1 - R$, this implies (b). Conversely, suppose that (b) is satisfied. Let $\delta = 1 - R - \varepsilon$ be given, where $\varepsilon > 0$. Since δ_n converges to $1 - R$, we have $\delta_n > 1 - R - \varepsilon$ for all sufficiently large n. Hence, (2.2) is satisfied for all sufficiently large n, which means that the sequence $(\lambda_n(x), \rho_n(x))$ is capacity-achieving

(a) $\iff$ (d): since (d) implies (c), we only need to show that (a) implies (d). First we show that $\lambda_n(x)$ converges uniformly to zero on $[0, \xi]$

as n goes to infinity. Let δ_n be the quantity defined in (b). It is proved in [10] that $\delta_n \leq (1 - R)(1 - R^{a_n})$, where $1/a_n = \int_0^1 \rho_n(x)dx$. Hence, for the sequence to be capacity-achieving it is necessary that a_n goes to infinity, i.e., that $\int_0^1 \rho_n(x)dx$ (and hence the same integral for $\lambda_n(x)$) goes to zero. Since $\lambda_n(x)$ is monotonically increasing (its derivative is obviously positive on $(0, \xi]$), this shows that $\lambda_n(x)$ converges to 0 uniformly. In the same way, we show that $1 - \rho_n^{-1}(1 - x)$ converges to zero uniformly, as this function is also monotonically increasing, and its integral from 0 to 1 equals that of $\rho_n(x)$. Let ε and δ be positive real numbers. There exists n_1 (depending on τ) such that for any $n \geq n_1$ we have

$$|(1 - R - \varepsilon)\lambda_n(x) - 1 + \rho_n^{-1}(1 - x)| < \tau.$$

Further, since $(\lambda_n(x), \rho_n(x))$ is capacity-achieving, there exists n_2 (depending on ε) such that for all $n \geq n_2$ and all $x \in (0, 1)$ we have $(1-R-\varepsilon)\lambda_n(x) - 1 + \rho_n^{-1}(1-x) < 0$. Setting $n_0 := \max(n_1, n_2)$ shows that (1) indeed implies (4) and finishes the proof. $\qquad\square$

In the following, we denote by $C^k[a, b]$ the space of all k times continuously differentiable functions on the interval $[a, b]$.

LEMMA 3.2. *Let ℓ be a fixed positive integer, $a < b$ be real numbers, and suppose that $(f_n(x))$ and $(g_n(x))$ are sequences of functions in $C^{\ell+1}[a, b]$ such that $f_n^{(k)}(x) > 0$ and $g_n^{(k)}(x) > 0$ on $[a, b]$ for all $0 \leq k \leq \ell + 1$, $g^{(k)}(x)$ is bounded by a real number c_k on $[a, b]$ for all n, and $f_n(x) - g_n(x)$ converges to zero uniformly for all $x \in [a, b]$. Then, for all $0 \leq k \leq \ell$, $f_n^{(k)}(x) - g_n^{(k)}(x)$ converges to zero uniformly for all $x \in [a, b]$.*

Proof. Let k be the smallest positive integer for which the claim is not true and assume that $k \leq \ell$. The assumptions imply that $k \geq 1$. Let $\alpha \in [a, b]$ be a point such that $f_n^{(k)}(\alpha) - g_n^{(k)}(\alpha)$ does not converge to zero. Hence, there exists $\tau > 0$ such that the absolute value of this difference is larger than τ for infinitely many n. Suppose first that $f_n^{(k)}(\alpha) - g_n^{(k)}(\alpha) > \tau$ for infinitely many n and consider the Taylor-expansion of $f_n - g_n$ around α:

$$f_n(x) - g_n(x) = \sum_{i=0}^{k-1} \left(f_n^{(i)}(\alpha) - g_n^{(i)}(\alpha) \right) \frac{(x - \alpha)^i}{i!} +$$
$$\left(f_n^{(k)}(\alpha) - g_n^{(k)}(\alpha) \right) \frac{(x - \alpha)^k}{k!} +$$
$$\left(f_n^{(k+1)}(\eta) - g_n^{(k+1)}(\mu) \right) \frac{(x - \alpha)^{k+1}}{(k + 1)!},$$

for some η and μ in $[a, b]$. The first term in the above sum can be made arbitrarily small by choosing n large enough. This is because $f_n^{(i)}(\alpha) - g_n^{(i)}(\alpha)$ converges to zero by the choice of k. The third term is bounded from below by $-c_{k+1}|x - x_0|^{k+1}/(k+1)!$, since $f_n^{(k+1)}(x)$ is positive on $[a, b]$. Hence, we can choose an x depending on τ and k only, so that the right hand side of the above expression is bounded from below by a positive quantity depending on τ and k only. This is a contradiction to the assumption that $f_n(\alpha) - g_n(\alpha)$ converges to zero. The case that $f_n^{(k)}(\alpha) - g_n^{(k)}(\alpha) < -\tau$ for infinitely many n is handled analogously by bounding the expression on the right hand side of the above equality from above.   □

With the help of the previous lemma, we can now show that the function $\delta_n \lambda_n(1 - \rho_n(1-x)) - x$ has to converge uniformly to zero on any compact subinterval of $[0, 1]$. This is done in two steps. In the first step, we prove this assertion for the function $\delta_n \lambda_n(x) - 1 + \rho_n^{-1}(1-x)$, see the next proposition. Then, using this result, we will prove the general assertion in Theorem 3.1 below.

PROPOSITION 3.1. *Let $(\lambda_n(x), \rho_n(x))$ be a capacity-achieving sequence of rate R, and suppose that there is an integer ℓ such that for all sufficiently large n all the derivatives of $1 - \rho_n^{-1}(1-x)$ up to the order $\ell+1$ are positive on $(0, 1)$. Then, for any $0 \le k \le \ell$, the kth derivative of $\delta_n \lambda_n(x) - 1 + \rho_n^{-1}(1-x)$ converges to zero uniformly on any closed subinterval of $(0, 1)$.*

Proof. In view of the previous lemma, it suffices to show that the kth derivative of $R_n(x) := 1 - \rho_n^{-1}(1 - x)$ and $\delta_n \lambda_n(x)$ are both bounded on $[a, b]$ for all $0 \le k \le \ell + 1$. The assertion is clear for $\delta_n \lambda_n(x)$, since $\lambda_n(x)$ is a polynomial with nonnegative coefficients. For the same reason, all the derivatives up to the order $\ell + 1$ of $\rho_n(x)$ are bounded from above on any compact subinterval of $(0, 1)$, if n is sufficiently large. Further, if this subinterval does not contain 0, all these derivatives are also bounded from below by a positive constant (depending possibly on the order of the derivative as well as the boundaries of the interval). This follows from the fact that the degrees of $\lambda_n(x)$ and $\rho_n(x)$ have to grow beyond bounds if these polynomials are from a capacity-achieving sequence [10, Theorem 1]. Note now that the kth derivative of $R_n(x)$ at any point α is obtained as a polynomial (with coefficients depending on k only) of the derivatives of $\rho_n(x)$ up to the order k at point $\rho_n^{-1}(1 - \alpha)$, divided by some power of $\rho_n'(\rho_n^{-1}(1 - \alpha))$. Since k is at most $\ell + 1$ and ℓ is fixed, and since the derivatives of $\rho_n(x)$ are bounded from above and below by constants independent of n, this shows that the derivatives of $\delta_n \lambda_n(x) - R_n(x)$ are bounded.   □

Now we can state and prove the "flatness condition."

THEOREM 3.1 (Flatness Condition). *Let $(\lambda_n(x), \rho_n(x))$ be a capacity-achieving sequence of rate R, and suppose that there is an integer ℓ such that for all sufficiently large n all the derivatives of $1 - \rho_n^{-1}(1-x)$ up to the order $\ell+1$ are positive on $(0, 1)$. Then, the kth derivative of $\rho_n(1 - \delta_n \lambda_n(x)) - 1 + x$ converges to zero uniformly on any closed subinterval of $(0, 1)$ for any $0 \le k \le \ell$.*

Proof. Let $u_n(x) := \rho_n(1 - \delta_n \lambda_n(x)) - 1 + x$. Using the previous proposition, it suffices to show that the derivatives of u_n up to the order $\ell + 1$ converge uniformly to zero if and only the same is true for $v_n(x) := \delta_n \lambda_n(x) - 1 + \rho_n^{-1}(1 - x)$. We first need a simple preliminary result, whose proof is left to the reader: for any $k \geq 0$ there exists a polynomial f_k in $2k + 2$ variables and with integer coefficients such that for any functions h, g which are at least k times continuously differentiable on an interval I, we have

$$(h(1 - g(x)) - (1 - x))^{(k)} = f_k\left(h, h', \ldots, h^{(k)}, g, g', \ldots, g^{(k)}\right).$$

Let $R_n(x) := 1 - \rho_n^{-1}(1 - x)$. Then we have $\rho_n(1 - R_n(x)) - (1 - x) = 0$. For any k, we thus have

(3.2) $$f_k\left(\rho_n, \rho_n', \ldots, \rho_n^{(k)}, R_n, R_n', \ldots, R_n^{(k)}\right) = 0.$$

Furthermore, we have

(3.3) $$\begin{aligned}(\rho_n(1 - \delta_n\lambda_n(x)) - (1 - x))^{(k)} = \\ f_k\left(\rho_n, \rho_n', \ldots, \rho_n^{(k)}, \delta_n\lambda_n, \delta_n\lambda_n', \ldots, \delta_n\lambda_n^{(k)}\right).\end{aligned}$$

By Proposition 3.1 we know that for any fixed $k \leq \ell$ the kth derivative of $\delta_n\lambda_n(x) - R_n(x)$ converges to zero uniformly on any closed subinterval of $(0, 1)$. But then, since f_k is a continuous function, this implies that for any x in that interval the absolute value of the difference of (3.2) and (3.3) can be made arbitrarily small by letting n grow large. This is the assertion to be proved. $\square$

4. Codes on other channels. The model of an erasure channel is rather simple compared to other channels like the Binary Symmetric Channel, or the Additive White Gaussian Noise Channel. However, many of the above results can be carried over to the case of these channels as well, as it is possible to design iterative decoding algorithms which, at each step, pass messages along the edges of the graph. The most powerful such "message passage decoder" is the belief propagation for a description of which we refer the reader to [7]. Richardson and Urbanke were able to analyze this decoder [9]. One of the main results of that paper is the derivation of a recursion for the (common) density functions of the message nodes at each iteration round of the algorithm. The analysis was further simplified in [8], and will be described below. First, we assume that the input alphabet is the set $\{\pm 1\}$. At each round, the algorithm passes messages from message nodes to check nodes, and then from check nodes to message nodes. We assume that at the message nodes the messages are represented as log-likelihood ratios

$$\log \frac{p(y|x = 1)}{p(y|x = -1)},$$

where y represents all the observations conveyed to the message node at that time. Now let f_ℓ denote the probability density function at the message nodes at the ℓth round of the algorithm. f_0 is then the density function of the error which the message bits are originally exposed to. It is also denoted by P_0. These density functions are defined on the set $\mathbb{R} \cup \{\pm\infty\}$. It turns out that they satisfy a *symmetry condition* [8] $f(-x) = f(x)\mathrm{e}^{-x}$. As a result, the value of any of these density functions is determined from the set of its values on the set $\mathbb{R}_{\geq 0} \cup \{\infty\}$. The restriction of a function f to this set is denoted by $f^{\geq 0}$. (The technical difficulty of defining a function at ∞ could be solved by using distributions instead of functions, but we will not further discuss it here.)

For a function f defined on $\mathbb{R}_{\geq 0} \cup \{\infty\}$ we define a *hyperbolic change of measure* γ via

$$\gamma(f)(x) := f(\ln \coth x/2)\mathrm{csch}(x).$$

If f is a function satisfying the symmetry condition, then $\gamma(f^{\geq 0})$ defines a function on $\mathbb{R}_{\geq 0} \cup \{\infty\}$ which can be uniquely extended to a function F on $\mathbb{R} \cup \{\pm\infty\}$. The transformation mapping f to F is denoted by Γ. It is a bijective mapping from the set of density functions on $\mathbb{R} \cup \{\pm\infty\}$ satisfying the symmetry condition to itself. Let f_ℓ denote the common density function of the messages passed from message nodes to check nodes at round ℓ of the algorithm. Suppose that the graph has a degree distribution given by $\lambda(x)$ and $\rho(x)$. Then we have the following:

$$(4.1) \qquad f_\ell = P_0 \otimes \lambda(\Gamma^{-1}(\rho(\Gamma(f_{\ell-1})))), \quad \ell \geq 1.$$

Here, $\otimes$ denotes convolution, and for a function f, $\lambda(f)$ denotes the function $\sum_i \lambda_i f^{\otimes(i-1)}$. In the case of the erasure channel, the corresponding density functions are two-point mass functions, with a mass p_ℓ at zero and a mass $(1 - p_\ell)$ at infinity. In this case, the iteration translates to

$$p_\ell = \delta\lambda(1 - \rho(1 - p_{\ell-1})),$$

where δ is the original fraction of erasures. This equality describes the progression of the erasure probability at each round of the decoding. Stated in this form, successful decoding translates to the condition (2.2).

The question now arises whether (4.1) can be used to obtain capacity-achieving sequences for the BSC or the AWGN channel. This is still an open question, and it seems that one would at least need an analytic condition to decide whether for a given initial noise function P_0, and given degree distributions $\lambda(x)$ and $\rho(x)$, the density functions converge to a Delta function at infinity.

It is conceivable that the flatness condition turns out to be helpful in designing capacity-achieving sequences. To argue for such an approach, let us take a closer look at how we would design capacity-achieving sequences of codes over the erasure channel. Suppose that we knew that there is a sequence of capacity-achieving right regular codes, say. This means that

all the nodes on the right hand side of the graph have the same degree, say
n. Then, we could find $\lambda(x)$ by differentiating

$$\delta\lambda(1 - (1 - x)^{n-1}) - x$$

and requiring that the derivatives be zero. This way we could recursively
solve for the various derivatives of $\lambda(x)$ at zero, and we would stop the
procedure as soon as we encounter a negative derivative. Then we would
increase n, and continue. Because of the flatness condition, this procedure
would give us sequences with better and better erasure protection.

We cannot really apply this procedure to codes on channels other than
the erasure channel. This is mainly because we do not have a flatness
condition on those channels, though I think that a similar assertion has to
hold there too. But it is remarkable that some minor parts of this approach
can indeed be carried over. For instance, if we let $f(x) = \delta\lambda(1-\rho(1-x))-x$,
then we know that for successful decoding, we need to have $f(x) \leq 0$ on
$[0, 1]$. In particular, this shows that $f'(0) \leq 0$, which gives us the condition

$$\lambda'(0)\rho'(1) \leq \frac{1}{\delta}.$$

(See also [10].) This condition can be generalized to other channels, and
is called the *stability condition* in [8]. capacity-achieving sequences have to
satisfy this condition with equality in case of codes over the erasure channel.
It is conjectured that the same is true for other channels if one replaces
this condition with the stability condition. If one could find analogues of
higher stability conditions, then this would presumably lead to candidates
for capacity-achieving sequences. On the practical side, this would lead to
a deterministic algorithm for the design of good low-density parity-check
codes.

REFERENCES

[1] P. ELIAS. Coding for two noisy channels. In *Information Theory, Third London Symposium*, pages 61–76, 1955.

[2] R.G. GALLAGER. *Low Density Parity-Check Codes*. MIT Press, Cambridge, MA, 1963.

[3] M. LUBY, M. MITZENMACHER, AND M.A. SHOKROLLAHI. Analysis of random processes via and-or tree evaluation. In *Proceedings of the 9th Annual ACM-SIAM Symposium on Discrete Algorithms*, pages 364–373, 1998.

[4] M. LUBY, M. MITZENMACHER, M.A. SHOKROLLAHI, AND D. SPIELMAN. Analysis of low density codes and improved designs using irregular graphs. In *Proceedings of the 30th Annual ACM Symposium on Theory of Computing*, pages 249–258, 1998.

[5] M. LUBY, M. MITZENMACHER, M.A. SHOKROLLAHI, AND D. SPIELMAN. Improved low-density parity-check codes using irregular graphs and belief propagation. In *Proceedings 1998 IEEE International Symposium on Information Theory*, page 117, 1998.

[6] M. LUBY, M. MITZENMACHER, M.A. SHOKROLLAHI, D. SPIELMAN, AND V. STEMANN. Practical loss-resilient codes. In *Proceedings of the 29th annual ACM Symposium on Theory of Computing*, pages 150–159, 1997.

[7] D.J.C. MACKAY. Good error-correcting codes based on very sparse matrices. *IEEE Trans. Inform. Theory*, **45**:399–431, 1999.

[8] T. RICHARDSON, M.A. SHOKROLLAHI, AND R. URBANKE. Design of provably good low-density parity-check codes. *IEEE Trans. Inform. Theory (submitted)*, 1999.

[9] T. RICHARDSON AND R. URBANKE. The capacity of low-density parity-check codes under message-passing decoding. *IEEE Trans. Inform. Theory (submitted)*, 1998.

[10] M.A. SHOKROLLAHI. New sequences of linear time erasure codes approaching the channel capacity. In *Proceedings of AAECC-13*, M. Fossorier, H. Imai, S. Lin, and A. Poli eds, number 1719 of Lecture Notes in Computer Science, pages 65–76, 1999.

[11] M. SIPSER AND D. SPIELMAN. Expander codes. *IEEE Trans. Inform. Theory*, **42**:1710–1722, 1996.

[12] D. SPIELMAN. Linear-time encodable and decodable error-correcting codes. *IEEE Trans. Inform. Theory*, **42**:1723–1731, 1996.

HYPERTRELLIS: A GENERALIZATION OF TRELLIS AND FACTOR GRAPH

WAI HO MOW*

Abstract. Factor graphs have recently been introduced as an efficient graphical model for codes to study iterative decoding algorithms. However, it is well-known that a factor graph generalizes only the time axis of a trellis, but omits the state transition representation. In this paper, a new graphical model, called the hypertrellis, is proposed to overcome this insufficiency of factor graphs. A hypertrellis is in essence a weighted hypergraph generalization of a traditional trellis. Its time topology, which extends the time axis of a trellis, can take the form of any factor graph. A key to this extension is the interpretation of a factor graph as a factor hypergraph. This is facilitated by introducing a "starfish" drawing representation for hypergraphs, which enhances the applicability of hypergraph models by enabling simpler drawing and easier visualization, relative to the traditional representation. The maximum likelihood decoding (MLD) problem is then formulated as a shortest hyperpath search on a hypertrellis. For hypertrellises with an acyclic time topology, a hyperpath-oriented MLD algorithm, called the one-way algorithm, is introduced. The one-way algorithm, as a hypertrellis generalization of the celebrated Viterbi algorithm, provides insights into efficient management of the surviving hyperpath history and various practical hyperpath-oriented simplifications. It is shown that as a MLD algorithm, the one-way algorithm has a lower minimal decoding delay. The computational complexity and the amount of storage needed are also better than with the well-known min-sum algorithm. Some connections between the hypertrellis and another recently proposed bipartite graph model, called the trellis formation, are also discussed.

Key words. Trellis, factor graph, hypergraph, hypertrellis, shortest hyperpath search, maximum likelihood decoder, iterative decoding, one-way algorithm.

1. Introduction. An important key to the great success of the Turbo coding schemes by Berrou *et al.* [3] is the "near-optimality" of the low-complexity iterative decoding algorithms. MacKay and Neal [17] subsequently observed that Gallager's classical iterative decoding algorithms [7] are also "near-optimal" when applied to low-density parity-check codes.[1] While some researchers have successfully extended and applied the Turbo principle [9] to more sophisticated detection problems, most iterative decoding algorithms were derived heuristically. In 1995, Wiberg *et al.* [27], [26] presented a systematic way to devise a class of iterative decoding algorithms based on the extended Tanner graphs (also called the TWL graphs) of codes. Subsequently, McEliece *et al.* [18] and others pointed out that Pearl's famous belief propagation technique [21] can be used to systematically derive iterative decoding algorithms based on Bayesian belief network

---

*Department of Electrical and Electronics Engineering, Hong Kong University of Science and Technology, Clear Water Bay, Kowloon, Hong Kong (Email: w.mow@ieee.org).

[1] Here the term "near-optimality" is used to describe the observation that the performance of a suboptimal decoder is only a fraction of a dB in signal-to-noise ratio worse than the optimal decoding performance at a specified error probability (e.g. at 10^{-5}) in the reported computer experiments.

representations of codes. The two approaches are equivalent and generate essentially the same class of iterative decoding algorithms. In 1997, the factor graph was collectively introduced by a group of researchers as a generalization of Tanner graphs and other graphical models (see [14]). The tutorial paper of Kschischang *et al.* [15] has demonstrated that the sum-product algorithms on factor graphs generalize many well-known algorithms in various research areas, among which iterative decoding algorithms are only a special case. It is clear that graphical modelling of codes plays a very important role in the theoretical understanding and practical applications of iterative decoding algorithms.

In the area of coding, the most frequently used graphical model is the trellis, which is a weighted layered graph representation for codes. It is well-known that the maximum likelihood decoding (MLD) problem can be formulated as a shortest path search on a trellis. The latter can be solved efficiently by the celebrated Viterbi algorithm (VA), which is closely related to the classical dynamic programming technique of Massé and Bellman (c.f. [10, Section 9]). It is also well-known that any linear convolutional or block code can be represented by a trellis. Determination of the least complicated trellis representation of codes has recently been a topic of intensive study. The complexity profile of a code trellis is typically measured by the numbers of nodes or edges along the time axis of the trellis. The VA has also been widely applied to many other detection problems in the wide areas of communications and signal processing, such as the demodulation of continuous phase modulation schemes, the sequence detection for intersymbol-interference channels, and the multiuser detection for asynchronous code-division multiple-access channels.

Because of its ubiquitousness, the trellis representation and the VA have already become essential textbook materials in the area of digital communications. While the VA is in essence a solution to the problem of maximum *a posteriori* (MAP) sequence detection for hidden Markov models [5], most textbooks introduce it as a trellis search algorithm. Although its other (such as probabilistic and algebraic) formulations are also available in the research literature, the shortest path search formulation is by far the simplest and most understandable interpretation for the VA. This is because the path search interpretation allows maximum visualization of the algorithmic details involved. It is indispensable for efficient implementation of the VA, especially when practical path-oriented simplifications (such as the sequential decoding, the M- and the T-algorithms) are required as in many real-world applications.

With the advent of the Turbo coding technique, the iterative decoding algorithms become a very attractive approach to achieve near-MLD performance. The efficient graphical modelling of codes, say, by using Bayesian belief networks or factor graphs, holds the key to systematically devise low-complexity iterative decoding algorithms. As a result, the problem of determining the least complicated graphical representation of codes becomes very important. However, the definitions and usefulness of a complexity measure depend crucially on the choice of an appropriate graphical model, which in turns depends on the targeted application. Currently, the graphical model that attracts most attention in the coding community is the factor graph.

One might expect that factor graphs will soon become as popular as, if not more popular than, trellises. However, factor graphs can never take the place of trellises since they essentially model codes at different "resolution" levels. A trellis is a model for Markov chains, whose factor graphs always have a chain structure. In other words, the useful information of the state transition structure, which can be visualized in a trellis, is completely omitted in the factor graph representation. In this sense, the factor graph only generalizes the time axis of a trellis. Roughly speaking, a factor graph models the topology of time variables, whose structure need not be a chain. The fact that factor graphs do not allow the visualization of any generalized state transition structure calls for a more sophisticated graphical model. The introduction of a new graphical model, called the *hypertrellis*, that enables visualization of both the time topology and the generalized state transition structure is the main objective of this work. In this sense, the hypertrellis "generalizes" both the trellis and the factor graph. It should however be noted that unlike the factor graph, the hypertrellis is a *hypergraph* (rather than a bipartite graph) model.[2]

With the hypergraph formulation, it is now possible to generalize some very well-known results from trellises to hypertrellises. This generalizing approach is adopted here. Specifically, MLD on a hypertrellis will be formulated as a shortest hyperpath search. Based on this formulation, a MLD algorithm, called the *one-way algorithm*, which is a generalization of the VA, will be introduced for a class of hypertrellises. The one-way algorithm is functionally equivalent to the well-known min-sum algorithm [26, Chapter 3]) (also called Pearl's belief revision algorithm [21]) in the sense that they produce exactly the same hard decoding decisions. Their relationship is similar to that between the VA and the forward-backward max-MAP algorithm. The one-way algorithm is computationally simpler than the min-sum algorithm, and has a smaller storage requirement. Besides, the min-sum algorithm is commonly formulated as a message-passing mechanism while the one-way algorithm will be presented based on the accumulation of hyperpath metrics. It should be noted that a similar algorithm on factor graphs was hinted in Section 3.3 of Reference [26], but no details or properties of the algorithm were given therein.[3] Besides, it is anticipated

---

[2] Strictly speaking, the hypertrellis is not a true generalization of the factor graph, since the hypertrellis is a hypergraph while the factor graph is a bipartite graph.

[3] While preparing the final version of this paper, it is recognized that the so-called one-way algorithm is an instance of non-serial dynamic programming [20], [4] when applied to optimization problems. It has also been discussed by Shenoy [24] in a general axiomatic setting. Therefore, the one-way algorithm is actually well-known at least in

that the hyperpath search formulation of the one-way algorithm can provide additional insights into various issues on efficient implementation and reduced-complexity modifications of the algorithm.

The rest of this paper is organized as follows. Section 2 presents some basics of factor graphs. Section 3 introduces the background and the "starfish" drawing representation of hypergraphs.[4] Based on this representation, the factor hypergraph is introduced as the hypergraph equivalent of the factor graph. The hypertrellis model, which enables extension of the time axis of a trellis to any factor hypergraph called the time topology, is proposed in Section 4. MLD on hypertrellis is formulated as a shortest hyperpath search in Section 5. Section 6 introduces the one-way algorithm as a hyperpath-oriented MLD algorithm for hypertrellises with an acyclic time topology. Some connections between the hypertrellis and the so-called trellis formation, which is the consequence of another recent attempt to extend trellises and factor graphs, are discussed in Section 7. Finally, the concluding remarks are presented in Section 8.

2. Basics of factor graphs. This section presents the basics of factor graphs necessary for the discussions in the forthcoming sections.

A factor graph model shows how a multivariate *global* function factorizes into a product of *local* functions. Let g be the global function of a $|S|$-variable set denoted by $X_S = \{x_1, x_2, \cdots, x_{|S|}\}$, where S is the index set $\{1, 2, \cdots, |S|\}$. The variable domain, called the *alphabet*, can take values from any specified discrete set and the function can be defined on any commutative semi-ring. Let f_j, for $j = 1, 2, \cdots, N$, be a local function of a $|S_j|$-variable set denoted by X_{S_j} with $S_j \subset S$ such that the collection $Q = \{S_1, S_2, \cdots, S_N\}$ of subsets of S corresponds to the factorization

$$(1) \qquad\qquad g(X_S) = \prod_{j=1}^{N} f_j(X_{S_j}).$$

The factor graph $G = (V, E)$ representing this factorization is a *bipartite graph* with its *node* (or *vertex*) set V being partitioned into a *variable node* set $V_x = S$ and a (local) *function node* set $V_f = Q$. The *edge* set E is defined by all variable-function node pair $(v_x, v_f) \in (V_x, V_f)$ with $v_x \in v_f$. Typically $S = S_1 \cup S_2 \cup \cdots \cup S_N$ and Q alone is sufficient to characterize the topology of G.

As an example, for the (7,4,3) Hamming code, $Q = \{\{1, 2, 3, 5\}, \{1, 2, 4, 6\}, \{1, 3, 4, 7\}\}$. Figure 1 shows the factor (or actually Tanner) graph

---

the area of optimization. However, to the best of our knowledge, its formulation as a shortest hyperpath search is new.

[4]While preparing the final version of the paper, it is recognized that the proposed "starfish" drawing representation of hypergraphs is not new, and was used to depict hypergraph-theoretic language grammars by Habel [8].

representing the following factorization of its *codeword indicator function*

(2) $g(x_1, \cdots, x_7) = (x_1 \oplus x_2 \oplus x_3 \oplus x_5) \times (x_1 \oplus x_2 \oplus x_4 \oplus x_6) \times (x_1 \oplus x_3 \oplus x_4 \oplus x_7)$,

which equals 1 if the arguments (i.e. $x_1, x_2, \cdots, x_7$ in this case) form a valid codeword and equals 0 otherwise. Here, all variables and local functions take only binary 0-1 values, "$\oplus$" denotes the mod-2 addition, and "$\times$" denotes the usual integer multiplication. Note that in this case, the multiplication and addition operators in the commutative semi-ring are specialized to mod-2 addition and integer multiplication operators respectively. In Figure 1, variable nodes are drawn as circles and function nodes as "$\oplus$". Every edge is drawn as a line joining the variable-function node pair it specifies.

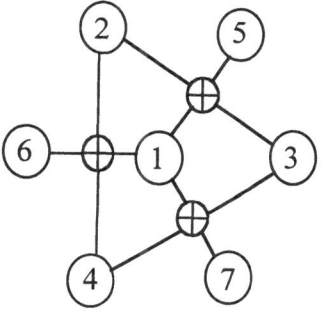

FIG. 1. *The factor graph of the (7,4,3) Hamming code.*

A trellis $T = (V, E)$ is a layered graph, where the node set V partitions into m disjoint classes of nodes denoted by $V_1, V_2, \cdots, V_m$ respectively, and the edge set E partitions into $m - 1$ disjoint classes of edges denoted by $E_1, E_2, \cdots, E_{m-1}$ respectively. For every edge $e_i \in E_i$ with $1 \leq i < m$, $e_i = \{v_i, v_{i+1}\}$ for some $v_i \in V_i$ and $v_{i+1} \in V_{i+1}$. The labels of node classes can be interpreted as discrete time indices, so that a node v_i represents a *state* at time i and an edge e_i represents a possible *state transition* from time i to time $i + 1$. Note that though it is typical that a trellis begins with a known state (i.e. $|V_1| = 1$), our definition of a trellis does not impose any restriction on the sizes of the node classes. As mentioned in Section 1, a factor graph in essence generalizes only the time axis of a trellis, but omits the representation of the state transition structure. Figure 2 exemplifies this insufficiency of the factor graph representation, which is the key motivation behind this work. Two different trellises corresponding to the same factor graph are shown in the figure. As a matter of fact, any trellis with the same time axis (or the same number of layers) corresponds to the same factor graph. This is of course the consequence of the lack of any state transition information in the factor graph representation. Further properties of factor graphs can be found in References [14] and [15].

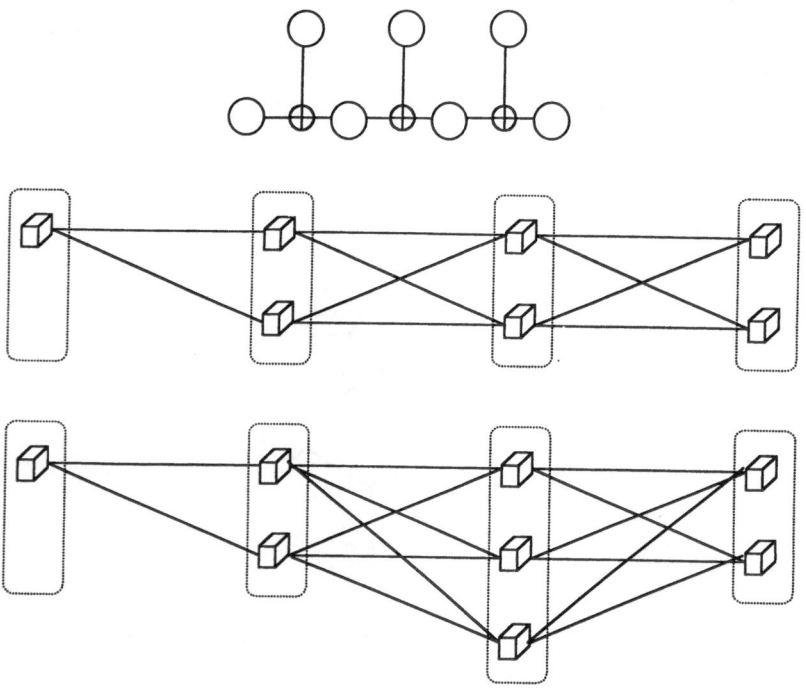

FIG. 2. *Two different trellises corresponding to the same factor graph.*

3. Hypergraph modelling. For some modelling problems, the hypergraph notion is required as the conventional graph concept becomes insufficient. This section introduces the hypergraph model, which is a key concept required for the introduction of the hypertrellis model. It should be noted that the "starfish" drawing representation to be presented is for general hypergraph models, though only factor hypergraphs and hypertrellises will be discussed in this paper.

3.1. Definitions. A hypergraph is a pair $H = (V, E)$, where the *node* set V is defined as $\{1, 2, \ldots, |V|\}$ and the *hyperedge* set $E = \{e_1, e_2, \ldots, e_{|E|}\}$ is a collection of subsets of V. Note that V is defined as an index set here for the sake of notational convenience, and can in fact be replaced by any other set with the same number of elements. Note also that by taking the hyperedge set as the second class of nodes, a hypergraph has also a description as a bipartite graph. The *degree* of a node is the number of hyperedges to which it belongs (or it is graphically joined). The *degree* of a hyperedge is simply its set size (or the number of nodes it graphically joins). An edge is a hyperedge of degree 1 or 2.[5] A hyperedge is said to be *proper* if it is not an edge. In words, a hypergraph is a generalization of a graph so that three or more nodes can be joined together by a single hyperedge. Clearly, if a hypergraph has no proper hyperedge (i.e. all hyperedges are edges), it reduces to a graph. Note that all hypergraphs considered in this section are *undirected*. A *directed* hypergraph can be defined by associating a direction to every hyperedge, as will be done in Section 4.

The following definitions will be useful in the forthcoming discussions. A *leaf* node is a node of degree one. A hypergraph is *weighted* if every hyperedge is associated with a value called a *weight*. A connected hypergraph H is said to be *cyclic* if it contains at least one hyperedge whose removal does not disconnect H. Only connected hypergraphs will be considered here. Any hypergraph that is not cyclic is said to be *acyclic*. The dual of a hypergraph is defined by exchanging the respective roles of the node set and the hyperedge set. Mathematically, the *dual hypergraph* $H^* = (V^*, E^*)$ of a hypergraph $H = (V, E)$ is defined by $V^* = \{1, 2, \cdots, |E|\}$ and $E^* = \{e_1^*, e_2^*, \ldots, e_{|V|}^*\}$, where $i \in e_j^*$ if and only if $j \in e_i$, for all $i \in V^*$ and $j \in V$. It is not difficult to see that the dual of H^* is H itself.

3.2. Traditional "closed-curve" drawing representation. Like a graph, a hypergraph is conventionally drawn as a set of circles representing the nodes, and a set of lines joining one or two nodes representing the edges. Every proper hyperedge of degree k (i.e. $k \geq 3$) is traditionally drawn as a simple closed curve encompassing the k nodes it joins. Refer to [2] and [16] for further background knowledge on hypergraphs.

As an example, the hypergraph in Figure 1 of [2] is reproduced here as Figure 3. In this example, there are eight nodes and six hyperedges, three out of which are edges. Namely, in our notation, $e_1 = \{3, 4, 5\}$, $e_2 = \{5, 8\}$, $e_3 = \{6, 7, 8\}$, $e_4 = \{2, 3, 7\}$, $e_5 = \{1, 2\}$, and $e_6 = \{7\}$. Even from this small example, it is not difficult to see that drawing and labelling

---

[5] A degree-1 edge is viewed here as a dangling edge connected to one node only and is in fact unpopular in the traditional graph representation.

hyperedges could be quite challenging in general, especially for hypergraphs having many high-degree hyperedges.

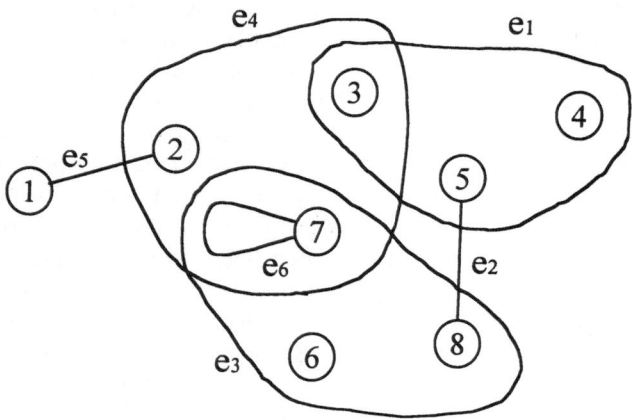

FIG. 3. *The traditional "closed-curve" representation of a hypergraph example.*

Almond [1] pointed out that hypergraphs are difficult to visualize and draw, and hence may be less favorable than its alternative graphical models. Indeed, difficulties in visualization and drawing may limit the application of hypergraph models to mostly simple unrealistic models.

3.3. The "starfish" drawing representation. It is observed that these disadvantages of hypergraph models are unessential, but are mainly due to the traditional manner of drawing proper hyperedges. To overcome this problem, we introduce the "starfish" drawing representation of proper hyperedges, which can be viewed as a consequence of the bipartite graph description of a hypergraph. A proper hyperedge of degree k is drawn as a starfish-like k-line segment, each of the k ends is connected to one node. Compared with the traditional "closed-curve" representation, it is in fact a more natural extension of the single-line representation of edges. Figure 4(a) shows the "starfish" representation of the same hypergraph shown in Figure 3. Though unessential, for the sake of clarity, a solid circle is added to the "starfish" center of every proper hyperedge. This representation avoids the unnecessary complication arising from drawing multiple overlapping closed curves and allows hyperedges to be labelled in a straightforward manner. As we shall see later, the "starfish" representation allows rather complicated hypergraphs to be drawn and labelled in a systematic manner and enables easy visualization.

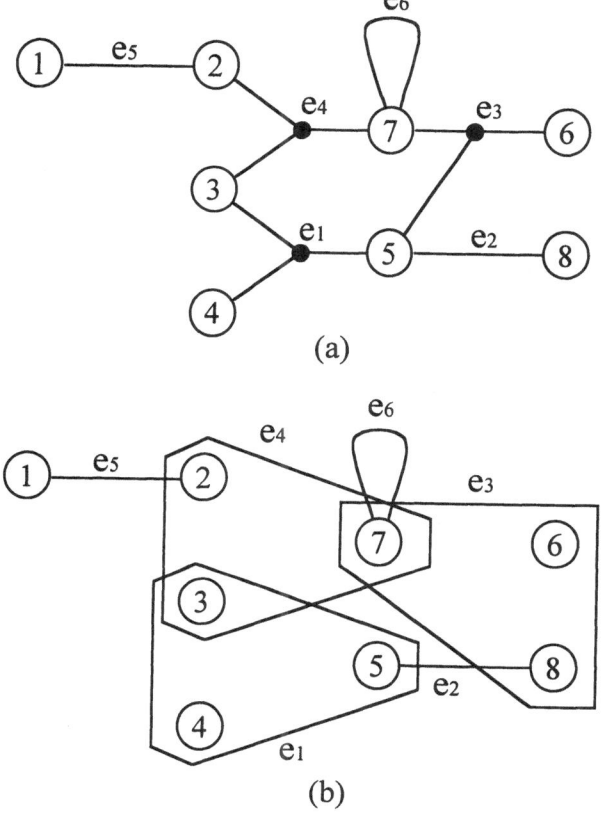

FIG. 4. *The hypergraph example in Figure 3 redrawn in: (a) the "starfish" representation; (b) the "closed-curve" representation after rearrangement.*

To show that the clarity of Figure 4(a) is not merely due to careful re-location of the nodes, the same hypergraph is redrawn in the "closed-curve" representation in Figure 4(b). While the latter is an obvious improvement of Figure 3 (thanks to the insight gained from Figure 4(a)), drawing and la-belling proper hyperedges are still non-straightforward. The relative merit of the "starfish" drawing representation will become even more obvious when practical hypergraph models of large sizes are represented.

3.4. Factor hypergraph. Without loss of generality, let the two classes of nodes in a bipartite graph be represented by circles and "$\oplus$" respectively. The existence of the aforementioned bipartite graph description for any hypergraph suggests that the two graphical models can be made mathematically equivalent. Furthermore, they are almost graphically identical with the "starfish" drawing representation of hypergraphs. The "starfish" of a proper hyperedge of degree k (with the central solid circle) resembles a "$\oplus$" node together with its k connecting edges in a bipartite graph. The only exception is when $k = 2$ and the hyperedge reduces to an edge, which is simply drawn as a single-line segment without any solid circle. It is noteworthy that the role of a solid circle in a hyperedge "starfish" is like that of a "soldering point" in the drawing of a circuit diagram. Its presence is for the purpose of clarity and, in some cases, may be omitted without ambiguity. For example, all solid circles in Figure 4(a) can be omitted without ambiguity. Figure 5 shows the factor graph corresponding to the hypergraph in Figure 4(a). Note however that this equivalence relationship depends on how the two node classes in the bipartite graph are labelled. More specifically, if every circle node together with its connecting edges in the bipartite graph is interpreted as a hyperedge, a hypergraph that is the dual of the previous one is resulted.

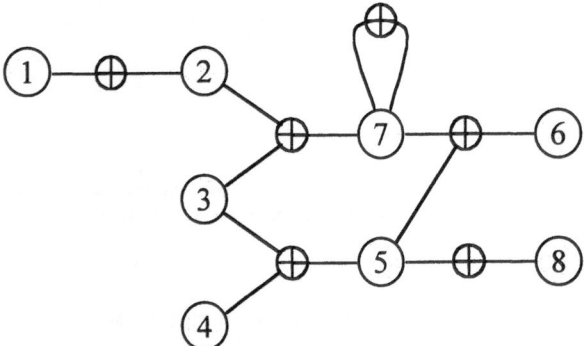

FIG. 5. *The factor graph corresponding to the hypergraph in Figure 4(a).*

Based on this graphical equivalence, any factor graph drawing can be interpreted as a factor hypergraph. A factor hypergraph $H = (V, E)$ has the node set $V = V_x$ and the hyperedge set $E = V_f$, where V_x and V_f are the respective variable and function node sets of the equivalent factor graph. There is no factor hypergraph counterpart corresponding to the definition of an edge set in a factor graph. In this sense, the mathematical definition of a factor hypergraph is simpler than that of a factor graph. Following the discussion, it is now unsurprising that any algorithm defined on a factor graph has a factor hypergraph equivalent, and vice versa.[6]

In the "starfish" drawing representation, the choice of using a solid circle to indicate the center of a proper hyperedge is arbitrary. In other words, the solid circle can be replaced by any other marker. If it is replaced by "$\oplus$", any hypergraph having only proper hyperedges has exactly the same appearance as a factor graph. For example, Figure 1 may as well be interpreted as the factor hypergraph of the (7,4,3) Hamming code. In subsequent sections, whether a graphical drawing should be interpreted as a factor graph or as a factor hypergraph will be clarified by the context of discussion. From here on, only the "starfish" drawing representation of hypergraphs will be used.

4. Hypertrellis modelling. The graphical similarity between factor hypergraphs and factor graphs implies that both graphical models also share similar shortcomings. In particular, both models have omitted the state transition information as discussed in Section 2. The discussion so far seems to suggest that factor hypergraphs are simply as good (or as bad) as factor graphs as a graphical model. Is there any significant advantage of factor hypergraphs over factor graphs besides its slightly simpler mathematical definition? Our introduction of factor hypergraphs is simply an intermediate step to the derivation of the hypertrellis model, which is a natural weighted hypergraph extension of trellises. In our opinion, the trellis-like extensions of other graphical models such as factor graphs and Bayesian networks are also possible. However, they appear to be less natural and more complicated without applying the notion of a weighted hypergraph. This topic will be further discussed in Section 7.

4.1. Definitions. A *hypertrellis* is a triple $T = (H, V, E)$, where the *time topology* $H = (V^H, E^H)$ with $V^H = \{1, 2, \cdots, m\}$ and $E^H = \{e_1^H, e_2^H, \cdots, e_n^H\}$ is a hypergraph, the node set V partitions into m disjoint classes of nodes denoted by $V_1, V_2, \cdots, V_m$ respectively, and the hyperedge set E partitions into n disjoint classes of hyperedges denoted by $E_1, E_2, \cdots, E_n$ respectively. Define the one-to-one mapping α by $\alpha(i) = V_i$, for $i = 1, 2, \cdots, m$. It maps the ith node of the time topology to the ith

---

[6]According to Almond [1], Kong [11] has already pioneered the use of hypergraphs to represent the factorization of a global belief function in as early as 1986. It is likely that the so-called factor hypergraph is just a rediscovery of Kong's factorization hypergraph.

node class of the hypertrellis. For $j = 1, 2, \cdots, n$, the jth hyperedge class E_j is a subset of the Cartesian product of all node classes in $\alpha(e_j^H)$, where α operates on e_j^H element-wise. Note that a hypertrellis T is itself a hypergraph with the node set V and the hyperedge set E. Only hypertrellises with a *connected* time topology are considered in this paper.

As a hypergraph, a hypertrellis is *weighted* if each of its hyperedges is assigned a value called a *hyperedge weight*. Adopting the notation in Section II, if a weighted hypertrellis T represents the factorization structure of a global function g, its time topology is the factor hypergraph of g (i.e. $(V^H, E^H) = (S, Q)$), the ith node class V_i is the alphabet of X_i, the jth hyperedge class E_j is the alphabet of X_{S_j}, and the hyperedge $e \in E_j$ representing the realization $X_{S_j} = e$ has a hyperedge weight $\lambda(e) = f_j(e)$.

A hypertrellis is said to be *t-acyclic* if its time topology is an acyclic hypergraph. Otherwise, it is said to be *t-cyclic*. Note that it is possible that a t-acyclic hypertrellis is a cyclic hypergraph.

A *complete hyperpath* $P = (V^P, E^P) = (\{v_1^P, v_2^P, \cdots, v_m^P\}, \{e_1^P, e_2^P, \cdots, e_n^P\})$ in T is a sub-hypergraph of T with $v_i^P \in V_i$ for $i = 1, 2, \cdots, m$, $e_j^P \in E_j$ for $j = 1, 2, \cdots, n$ and $e_1^P \cup e_2^P \cdots \cup e_n^P = V^P$. Any complete hyperpath in T is isomorphic to its time topology. Any connected sub-hypergraph of a complete hyperpath is called a *hyperpath*. If a hyperpath is not complete, it is said to be *partial*. A hyperpath is said to be *weighted* if each of its hyperedge is assigned a hyperedge weight. The sum of all hyperedge weights associated with a hyperpath is called the *hyperpath metric*. It is assumed that every hyperedge in T belongs to at least one complete hyperpath. Therefore, a hypertrellis is essentially equivalent to the union of all of its complete hyperpaths.

A hypertrellis specializes to a trellis if its time topology reduces to a time axis, that is isomorphic to a chain graph. In our notations, it means that after a proper re-labelling of the edges in the time topology, $e_j^H = \{j, j+1\}$, for $j = 1, 2, \cdots, n = m - 1$. In words, a hypertrellis is the generalization of a trellis, whose time axis is extended to a time topology represented by the equivalent factor hypergraph. A trellis is conventionally viewed as a two-dimensional graph, whose nodes are located according to the time and state axes. From this perspective, a hypertrellis is a generalized trellis whose time axis has been extended from a chain graph to a hypergraph. It is also an extension of the factor hypergraph so that the generalized state transition structure can also be visualized.

4.2. Drawing representation. As a hypertrellis is also an m-partite hypergraph, the "starfish" drawing representation can be applied. For practical applications, it is important to find a systematic way of drawing hypertrellises that facilitates the visualization of the relationships among hyperpaths. To achieve this, one can imagine a hypertrellis as a 3-dimensional network model with all nodes in V_i grouped together in the same horizontal but different vertical position. The hypertrellis can then be drawn as a projected 2-dimensional view of the 3-dimensional network model.

In Figures 6 to 10, the following notations for drawing hypertrellises will be adopted. Each node is drawn as a cube. All nodes encircled by a dotted line belong to the same node class. To highlight the all-zero hyperpath, all of its hyperedges are drawn as bold lines. The upper and lower nodes of the same class represent the bits 0 and 1 respectively.

Figure 6 shows the hypertrellis and the factor hypergraph of the (3,2) single parity check code. Node classes in the hypertrellis are in one-to-one correspondence with nodes in the time topology. As there is only one hyperedge in the time topology, it corresponds to a single *hypertrellis section*. There are a total of four hyperedges in the section. Every three nodes joined together by a hyperedge specifies a valid codeword.

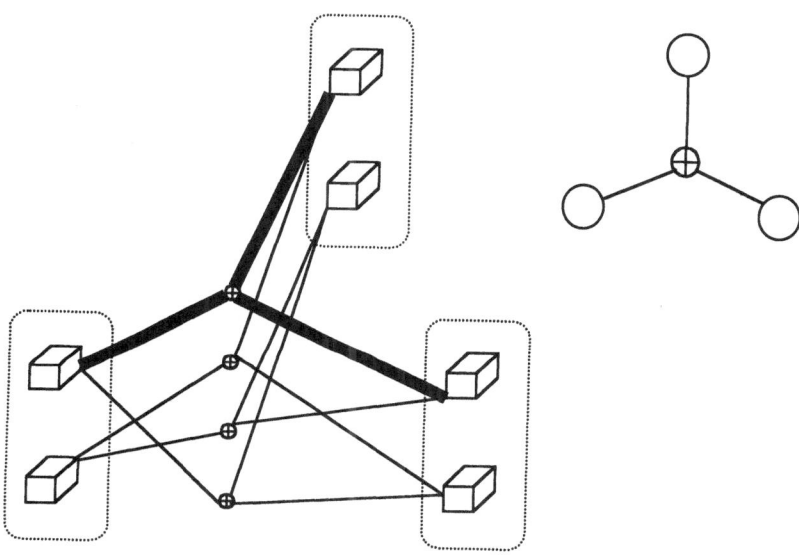

FIG. 6. *The hypertrellis and the factor hypergraph of the (3,2) single parity check code.*

Figure 7 shows the hypertrellis and the factor hypergraph of a (7,4,2) linear binary code. There are seven node classes and three hypertrellis sections. Hyperedges in the same section belong to the same class. Node classes in the hypertrellis are in one-to-one correspondence with nodes in the time topology. Hypertrellis sections in the hypertrellis are in one-to-one correspondence with hyperedges in the time topology. Hypertrellis

sections are joined together via common nodes. Nodes joined together by
a hyperedge specify a valid codeword segment for the corresponding variables. In other words, every hyperedge specifies a possible choice of the
corresponding codeword segment. Consequently, *every valid codeword corresponds to a unique complete hyperpath in the hypertrellis.* A rigorous way
to derive this one-to-one correspondence between codewords and complete
hyperpaths is to set all hyperedge weights (i.e. all local function values
$f_j(.)$'s) to 1 so that the global function g becomes the codeword indicator
function. Therefore, a code hypertrellis, even without specifying weights,
allows graphical visualization of the codeword indicator function.

Figure 8 shows some possible factor graph and hypertrellis representations of a 3-section 2-state unterminated trellis. It is widely accepted in
the literature that the factor graph in Figure 8(a) corresponds to the time
axis of the trellis in Figure 8(d). Contrary to the conventional wisdom, the
introduced hypertrellis model suggests that the factor graphs in Figures
8(a) and (b) should correspond to the time topology of the hypertrellis in
Figure 8(c) and the time axis of the trellis in Figure 8(d) respectively.

Figure 9 shows the hypertrellis of the (7,4,3) Hamming code, whose
factor graph was shown in Figure 1 and whose codeword indicator function
was expressed in (2). The hypertrellis and the factor hypergraph of a Turbo
code with unit-memory recursive systematic convolutional components and
a 3-bit interleaver are shown in Figure 10.

Note that there is no degree-1 edge in the time topologies of the hypertrellises shown in Figures 6 to 10 because all the local functions have
more than one variables as their arguments. In general, a local function
of one variable specifies the weights assigned to a class of dangling edges,
each of which connects to a distinct node in a node class. For notational
simplicity, all degree-1 edges are omitted and their weights are shown next
to their associated nodes. As an example, Figure 11 shows the *decoding
hypertrellis* equivalent to the factor (actually Tanner) graph example of the
(7,4,2) binary linear code in Figure 3 of Reference [6], which is actually a
polished version of Wiberg's example presented in Figure 3.3 of Reference
[26]. The hyperedge weights therein are given by the correlation metrics
derived from the outputs of an additive white Gaussian noise channel.

4.3. Complexity measures. The efficiency or complexity of a specific graphical model for a given code is reflected in some combinatorial
properties of the model in use. For the hypertrellis model, two natural
complexity measures are the *node-complexity profile* and the *hyperedge-
complexity profile.* The node-complexity profile (NCP) is defined by the
sequence $(|V_1|, |V_2|, \cdots, |V_m|)$, where $|V_i|$ represents the number of nodes in
the ith node class. The hyperedge-complexity profile (ECP) is defined by
the sequence $(|E_1|_{|e_1|}, |E_2|_{|e_2|}, \cdots, |E_n|_{|e_n|})$, where $|e_j|$ and $|E_j|$ represents
the degree and the number of hyperedges in the jth hyperedge class respec-
tively. They are reasonably good estimates of the complexity of decoding

algorithms with a hypertrellis formulation and may be used as efficiency criteria for comparing different hypertrellis representations of the same code.

As an example, consider the hypertrellis of the (7,4,3) Hamming code in Figure 9. Its NCP and ECP are $(2,2,2,2,2,2,2)$ and $(8_4, 8_4, 8_4)$ respectively.

5. Maximum likelihood decoding as a shortest hyperpath search. For MLD over memoryless channel, the commutative semi-ring, on which the global function g is defined, is typically taken as the min-sum real-number ring. That is, the multiplication and addition operators in the general semi-ring are specialized to the real-number addition and minimum operators respectively. Adopting the mathematical notations in Section 2, the factorization in (1) becomes

$$g(X_S) = \sum_{j=1}^{N} f_j(X_{S_j}).$$

The goal of MLD is to find the most likely *information symbols* corresponding to the value of X_S (i.e. the most likely codeword) with

$$\min_{X_S} g(X_S),$$

where the minimization is taken over all possible values of X_S. As discussed in Section 4.2, every value of X_S corresponds to a unique complete hyperpath in the code hypertrellis. Among all possible hyperpaths in the decoding hypertrellis, the one with minimum hyperpath metric gives the most likely codeword. Such a hyperpath is called the *shortest hyperpath*. Therefore, MLD can be viewed as a shortest hyperpath search on the decoding hypertrellis. This is a hypertrellis generalization of the well-known result that MLD can be interpreted as a shortest path search on the decoding trellis.

Remark. Strictly speaking, a shortest hyperpath search only finds the most likely codeword instead of information symbols as required by the MLD. This is not a problem for systematic codes, since information symbols are simply part of the codeword. In the case of non-systematic codes, the code should be first expanded into a systematic one by including all information symbol variables before being modelled by a hypertrellis. Typically, the information symbol variables are leaf nodes in the time topology and do not increase the ECP of the hypertrellis. For example, an MLD algorithm based on the hypertrellis in Figure 8(c) or the trellis in Figure 8(d) has the same complexity since the only function of the information bits (or the leaf nodes) is to label the hyperedges.

6. One-way algorithm for T-acyclic hypertrellises. The shortest hyperpath search interpretation of MLD on hypertrellis discussed in Section 5 is valid whether the hypertrellis is t-cyclic or t-acyclic (as defined in Section 4.1). It is well-known that the min-sum algorithm performs MLD on acyclic factor graphs [26]. Likewise, the *one-way algorithm* to be presented is an MLD algorithm on t-acyclic hypertrellises (i.e. also on acyclic factor graphs or hypergraphs). For this reason, we shall only consider t-acyclic hypertrellises in this section.

Note however that the one-way algorithm is not identical to the min-sum algorithm. The min-sum algorithm specializes to the Max-MAP algorithm,[7] while the one-way algorithm specializes to the VA.

Consider the time topology of a t-acyclic hypertrellis, which is essentially a factor hypergraph with a tree structure. The computation of the one-way algorithm progresses according to a *directed* version of the time topology called the *computational time topology*. By selecting a variable node as the root node, all variable nodes can be classified according to which levels of the tree they belong to. Let us adopt the convention that the root node is in level 0 and the level numbers are in increasing order. As a consequence of the tree structure of the acyclic hypergraph, all nodes connected by the same hyperedge have the same level number except one, whose level number is exactly one less than the others. A direction can be defined for every hyperedge by assigning the connected node with a smaller level number as a *sink* node and other connected nodes as *source* nodes. In this way, a directed acyclic hypergraph is resulted for every choice of the root node. Figure 12 shows an example of how to obtain a computational time topology from the acyclic factor hypergraph in Figure 7. In the figure, an arrow head is added to point to the sink node joined by every directed hyperedge.

The computational procedure follows the hyperedge directions of the computational time topology, namely, from leaf nodes to the root node. The set of level-k hyperedges are those joining level-k nodes and level-$(k+1)$ nodes. The computation for nodes and hyperedges on the same level may proceed in any order. This essentially imposes a *partial ordering constraint* in the computational procedure. For notational convenience, the one-way algorithm will be presented in decreasing levels of nodes and hyperedges. It is important to remember that the actual computational time schedule needs only to satisfy the aforementioned partial ordering constraint.

Denote the $(K+1)$-level partition of node classes $\{V_1, V_2, \cdots, V_m\}$ by $\{V_0^L, V_1^L, \cdots, V_K^L\}$, and the K-level partition of hyperedge classes $\{E_1, E_2, \cdots, E_n\}$ by $\{E_0^L, E_1^L, \cdots, E_{K-1}^L\}$. Note that $V_0^L = \{V_{\text{root}}\}$ contains only the root node class corresponding to the root node in the time topology. Define by $E(v) = \{e \in E : v \in e\}$ the set of all hyperedges joining node v, and recall that $\lambda(e)$ denotes the weight of the hyperedge e.

---

[7]Strictly speaking, it is the max-sum (rather than the min-sum) algorithm that specializes to the Max-MAP algorithm. However, the max-sum and the min-sum algorithms are algorithmically equivalent since the minimum operator becomes the maximum operator (and vise versa) if all of its operands are negated and the output is negated again.

The one-way algorithm recursively computes the so-called *node metric* $\Gamma(v)$ as follows:

Step 1: For all $v^L \in V_K^L$ and for all $v \in v^L$, set $\Gamma(v) = \lambda(\{v\})$.

(*Remark.* This step initializes all level-K node metrics as the corresponding degree-1 edge weights).

Step 2: For $k = K - 1, K - 2, \cdots, 0$, and for all $v^L \in V_k^L$ and for all $v \in v^L$,

$$(3) \qquad \Gamma(v) = \lambda(\{v\}) + \sum_{e^L \in E_k^L(v)} \left\{ \min_{e \in e^L} \left[\lambda(e) + \sum_{v' \in e - \{v\}} \Gamma(v') \right] \right\}$$

where $E_k^L(v) = \{e^L \cap E(v) : e^L \in E_k^L\}$ is the collection of all level-k hyperedge classes, which consist only of the hyperedges joining node v. After computing the minimum in (3) for each hyperedge class e^L, the *surviving* hyperedge in e^L that achieves the minimum is memorized.

(*Remark.* This step calculates, in a node-by-node level-by-level manner, the node metric $\Gamma(v)$ as the hyperpath metric of the *shortest partial hyperpath* terminating at node v, following the hyperedge directions in the computational time topology. It is noteworthy that the *add-compare-select-add* operations in (3) generalizes the well-known *add-compare-select* operations in the Viterbi algorithm.)

Step 3: Compute $\min_{v \in V_{\text{root}}} \Gamma(v)$ as the overall shortest complete hyperpath metric and identify the node $v_{\min}$ that achieves the minimum. The shortest hyperpath is then obtained by tracing back all surviving hyperedges starting from node $v_{\min}$. Each information symbol associated with the most likely codeword is decoded according to which node in the corresponding node class belongs to the shortest hyperpath.

An example will clarify the details of the algorithm. Figure 13 presents a detailed application of the algorithm to the decoding hypertrellis in Figure 11. Figures 13(a) and (b) show the surviving hyperedges and node metrics after the respective level-1 and level-2 computations in Step 2. As weights in the example are based on the correlation metric instead of the Euclidean metric, the minimum operation in (3) is replaced by the maximum operation (see also Footnote 7). In the figure, the node metrics $\Gamma(v)$'s and the hyperedge values calculated by the expression in the square brackets of (3) are shown. Surviving hyperedges are drawn as bold lines. From Figure 13(b), it can be seen that the "shortest" hyperpath metric is 15, and the overall "shortest" complete hyperpath is shown as a bold dotted line. The seven decoded bits, four of which are information bits, are also shown as node labels in a factor hypergraph. Unsurprisingly, the decoded bits in the example are exactly the same as those in Figure 3 of [6].

The example sheds light on some general implementation issues of the presented algorithm. Let a node v in a hypertrellis belong to the node class $\alpha(v^H)$, where v^H is a variable node in the time topology. (Recall the definition of the mapping α in Section 4.1.) The number of surviving hyperedges needed to be memorized at node v is $\deg(v^H) - 1$, where $\deg(v^H)$ denotes the degree of the variable node v^H. In particular, if v^H is a leaf node, it is unnecessary for node v to memorize any surviving hyperedge. The storage requirement depends on the number of hyperedges in the hyperedge class E_j joining v. For the example in Figure 13, there are only 4 out of 14 nodes, which need to memorize surviving hyperedges. Each of the two nodes in the center needs to memorize two hyperedges, and each of the other two nodes needs to memorize one hyperedge. There are a total of six selections of surviving hyperedges, each of which requires only 1 bit. Hence, a total of 6 bits are sufficient to store the whole *hyperpath history*. In addition, the number of node metrics to be stored is upper-bounded by the maximum number of nodes in one level of hypertrellis excluding the highest level K, i.e. $\max_{k=0}^{K-1} \sum_{v^L \in V_k^L} |v^L|$. The actual storage requirement is typically much less.

With the same performance, the computational complexity of the one-way algorithm is approximately half of that of the min-sum algorithm. Following the discussion above, the storage requirement of the former is also much lower than that of the min-sum algorithm, which typically requires to store the node metric for every node. Therefore, the presented one-way algorithm can achieve the same performance at much lower computational and storage complexities relative to the well-known min-sum algorithm.

It should be clear from the procedure that it is a hypertrellis generalization of the VA. If the central node of the computational time axis of a trellis is chosen as the root node, the one-way algorithm specializes to the bidirectional VA, which is well-known for halving the decoding delay by doubling the hardware complexity. For general hypertrellises, different choices of the root node in the computational time topology may result in different decoding delays. For delay-constrained applications, a high degree of parallelism in the decoding algorithm is desirable. Assuming maximal parallelism, it is not difficult to see that the minimum achievable delay in terms of the number of hyperedges is equal to approximately half of the diameter of the time topology. The minimum decoding delay achievable by the min-sum algorithm is twice as long.

7. Connections with trellis formations. Independent of our work on hypertrellises [19], there is another graphical generalization of factor graphs and trellises, called *trellis formations*, proposed by Kötter and Vardy [13]. As suggested by its name, it is a graphical structure based on trellis sections. In general, a factor graph has to be properly *normalized* before it can be converted into a trellis formation. A normalized factor graph satisfies the constraint that each of its function nodes must join no more than two state nodes [12] so that it corresponds to a trellis section.[8] In our terminology, the normalization process ensures that the factor hypergraph must be a graph.

There is another type of normalized factor graphs which are equivalent to Forney's generalized state realization graphs. This second type of normalized factor graphs satisfy the constraint that each state node must be joined by no more than two edges [12]. In our terminology, this second type of normalization ensures that the dual of the factor hypergraph must be a graph (c.f. Section 3.1 for the definition of a dual hypergraph).

Since the time topology of a hypertrellis can be an arbitrary hypergraph, trellis formations are equivalent to a subclass of hypertrellises whose time topologies satisfy the aforementioned graph constraint. From our viewpoint, the graph constraint on factor hypergraphs is necessary simply because trellis formations are constructed based on trellis sections. With the concept of hypertrellis sections, any factor graph (or its equivalent factor hypergraph) can be expanded into hypertrellis in a systematic manner as discussed.

Finally, the differences in the respective complexity measures for trellis formations and hypertrellises deserve some remarks. As an example, consider the (7,4,3) Hamming code again. The node-complexity and edge-complexity profiles of its trellis formation are $(2, 2, 2, 2, 2, 2, 2, 8, 8, 8) = (2^7 8^3)$ and $(8, 8, 8, 8, 8, 8, 8, 8, 8, 8, 8, 8) = (8^{12})$ respectively, while the node-complexity and hyperedge-complexity profiles of its hypertrellis are (2^7) and (8_4^3) respectively, as given in Section 4. The complexity measure of a code depends on the graphical model in use. In [12], the product of all elements in the node-complexity profile of a trellis formation was used to measure the complexity of a particular code representation. For a given code, the simplest representation under the trellis formation modelling and that under the hypertrellis modelling could be very different. As no normalization process of the factor graph is involved and trellis formations are equivalent to a subclass of hypertrellises, we believe that the complexity measure defined on hypertrellises should give a more accurate measure of the actual decoding complexity.

8. Concluding remarks. A "starfish" drawing representation of hypergraphs that enables simpler drawing and easier visualization than the traditional "closed-curve" representation has been introduced to enhance the applicability of hypergraph models. A hypergraph approach to modelling the function factorization was investigated. The hypergraph equivalents of factor graphs, called factor hypergraphs, were introduced. Using the "starfish" representation, the graphical appearance of the factor hy-

---

[8]A reviewer pointed out that Kötter introduced the constraint simply for easier mathematical handling.

pergraph was made almost identical to that of the equivalent factor graph. Hypertrellises have been proposed as a weighted hypergraph extension of trellises and factor graphs so that both the generalized time and state transition structures can be visualized graphically. It was pointed out that the recently proposed trellis formations are equivalent to a subclass of hypertrellises and the complexity measures defined on hypertrellises are likely to be more accurate.

MLD was formulated as a shortest hyperpath search on hypertrellises. For t-acyclic hypertrellises, the one-way algorithm was introduced as a solution to the shortest hyperpath search problem. It is more attractive than the well-known min-sum algorithm in terms of the computational complexity, the storage requirement and the minimal decoding delay.

The role of the one-way algorithm for hypertrellises is similar to that of the famous VA for trellises. It is anticipated that hyperpath-oriented simplifications of the one-way algorithm can be derived in a way similar to those well-known path-oriented simplifications of the VA such as the M-algorithm and the T-algorithm. This is one of our future research topics.

Finally, it should be remarked that while the hypertrellis model and the one-way algorithm were discussed in the context of decoding, its application areas beyond coding are numerous and are as wide as those of factor graphs [15].[9]

Acknowledgements. This author is deeply indebted to Prof. Joachim Hagenauer and his research team, Technische Universität München, for their hospitality and many stimulating activities during his visit in 1996 sponsored by the Humboldt Foundation. Without them, the motivation behind this work would not have existed. He sincerely thank Prof. Raymond Yeung, The Chinese University of Hong Kong, who noted what was originally called the "joint-oriented graph" is in fact the well-known hypergraph and brought his attention to Kötter's works in Reference [12]. He would also like to acknowledge Prof. Susumu Yoshida, Kyoto University, for providing an excellent working environment, in which the first draft of this paper was completed, during his 1-month visit in 2000 sponsored by the Telecommunications Advancement Organization of Japan. He is grateful to Prof. P. P. Shenoy, University of Kansas, for providing information about his works on valuation networks. Finally, he would like to thank the anonymous reviewer and the editors, Dr. J. Rosenthal and Dr. B. Marcus, for their efforts in detailed proofreading and improving the presentation of this paper.

---

[9]While preparing the final version of this paper, it is recognized that the factor graph is in fact a special case of Shenoy's valuation network [23] and the sum-product algorithm is an instance of the Shenoy-Shafer local computation architecture [25]. The latters are well-known in the area of probabilistic expert systems [22]. Therefore, the proposed hypertrellis model can as well be considered as an extension of the valuation network.

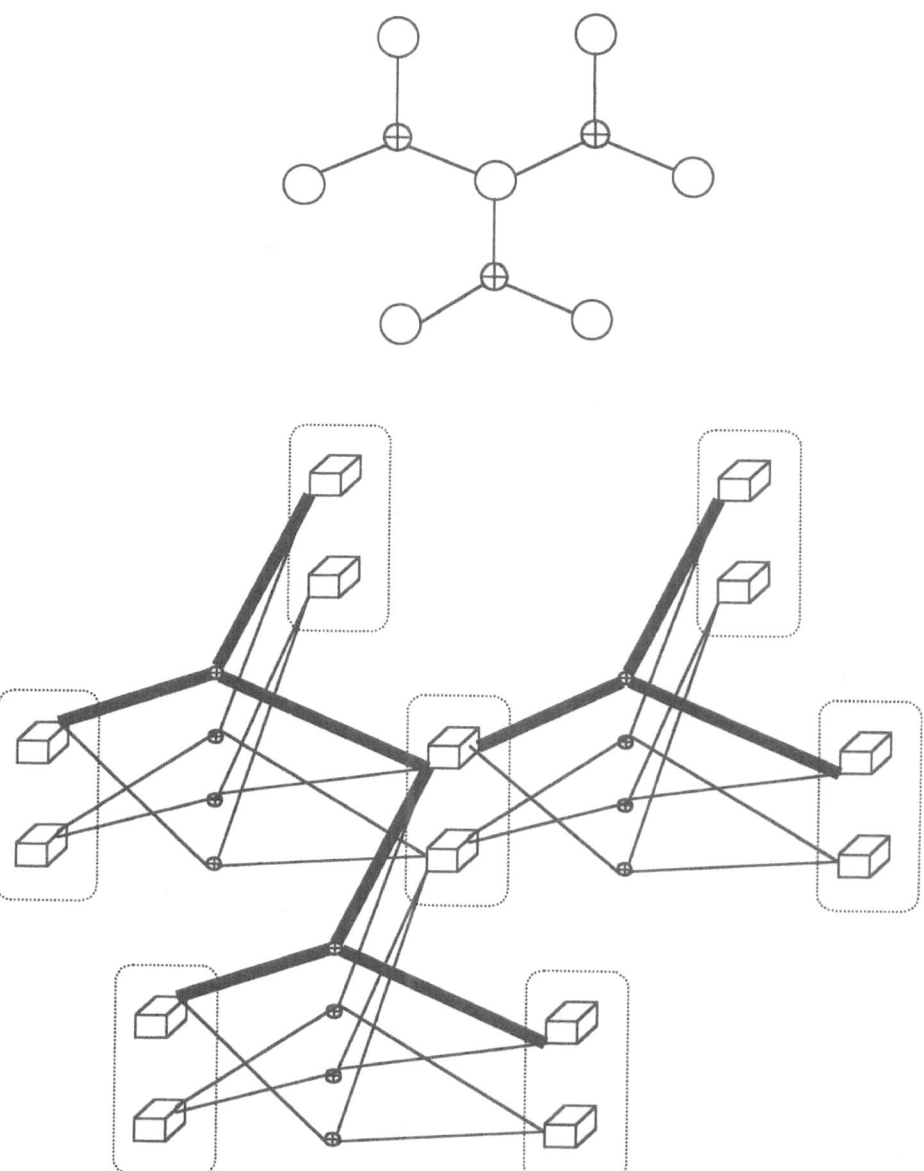

FIG. 7. *The hypertrellis and the factor hypergraph of the (7,4,2) linear binary code.*

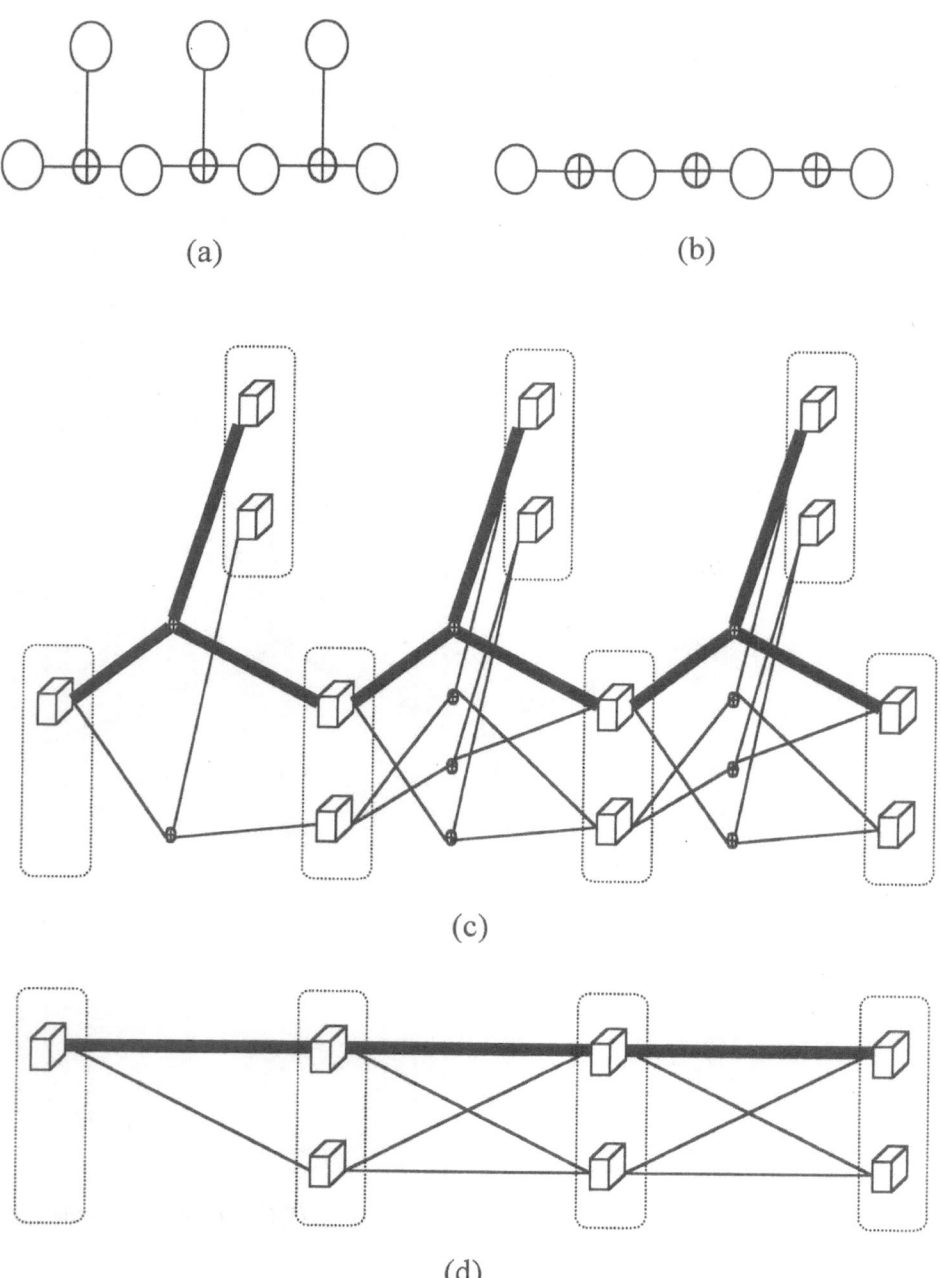

FIG. 8. *Various factor graph and hypertrellis representations for a 3-section 2-state unterminated trellis.*

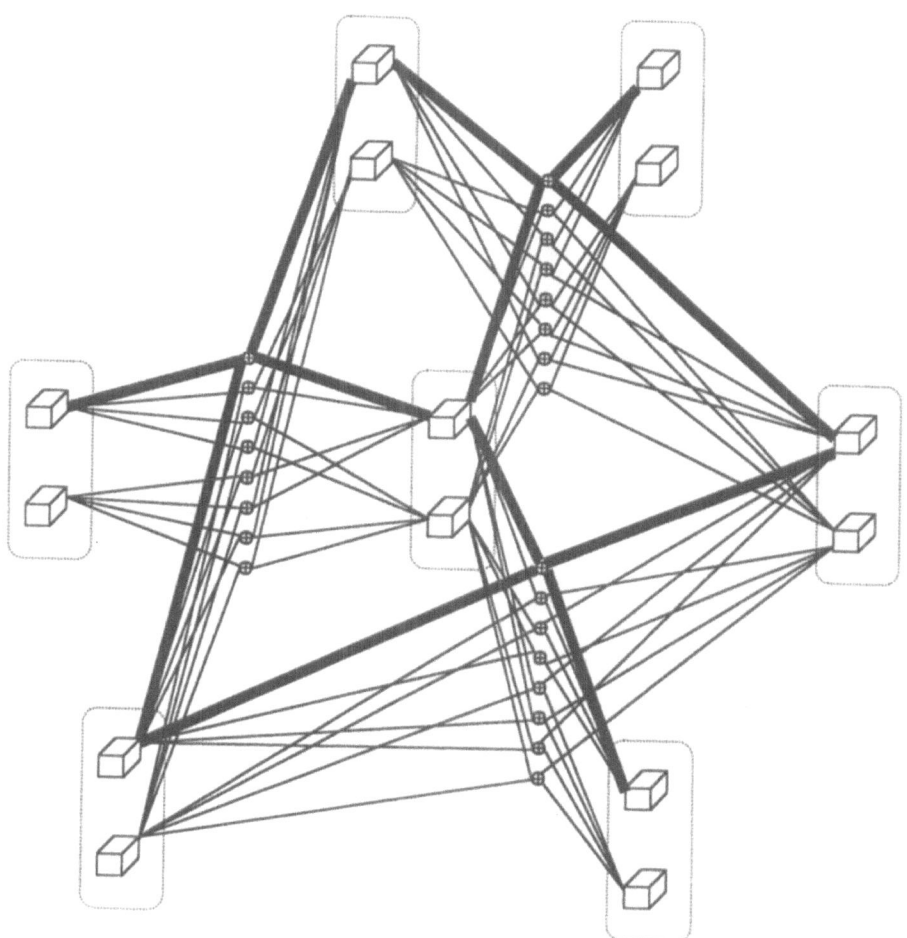

FIG. 9. *The hypertrellis of the (7,4,3) Hamming code.*

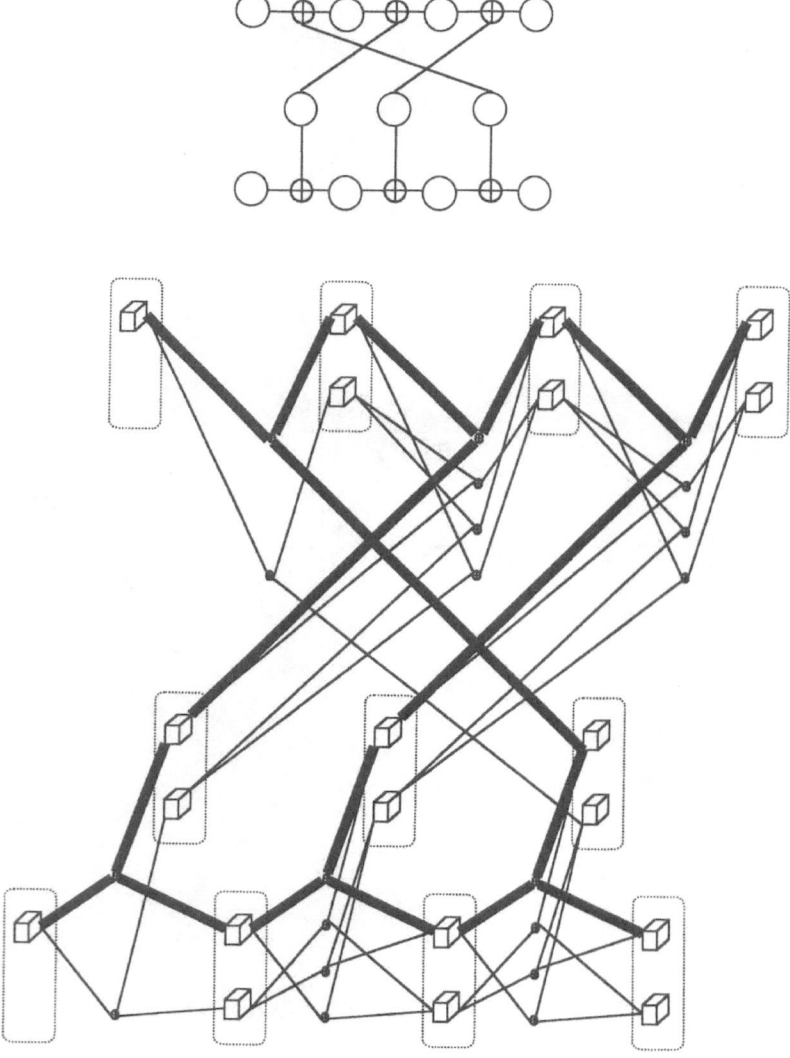

FIG. 10. *The hypertrellis and the factor hypergraph of a Turbo code with 2 unit-memory recursive systematic convolutional component codes and a 3-bit interleaver.*

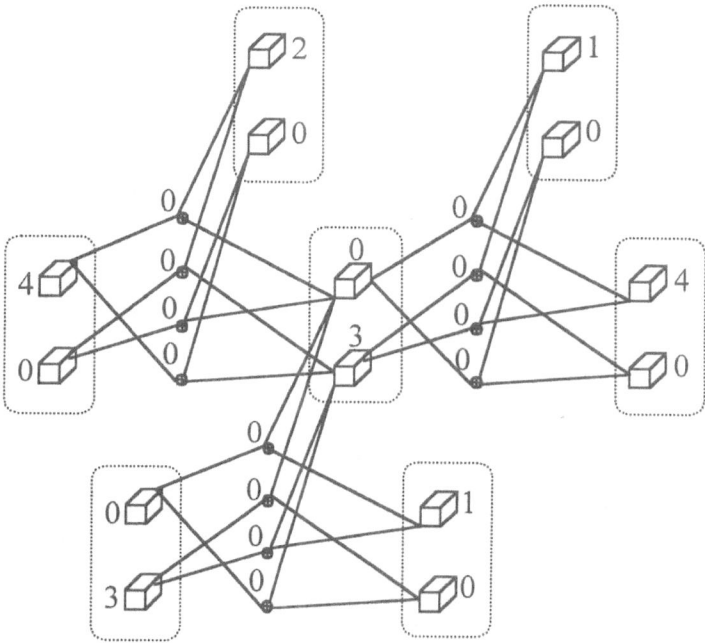

FIG. 11. *The decoding hypertrellis equivalent to the factor graph example of the (7,4,3) linear binary code in Figure 3.3 of Reference [6].*

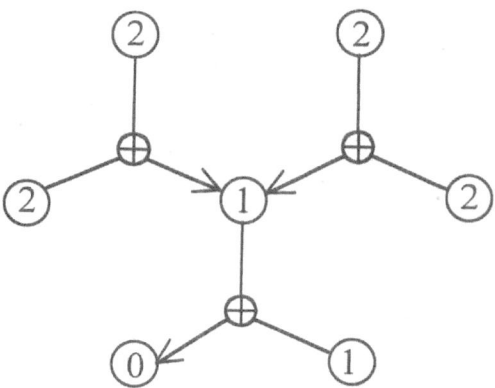

FIG. 12. *A computational time topology with tree level numbers as node labels, where the root node is labeled with 0.*

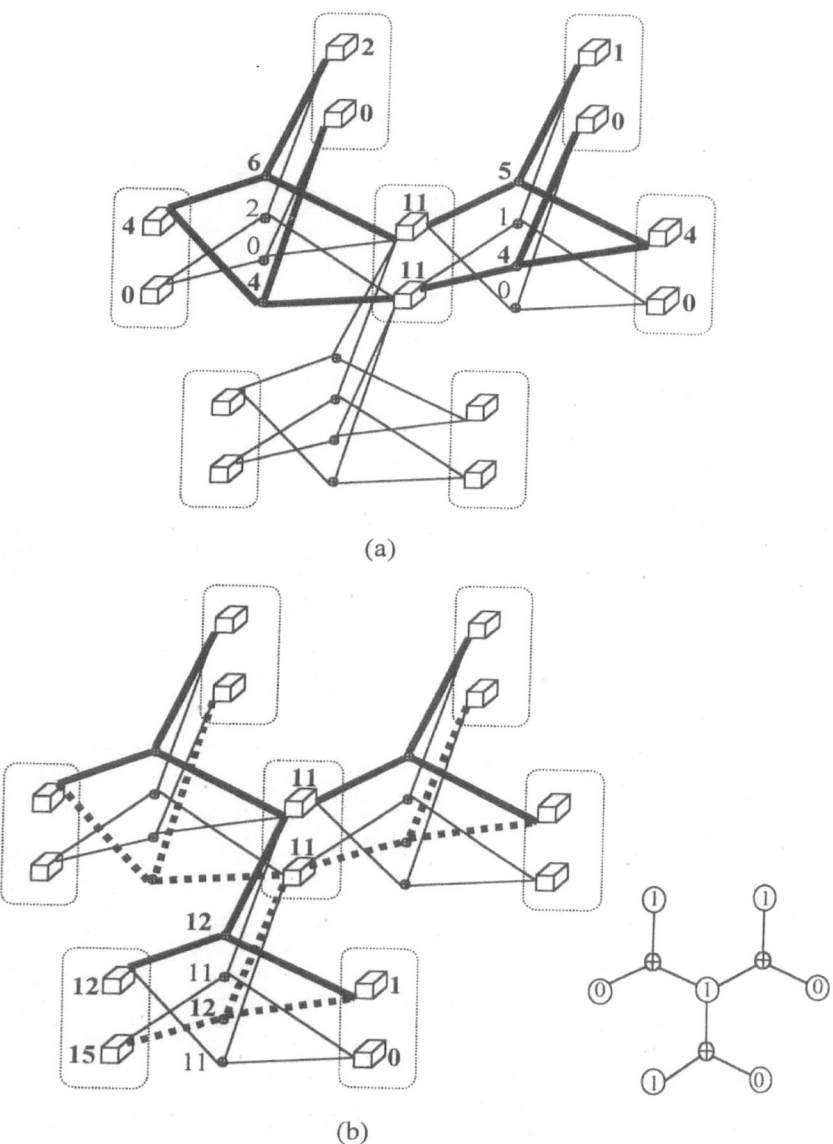

FIG. 13. *The one-way algorithm applied to the weighted hypertrellis in Figure 11. (a) Level-1 computation. (b) Level-0 computation and the decoded codeword.*

REFERENCES

[1] R.G. ALMOND, *Graphical Belief Modelling*, Chapman and Hall: London, 1995.

[2] C. BERGE, *Hypergraphs*, North-Holland, 1989.

[3] C. BERROU, A. GLAVIEUX, AND P. THITIMAJSHIMA, "Near Shannon limit error-correcting coding and decoding: Turbo codes," *Proc. IEEE Int. Conf. Commun. (ICC'93)*, Geneva, Switzerland, 1993, pp. 1064–1070.

[4] U. BERTELÈ AND F. BRIOSCHI, *Nonserial Dynamic Programming*, New York: Academic Press, 1972.

[5] G.D. FORNEY, "The Viterbi Algorithm," *Proc. IEEE*, **61**, pp. 268–278, Mar. 1973.

[6] G.D. FORNEY, "On Iterative Decoding and the Two-Way Algorithm," Proc. International Symposium on Turbo Codes, Brest, France, 1997.

[7] R.G. GALLAGER, *Low-Density Parity-Check Codes*, Cambridge, MA: MIT Press, 1966.

[8] A. HABEL, *Hyperedge Replacement: Grammars and Languages*, Berlin: Springer-Verlag, 1992.

[9] J. HAGENAUER, "The Turbo Principle: Tutorial Introduction and State of the Art," *Proc. International Symposium on Turbo Codes*, Brest, France, 1997.

[10] A. KAUFMANN, *Graphs, Dynamic Programming, and Finite Games*, translated by H.C. Sneyd, Academic Press, 1967, *Méthodes et Modèles de la Recherche Opérationnelle*, originally published in 1964 (in French).

[11] C.T.A. KONG, "Multivariate Belief Functions and Graphical Models," PhD thesis, Technical Report S-107, Harvard University, Department of Statistics, 1986.

[12] R. KÖTTER, "Factor Graphs, Trellis Formations, and Generalized State Realizations," presented at *the IMA Summer Program on Codes, Systems and Graphical Models*, August 2–6, 1999. Downloadable from *http://www.ima.umn.edu/talks/workshops/aug2-13.99/8-2-13.99.html*.

[13] R. KÖTTER AND A. VARDY, "Factor Graphs: Constructions, Classification, and Bounds," *Proc. International Symposium on Information Theory (ISIT'98)*, Cambridge, MA, USA, August 1998.

[14] F.R. KSCHISCHANG AND B.J. FREY, "Iterative Decoding of Compound Codes by Probability Propagation in Graphical Models," *IEEE J. Selected Areas in Commun.*, **16**, pp. 219–230, 1998.

[15] F.R. KSCHISCHANG, B.J. FREY, AND H.-A. LOELIGER, "Factor Graphs and the Sum-Product Algorithm," submitted to *IEEE Trans. Inform. Theory*, July 27, 1998. Downloadable from *http://www.comm.utoronto.ca/frank/factor/* .

[16] S.L. LAURITZEN, *Graphical Models*, Oxford University Press, 1996.

[17] D.J.C. MACKAY AND R.M. NEAL, "Near Shannon limit performance of low density parity check codes," *Electron. Lett.*, **32**(18), pp. 1645–1646, 1996.

[18] R.J. MCELIECE, D.J.C. MACKAY AND J.-F. CHENG, "Turbo decoding as an instance of Pearl's belief propagation algorithm," *IEEE J. Selected Areas in Commun.*, **16**, pp. 140–152, 1998.

[19] W.H. MOW, "Hypertrellis: a Generalization of Trellis and Factor Graph," material presented at the Hong Kong University of Science and Technology, July 7, 1999. Accepted by (though not presented at) *the IMA Summer Program on Codes, Systems and Graphical Models*, August 2–6, 1999. Downloadable from *http://www.ima.umn.edu/talks/workshops/aug2-13.99/8-2-13.99.html* .

[20] G.L. NEMHAUSER, *Introduction to Dynamic Programming*, New York: John Wiley & Sons, 1966.

[21] J. PEARL, *Probabilistic Reasoning in Intelligent Systems: Networks of Plausible Inference*. San Mateo, CA: Morgan Kaufmann, 1988.

[22] G. SHAFER, *Probabilistic Expert Systems*. CBMS-NSF Regional Conference Series in Applied Mathematics, **67**, PA: SIAM, 1996.

[23] P.P. SHENOY, "A Valuation-based Language for Expert Systems," *International Journal of Approximate Reasoning*, 3(5), pp. 383–411, 1989.

[24] P.P. SHENOY, "Valuation-based Systems for Discrete Optimization," in *Uncertainty in Artificial Intelligence*, **6**, pp. 385–400, 1991.

[25] P.P. SHENOY AND G. SHAFER, "Axioms for Probability and Belief-function Propagation," in *Uncertainty in Artificial Intelligence*, **4**, pp. 169–198, 1990.

[26] N. WIBERG, *Codes and Decoding on General Graphs*, Linköping Studies in Science and Technology. Dissertations, No. 440, 1996.

[27] N. WIBERG, H.-A. LOEGLIGER AND R. KÖTTER, "Codes and Iterative Decoding on General Graphs," *European Transactions on Telecommunications*, **6**, pp. 513–525, 1995.

Part 3. Decoding techniques

BSC THRESHOLDS FOR CODE ENSEMBLES BASED ON "TYPICAL PAIRS" DECODING[*]

SRINIVAS AJI[†], HUI JIN[†], AAMOD KHANDEKAR[†], DAVID J.C. MACKAY[‡], AND ROBERT J. MCELIECE[†]

Abstract. In this paper, we develop a method for closely estimating noise threshold values for ensembles of binary linear codes on the binary symmetric channel. Our method, based on the "typical pairs" decoding algorithm pioneered by Shannon, completely decouples the channel from the code ensemble. In this, it resembles the classical union bound, but unlike the union bound, our method is powerful enough to prove Shannon's theorem for the ensemble of random linear codes. We apply our method to find numerical thresholds for the ensembles of low-density parity-check codes, and "repeat-accumulate" codes.

1. Introduction. In this paper, we consider the performance of ensembles of codes on the binary symmetric channel. Our particular focus is on the question as to whether or not a given ensemble is "good," in the sense of MacKay [7]. In short, an ensemble of codes is said to be good, if there is a $p > 0$ such that the ensemble word error probability (with maximum-likelihood decoding) on a BSC with crossover probability p approaches zero as the block length approaches infinity. The largest such p for a given ensemble is called the (noise) *threshold* for the ensemble. Our main result (Theorem 4.1) is a technique for finding a lower bound on the ensemble threshold, which is based on the ensemble's weight enumerator.

Of course the classical union bound provides one way of using weight enumerators to estimate ensemble thresholds, but the estimates are poor. Gallager [4, Chapter 3] gave a variational method for upper bounding the probability of maximum-likelihood decoding error for an arbitrary binary code, or ensemble of codes (given an expression for the average weight-enumerator function) on a general class of binary-input channels. Gallager's technique, however, is quite complex, and even in the special case of the BSC it is difficult to apply to the problem of finding ensemble thresholds.[1]

In this paper we abandon the full maximum-likelihood decoder, and instead focus on a slightly weaker decoding algorithm, which is much easier to analyze, the *typical pairs* decoder. This technique was pioneered by

---

[*]The work of Aji, Jin, Khandekar, and McEliece on this paper was supported by NSF grant no. CCR-9804793, and grants from Sony, Qualcomm, and Caltech's Lee Center for Advanced Networking. David MacKay's work is supported by the Gatsby Charitable Foundation.

[†]Department of Electrical Engineering (136-93), California Institute of Technology, Pasadena, CA 91125.

[‡]Department of Physics, University of Cambridge, Cavendish Laboratory, Madingley Road, Cambridge, CB3 0HE, United Kingdom.

[1]We have been able to show that the thresholds obtained by our method are the same as the best obtainable by the Gallager methodology.

Shannon [11, Theorem 11], but as far as we can tell was not used to analyze ensembles other than the ensemble of all codes (which we call the Shannon ensemble in Section 5, below) until the 1999 paper of MacKay [7], in which it was used to analyze certain ensembles of low-density-parity check codes. In brief, when presented with a received word $\mathbf{y}$, the typical pairs decoder seeks a codeword $\mathbf{x}$ such that the pair $(\mathbf{x}, \mathbf{y})$ belongs to the set T of "typical pairs." (We give a precise definition of T in section 2, which follows.) In Section 3, we develop an upper bound on the typical-pairs decoder's error probability (Theorem 3.1) which, like the classical union bound, decouples the code's weight enumerator from the channel, but unlike the union bound, when combined with the law of large numbers, gives good estimates for code thresholds (Theorem 4.1).

We then apply Theorem 4.1 to three families of binary code ensembles: (1) The Shannon ensemble, consisting of all linear codes of rate R; (2) the Gallager ensemble, consisting of (j, k) low-density parity-check codes; and (3) the ensemble of Repeat-Accumulate codes introduced by Divsalar, Jin and McEliece [2]. In the case of the Shannon ensembles, we show that our method yields thresholds identical to those implied by Shannon's theorem. Thus the typical sequence method, despite its suboptimality, loses nothing (in terms of coding thresholds) for the Shannon ensemble.

Finally, we compare our thresholds to the *iterative* thresholds for the Gallager and RA ensembles recently obtained by Richardson and Urbanke [10], in order to estimate the price paid in coding threshold for the benefits of iterative decoding. In most cases, this loss is quite small, and in the case of $j = 2$ LDPC codes, there appears to be no penalty at all.

The method described in this paper can be readily extended to many other channel models, including channels with memory (cf. [7, Section II]). This extension will be developed in a forthcoming paper, where the emphasis will be on the binary erasure channel and the additive Gaussian noise channel.

2. Typical pairs. Let T be a set of binary vectors of length n which is closed under coordinate permutations, and let $\mathbf{Z} = (Z_1, Z_2, \ldots, Z_n)$ be the BSC noise vector, i.e., the Z_i's are i.i.d. random variables with common density

$$\Pr\{Z = 0\} = 1 - p, \qquad \Pr\{Z = 1\} = p.$$

If we define the set T to be a set of "typical" noise vectors, then T represents the typical channel outputs if the zero-word is transmitted, and the $T + \mathbf{x}$ represents the set of typical channel outputs if the codeword $\mathbf{x}$ is transmitted. In the typical-pairs decoder (to be defined shortly), decoder errors can result if the channel output is in the typical set of more than one codeword. We are therefore interested in the quantity $\Pr\{\mathbf{Z} \in T \cap (T + \mathbf{x})\}$.

If T is invariant under coordinate permutations, the probability $\Pr\{\mathbf{Z} \in T \cap (T + \mathbf{x})\}$ depends only on the weight of $\mathbf{x}$. Thus we define, for $h = 0, 1, \ldots, n$,

(2.1) $$P_h(T) = \Pr\{\mathbf{Z} \in T \cap (T + \mathbf{x})\},$$

where $\mathbf{x}$ is any vector of weight h. The quantity $P_h(T)$ is then the probability of error in a typical-set decoder in the case of a code having only two codewords separated by a Hamming distance h.

For example,

(2.2) $$P_0(T) = \Pr\{\mathbf{Z} \in T\}.$$

Since any set T which is invariant under coordinate permutations must consist of all vectors of weight $k \in K$, where K is a subset of $\{0, 1, \ldots, n\}$, the probabilities $P_h(T)$ depend only on the set K. A short combinatorial calculation gives

(2.3) $$P_h(T) = \sum_{k_1 \in K} p^{k_1} (1-p)^{n-k_1} \sum_{k_2 \in K} \binom{h}{(h+k_1-k_2)/2} \binom{n-h}{(k_1-h+k_2)/2}.$$

This is because a vector of weight k_1 has probability $p^{k_1}(1-p)^{n-k_1}$, and there are exactly $\binom{h}{(h+k_1-k_2)/2} \binom{n-h}{(k_1-h+k_2)/2}$ vectors of weight k_1, which have the property that when the first h components are complemented,

i.e., the vector $\mathbf{x} = (\overbrace{11\cdots1}^{h} \overbrace{00\cdots0}^{n-h})$ is added, the resulting vector has weight k_2. Applying (2.3) to the case $h = 0$, we obtain

$$P_0(T) = \sum_{k \in K} p^k (1-p)^{n-k} \binom{n}{k},$$

in agreement with (2.2).

In our main application (Theorem 4.1) the set T will be the "typical sequences" of length n and so will be denoted by T_n. The definition of T_n is

(2.4) $$T_n = \left\{ \mathbf{z} : \left| \frac{\mathrm{wt}(\mathbf{z})}{n} - p \right| \le \epsilon_n \right\},$$

where ϵ_n is a sequence of real numbers approaching zero more slowly than $n^{-1/2}$, i.e., $\epsilon_n \sqrt{n} \to \infty$. Then by a straightforward extension of the weak law of large numbers,

(2.5) $$\lim_{n \to \infty} \Pr\{\mathbf{Z} \in T_n\} = 1.$$

Furthermore, by defining $K_n = \{k : n(p - \epsilon_n) \le k \le n(p + \epsilon_n)\}$, and using the formula (2.3), it is relatively easy to prove that for any δ in the range $0 \le \delta \le 2p$, we have

(2.6) $$\lim_{n \to \infty} -\frac{1}{n} \log P_{\delta n}(T_n) = K(\delta, p),$$

where $K(\delta, p)$ is given by the equivalent formulas

$$(2.7) \qquad K(\delta, p) = H(p) - \delta \log 2 - (1 - \delta)H\left(\frac{p - \delta/2}{1 - \delta}\right)$$

$$(2.8) \qquad = H(\delta) - pH\left(\frac{\delta}{2p}\right) - (1 - p)H\left(\frac{\delta}{2(1 - p)}\right),$$

where $H(x)$ is the entropy function, i.e., $H(x) = -x \log x - (1-x) \log(1-x)$. (These formulas are true only for $\delta < 2p$; for $\delta \geq 2p$, $K(\delta, p)$ is infinite, since $P_h(T_n) = 0$ for $h > 2n(p + \epsilon_n)$.) In Figure 1, we have plotted the function $K(\delta, p)$ for several values of p.[2]

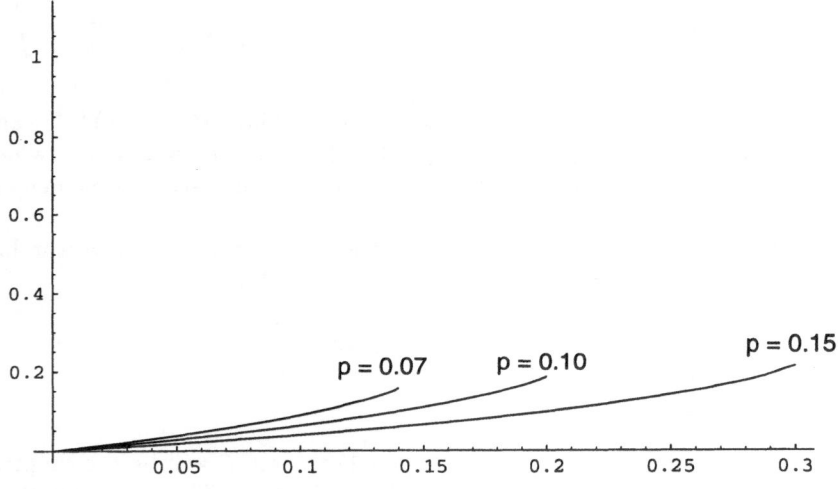

FIG. 1. *The Function* $K(\delta, p)$ *for* $p = 0.07, 0.10., 0.15$.

In fact, a closer examination of the limit in (2.6) shows that for a fixed value of p, the limit is *uniform*. That is, for a fixed p, there exists a sequence of positive numbers $\beta_n \to 0$, such that

$$(2.9) \qquad \left| -\frac{1}{n} \log P_{\delta n}(T_n) - K(\delta, p) \right| < \beta_n \quad \text{for all } 0 < \delta < 2p.$$

Alternatively, we can write (2.9) as

$$(2.10) \qquad P_h(T_n) = e^{-n(K(\delta, p) + o(1))},$$

where $\delta = h/n$.

---

[2]In Figure 1, and all the other figures in the paper, computations using logarithms use natural logarithms.

3. The typical pairs decoding method. Suppose C is an (n, k) binary linear code, with weight enumerator $(A_0, A_1, \ldots, A_n)$, i.e., C contains exactly A_h words of Hamming weight h, for $h = 1, \ldots, n$. We suppose that at the transmitter, a codeword $\mathbf{x} \in C$ is selected at random, transmitted over a BSC with crossover probability p, and received at the destination as $\mathbf{y}$. The T-decoder tries to infer $\mathbf{x}$ based on knowledge of the code C, the noisy codeword $\mathbf{y}$, and the channel noise parameter p. The T-decoder works as follows.

For every codeword $\mathbf{x}_i$, the ith "pseudonoise" $\mathbf{z}_i = \mathbf{y} - \mathbf{x}_i$ is computed. If there are no indices i for which $\mathbf{z}_i \in T$, the decoder fails. Otherwise, among those indices such that $\mathbf{z}_i \in T$, the decoder choose one for which the Hamming weight $w(\mathbf{z}_i)$ is smallest. In short, the decoder chooses the *most likely codeword for which $\mathbf{z}_i$ is typical.* In what follows we do not distinguish between decoder error and failure, and denote the probability of decoder error (or failure) by P_E.

THEOREM 3.1. *If P_E denotes the probability that the T-decoder does not correctly identify the transmitted codeword, then*

$$(3.1) \qquad P_E \leq (1 - P_0(T)) + \sum_{h=1}^{n} A_h \min(\beta^h, P_h(T)),$$

where $\beta = 2\sqrt{p(1-p)}$ is the channel Bhattacharyya parameter.[3]

Proof. Let $(\mathbf{x}_0, \mathbf{x}_1, \ldots, \mathbf{x}_{M-1})$ be an ordering of the code with $\mathbf{x}_0$ being the all-zeros word, and suppose $\mathbf{x}_0$ is transmitted. For $i = 0, 1, \ldots, M - 1$, define the following events:

$$
\begin{aligned}
T_i &= \{\mathbf{z}_i \in T\} & \text{($\mathbf{z}_i$ is typical)} \\
V_i &= \{w(\mathbf{z}_i) \leq w(\mathbf{z}_0)\} & \text{($\mathbf{z}_i$ is more likely than $\mathbf{z}_0$)} \\
S_i &= T_i \cap V_i & \text{($\mathbf{z}_i$ is typical and is more likely than $\mathbf{z}_0$)}
\end{aligned}
$$

Then the T-decoder will fail only if at least one of the events $T_0', S_1, \ldots, S_{M-1}$ occurs. Thus if $\mathcal{E}$ denotes the event "T-decoder fails, given that $\mathbf{x}_0$ was transmitted," we have (here T_0' denotes the complement of T_0)

$$
\begin{aligned}
(3.2) \qquad \mathcal{E} &= T_0' \cup \left(\bigcup_{i=1}^{M-1} S_i \right) \\
&= T_0' \cup \left(T_0 \cap \bigcup_{i=1}^{M-1} S_i \right) \\
&= T_0' \cup \left(\bigcup_{i=1}^{M-1} (T_0 \cap S_i) \right).
\end{aligned}
$$

Therefore the probability of T-decoder error, given that $\mathbf{x}_0$ was transmitted, can be upper bounded as follows:

$$(3.3) \qquad \Pr\{\mathcal{E}|\mathbf{x}_0\} \leq \Pr\{T_0'|\mathbf{x}_0\} + \sum_{i=1}^{M-1} \Pr\{T_0 \cap S_i|\mathbf{x}_0\}.$$

---

[3]The term β^h in (3.1) is present for technical reasons, e.g., the proof of Theorem 4.1. Normally, it will be smaller than the term $P_h(T)$ only for very small values of h.

But $\Pr\{T_0'|\mathbf{x}_0\} = 1 - \Pr\{T_0|\mathbf{x}_0\}$, and $\Pr\{T_0|\mathbf{x}_0\} = \Pr\{\mathbf{Z} \in T\} = P_0(T)$, from (2.2). Thus

$$(3.4) \qquad\qquad \Pr\{T_0'|\mathbf{x}_0\} = 1 - P_0(T).$$

Also, since $S_i = T_i \cap V_i$, it follows that

$$\Pr\{T_0 \cap S_i|\mathbf{x}_0\} \le \min(\Pr\{V_i|\mathbf{x}_0\}, \Pr\{T_0 \cap T_i|\mathbf{x}_0\}).$$

By the familiar union bound argument [8, Theorem 7.5], we have

$$\Pr\{V_i|\mathbf{x}_0\} \le \beta^{h_i},$$

where h_i is the Hamming weight of $\mathbf{x}_i$.

Also note that by definition $\mathbf{z}_i = \mathbf{y} - \mathbf{x}_i$, and so we have, for $i = 1, \ldots, M - 1$, $\mathbf{z}_i = \mathbf{z}_0 + (\mathbf{x}_0 - \mathbf{x}_i) = \mathbf{z}_0 - \mathbf{x}_i$, since $\mathbf{x}_0$ is the all-zeros word. Thus $T_i = \{\mathbf{z}_0 \in T + \mathbf{x}_i\}$, and so

$$\begin{aligned} \Pr\{T_0 \cap T_i|\mathbf{x}_0\} &= \Pr\{\mathbf{Z} \in T \cap (T + \mathbf{x}_i)\} \\ &= P_{h_i}(T) \end{aligned}$$

where h_i is the Hamming weight of $\mathbf{x}_i$. Hence

$$(3.5) \qquad \begin{aligned} \sum_{i=1}^{M-1} \Pr\{T_0 \cap S_i|\mathbf{x}_0\} &\le \sum_{i=1}^{M-1} \min(\beta^{h_i}, P_{h_i}(T)), \\ &= \sum_{h=1}^{n} A_h \min(\beta^h, P_h(T)), \end{aligned}$$

since there are exactly A_h words of Hamming weight h in $\mathcal{C}$. Combining (3.3) with (3.4) and (3.5), gives (3.1). $\qquad\square$

4. Code ensembles. By an *ensemble* of linear codes we mean a sequence $\mathcal{C}_{n_1}, \mathcal{C}_{n_2}, \ldots$ of sets of linear codes of a common rate R, where $\mathcal{C}_{n_i}$ is a set of (n_i, k_i) codes with $k_i/n_i = R$. We assume that the sequence $n_1, n_2, \ldots$ approaches infinity. If C is an (n, k) code in the ensemble, we denote the weight enumerator of C by the list $A_0(C), A_1(C), \ldots, A_n(C)$. The *average weight enumerator* for the set $\mathcal{C}_n$ is defined as the list

$$\overline{A}_0^{(n)}(C), \overline{A}_1^{(n)}(C), \ldots, \overline{A}_n^{(n)}(C),$$

where

$$(4.1) \qquad \overline{A}_h^{(n)} \triangleq \frac{1}{|\mathcal{C}_n|} \sum_{C \in \mathcal{C}_n} A_h(C) \qquad \text{for } h = 0, 1, \ldots, n.$$

We define, for each n in the sequence $n_1, n_2, \ldots$, the function

$$(4.2) \qquad r_n(\delta) \triangleq \frac{1}{n} \log \overline{A}_{\lfloor \delta n \rfloor}^{(n)} \qquad \text{for } 0 < \delta < 1,$$

Also, we define the *ensemble spectral shape* :

(4.3) $$r(\delta) \triangleq \lim_{n\to\infty} r_n(\delta) \qquad \text{for } 0 < \delta < 1,$$

assuming that the limit exists. In this case, we may write

(4.4) $$\overline{A}_h^{(n)} = e^{n(r(\delta)+o(1))},$$

where $\delta = h/n$.

Now we apply Theorem 3.1, using the set T_n, defined in (2.4), to a code $C \in \mathcal{C}_n$:

(4.5) $$P_E \le \eta_n + \sum_{h=1}^{n} A_h(C)P_h(T_n),$$

where $\eta_n = \Pr\{T'_n\} \to 0$ by (2.5). If we average (4.5) over all codes in the ensemble $\mathcal{C}_n$, we obtain the following upper bound on $\overline{P}_E^{(n)}$, the ensemble decoder error probability:

(4.6) $$\overline{P}_E^{(n)} \le \eta_n + \sum_{h=1}^{n} \overline{A}_h^{(n)} P_h(T_n).$$

Replacing $\overline{A}_h^{(n)}$ with the right side of (4.4), and $P_h(T_n)$ with the right side of (2.10), (4.6) becomes

(4.7) $$\overline{P}_E^{(n)} \le \eta_n + \sum_{h=1}^{n} e^{-n(K(\delta,p)-r(\delta)+o(1))}.$$

It now appears that if p is chosen so that the function $K(\delta,p) - r(\delta)$ is positive for all $0 < \delta < 1$, so that the exponent in the sum in (4.7) is always negative, the ensemble word error probability $\overline{P}_E^{(n)}$ will approach zero, as $n \to \infty$. This is in fact true, provided we make the following two technical assumptions about the behavior of $\overline{A}_h^{(n)}$, for $h = o(n)$.

• *Assumption 1.* There exist a sequence of integers d_n such that $d_n \to \infty$ and

(4.8) $$\lim_{n\to\infty} \sum_{h=1}^{d_n} \overline{A}_h^{(n)} = 0.$$

(This assumption says, roughly, that the minimum distance of the ensemble is at least d_n.)

• *Assumption 2.* There exist a sequence of real numbers $\theta_n \ge 0$ such that

(4.9) $$r_n(\delta) \le r(\delta) + \theta_n, \qquad \text{where} \quad \lim_{n\to\infty} \frac{n\theta_n}{d_n} = 0.$$

We now state our main result:

THEOREM 4.1. *Suppose the code ensemble has spectral shape $r(\delta)$, and also that it satisfies Assumptions 1 and 2. Then if the crossover probability $p < 1/2$ of the channel satisfies*

$$K(\delta, p) > r(\delta) \qquad for \quad 0 < \delta < 2p,$$

then $\overline{P}_E^{(n)} \to 0$ as $n \to \infty$.

There is a slightly weaker version of Assumption 1 that guarantees that the ensemble *bit* error probability approaches zero:

- *Assumption 1'.* There exist a sequence of integers d_n such that $d_n \to \infty$ and

$$(4.10) \qquad \lim_{n \to \infty} \sum_{h=1}^{d_n} \frac{h}{n} \overline{A}_h^{(n)} = 0.$$

The corresponding modification of Theorem 4.1 follows.

THEOREM 4.2. *Suppose the code ensemble has spectral shape $r(\delta)$, and also that it satisfies Assumptions 1' and 2. Then if the crossover probability $p < 1/2$ of the channel satisfies*

$$K(\delta, p) > r(\delta) \qquad for \ 0 < \delta < 2p,$$

then $\overline{P}_b^{(n)} \to 0$ as $n \to \infty$, where P_b denotes the T-decoder's bit error probability.

(A proof of Theorem 4.1 will be found in the Appendix. The proof of Theorem 4.2 is similar and is omitted.)

In the following three sections, we will apply Theorem 4.1 to three different ensembles of binary linear codes: (1) The Shannon ensemble, consisting of all linear codes of rate R; (2) the Gallager ensemble, consisting of (j, k) low-density parity-check codes; and (3) the ensemble of Repeat-Accumulate codes introduced by Divsalar, Jin and McEliece [2].

5. The Shannon ensemble. For the set of random linear codes of rate R, we have

$$(5.1) \qquad \overline{A}_h^{(n)} = \binom{n}{h} 2^{-n(1-R)},$$

from which it follows via a routine calculation that

$$(5.2) \qquad r(\delta) = H(\delta) - (1 - R) \log 2.$$

This function is shown for $R = 1/3$ in Figure 2.

To apply Theorem 4.1 to the Shannon ensemble,[4] for a given rate R we must find the largest p such that $K(\delta, p) > H(\delta) - (1 - R) \log 2$ for all $0 < \delta < 2p$.

---

[4]Assumptions 1 and 2 are satisfied with $d_n = Kn$ for a suitable positive constant $K = K(R)$, and $\theta_n = 0$.

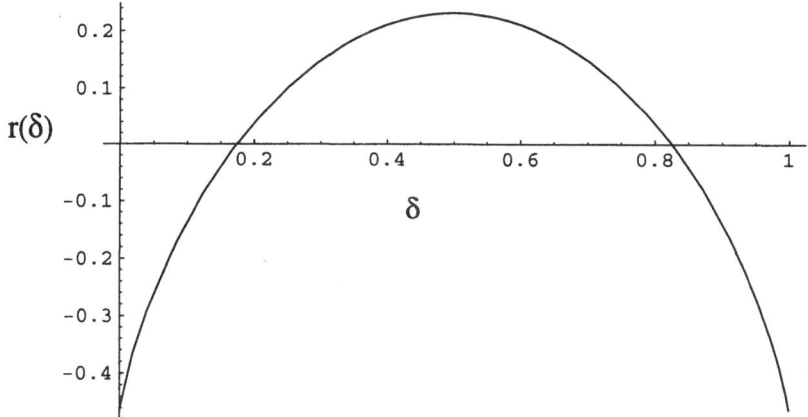

FIG. 2. *The function $r(\delta)$ for the ensemble of $R = 1/3$ linear codes.*

Using (5.2) and (2.8), this inequality becomes

$$(5.3) \qquad pH\left(\frac{\delta}{2p}\right) + (1-p)H\left(\frac{\delta}{2(1-p)}\right) < (1-R)\log 2.$$

The maximum of the left side of (5.3) in the range $0 < \delta < 2p$ occurs at $\delta = 2p(1-p)$, and is $H(p)$. Thus the inequality $K(\delta,p) > H(\delta)-(1-R)\log 2$ required by Theorem 4.1 becomes simply $H(p) < (1-R)\log 2$, or $H_2(p) < 1 - R$, where $H_2(p)$ is the binary entropy function. Thus we have proved

THEOREM 5.1. *The ensemble of random linear codes of rate R is good on a BSC with crossover probability p if $H_2(p) < 1 - R$.*

The idea of the proof is illustrated in Figure 3, where we see the function $K(\delta, 0.174)$ just touching the $r(\delta)$ curve of Figure 2. This shows that the threshold for the ensemble of $R = 1/3$ linear codes is $p = 0.174$, which reflects the fact that $H_2(0.174) = 1 - 2/3$.

Of course, Theorem 5.1 is just Shannon's theorem for linear codes on the BSC. We have included it only to demonstrate that Theorem 4.1 is powerful enough to reproduce Shannon's theorem. In the next two sections we will apply it to more interesting ensembles.

6. The Gallager ensemble. In this section, we discuss the application of Theorem 4.1 to the ensemble of (j, k) low-density parity-check codes defined by Gallager [4].[5] In brief, every code in Gallager's (j, k) ensemble is defined by a parity-check matrix which has j ones in each column

---

[5]There are numerous ways to define this ensemble. The definition we follow was given by Gallager [4, Section 2.2], and differs, e.g. from the ensemble analyzed by MacKay in [7, Section II].

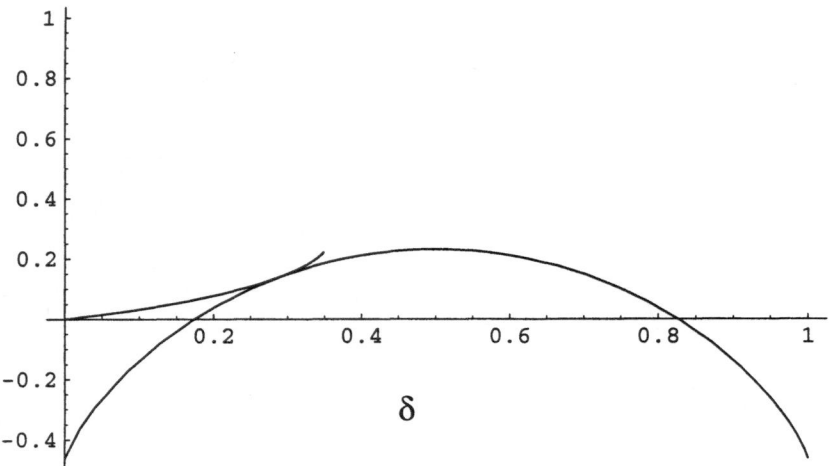

FIG. 3. *The function $r(\delta)$ for the ensemble of $R = 1/3$ linear codes, together with the function $K(\delta,p)$ for $p = 0.174$.*

and k ones in each row. The rate of each code in the ensemble is at least $R_{j,k} = 1 - (j/k)$.

The spectral shape $r_{j,k}(\delta)$ for the (j, k) ensemble was determined by Gallager. It can be expressed in parametric form, as follows:

$$\delta_{j,k}(s) = \frac{1}{k}\frac{\partial \mu}{\partial s}(s, k)$$

$$r_{j,k}(s) = \frac{j}{k}\left(\mu(s,k) - s\frac{\partial \mu}{\partial s}(s,k) + (k-1)\log 2\right) - (j-1)H\left(\frac{1}{k}\frac{\partial \mu}{\partial s}(s,k)\right)$$

where the parameter s ranges from $-\infty$ to $+\infty$, and the function $\mu(s, k)$ is defined by

$$\mu(s, k) \triangleq \log \frac{(1 + e^s)^k + (1 - e^s)^k}{2^k}.$$

Figure 4 shows the function $r_{j,k}$ for $(j, k) = (3, 6)$.

Given the spectral shape, it is an easy task to apply Theorem 4.1 to find the corresponding BSC ensemble thresholds.[6] A short table of these thresholds, together with the corresponding Shannon limit, is given below.

---

[6] To satisfy Assumptions 1 and 2 for $j \geq 3$, we can take $d_n = Kn$ for a suitable constant $K = K(j)$, and $\theta_n = 0$. For $j = 2$, we can prove the existence of a sequence of d_n's which satisfy Assumptions 1' and 2 with $\theta_n = 0$, though we do not have an explicit expression for them.

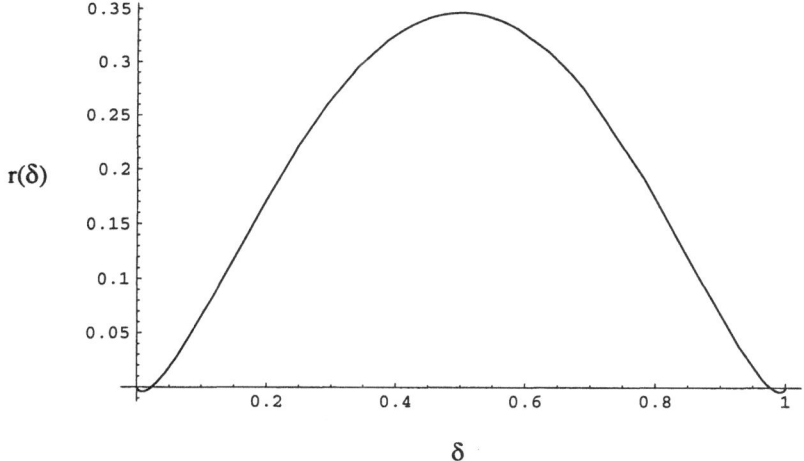

FIG. 4. *The function* $r(\delta)$ *for the ensemble of* $(3,6)$ *LDPC codes.*

(j,k)	$R_{j,k}$	$p_{j,k}$	RU limit	Shannon limit
(3,6)	1/2	0.0915	0.084	0.109
(3,5)	2/5	0.129	0.113	0.145
(4,6)	1/3	0.170	0.116	0.174
(3,4)	1/4	0.205	0.167	0.214
(2,3)	1/3	0.0670	0.0670	0.174
(2,4)	1/2	0.0286	0.0286	0.109

For example, consider the "$(3,5)$" line in the table. The corresponding Gallager ensemble consists of codes which have parity-check matrices with 3 ones per column and 5 ones per row. The rate of all codes in this ensemble at least $R_{3,5} = 1 - (3/5) = 2/5$. Using Theorem 4.1, it is calculated that for any BSC with crossover probability $p < 0.129$, the $(3,5)$ ensemble is good, i.e., the average word error probability of the T-decoder approaches 0, as $n \to \infty$. This should be compared to the Shannon limit for the ensemble of all linear codes of rate 2/5 (cf. Theorem 5.1), which is $p = 0.145$, which indicates the price which is paid for having the $(3,5)$ structure. Finally, we note that the Richardson-Urbanke limit [10] for the $(3,5)$ ensemble is $p = 0.113$, i.e., with belief propagation–style iterative decoding, the ensemble decoder error probability approaches 0 if and only if $p < 0.113$.

(The values $p_{j,k}$ for $(j,k) = (3,6)$, $(3,5)$, $(4,6)$, and $(3,4)$ given in the above table appear to agree with the values given by Gallager [4] in his Figure 3.5, although he gave no numerical values. However, as we mentioned above, we have been able to show that the thresholds obtained from our Theorem 4.1 are the same as the best obtainable using Gallager's methodology, so our threshold values are at least as good as Gallager's.)

We conclude this section with some remarks on the ensemble of $(2, k)$ LDPC codes. Originally dismissed by Gallager because their minimum distance is $O(\log n)$ [4, Theorem 2.5], they are nevertheless quite interesting, and are variously called "graph-theoretic," "circuit," or "cycle" codes [9, Section 5.8], [6] because of their close connection to finite undirected graphs. Using Theorem 4.2, we can show that for $p < p^*(k)$, the *bit error probability* for T-decoding of the $(2, k)$ ensemble approaches zero, where $p^*(k)$ is given by the exact formula

$$(6.1) \qquad p^*(k) = \frac{1}{2}\left(1 - \sqrt{1 - \frac{1}{(k-1)^2}}\right).$$

(The ensemble word error probability does not approach zero for any $p > 0$.)

Furthermore, Wiberg [12, Example 5.1] showed that with *iterative* decoding, the ensemble of $(2, k)$ cycle codes has ensemble bit error probability approaching zero for $p < p^*(k)$. Numerically, the Richardson-Urbanke method appears to give the same value, so it seems safe to say that (6.1) gives the exact iterative threshold for the Gallager $(2, k)$ ensemble.[7]

Finally, it was shown by Decreusefond and Zémor [3] that for an "expurgated" ensemble of $(2, k)$ cycle codes, the *exact* maximum-likelihood BSC coding threshold is equal to $p^*(k)$. Since as we have seen, the threshold for the unexpurgated ensemble is at least this good, it seems very likely that $p^*(k)$ is the exact ML threshold for the unexpurgated ensemble as well. These results strongly suggest that that for $(2, k)$ cycle codes, the iterative and maximum-likelihood thresholds are the same, and are given by the formula (6.1).

7. The ensemble of repeat-accumulate codes. In brief, for an integer $q \geq 2$, the ensemble of q-repeat accumulate codes consists of those codes which can be encoded by the serial concatenation of a q-ary repetition encoder, followed by a pseudorandom permutation, followed by a rate 1 code with (square) generator matrix of generic shape

$$G = \begin{pmatrix} 1 & 1 & 1 & 1 & 1 \\ 0 & 1 & 1 & 1 & 1 \\ 0 & 0 & 1 & 1 & 1 \\ 0 & 0 & 0 & 1 & 1 \\ 0 & 0 & 0 & 0 & 1 \end{pmatrix}.$$

The basic combinatorial fact about the ensemble of (qk, k) RA codes is the following formula for the average number of input words of weight w

---

[7]For a survey of iterative decoding of cycle codes, see [5].

which are encoded into output words of weight h [2, Eq. (5.4)]:

(7.1)
$$\overline{A}_{w,h}^{(qk)} = \frac{\binom{k}{w}\binom{qk-h}{\lfloor qw/2 \rfloor}\binom{h-1}{\lceil qw/2 \rceil - 1}}{\binom{qk}{qw}}.$$

It follows then that if $\overline{A}_h^{(qk)}$ denotes the average number of words of weight h in the ensemble,

(7.2)
$$\overline{A}_h^{(qk)} = \sum_{w=1}^{N} \overline{A}_{w,h}^{(qk)}.$$

From (7.1) and (7.2), it can be shown that the spectral shape $r(\delta)$ for the ensemble of q-RA codes is as follows :

(7.3) $\quad r(\delta) = \max_{0 \le x \le 1/q} \left\{ -\frac{q-1}{q} H(qx) + (1-\delta)H(\frac{qx}{2(1-\delta)}) + \delta H(\frac{qx}{2\delta}) \right\}.$

Figure 5 shows the $r(\delta)$ curve for the ensemble of $q = 3$ RA codes.[8]

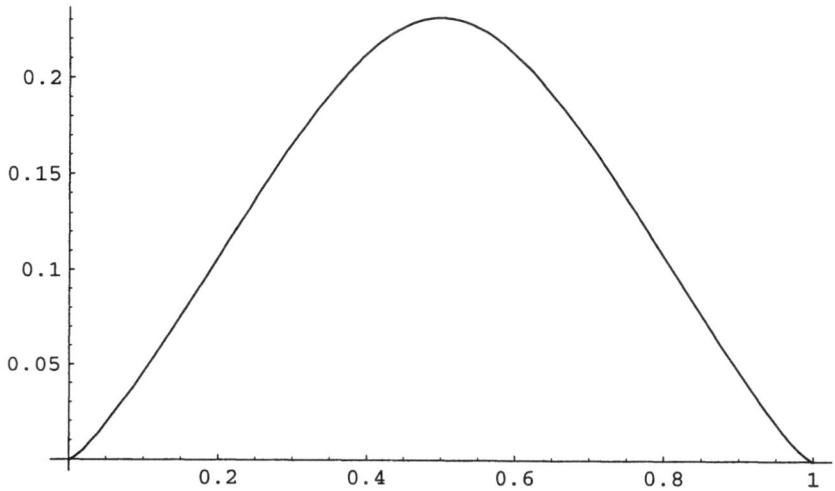

FIG. 5. *The function $r(\delta)$ for the ensemble of $R = 1/3$ RA codes.*

Combining (7.3) with Theorem 4.1, it is a straightforward computation to obtain the thresholds in the following table.

---

[8]To satisfy Assumptions 1 and 2 for $q \ge 3$, we can take $d_n = \log^2 n$ and $\theta_n = (K \log n)/n$ for suitable constants $K = K(q)$. For $q = 2$, we can only show the existence of a sequence d_n satisfying Assumptions 1' and 2 by taking $d_n = 2$ and $\theta_n = (K \log n)/n$.

q	R_q	p_q	RU limit	Shannon limit
2	1/2	0.029*	–	0.109
3	1/3	0.132	0.142	0.174
4	1/4	0.191	0.188	0.215
5	1/5	0.228	0.216	0.243
6	1/6	0.254	0.235	0.264
7	1/7	0.274	0.250	0.281

For example, consider the $q = 3$ line of the table. It indicates that the common rate for all $q = 3$ RA codes is $R = 1/3$, and that this ensemble is good on any BSC with crossover probability $p < 0.132$. By way of comparison, the Shannon threshold for the ensemble of all rate 1/3 linear codes is seen to be $p < 0.174$. Finally, the Richardson-Urbanke iterative decoding threshold [Richardson and Urbanke, private commmunication] is $p < 0.142$. Since we can show that the T-decoding algorithm always gives the same ensemble threshold as does maximum-likelihood decoding, which must be at least as good as the iterative threshold, this apparently shows either that the thresholds given in Theorem 4.1 are not always the best possible for T-decoding, or that the R-U theorem is not correct for this ensemble. A resolution of this paradox would be very welcome.

Finally we note that for the ensemble of $q = 2$ RA codes, the word error probability for T-decoding does not approach zero for any $p > 0$, but, again by using Theorem 4.2, we can show that the ensemble *bit* error probability approaches zero for $p < 0.029$.

APPENDIX

A. Proof of Theorem 4.1. We first define the *ensemble threshold* as follows:

$$(A.1) \qquad p_0 = \sup\{p : K(\delta, p) > r(\delta), 0 < \delta < 2p\}.$$

LEMMA A.1. *If $p < p_o$, then there exist real numbers $\alpha_0 > 0$ and $\gamma_0 > 0$, and a positive integer N_0, such that for $n \geq N_0$,*

$$\sum_{h=d_n}^{\alpha_0 n} \overline{A}_h^{(n)} \beta^h = O(e^{-d_n \gamma_0}),$$

where $\beta = 2\sqrt{p(1-p)}$.

Proof. Using the definition (2.7), It is straightforward to show that

$$\lim_{\delta \to 0} \frac{K(\delta, p_0)}{\delta} = \frac{\partial K(0, p_0)}{\partial \delta} = -\log \beta_0,$$

where $\beta_0 = 2\sqrt{p_0(1 - p_0)}$. Hence for $p < p_0$, we have

$$\limsup_{\delta \to 0} \frac{r(\delta)}{\delta} \leq \lim_{\delta \to 0} \frac{K(\delta, p_0)}{\delta}$$
$$= -\log \beta_0 = 2\sqrt{p_0(1 - p_0)}$$
$$< -\log \beta = 2\sqrt{p(1 - p)}.$$

This, together with Assumption 2, implies that there exists $\alpha_0 > 0$, $\gamma_0 > 0$, and a positive integer N_0 such that for $n \geq N_0$, we have

$$\sup_{d_n/n \leq \delta < \alpha_0} \frac{r_n(\delta)}{\delta} < \frac{n\theta_n}{d_n} + \sup_{0 \leq \delta < \alpha_0} \frac{r(\delta)}{\delta} < -\log \beta - \gamma_0.$$

Hence we have, for $n \geq N_0$,

$$\sum_{h=d_n}^{\alpha_0 n} \overline{A}_n^{(n)} \beta^h = \sum_{h=d_n}^{\alpha_0 n} e^{-h(\log \beta - r_n(\delta)/\delta)} < \sum_{h=d_n}^{\alpha_0 n} e^{-h\gamma_0}$$
$$< \sum_{h=d_n}^{\infty} e^{-h\gamma_0} = O(e^{-d_n \gamma_0}),$$

which completes the proof. $\qquad\square$

Now we can give the proof of Theorem 4.1. With the notation being as established above, we have, by Theorem 3.1, for $p < p_0$,

$$(A.2) \qquad P_E \leq \sum_{h=1}^{d_n} \overline{A}_h^{(n)} + \sum_{h=d_n}^{\alpha_0 n} \overline{A}_h^{(n)} \beta^h + \sum_{h=\alpha_0 n}^{n} \overline{A}_h^{(n)} P_h(T) + o(n).$$

The first sum in (A.2) approaches zero by Assumption 1, the second sum approaches zero by Lemma A.1 together with the fact that $d_n \to \infty$. The third sum is

$$(A.3) \qquad \sum_{h=\alpha_0 n}^{n} \overline{A}_h^{(n)} P_h(T) = \sum_{h=\alpha_0 n}^{n} e^{-n(K(\delta,p) - r(\delta) + o(1))}$$
$$\leq \sum_{h=\alpha_0 n}^{n} e^{-n(K(\delta,p) - K(\delta,p_0) + o(1))},$$

where the first line follows from (2.10) and Assumption 2, and the second line follows from the definition (A.1) of p_0.

Finally, let ϵ be such that

$$K(\delta, p) - K(\delta, p_0) \geq \epsilon \qquad \text{for } \alpha_o \leq \epsilon \leq 2p.$$

Then for n sufficiently large, the exponent in (A.3) will be $\geq \epsilon/2$, and so the sum will be upper bounded by $n \cdot e^{-n\epsilon/2}$, which goes to zero as $n \to \infty$.
$\square$

Acknowledgement. This paper is an outgrowth of conversations that took place at the Institute for Mathematical Analysis, Minneapolis, Minnesota, during the workshop on Graphical Models and Iterative Decoding that took place in August, 1999. The authors wish to thank IMA for its hospitality and conducive work environment.

REFERENCES

[1] T.M. COVER AND J.A. THOMAS, *Elements of Information Theory*. New York: John Wiley and Sons, 1991.

[2] D. DIVSALAR, H. JIN, AND R. MCELIECE, "Coding Theorems for 'Turbo-Like' Codes." Proc. 1998 Allerton Conf., pp. 201–210.

[3] L. DECREUSEFOND AND G. ZÉMOR, "On the error-correcting capabilities of cycle codes of graphs," *Combinatorics, Probability, and Computing*, vol. 6 (1997), pp. 27–38.

[4] R. GALLAGER, *Low-Density Parity-Check Codes*. Cambridge, Mass.: The M.I.T. Press, 1963.

[5] G.B. HORN, "The iterative decoding of cycle codes," submitted to *IEEE Trans. Inform. Theory*.

[6] D. JUNGNICKEL AND S.A. VANSTONE, "Graphical codes revisited," *IEEE Trans. Inform. Theory*, vol. IT-43 (Jan. 1997), pp. 136–146.

[7] D.J.C. MACKAY, "Good error-correcting codes based on very sparse matrices," *IEEE Trans. Inform. Theory*, vol. IT-45 (March 1999), pp. 399–431.

[8] R.J. MCELIECE, *The Theory of Information and Coding*. Reading, Mass.: Addison-Wesley, 1977.

[9] W.W. PETERSON AND E.J. WELDON, JR., *Error-Correcting Codes, 2nd. ed.* Cambridge, Mass.: The MIT Press, 1972.

[10] T. RICHARDSON AND R. URBANKE, "The capacity of low-density parity-check codes under message-passing decoding," submitted to *IEEE Trans. Inform. Theory*.

[11] C.E. SHANNON, *The Mathematical Theory of Information*. Urbana, IL: University of Illinois Press, 1949 (reprinted 1998).

[12] N. WIBERG, *Codes and Decoding on General Graphs*. Linköping Studies in Science and Technology. Dissertation, no. **440**. Linköping University, Linköping, Sweden, 1996.

PROPERTIES OF THE TAILBITING BCJR DECODER

JOHN B. ANDERSON* AND KEMAL E. TEPE†

Abstract. The tailbiting BCJR algorithm extends the maximum a posteriori (MAP) decoder of Bahl et al. to the case of tailbiting trellis codes. The algorithm consists of forward and backward recursions that start from the left and right principal eigenvectors of the product of the trellis gamma matrices. The result is a slightly suboptimal symbol-by-symbol MAP decoder that performs much less computation than the true MAP decoder. The decoder has both iterative and non-iterative realizations. We formally justify the algorithm and develop its properties. Storage of the entire recursion outcome is not required and we relate the needed length to the encoder memory and the encoder decision depth parameter. By tests of actual decoders, the bit error rate of the algorithm is compared to that of true MAP, maximum likelihood, and circular Viterbi decoders. For a given encoder, the BER of these decoders depends on the ratio of the tailbiting circle size to the encoder memory. We argue that there exists a certain practical optimum ratio of circle size to memory, and at this ratio the BER of the tailbiting BCJR decoder is essentially that of the true MAP decoder.

AMS(MOS) subject classifications. 94B10 Convolutional codes; 94B35 Decoding.

1. Introduction. We study a decoding algorithm for tailbiting trellis codes that is based on the idea of *a posteriori* probability, or APP, decoding. An APP decoder is one that computes the probability of a transmitted data symbol or encoder state or encoder transition, given the observed channel outputs and any *a priori* probability known about the encoder states. A MAP (maximum *a posteriori* probability) decoder is an APP decoder that makes a hard decision about a data symbol or state by choosing the one with highest APP. The classical BCJR algorithm, named for the authors of an early paper [1], computes the probabilities of states and transitions of a Markov source and has been applied to APP and MAP decoding of terminated trellis codes. The algorithm and its notation trace back to the statistical literature of the 1960s. In an earlier paper [2] we extended the BCJR idea to the decoding of codes that have a tailbiting trellis structure. In this paper we give a formal justification for the algorithm's structure and a detailed rendering of its properties as a decoder for binary convolutional tailbiting codes. Both BSC and AWGN coding are considered.

Tailbiting (TB) convolutional codes trace back to works by Solomon and Tilborg [3] and Ma and Wolf [4], among others. They are defined by the fact that the encoder begins at the same state in which it will later end. The paper focuses on rate $1/2$ time-invariant feedforward convolutional encoders, and a notation for these is defined with the aid of Fig. 1. The encoder is defined by two shift register tap sets g_1 and g_2 expressed as left-

---

*Dept. Information Technology, Box 118, Univ. of Lund, SE-221 00 Lund, SWEDEN.
†Elec., Computer and Systems Eng. Dept., Rensselaer Poly. Institute, Troy, NY 12180, USA.

justified octals; for example $(g_1, g_2) = (11101, 11001)$ is denoted (72,62). The tap sets are of length $m + 1$, where m is the memory of the encoder. To start the encoder in the TB mode, the last m data bits are used to initialize the shift register. Another useful class of encoders is the recursive systematic class. Recent work shows that they may be used in the TB mode [6, 19]. We have studied them extensively as well, but since results of the type in this paper hardly differ from the feedforward class, we will not discuss them separately.

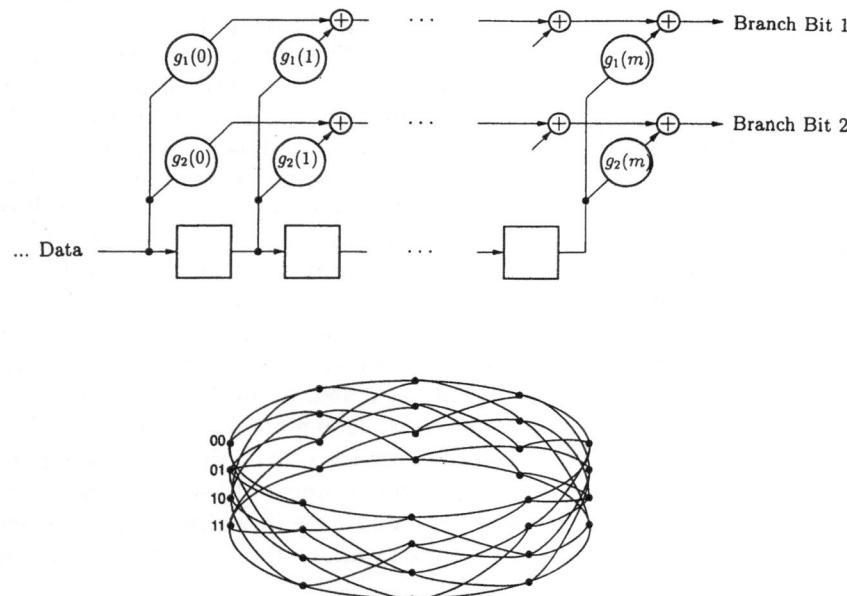

FIG. 1. *(a) (Above) General rate 1/2 memory m feedforward encoder with generators (g_1, g_2). (b) (Below) 8-stage circular tailbiting trellis for encoder (111,101).*

Because of the TB condition, the trellis generated by the encoder is circular, and Fig. 1 gives an 8-stage example for the (7,5) encoder. Define an (N, L, d) binary tailbiting code to have N-bit codewords comprising L stages, whose minimum Hamming distance is d. Its rate is $R = L/N$ bits per channel use. In [5], we give an exhaustive list of TB encoders that optimize d at all short and medium L and rates $1/4 - 2/3$.

Convolutional tailbiting codes are important for several reasons. First, the codes are very powerful. Reference [5] shows that (N, L, d) TB codes achieve the minimum distance of most of the best known block codes at short and medium N and that they do so at a relatively small trellis complexity. Second, as convolutional codes they do not require terminating bits in order to avoid a higher error rate at the ends of words. There is thus no rate reduction on the coding and no accompanying loss of energy efficiency. Finally, as trellis codes these TB codes allow the use of trellis algorithms that have computational advantages and that easily adapt to soft channel input and soft decoder output. The BCJR decoder in this paper is one such decoder. Soft output APP decoders are essential in iterative decoding methods such as turbo coding. As soft input decoders, MAP decoders can take account of *a priori* data probabilities. When these exist the improvement in, e.g., bit error rate over distance-minimizing decoders can be dramatic [2, 7].

One result of this paper is that there is a subtle difference between the MAP and BCJR decoders, and so we next define carefully what is meant by each. The *sequence* MAP decoder solves for the *a posteriori* most likely sequence of state transitions in the encoder trellis. Its output is an entire codeword. The *symbol-by-symbol* MAP decoder solves for the most likely value of the encoder state S_t or transition or data symbol at time t. The sequence of these may not correspond to states or data in the sequence MAP output. The true symbol-by-symbol MAP decoder would exhibit all sequences of states and their probabilities and sum the probabilities of all state sequences that contain the desired state or transition or symbol at time t. Our colleague K. Zigangirov has proposed such a MAP decoder for tailbiting codes [18, Chap. 7]. It can be thought of as two recursions that each produce and store a sequence of $2^m \times 2^m$ matrices.

In contrast, decoders of the BCJR type do not work with probabilities of whole state sequences. Let us formally take a BCJR-type decoder as one with two recursions that produce and store a sequence of size-2^m *vectors* instead of $2^m \times 2^m$ matrices, with each vector relating to the 2^m state probabilities at a time t. Its reason for being is that it is simpler by a factor 2^m. It is well known that the BCJR decoder is true symbol-by-symbol MAP for *terminated* codes. We will find that this is not quite so for tailbiting codes.

An important aspect of APP-based decoders is that their recursions can be viewed as matrix and vector operations. This can be a great advantage: Objects such as eigenvectors in fact have physical meaning, and when expressed in a vector programming language the recursions become startlingly simple. The dynamic programming in a Viterbi algorithm is a less natural operation to an arithmetic-based processor and matrix properties have less physical meaning. But the vector nature of APP decoders can sometimes also mislead: Certain matrices in APP recursions are sparse and multiplications with them do not have the complexity of full matrix operations. Both BCJR and true MAP decoders can be simplified by taking account of sparsity and trellis structure. We will not pursue this in detail.

APP-based decoders form a subset of the more general class of "belief propagation" algorithms, a field that has attracted much recent interest. See for example refs. [8, 9] and other papers in this book. Belief propagation around a tailbiting circular code trellis is a very special simple case. Sensible code trellises are irreducible – meaning that any state may be reached from any other – and in addition belief can be considered to propagate in just two directions, clockwise and counter-clockwise. The sections that follow display many properties that we suspect do not apply beyond this bidirectional propagation case.

A formal justification of the tailbiting BCJR (henceforth denoted TB-BCJR) decoder appears in Section 2. The critical difference from a terminated BCJR decoder is the appearance of principal eigenvectors in the decoder and this leads to interesting properties. Section 3 explores how to find the eigenvectors. It turns out that finding them is a fixed overhead that depends on the convolutional tap set and not on the tailbiting circle size; similarly, the storage of recursion outcomes can be limited. The rest of the paper turns from a formal approach to one of exploring decoder behavior through measurements of bit error rate (BER). Section 4 studies the TB-BCJR error rate and compares it to that of tailbiting Viterbi and true MAP decoders. It turns out that no more than a certain encoder memory m is needed for a fixed circle size L and only a certain L is needed for a given m; put another way, the ratio L/m should assume a certain value. This value depends on the *decision depth* parameter of the convolutional encoder. This parameter also gives an insight into when the TB-BCJR and true MAP decoders have similar BERs.

2. Formal justification of the tailbiting BCJR. In this section we adapt the BCJR algorithm to tailbiting and show that under tailbiting an algorithm of the BCJR type should be initialized with the principal eigenvectors of a certain matrix. These are unique and component-wise positive vectors. Properties of the forward and backward BCJR recursions are developed, including particularly the fact that the recursions converge to these eigenvectors. It will turn out that the TB-BCJR, whether based on eigenvectors or converging recursions, is not quite a true MAP decoder.

The BCJR for Terminated Trellis Codes. The BCJR is known also as the forward-backward or sum-product algorithm. It solves for the set of probabilities

(2.1)
$$P\{S_t = i \mid Y_1^L\}$$

that the encoder state S_t at trellis stage t was i, given that the codeword $Y_1, \ldots, Y_L$ was received. Here each Y_t is the set of c channel outputs corresponding to stage t when the code is convolutional with rate b/c; the notation Y_1^L is shorthand for the c-groups 1 through L. Actually, the algorithm solves a system of equations in the variables

(2.2)
$$\lambda_t(i) = P\{S_t = i, Y_1^L\}$$

which are simply the probabilities in (2.1) scaled by the fixed number $P\{Y_1^L\}$. It is convenient to group together $\lambda_t(i), i = 1, \ldots, 2^m$, into a vector $\boldsymbol{\lambda}_t$, one for each trellis stage.

To find these probabilities, the procedure defines three sets of working probabilities. Define for each time t the row vector $\boldsymbol{\alpha}_t$ with ith element

(2.3)
$$\alpha_t(i) = P\{S_t = i, Y_1^t\}, \quad t = 1, \ldots, L.$$

Define the column vector $\boldsymbol{\beta}_t$ with jth element

(2.4)
$$\beta_t(j) = P\{Y_{t+1}^L \mid S_t = j\}, \quad t = 1, \ldots, L - 1.$$

Finally, define the matrix $\boldsymbol{\Gamma}_t$ with i, j element

(2.5)
$$\Gamma_t(i, j) = P\{S_t = j, Y_t \mid S_{t-1} = i\}, \quad t = 1, \ldots, L.$$

The matrix set here comprises the input to the algorithm; it contains the channel outputs, the encoder transition pattern and the *a priori* probabilities. Given $\boldsymbol{\Gamma}_1, \ldots, \boldsymbol{\Gamma}_L$, the BCJR works as follows.

(i) Form the set of row vectors $\boldsymbol{\alpha}_1, \ldots, \boldsymbol{\alpha}_L$ by the *forward recursion*

(2.6)
$$\boldsymbol{\alpha}_t = \boldsymbol{\alpha}_{t-1} \boldsymbol{\Gamma}_t, \quad t = 1, \ldots, L.$$

(ii) Form the set of column vectors $\boldsymbol{\beta}_{L-1}, \ldots, \boldsymbol{\beta}_1$ by the *backward recursion*

(2.7)
$$\boldsymbol{\beta}_t = \boldsymbol{\Gamma}_{t+1} \boldsymbol{\beta}_{t+1}, \quad t = L - 1, \ldots, 0.$$

(iii) Form the output set $\boldsymbol{\lambda}_1, \ldots, \boldsymbol{\lambda}_L$ by the operation

(2.8)
$$\boldsymbol{\lambda}_t = \boldsymbol{\alpha}_t \bullet \boldsymbol{\beta}_t',$$

where '$\bullet$' means component-wise multiplication. Observe that with a vector processing algorithm, (2.6),(2.7) and (2.8) are each a single line.

The relationships (2.6)–(2.8) here hold if the encoder outputs are Markov and the channel is memoryless. Under the same conditions, (2.5) can be written as

(2.9)
$$\Gamma_t(i, j) = P\{Y_t \mid i \to j\} P\{i \to j\},$$

where $i \to j$ is the tth state transition. For the BSC, Y_t consists of symbols 0 and 1, and the probabilities $P\{Y_t \mid i \to j\}$ stem from the crossover probability p. For the AWGN channel, the data-bearing values $+\sqrt{E_s}$ and $-\sqrt{E_s}$ are corrupted by additive Gaussian noise with variance $N_0/2$ to form the c variables in each Y_t. Now $P\{Y_t \mid i \to j\}$ should be taken as a Gaussian density that depends on the signal-to-noise ratio $E_b/N_0 = E_s/RN_0$, and corresponding changes should be made in (2.3)–(2.5).

For terminated coding the encoder starts before transition 1 and ends after transition L in state 0. Some contemplation of recursions (2.6) at $t = 1$ and (2.7) at $t = L - 1$ shows that they produce the right outputs when they are initialized by $\alpha_0 = (1, 0, \ldots, 0)$ and $\beta_L = (1, 0, \ldots, 0)'$. Since successive α and β become rapidly smaller, it is convenient to normalize each to unit sum as it appears. This will have no formal mathematical effect on λ_t, since it will always be normalized by $\sum_i \lambda_t(i)$. Because of this normalization, we will scale α and β at will through the rest of the paper.

For a feedforward encoder it turns out that the APP that data symbol d_t is 0 ($+\sqrt{E_s}$ in the AWGN case) is the sum

$$(2.10) \qquad \sum_{i \varepsilon S} \lambda_t(i),$$

where λ_t has been normalized and S is the set of states to which data 0 transitions lead. For a feedback encoder the same APP is

$$\sum_{i \to j \varepsilon \mathcal{T}} \sigma_t(i \to j), \quad \sigma_t(i \to j) = \alpha_{t-1}(i)\Gamma_t(i, j)\beta_t(j);$$

here $\sigma_t(i \to j)$ is the normalized probability of a transition $i \to j$ at time t and $\mathcal{T}$ is the set of such transitions caused by data 0. In the sequel we will use only (2.10). These expressions are probabilities; the BCJR decoders in this paper are MAP decoders that make a hard decision on d_t, by putting out the symbol with the highest probability.

The Extension to Tailbiting – Forward Recursion. In the case of tailbiting, the start and end states of the encoder, S_0 and S_L, are equal but unknown and must somehow be estimated by the decoder. For the BCJR algorithm the problem comes down to initiating the recursions: The forward recursion at $t = 1$ starts from some S_0 and the backward one at $t = L - 1$ from the same S_0.

In what follows we will equate not the states S_0 and S_L, but their *distributions*; that is, we will set $P\{S_0 = i, Y_1^L\} = P\{S_L = i, Y_1^L\}$, all i. A true TB decoder enforces $S_o = S_L$. The implies but is not implied by equal distribution. The rest of the decoder is (2.6)–(2.8), but the decoder begins from an assumption weaker than tailbiting, and it therefore may not solve for the exact probabilities (2.1)–(2.4). It does nonetheless produce probability distributions. These solutions have special properties, the most interesting of which is that they apparently tend in practical situations to the true tailbiting APP decoder solutions. To distinguish the new solutions we will call them tailbiting BCJR (TB-BCJR) solutions and denote them by $Q\{S_t = i, Y_1^L\}, Q\{Y_{t+1}^L = i, S_t = j\}$, etc., rather than with P as in for example (2.1)–(2.4).

We will take first the forward recursion and give an argument that produces the fundamental equation of the TB-BCJR algorithm. Strictly speaking, the equation follows from the Perron-Frobenius theorem and the fact that the algorithm consists of recursions, but the argument will give us a useful insight. From (2.6) at $t = 1$,

$$Q\{S_1 = j, Y_1\} = \alpha_1(j) = \sum_i \alpha_0(i)\Gamma_1(i,j) =$$

$$\sum_i \alpha_0(i)Q\{S_1 = j, Y_1 \mid S_0 = i\}.$$

From the laws of probability it is clear that $\alpha_0(i)$ in this expression must be a probability distribution on the outcome of S_0. We can select any distribution, and with the equal distribution assumption in mind, we choose

$$(2.11) \quad \alpha_0(i) = \frac{Q\{S_L = i, Y_1^L\}}{Q\{Y_1^L\}} = Q\{S_L = i \mid Y_1^L\}, \quad i = 1, \ldots, 2^m.$$

Now iterate (2.6) $L - 1$ times to obtain $\alpha_L = \alpha_0\Gamma_1\ldots\Gamma_L$. From (2.11),

$$\alpha_L(i) \overset{\triangle}{=} Q\{S_L = i, Y_1^L\} = \alpha_0(i)Q\{Y_1^L\}, \quad i = 1, \ldots, 2^m.$$

Consequently,

$$(2.12) \qquad\qquad \alpha_0 = \alpha_0\Gamma_1 \cdots \Gamma_L / Q\{Y_1^L\};$$

that is, the α_0 selected is a left eigenvector of the matrix $\Gamma_1 \ldots \Gamma_L / Q\{Y_1^L\}$.

In fact, α_0 is the unique, positive principal[1] left eigenvector of the matrix in (2.12), henceforth abbreviated as the l.p.e.v. This is a consequence of the century-old Perron-Frobenius Theorem. A classic description[2] of this theorem of linear algebra appears in Bellman [11]. It states that a positive matrix G has a unique (to a constant) positive l.p.e.v. and the eigenvalue corresponding is positive. A positive vector or matrix iz one whose components are all greater than zero. Moreover, all other eigenvectors have at least one component with opposite sign. The matrix in (2.12) will be positive if the encoder in the absence of the tailbiting condition can reach

---

[1] A principal eigenvector, whether left or right, is the one with largest eigenvalue. The principal eigenvalue is the largest eigenvalue; the principal left and right eigenvectors correspond to the same principal eigenvalue.

[2] We use Bellman's terminology. A recent treatment of the theorem appears in [21].

every state at L from every state at 1. We will restate the theorem as a property of the tailbiting matrix product $G = \Gamma_1 \cdots \Gamma_L / Q\{Y_1^L\}$.

Property 2.1 Let α satisfy $\alpha = \alpha G$, where G is the positive tailbiting matrix just defined. Then there exists precisely one α that is a probability distribution and it is the l.p.e.v. of G.

Another implication of Perron-Frobenius is that a value $Q\{Y_1^L\} > 0$ exists (it is the l.p.e.v. of $\Gamma_1 \cdots \Gamma_L$); since a positive α_o exists as well, the Q distribution used in (2.11)–(2.12) exists.

We turn next to a second fact of matrix algebra, that for any matrix A with principal eigenvalue λ and an initial u not orthogonal to the l.p.e.v. v, the l.p.e.v. satisfies

$$(2.13) \qquad v = \lim_{n \to \infty} \frac{uA^n}{\lambda^n}$$

If now G is positive and $\alpha = \alpha G$, then $\lambda = 1$ and uG^n converges to α for any positive u. The last follows because for $u \neq 0$, uG has no zero components and hence cannot be orthogonal to a positive l.p.e.v. We summarize as:

Property 2.2 Let the tailbiting G be positive. The l.p.e.v. α satisfying $\alpha = \alpha G$ satisfies $uG^n \to \alpha$ with n for any nonnegative nonzero starting distribution u. The recursion (2.6), iterated nL times and normalized each cycle to unit sum, *necessarily* converges to the l.p.e.v. of G.

There are important facts here for a physical tailbiting encoder. First and foremost, Property 2.2 says that iterating the forward BCJR recursion for n cycles always leads to the same result in the limit, namely the α_0 that stems from the equal distribution assumption. Whether we first make that assumption ourselves and proceed as in (2.11)–(2.12) is actually irrelevant. Such an ongoing recursion is a natural way to construct a decoder: Properties 2.1 and 2.2 imply that the strategy indeed converges, but that it converges to the "Q" solution set, not the tailbiting one.

A second fact is that the key condition for both properties is that G be positive. That will be so if, (i), an L-step encoder transition exists from state i to state j for all i, j; (ii), $p > 0$ in the BSC and $E_b/N_o < \infty$ in the AWGN channel; and (iii), all data sequences have nonzero probability. These are sufficient but not necessary conditions for Properties 2.1 and 2.2. They are an obvious set of regularity conditions for a TB coding system. If the TB trellis contains two noncommunicating subsections, for example, different positive starting u can lead to different limits in (2.13), but we take this to be an irregular code.

Forney and Horn and others [10, 20] have advocated the idea of pseudocodewords as a way to illustrate the difference between the "P" and "Q" solutions here. Pseudocodewords are multicycle words that start at a state and end in the same state after n cycles, but not necessarily after fewer than n.

Tailbiting Symmetry. As just explained, a set $\alpha_1, \ldots, \alpha_L$ of state distributions, one for each stage, is produced via solution of (2.12) and $L - 1$ applications of (2.6). Suppose the equal distribution assumption is applied at a different stage and the solution performed from there. By symmetry, the same essential distributions should be produced for a given channel output Y_1^L, no matter which trellis stage is labeled as time 0. We can formalize this notion as the following Tailbiting Symmetry Principle: A solution for the data symbol (or encoder state or encoder transition) probabilities cannot depend on the starting stage of the analysis. We look now at several properties that stem from this principle.

As a preliminary, define the notation $\sim$ to mean that two positive distributions are alike except for a scalar multiple. For example, the distributions α_0 and α_L in the TB-BCJR solution satisfy $\alpha_0 \sim \alpha_L$. The scale factor is in fact $1/Q\{Y_1^L\}$.

In what follows it is useful to think of α_0 not as a state distribution at $t = 0$ on a tailbiting circle, but as a fortuitous choice of starting distribution for a straight-line trellis, as shown in Fig. 2. The solution for the state distribution at the following stages will be unaffected. Yet it is often easier to think about the solution.

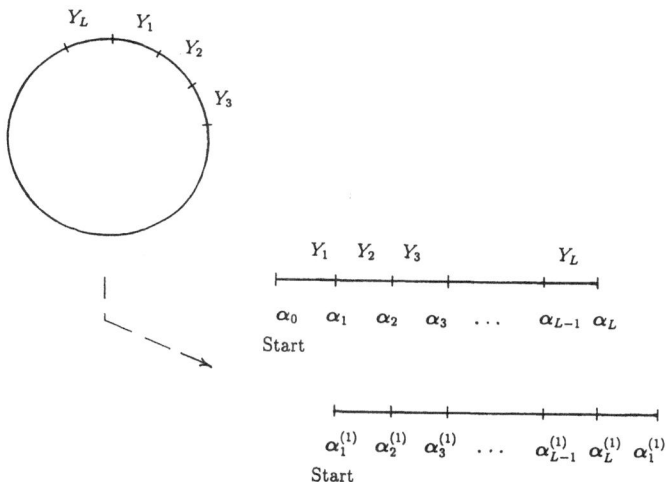

FIG. 2. *Two straight line portrayals of the same circular TB trellis word, one beginning one stage later. Subscripts indicate an arbitrary absolute time. $Y_1, \ldots, Y_L$ are arriving channel outputs. BCJR forward recursions as indicated for the two cases.*

We have first the following simple property, which expresses the fact that consistent l.p.e.v.'s are obtained when the equal distribution assumption is applied, no matter where the TB circle is opened out into a straight trellis.

Property 2.3. Consider the sequence of vectors $v_1, \ldots, v_L$ and matrices $A_1, \ldots A_L$, where the v_ℓ satisfy $v_\ell = v_{\ell-1} A_\ell$ and $v_0 = v_L$. Furthermore, all cyclic products of the matrices are positive. Then one v_k is the l.p.e.v. for cyclic shift product $A_{k+1} \ldots A_L A_1 \ldots A_k$ with eigenvalue λ, if and only if the remaining vectors v_ℓ are the l.p.e.v.'s with the same λ for the cyclic products $A_{\ell+1} \ldots A_L A_1 \ldots A_\ell$. That is, all the vectors succeed as principal eigenvectors or none of them do.

Proof. By hypothesis, $\lambda v_k = v_k A_{k+1} \ldots A_L A_1 \ldots A_k$. Right multiply by A_{k+1} to obtain $\lambda v_k A_{k+1} = v_k A_{k+1} \ldots A_L A_1 \ldots A_k A_{k+1}$. Since $v_{k+1} = v_k A_{k+1}$, we have $\lambda v_{k+1} = v_{k+1} A_{k+2} \ldots A_L A_1 \ldots A_{k+1}$, which proves the result for v_{k+1}. Repeat to show for all v_ℓ. This proves the forward implication. To prove the reverse implication, we show that if any vector in the chain fails to be an eigenvector then all the rest fail as well. Suppose $v_{\ell-1}$ succeeds as an eigenvector for λ at some $\ell - 1$ and lies next to a v_ℓ that fails. This must happen somewhere if some but not all of the vectors fail as eigenvectors. Then the same argument as above shows that v_ℓ succeeds as an eigenvector, contra the assumption. $\square$

Returning now to the forward BCJR recursion, we have from (2.12) that $\alpha_0 = \alpha_0 \Gamma_1 \cdots \Gamma_L / Q\{Y_1^L\}$. The l.p.e.v. of $\Gamma_1 \cdots \Gamma_L / Q\{Y_1^L\}$ is the same as that of $\Gamma_1 \cdots \Gamma_L$; only the eigenvalue is scaled by $1/Q\{Y_1^L\}$. It is reasonable to add to the aforementioned TB trellis regularity conditions that all cycles of the Γ-matrices are positive. The property thus applies to the cyclic rotations of $\Gamma_1, \ldots, \Gamma_L$ and their l.p.e.v.'s.

Consider a forward recursion that begins one stage to the right, as illustrated in Fig. 2. The superscripts indicate the starting point of the recursions in units to the right, and the subscripts denote absolute time. By the property, the l.p.e.v. of $\Gamma_2 \cdots \Gamma_L \Gamma_1$, which will be the starting vector $\alpha_1^{(1)}$, must be a scalar multiple of $\alpha_1^{(0)} = \alpha_0^{(0)} \Gamma_1$, which is an α that stems from the recursion that begins at time 0. It is the logic leading to (2.12) — the equating of distributions — that specifies $\alpha_1^{(1)}$; the Symmetry Principle says that this start vector should lead to a set of vectors that are consistent with the set produced by a recursion that starts at time 0. The property confirms that no other vector can do this.

The statement $\alpha_1^{(1)} \sim \alpha_1^{(0)}$ now says that the distributions $Q\{S_1 = i | Y_1^L\}$ and $Q\{S_1 = i, Y_1\}, i = 1, \ldots, 2^m$, are alike to a scalar. Repeating the arguments one shift left shows that $\alpha_1^{(-1)} \sim \alpha_1^{(0)}$, so that $Q\{S_1 = i, Y_1\}$ and $Q\{S_1 = i, Y_L, Y_1\}$ are similarly alike to a scalar. Now scale these second two by scalars $1/Q\{Y_1\}$ and $1/Q\{Y_L, Y_1\}$, respectively; all three distributions are now valid probability distributions and so they are not just scalings, but are in fact equal:

$$Q\{S_1 = i \mid Y_1^L\} = Q\{S_L = i \mid Y_1\} = Q\{S_L = i \mid Y_L, Y_1\}, \quad i = 1, \ldots, 2^m.$$

Repetitions of the argument at all shifts and all S_t demonstrates

Property 2.4. Let time run mod L. Then at any stage t and for any state i,

(2.14)
$$\begin{aligned}
Q\{S_t = i \mid Y_t\} &= Q\{S_t = i \mid Y_{t-1}, Y_t\} = \cdots \\
&= Q\{S_t = i \mid Y_{t+1}, \ldots, Y_L, Y_1, \ldots, Y_t\}.
\end{aligned}$$

That is, the state Q distribution at t given the observation Y_t does not depend on earlier observations. This is a consequence of the equal distribution that leads to (2.12). We stress once again that the probabilities here are those that stem from (2.12), not those in true tailbiting.

The Backward Recursion. It remains to investigate the backward TB-BCJR recursion. The recursion (2.7) continues to hold and as with the forward one, the problem is how to start the recursion. We will show that as a consequence of the equal distribution assumption, the backward β_t are the *right* principal eigenvectors (r.p.e.v's) of the Γ-products.

First we cite another useful property of the TB-BCJR recursions. It states that every backward recursion is the forward recursion of some other TB trellis, and vice versa. Without loss of generality, let the forward direction in the "first" trellis be left to right for straight-line trellises and clockwise for a circular one.

Property 2.5. Suppose $\beta_0, \beta_1, \ldots, \beta_L$ are the outcomes of a backward recursion starting from β_L. Here the channel outcomes $Y_1, \ldots, Y_L$ give rise to $\Gamma_1, \ldots, \Gamma_L$ and $\beta_t = \Gamma_{t+1}\beta_{t+1}$, $t = 1, \ldots, L-1$. Then a forward recursion $\alpha_0, \alpha_1, \ldots, \alpha_L$ exists whose vectors are identical to $\beta_0, \beta_1, \ldots, \beta_L$ taken in reverse order (i.e., $\alpha_t = \beta_{L-t}$). For this recursion, the forward direction is right to left (counterclockwise). The channel outcomes from first to L-th are $Y_L, \ldots, Y_1$ and the gammas are $\Gamma_L', \ldots, \Gamma_1'$, where Γ_t stems from Y_t for the first trellis. The starting α_0 in the second trellis is β_L from the first. The data symbols in the second numbered according to $\alpha_1, \ldots, \alpha_L$ in the second are the symbols $d_{L-1}, \ldots, d_1, d_L$ in the first.

These relationships are all sketched in Fig. 3, which shows the structure of both recursions. At stage t, the first backward recursion is $\beta_{t-1} = \Gamma_t\beta_t$. Complementing both sides, we get $\beta_{t-1}' = \beta_t'\Gamma_t'$; this shows that the new α_ℓ is $\beta_{L-\ell}', \ell = 1, \ldots, L$, provided that the new α-recursion starts with $\alpha_0 = \beta_L$. The new Γ-matrices are evidently transposes of the old. From the properties of any forward recursion, the new α_0 must satisfy $\alpha_0 \sim \alpha_L$. Thinking instead in reverse, we see that there must exist a backward recursion for every proper circular trellis for which $\beta_L = \Gamma_1 \cdots \Gamma_L\beta_L/Q\{Y_1^L\}$. This follows because in the matching α-recursion in the reversed trellis, it must be that $\alpha_0 = \alpha_0\Gamma'_L \cdots \Gamma'_1/Q\{Y_1^L\}$; complementing both sides and taking $\beta_L = \alpha'_0$ gives

$$(2.15) \quad \beta_L = \alpha'_0 = (\Gamma'_L \cdots \Gamma'_1)'\alpha'_0/Q\{Y_1^L\} = \Gamma_1 \cdots \Gamma_L\beta_L/Q\{Y_1^L\}$$

β_L here is evidently a right principal eigenvector (denoted r.p.e.v.), which satisfies the Perron-Frobenius Property 2.1, with right substituted for left.

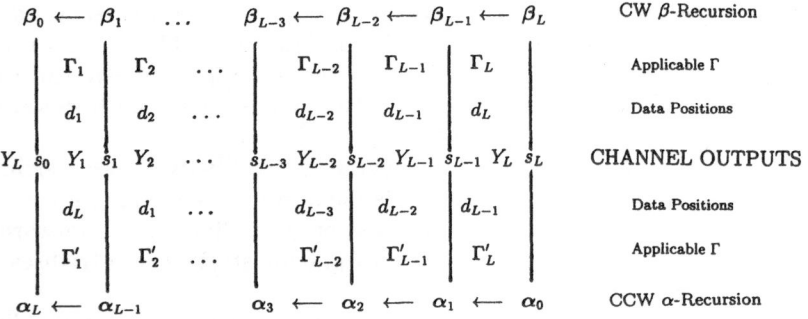

FIG. 3. *Equivalence of clockwise backward recursion (top) to counterclockwise forward recursion (bottom). At the middle is the received $Y_1, \ldots, Y_L$. All subscripts show absolute time except the bottom α-row, which is subscripted backwards. Top β-set and bottom α-set will be identical if $\alpha_0 = \beta'_L$ and Y and Γ are used as indicated.*

Clearly, the property does not depend on which direction is taken as forward, and we could as well have begun with a forward recursion and ended with a backward one. The property expands on the well known fact about terminated trellises (i.e., those with $\alpha_0 = \beta'_L = (1, 0, \ldots, 0)$) that they can equally well be decoded in either direction.

To summarize, we have shown that a backward recursion exists beginning from a positive β_L that is the r.p.e.v. of $\Gamma_1 \cdots \Gamma_L$, for which $\beta_0 = \Gamma_1\beta_1$ and $\beta_0 \sim \beta_L$. Furthermore, Properties 2.1–2.4 apply; that is, (i), there is only one such r.p.e.v. and one such recursion outcome; (ii), in the limit of many cycles the recursion converges to this β_L for any nonnegative nonzero starting vector; (iii), if a backward recursion starts instead from position ℓ, with β_ℓ generated as in (i), it will generate a set of vectors that are the same to a scalar multiple; and (iv), a backward version of (2.14) will hold.

The discussion here shows that a circular backward recursion exists based on the stage to stage relations (2.6)–(2.7) and that solving for it, by iteration or otherwise, produces a unique, inescapable result. When combined in step (2.8) with the unique forward outcome, the result is a unique TB-BCJR outcome set $\lambda_1, \ldots, \lambda_L$.

3. Implementation aspects. It is convenient to characterize tailbiting coding by its circle size relative to the encoder memory m, since the decoder error performance and much other behavior organize themselves this way. There are three cases: We will call these short, medium and long tailbiting. In short tailbiting at rate 1/2, the circle size is roughly 1–2 times m; or alternately m is between $L/2$ and L. These codes tend to behave like short block codes and here the TB-BCJR may have a somewhat higher bit error rate than a true MAP decoder.

In medium tailbiting, the circle is $2m$–$4m$, or m lies in $(L/4, L/2)$. Decoders behave like trellis decoders and the TB-BCJR has essentially the true MAP error rate. In ref. [5] we show that the minimum distance of the best TB codes of size L does not grow once m approximately exceeds $L/2$; this says that the minimum distance of a given convolutional generation is circle-size limited when tailbiting is short and achieves its full size only with medium tailbiting. Extensive tests of the bit error rate of real decoders [12] has shown that the same general conclusion applies to BER. For a given L, medium tailbiting seems to be ideal in the sense that lowering m tends to decrease performance and raising m increases trellis complexity unnecessarily. In long tailbiting, m is less than $L/4$ and the decoder behavior is similar to a terminated decoder's with two differences: There is no rate reduction and if the circle is traversed only once there is a brief period of raised BER during which the TB decoder decides the starting state.

The L/m ratios that mark the cases depend on the rate, and the discussion here focuses on rate 1/2.

Finding the Eigenvector: The Power and Wrap Algorithms. The TB-BCJR algorithm consists of steps (2.6)–(2.8), preceeded by finding the right and left principal eigenvectors of $\Gamma_1 \cdots \Gamma_L$, or at least something close to them. Different ways of performing the eigen calculation lead to several practical versions of the TB-BCJR. In many applications, it will be unnecessary to form the full matrix $\Gamma_1 \cdots \Gamma_L$, and in no case is it necessary to perform the full solution for all 2^m eigenvectors.

The baseline procedure is to find the principal eigenvectors by one of the well-known accurate procedures. We emphasize that this method is not iterative: It directly computes the exact solution to a set of equations. It is interesting to try out initial vectors for the recursions other than the l.p.e.v and r.p.e.v., in an effort to find ones that are somehow better, or perhaps better embody the tailbiting condition. We have tried many and compiled histograms that show observed decoder error rate as a function of trellis stage. An example is shown in Fig. 4 for the code (72,62), an AWGN channel with $E_b/N_0 = 2.5$ dB and $L = 20$ stages. The figure shows the four cases of α starting with the uniform and left p.e.v. and β starting with the uniform and right p.e.v. The uniform vectors cause a starting and/or ending transient in the BER which dies out as the decoding moves away from the incorrect start or end. We have observed that only the p.e.v. starts seem to give a flat histogram. No lower histogram than this flat one has ever been observed. In particular, starting the backward recursion with the forward α at that stage leads to a startup transient and a poorer BER histogram.

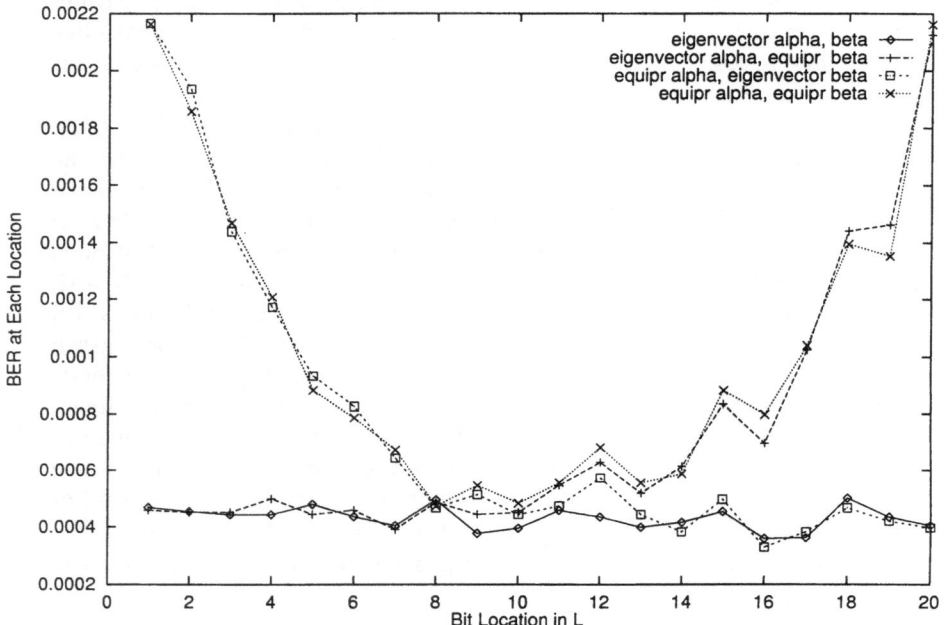

FIG. 4. *Histogram showing bit error rates divided by L at each circle position, with equiprobable and principal eigenvector starting vectors for the recursions. Four combinations shown. Only the case with both PEVs is flat. Encoder (72,62), AWGN, $E_b/N_o = 2.5dB$, $L = 20$. To find overall BER integrate the histogram.*

We turn next to an algorithm that *is* iterative. Property 2.2 states that the result of simply iterating the forward and backward recursions around and around the trellis circle must eventually converge to the α_0 and β_L that correspond to the equal distribution assumption. This occurs for any nonnegative starting vector. When nothing is known about the starting distribution, we may as well start with the equiprobable one, $e \triangleq 2^{-m}(1, 1, \ldots, 1)$. In formal terms we have for w complete iterations

$$(3.1) \qquad \alpha_0 = \lim_{w \to \infty} eG^w$$

where $G = C\Gamma_1 \cdots \Gamma_L$ is the Γ-product scaled by C so that it has unit principal eigenvalue. The calculation is said to "wrap" $w - 1$ times. We have called this method the wrap algorithm in [2].

The flow of a practical wrap algorithm is shown in Fig. 5. The figure imagines two w-fold repetitions of the trellis and the $Y_1, \ldots, Y_L$ sequence in two straight runs. The recursions execute along the arrows and the last sets of $\alpha_1, \cdots, \alpha_L$ and $\beta_L, \cdots, \beta_1$ are held as output. The present sets can overwrite the previous ones, so that the total storage is $2L$ vectors; in addition the Γ-set needs to be either stored or regenerated as needed. The storage and the last of the w wraps are unchanged from the baseline algorithm.

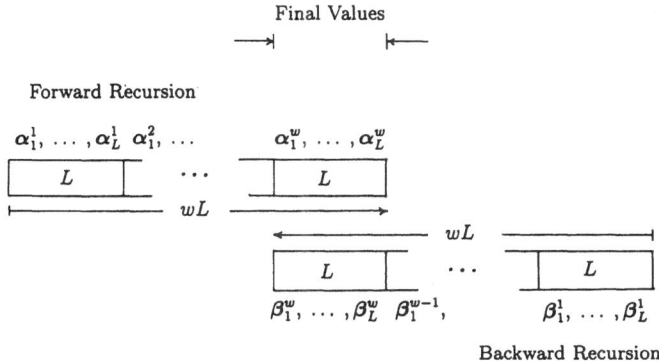

WRAP ALGORITHM

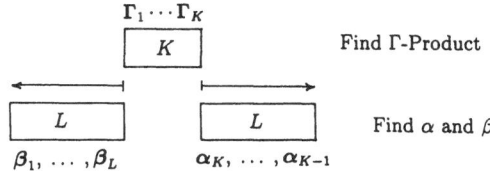

POWER ALGORITHM

FIG. 5. *The wrap and power algorithms for finding α- and β-sets, envisioned as working on repeated copies of the trellis and received signal.*

Another method, called the power algorithm, is based on the so-called power method for finding principal eigenvectors. In this method the recursions themselves are not iterated. We achieve (3.1) by forming $G = C\Gamma_1 \cdots \Gamma_L$ directly as a matrix product; then G is raised to a power; finally, $\alpha_L \sim \alpha_0$ is taken as eG^v and β_1 as $G^v e'$, where v is an integer. Only then are the BCJR recursions performed, one cycle only to produce the α- and β-sets. Storage is again that of the baseline and wrap algorithms.

The important issue in the power scheme is the size of v. We find that $v=2$–3 is almost always enough. Figure 6 shows the BER for the code (72,62) with circle 20, an AWGN channel, and the power algorithm with $v = 1, 2, 3, 50$. The power 2 achieves virtually the baseline algorithm BER. We find similar behavior with the BSC and in general with codes that employ medium tailbiting. Codes with short tailbiting need a little higher power. Codes with long tailbiting can use a power *smaller* than 1; that is, the product $\Gamma_1 \cdots \Gamma_L$ need not be fully carried out. We turn to this case shortly.

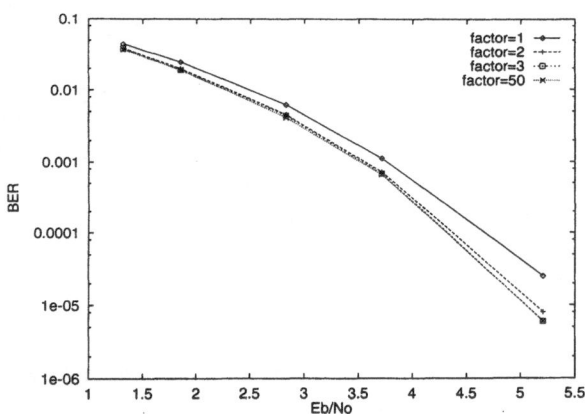

FIG. 6. *Bit error rate of (72,62), $L = 20$, over AWGN channel, with the decoder using the outcome of the power algorithm calculation of the α- and β-sets. Power $v = 1, 2, 3, 50$.*

As is well known in matrix theory, if the principal and next-largest eigenvalues of $\Gamma_1 \cdots \Gamma_L$ have ratio close to 1, the convergence of the power algorithm with v will be slow. By solving for the ratio and trapping such cases during decoder operation, we have observed that slow convergence is not common. It happens almost exclusively in good channels and then only when the received sequence has a similar probability of being sent in two conflicting ways. A power algorithm with small v will now produce somewhat distorted vectors, but since there is no clear decoder decision to take and since the tie event is unlikely in a good channel, the distortion's effect on the overall BER is negligible.

When the power v and the wrap cycles w are the same, the two procedures will in principle lead to the same estimate for the eigenvector. In practice they generally do so, but the power algorithm is the more stable one mathematically and it is simpler to implement in terms of software steps. In Section 4 we use the power algorithm. However, a strict count of arithmetic operations shows that the wrap algorithm is the only truly efficient procedure. In fact, the power algorithm under short and medium tailbiting leads to the same order of computation as the more complex true MAP decoder. We intend to return to this subject in a future paper.

The Convergence Depth of $\Gamma_1 \cdots \Gamma_K$. With long tailbiting only a fraction of the total product $\Gamma_1 \cdots \Gamma_L$ needs to be computed for an accurate estimate of one α and β. A rough statement of the convergence problem is: After how many stages does the trellis decoding "forget" its starting conditions? For a more precise statement we note that element i, j of the product $\boldsymbol{H} = \Gamma_1 \cdots \Gamma_K$ is the probability that Y_1^K was the channel output and the encoder moved from state i to state j in K steps. If the rows of $\boldsymbol{H}$ are essentially the same, then the distribution $Q\{S_K = j, Y_1^K \mid S_0 = i\}$ does not depend on the starting state and we have an estimate of $Q\{S_K = j, Y_1^K\}$ instead. At which K does this happen?

With reference to Fig. 7a, consider now the problem for the BSC(p) when all zeros are sent. This will serve to illustrate ideas and motivate an algorithm to estimate the convergence depth. The $i,1$ element of $\boldsymbol{H}$ is made up of a number of path terms of the form $pp \cdots p(1-p)(1-p) \cdots (1-p)C_o$, in which the number of p's is equal the Hamming distance d of the path from state i to 1 and the number of $(1-p)$'s is $Kc - d$; C_o is set by the *a priori* data probabilities for the path and we will assume that C_o exceeds 0 by a reasonable margin. As p tends to 0, the path $i \to 1$ with smallest d dominates $H(i, 1)$. This path is heavier in Fig. 7a.

If all paths to other states are heavier than d after K stages, then entry $H(i, j)$, which represents the path $i \to j, j \neq 1$, is much smaller than $H(i, 1)$. Such a K must eventually be reached, because the other paths out of i all continually gain weight; furthermore, it will also be true at stages $K + 1, K + 2, \ldots$ because the least-weight path to state 1 continues to have weight d while the weights to other states can only grow. Let K^* be the deepest such stage for all the starting states i. Then for $\boldsymbol{H}$ that includes K^* Γs, we have $H(i, 1) \gg H(i, j), \ j \neq 1$, for all starting states i, as p tends to 0.

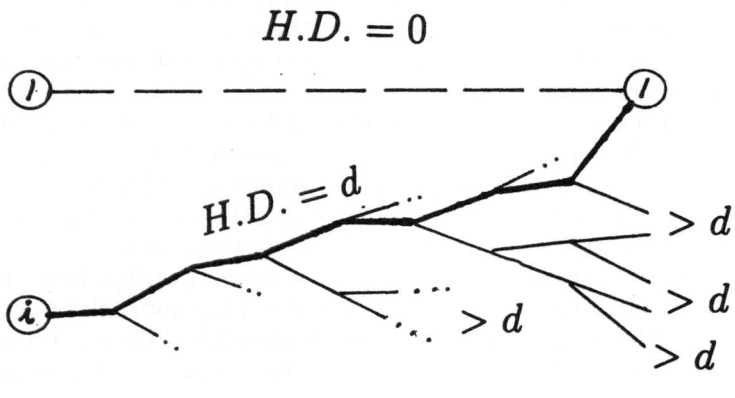

(a)

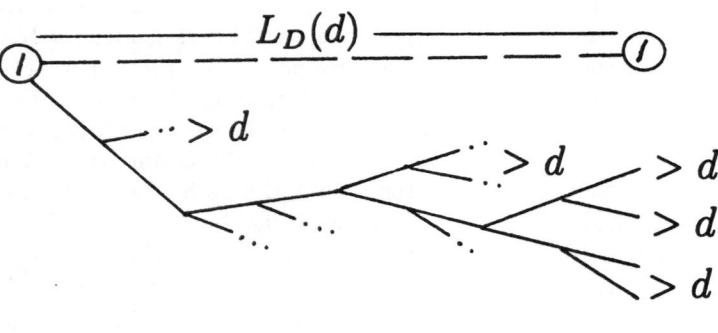

(b)

FIG. 7. *(a) Illustration of the problem of finding convergence time for a BSC with no errors. d is least Hamming distance (H.D.) of any path from state i to state 1. (b) Illustration of finding decision depth function at distance d.*

To conclude, let α_0 be any starting distribution. Then after K^* stages, $\alpha_0 H$ is a vector dominated by its first element. This is a statement that $P\{S_{K^*} = j \mid Y_1^{K^*}\}$ is virtually 1 for $j = 1$ and 0 otherwise.

We are led to the following algorithm to find the convergence depth at small p or high E_b/N_o. Since the tailbiting code is linear, we can continue to assume that all zeros are sent. Note that slightly different answers may be obtained with forward and backward searches.

Convergence Depth Algorithm

(i) Repeat for starting states $i = 1, \ldots 2^m$:
Execute a dynamic program (i.e., a VA) on the tailbiting trellis from state i (set the starting weight of node i to 0 and of the others to ∞), in order to find the least-weight trellis path to each node j at stage K.
Stop when weight(node 1) < weight(node j), all $j \neq 1$. Record K.
(ii) The largest K is the convergence depth.

We have carried out the algorithm for all the codes in this paper, and for many others as well, and in all cases the observed depth is very close to a tabulated parameter called the *decision depth*, L_D. This depth measures the channel observation width in stages needed by a trellis decoder that starts from a known state (a so called "first decision" decoder), if its asymptotic error probability is to be that predicted by the encoder free distance. Decision depths for most feedforward convolutional encoders are listed in [13]. The depth approximately satisfies

$$(3.2) \qquad L_D(d_f) \approx \frac{d_f}{ch_B^{-1}(1 - R)},$$

where d_f is the free distance of a convolutional code of rate R with c bits on each trellis branch, and $h_B^{-1}(\ )$ is the inverse of the binary entropy function. For example, rate 1/2 decoders need to satisfy $L_D(d_f) \approx 4.54 d_f$. A formal definition is:

Definition. Consider all paths diverging from the all zero state of a convolutional code, which do not later merge to the all zero state (our state 1). The decision depth is the first depth at which all such paths have weight $> d_f$.

The calculation is illustrated in Fig. 7b.

Figure 8 shows the convergence behavior as a function of trellis depth K for several rate 1/2 best-d_f codes over the AWGN channel. The vertical axis of both plots is the measured standard deviation in the estimate of α_K after K stages forward and β_0 after K stages backward, according to the following scheme. First, the all $+\sqrt{E_s}$ data sequence is transmitted through noise with variance $N_o/2$ and both recursions begin from the uniform vector, which is the best starting vector in the absence of any source

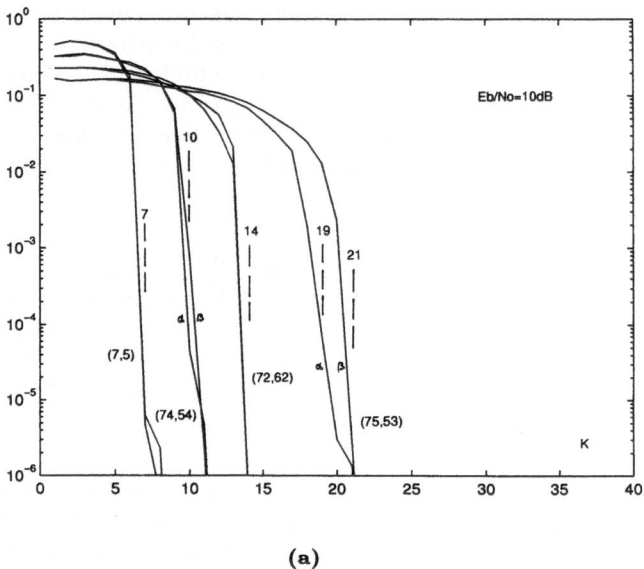

(a)

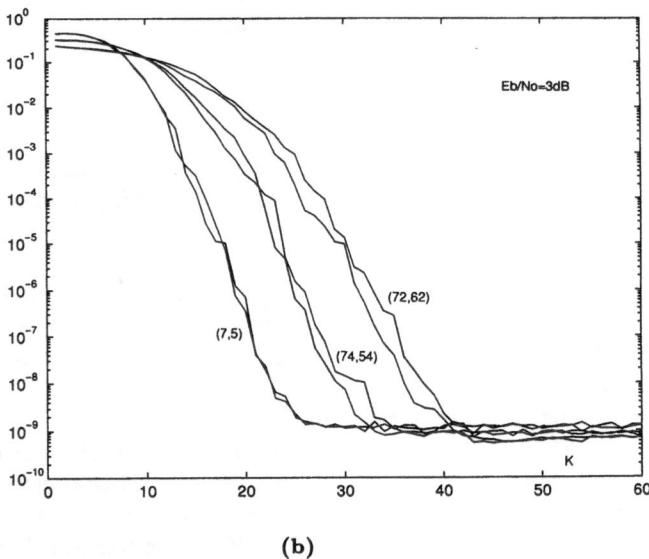

(b)

FIG. 8. (a) rms deviation of α and β components after trellis depth K. Average over 60 trials. AWGN, $E_b/N_o = 10dB$; codes (7,5), (74,54), (72,62), (75,53). Convergence depths from algorithm A4 are shown. (b) Same, but $E_b/N_o = 3dB$. Codes (7,5), (74,54), (72,62).

or channel information. Then the recursions are repeated for each of the 2^m starting vectors of the form $(0, \ldots, 1, 0, \ldots, 0)$. The standard deviation among the 2^m final α or β components relative to their uniform-start estimate is computed. Sixty repetitions of the entire experiment are performed and the average of the outcomes appears in the plots. They thus show the rms outcome over the 2^m most extreme starting vectors.

Figure 8a, for $E_b/N_o = 10$ dB, shows a precipitous drop for each code at close to the code's decision depth. Table 1 lists the decision depths for both the forward and backward trellises, as well as the outcomes of the foregoing algorithm for the precise depth. Figure 8b is a repeat at $E_b/N_o = 3$ dB. Now the decline in average standard deviation is slower with depth but it has still fallen to 10% after one decision depth of stages. The falloff is even slower at very poor E_b/N_o, but it is clear that at least for this rate and these codes and for reasonable E_b/N_o, one α_t and β_t may be considered known after L_D stages.

TABLE 1

Standard decision depths and convergence depth algorithm estimates for clockwise and counterclockwise trellises for some generators used in this paper. "df" indicates free distance. Forward (α) recursion is in the clockwise direction.

| Encoder | df | Convergence Depth Alg. | | Measured Decision Depth | |
		Alpha Depth	Beta Depth	CW (Alpha)	CCW (Beta)
(7,5)	5	7	7	8	8
(74,54)	6	10	10	10	11
(72,62)	7	14	14	15	19
(75,53)	8	19	21	19	23
(414,730)	8	23	26	18	24
(554,744)	10	28	29	28	27
(712,476)	10	30	31	29	28
(561,753)	12	39	37	37	33

From the definitions of decision depth and convergence depth, it seems clear that adding decoder path memory significantly beyond L_D cannot improve the decoder bit error rate in a reasonable channel. It thus appears that (3.2) is the approximate start of what we have called the long tailbiting range. At rate 1/2, in fact $L/m \approx 4$–5 has been observed experimentally as the starting point; (3.2) yields $4.5m$. It also makes sense that circle size (3.2) is related to the point where the TB-BCJR and true MAP tailbiting decoders have similar BER. A rough proof of an estimate goes as follows. The true MAP tailbiting decoder may be realized as a weighted sum of 2^m applications of the terminated BCJR, one for each of the 2^m starting states[3]. Once the recursions proceed beyond the convergence depth, their outcomes are essentially the same, regardless of starting state. For both the forward and backward recursions to be independent of starting state,

---

[3]This algorithm was communicated to us by L. Wei of Australian National Univ.

the TB circle must exceed the sum of the α- and β-depths. If we accept that these depths are both approximately (3.2), then the TB-BCJR and true MAP decoders have similar outcomes when the circle is larger than about twice (3.2).

The discussion here leads to the following conclusions about the TB-BCJR algorithms.

(i) In short and medium tailbiting, the recursions do not "forget" their origin in one cycle. The transition to long tailbiting occurs after one decision depth of stages.

(ii) For short and medium tailbiting, the computation needs more than one trip around the circle. 2–3 trips are sufficient if BER is the performance measure. The wrap algorithm is most efficient.

(iii) For long tailbiting, the computation of a single α or β, essentially the p.e.v. calculation, takes less than one TB circle; as the circle size grows for a fixed encoder, the computation in this step dwindles in relation to the whole.

(iv) For circles larger than about $2d_f/ch_B^{-1}(1-R)$, the true MAP and TB-BCJR algorithms make similar decisions.

4. BER performance and comparisons to non-BCJR decoding. A major issue in the performance of the TB-BCJR is how the error rate compares to ordinary minimum distance decoding. A second issue is how close the TB-BCJR comes to the true MAP decoder. This section presents the results of decoder tests over simulated AWGN and BSC channels. We need to make comparisons among the design paradigms introduced in Section 1: maximum *a posteriori* versus maximum likelihood decoding, symbol-by-symbol versus sequence decoding, and decoding based on probability versus distance. Here is a list of decoders that will be tested in this section together with their identifying acronyms.

(i) ML will denote the sequence maximum likelihood tailbiting decoder. One implementation of the ML is based on the usual Viterbi algorithm, started successively at each trellis state $S_i, i = 1, \ldots, 2^m$. Making one trip around the circle, the decoder records the minimum distance path starting at S_i; the calculation is repeated for each S_i; the decoder output is the minimum distance path for the 2^m starts so recorded. With both the BSC and the AWGN channels, the distance-minimizing decoder is also a sequence ML decoder. When the source data are i.i.d., the decoder is also sequence MAP. The ML decoder will constitute our baseline performance.

(ii) sbsMAP will denote the true symbol-by-symbol tailbiting MAP decoder of Zigangirov [18, Chap. 7].

(iii) TB-BCJR denotes the power TB-BCJR. This algorithm approximates the sbsMAP.

(iv) CVA denotes the circular Viterbi algorithm, a distance-minimizing tailbiting decoder that works in one pass. Different versions of the CVA have been proposed [14, 15] which have in common that the usual VA traverses a number of times around the TB circle, comparing outcomes from earlier circles until a stopping condition is met. Our CVA traverses $U + L + U$ stages, where L is the circle size, and puts out the middle L branches of the minimum-distance length $U + L + U$ path. The algorithm and the optimal choice of U are described in a forthcoming paper [16]. In brief, U need not be long, and the scheme has virtually the same error performance as the ML decoder under medium and long tailbiting.

Before proceeding to the tests, we emphasize several points about these schemes. The ML and sbsMAP are 2^m-fold more complex, more or less, than the TB-BCJR and CVA, and their significance is mostly as baselines. The practical schemes are these last two. Second, the CVA, being a straightforward VA in essence, is simpler than the TB-BCJR and is surely preferred when the data are i.i.d. When they are not, the TB-BCJR, which alone uses the data probabilities, can have a large advantage. Finally, if a probability rather than a symbol is to be the output (an APP decoder), then the CVA and the ML cannot be used. Finally, we remark that some tests are performed for the BSC and others for the AWGN channel; both have similar behavior with the one difference that the AWGN needs 2–2.5 dB less E_b/N_o than the BSC for the same error rate.

The first set of tests applies to the BSC and a circle of size $L = 10$. Figure 9 shows the BER versus equivalent E_b/N_o when a good encoder of memory 2–5 is employed with this TB circle size. (These best-BER encoders were found by an exhaustive procedure described in [12], which also presents the semi-analytic method which was used to compute the BER; these encoders are similar to but not always the same as the best minimum distance encoders). The transition from long to short tailbiting is clear here. The $m = 2$ encoders are in the long mode while the $m = 5$ encoder is in the short mode; the extra memory does not much lower the BER because the $L = 10$ circle is too short for the longer memories.

Figure 10 shows the opposite case, a fixed encoder (66,62) with a circle size that progresses through 5,8,16,24,40. The channel is AWGN. The first two cases are medium tailbiting and the later ones are long. To get nearly maximum performance out of (66,62), the tailbiting needs to be long, about 16 in this case, but once it is indeed long a further lengthening does not save much on the E_b/N_o axis.

The TB-BCJR, CVA and the ML baseline BERs are compared in Fig. 11 for the AWGN channel. The encoder is (72,62), a memory 4 encoder with $d_f = 7$, the best available distance at this memory; tests [12] show that it also has close to optimal BER for $m = 4$. Circle sizes are 6 and 20 stages. At size 20 the coding is solidly in the medium tailbiting mode and all the decoders have the same performance. At size 6 the tailbiting is short: Now the TB-BCJR and the CVA are still the same, but both are a small but statistically significant margin worse than ML decoding.

A second AWGN comparison of these decoders appears in Fig. 12. This time the encoder is quite a special one, the time-invariant tailbit-

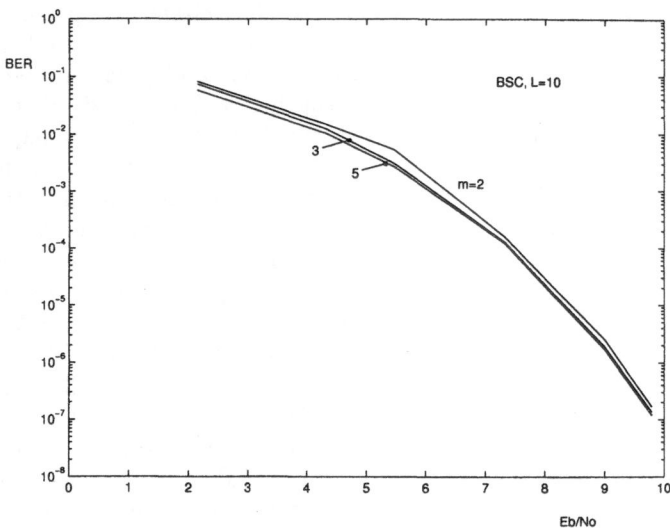

FIG. 9. *Measured bit error rate versus E_b/N_o in dB for the best BSC codes at memory 2,3,5 and circle size 10. Best $m = 4$ code lies in between 3 and 5 curves.*

ing representation of the (24,12) extended Golay code, with generators (414,730)[4]. In earlier work [5, 18] it has been shown that these generate the best minimum distance rate 1/2 memory 6 code with circle 12. No other $m = 6$, $L = 12$ code has its minimum distance of 8; however, it has a huge number, 759, of nearest neighbors and there exist considerably better $m = 6$ codes for larger circles, both in terms of distance and neighbors. Figure 12 shows that the CVA and the TB-BCJR at circle 12 are significantly worse than ML decoding, about 0.5 dB. Work by P. Ståhl shown in [18, p. 334] plots the sbsMAP error rate; it is almost identical to the ML curve here. In short, the ML and sbsMAP decoders have one performance and the CVA and TB-BCJR have another. At circle 48, the code is in the medium mode, the BER curve improves by more than a dB, and all three algorithms perform the same. The optimum-distance $m = 6$ encoder at $L = 48$, incidentally, is (564,634) with $d_{min} = 10$ and 528 neighbors [5, 18]; this encoder will contribute a *second* dB of gain.

We have observed the behavior in Figs. 11 and 12 with many codes, both for BSC and AWGN channels. We can summarize it as follows. When the tailbiting is short, the ML and probably the sbsMAP have a better BER than the CVA and TB-BCJR decoders. The Golay code is the most pernicious example we have seen and the difference is otherwise much smaller. As the tailbiting grows to medium, the BER differences among all these

---

[4]For a full discussion of this code's interesting properties, see Horn [20] and references therein.

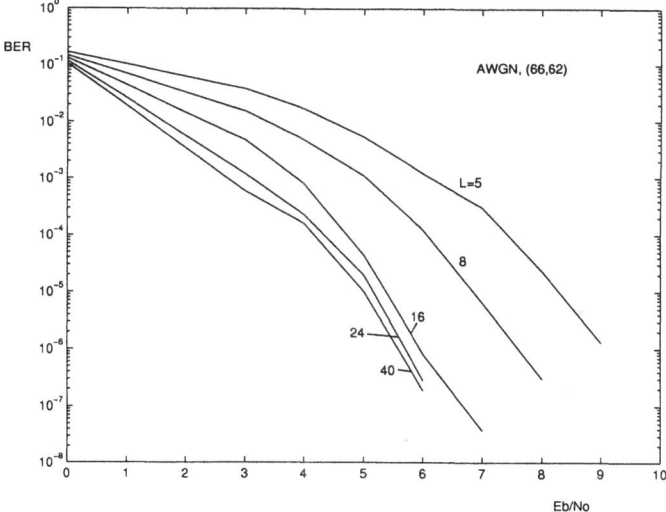

FIG. 10. *Measured bit error rate versus E_b/N_o in dB for code (66,62) over AWGN, at circle sizes 5,8,16,24,40. Note that an $L = 10$ curve here would lie approximately 2.5 dB left of those in Fig. 9, because of the AWGN channel.*

decoders disappear. To extract the best performance from a given encoder memory, the tailbiting should be at least medium. This is forcefully shown by the $m = 6$ case in Fig. 12. We doubt in fact that there is any practical use for short tailbiting in BER-criterion decoding.

5. Conclusions. By a study of rate 1/2 codes we have illustrated how tailbiting decoder BER depends on the length of the tailbiting circle. Short, medium and long modes are set approximately by the ratio of two encoder parameters, the memory and the decision depth. We have focused on a new efficient decoder called the tailbiting BCJR, which is based on the forward and backward vector recursion principle of Bahl et al. in [1] and earlier works. We have given a formal justification of this decoder, detailed a number of properties and shown in particular how the decoder design stems from certain principal eigenvectors. On the practical side, we have shown that the eigenvectors are essentially found after a certain number of trellis stages.

We have compared the new decoder to a number of related decoders and found evidence that its performance may fall a little short of optimal when the tailbiting circle is short. Unfortunately, this shortfall is inevitable since an incorrect solution is the limit of the standard BCJR recursions. The incorrect solution may be viewed as stemming from equating the TB state distribution rather than the states themselves. We can say at this writing, however, that few practical applications exist for which this shortfall is significant. In return for the slightly suboptimal performance comes a 2^m-fold saving in computation.

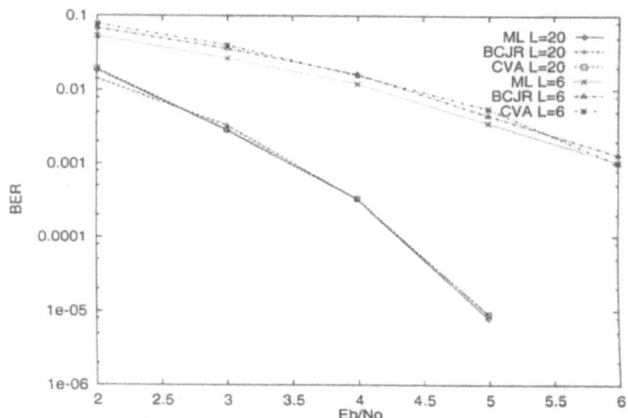

FIG. 11. *Measured bit error rate versus E_b/N_o in dB for code (72,62) over AWGN and circles 6 and 20, comparing ML, CVA and TB-BCJR decoders.*

It should be mentioned that the TB-BCJR itself may be made more efficient by modifications in the style of those in sequential, or "reduced search", decoders. Some of these appear in ref. [17]. Another point is that when the data are i.i.d. and hard output is acceptable, the dynamic program based ML and CVA schemes may apparently be used without much harm to the bit error rate. The family of circular VA algorithms are simpler than the linear algebra based BCJR family and are thus worth further attention. When the soft output is the object or the data probabilities are skewed, one must use the TB-BCJR, or if the last degree of optimality is the goal, the true symbol-by-symbol MAP.

6. Acknowledgments. We wish to acknowledge informative discussions with K. Zigangirov and R. Johannesson of Lund University, L. Wei of Australia National University, G.D. Forney, Jr. of Motorola, and S. M. Hladik of General Electric Research and Development Center, as well as a dedicated anonymous reviewer. Further, the following researchers shared results from their own implementations of the TB-BCJR algorithm, which were important to verifying the work presented here: P. Ståhl and M. Handlery of Lund University, E. Offer of Technical Univ. Muenchen, and S.M. Hladik.

This work was supported in part by a gift to Rensselaer Polytechnic Institute by L.M. Ericsson Inc., Research Triangle Park, NC 27709 USA.

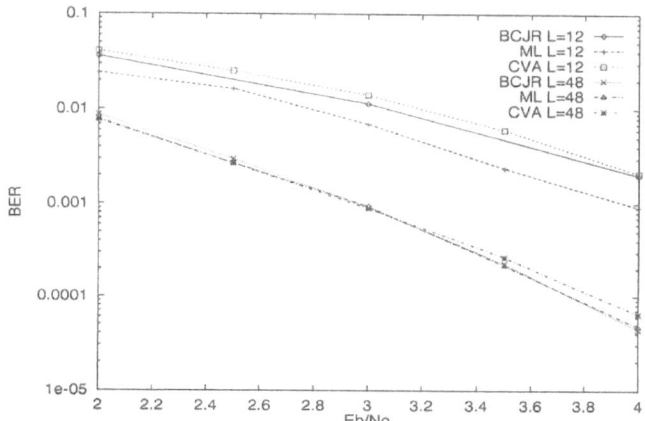

FIG. 12. *Measured bit error rate versus E_b/N_o in dB for code (414,730) over AWGN and circles 12 and 48, comparing ML, CVA and TB-BCJR. The optimal-BER code at $L = 48$ has asymptotically 2dB better performance than the $L = 48$ result here.*

REFERENCES

[1]   L.R. BAHL, et al., "Optimal decoding of linear codes for minimizing symbol error rate," *IEEE Trans. Information Theory*, **20**, pp. 284–287, Mar. 1974.

[2]   J.B. ANDERSON AND S.M. HLADIK, "Tailbiting MAP decoders," *IEEE J. Sel. Areas Commun.*, **16**, pp. 297–302, Feb. 1998.

[3]   G. SOLOMON AND H.C.A. VAN TILBORG, "A connection between block and convolutional codes," *SIAM J. Appl. Math.*, **37**, Oct. 1979.

[4]   H.H. MA AND J.K. WOLF, "On tailbiting convolutional codes," *IEEE Trans. Communications*, **34**, pp. 104–111, Feb. 1986.

[5]   P. STÅHL, J.B. ANDERSON AND R. JOHANNESSON, "Optimal and near-optimal encoders for short and moderate-length tailbiting trellises," *IEEE Trans. Information Theory*, **45**, pp. 2562–2571, Nov. 1999.

[6]   J.B. ANDERSON AND R. JOHANNESSON, "A condition for feedback tailbiting convolutional encoders and a short list of allowed feedback polynomials," *Proceedings*, Conf. on Information Systems and Sciences, Princeton Univ., Princeton, NJ, Mar. 1998.

[7]   J. HAGENAUER, "Source-controlled channel decoding," *IEEE Trans. Communications*, **43**, pp. 2449–2457, Sept. 1995.

[8]   S.M. AJI, G.B. HORN AND R.J. MCELIECE, "On the convergence of iterative decoding on graphs with a single cycle," *Proceedings*, Conf. on Information Systems and Sciences, Princeton Univ., Princeton, NJ, Mar. 1998.

[9]   R.J. MCELIECE, D.J.C. MACKAY AND J.-F. CHENG, "Turbo decoding as an instance of Pearl's 'Belief Propagation' Algorithm," *IEEE J. Sel. Areas Commun.*, **16**, pp. 140–152, Feb. 1998.

[10] G.D. FORNEY, JR., F.R. KSCHISCHANG AND B. MARCUS, "Iterative decoding of tail-biting trellises," unpublished manuscript (25pp), July 1998; summary in *Proceedings*, IEEE Information Theory Society Workshop, San Diego, CA, Feb. 1998.

[11] R. BELLMAN, *Introduction to Matrix Analysis*, 2nd Ed., McGraw-Hill, New York, 1970.

[12] J.B. ANDERSON, "Best short rate 1/2 tailbiting codes for the bit error rate criterion," *IEEE Trans. Communications*, **48**, Apr. 2000.

[13] J.B. ANDERSON AND K. BALACHANDRAN, "Decision depths of convolutional codes," *IEEE Trans. Information Theory*, **35**, pp. 455–459, Mar. 1989.

[14] K. SH. ZIGANGIROV AND V.V. CHEPYZHOV, "Study of tailbiting convolutional codes," *Proceedings*, 4th Joint Swedish-Soviet Intern. Workshop Information Theory, Gotland, Sweden, pp. 52–55, 1989.

[15] R.V. COX AND C.-E. SUNDBERG, "An efficient adaptive circular Viterbi algorithm for decoding generalized tailbiting convolutional codes," *IEEE Trans. Vehicular Tech.*, **43**, pp. 57–68, Feb. 1994.

[16] J.B. ANDERSON AND S.M. HLADIK, "An optimal circular Viterbi decoder," in submission, *IEEE J. Sel. Areas Commun.*, Mar. 2000.

[17] V. FRANZ AND J.B. ANDERSON, "Concatenated decoding with a reduced-search BCJR algorithm," *IEEE J. Sel. Areas Commun.*, **16**, pp. 186–195, Feb. 1998.

[18] R. JOHANNESSON AND K.SH. ZIGANGIROV, *Introduction to Convolutional Coding*, IEEE Press, Piscataway, NJ, 1999.

[19] R. RAMESH, E. WONG AND H. KOORAPATHY, "On tailbiting recursive systematic convolutional encoders," in submission, *IEEE Trans. Communications*, 1998.

[20] G. HORN, Ph.D. Thesis, "Iterative decoding and pseudo-codewords," Dept. Electrical Eng., Calif. Inst. Technology, Pasadena, May 1999.

[21] P. LANCASTER AND M. TISMENETSKY, *The Theory of Matrices*, 2nd Ed., Academic Press, San Diego, 1985.

ITERATIVE DECODING OF TAIL-BITING TRELLISES AND CONNECTIONS WITH SYMBOLIC DYNAMICS*

G. DAVID FORNEY, JR.[†], FRANK R. KSCHISCHANG[‡], BRIAN MARCUS[§], AND SELIM TUNCEL[¶]

Abstract. The sum-product and min-sum algorithms are used to decode codes defined by trellises. In this paper, we discuss the behavior of these and related algorithms on tail-biting (TB) trellises.

The convergence of the sum-product algorithm on tail-biting trellises was analyzed recently by Anderson and Hladik [2] and was shown to give approximate *a posteriori* probabilities.

We introduce and analyze generating function versions of both algorithms, each involving a parameter x. This involves the minwps graph, a tool borrowed from the symbolic dynamics of Markov chains. We determine the limiting behavior as $x \to 0$ of the result of the generating-function sum-product algorithm and show how this relates to maximum-likelihood sequences and the generating-function min-sum algorithm.

Key words. Tail-biting trellises, pseudocodewords, generating-function sum-product algorithm, generating-function min-sum algorithm, weight-per-symbol, minwps graph.

AMS(MOS) subject classifications. 94B99, 37B10.

1. Introduction. The iterative algorithms used for decoding capacity-approaching codes such as turbo codes and low-density parity-check codes are instances of a general class of decoding algorithms for codes defined on graphs, called the sum-product algorithm over a semi-ring [19, 5, 10, 13]. Special cases include the forward-backward (APP, BCJR, "MAP," belief propagation) algorithm and the min-sum (bidirectional Viterbi, belief revision) algorithm.

The behavior of these algorithms on cycle-free graphs is well understood. However, little is known about their theoretical performance and behavior on graphs with cycles, although empirically they often work very well. The simplest graph with cycles is a single cycle, which in coding corresponds to a tail-biting (TB) trellis.

Anderson and Hladik [2] showed that the sum-product algorithm over the real numbers converges on TB trellises. In fact, Perron-Frobenius theory provides a very strong convergence result. We review the result of [2] in Section 5 (see Proposition 5.1).

The convergence behavior of the min-sum algorithm for TB trellises is governed by the trellis path(s) with the minimum average weight per

---

*A preliminary version of this paper was presented at the IEEE Information Theory Workshop, San Diego, CA, February 9-11, 1998.

[†]MIT, Cambridge, MA 02139.

[‡]ECE Department, University of Toronto, Toronto, Ont., M5S 3G4, CANADA.

[§]IBM Almaden Research Center 650 Harry Rd., San Jose, CA 95120.

[¶]Mathematics Department, University of Washington, Seattle, WA 98195.

cycle, called the "dominant pseudocodeword(s)." Thus, if the dominant pseudocodeword(s) is not a codeword, i.e., if its length is more than one cycle, then the min-sum algorithm will tend to detect a non-codeword.

The dominant pseudocodewords can be assembled into a graph, called the minwps graph, which is a special case of an object used in the analysis and construction of codes between Markov chains ([14, 17, 15, 4]). We rely heavily on the minwps graph in our analysis of the algorithms. In particular, we show that the maximum likelihood paths spend all but a tiny fraction of time in the minwps graph (Proposition 6.1 of Section 6).

In Section 7, we introduce a generating-function version of the sum-product algorithm involving a parameter x, which more explicitly reveals the dominant pseudocodewords and helps to clarify the relationship between the ordinary sum-product and min-sum algorithms on TB trellises. We show that the generating-function sum-product algorithm converges for each $x > 0$ (Proposition 7.1), and in Section 8 we determine its limiting behavior as $x \to 0$ (Proposition 8.2). We show how this can be used to detect some dominant pseudocodewords (Proposition 8.3), but that it may fail to detect all such words (Example 4 of Section 9). In Section 10, we introduce a generating-function version of the min-sum algorithm and show that it converges over a particular subsequence of the integers (Proposition 10.2).

Leading terms of eigenvalue and eigenvector functions play an important role in determining the limiting behavior of the generating-function sum-product algorithm. This is discussed in detail in Section 8 and illustrated with examples in Section 9. Proposition 10.1 shows how these leading-term functions represent eigenvectors over a semiring of monomial functions.

A Note on notation: We use $\sim$ to denote "equal up to scale" and $\star$ to denote the componentwise product of two vectors.

2. Conventional and tail-biting trellises. A **series conventional finite-length trellis** T is defined on an ordered discrete index set ("time axis") $I \subseteq Z$ that may be identified with an interval $[0, n] = \{0, 1, \ldots, n\}$, where n is the **trellis length.**

Given the time axis, finite state spaces Σ_k, $0 \le k \le n$, and output alphabets $A_k, 0 \le k < n$, are defined, with $|\Sigma_0| = |\Sigma_n| = 1$. Further, a **trellis section** $T_k \subseteq \Sigma_k \times A_k \times \Sigma_{k+1}$ is defined for $0 \le k < n$ as the set of allowed state transitions (s_k, a_k, s_{k+1}) from state s_k to state s_{k+1} with label a_k. The elements of T_k are called **branches.** The indicator function of T_k is:

$$\phi_k(s_k, a_k, s_{k+1}) = 1, \text{ if } (s_k, a_k, s_{k+1}) \in T_k;$$
$$= 0, \text{ otherwise.}$$

A valid trellis path is a state/output sequence $(s, a) \in (\prod_k \Sigma_k) \times (\prod_k A_k)$ that satisfies all local constraints $(s_k, a_k, s_{k+1}) \in T_k$ for all $0 \le$

$k < n$. The indicator function

$$\phi(s,a) = \prod_k \phi_k(s_k, a_k, s_{k+1})$$

equals 1 if (s,a) is a valid trellis path, and equals 0 otherwise. The code defined by the trellis is the set of all output sequences a that correspond to valid trellis paths.

A **tail-biting (TB) trellis** T is defined on an index set I that is identified with a cyclic group $Z_n = \{0, 1, \ldots, n-1\}$, where n is the TB trellis length [3]. All index arithmetic is performed modulo n.

Again, for each $k \in Z_n$, a finite state space Σ_k and output alphabet A_k are defined, with $\Sigma_n = \Sigma_0$. A trellis section T_k is defined for each $k \in Z_n$ as the set of allowed triples (s_k, a_k, s_{k+1}), and $\phi_k(s_k, a_k, s_{k+1})$ is defined as the indicator function of T_k. Again a valid TB trellis path is a state/output sequence (s, a) that satisfies all local constraints $(s_k, a_k, s_{k+1}) \in T_k$ for all $k \in Z_n$, <u>and</u> $s_n = s_0$. For $j = 0, \ldots, n-1$, the branches beginning in Σ_j constitute the jth **phase** of the TB trellis. Figure 1 shows a TB trellis for a linear block code.

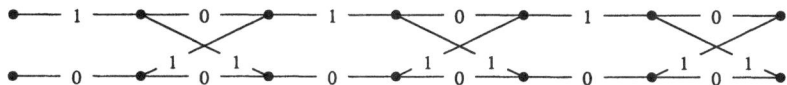

FIG. 1. *TB trellis for a linear block code.*

Note that a conventional trellis of length n may be regarded as a TB trellis with $|\Sigma_0| = |\Sigma_n| = 1$.

3. The sum-product algorithm on conventional trellises. On a conventional trellis, the **sum-product algorithm** involves a forward and a backward recursion, and is sometimes called the forward-backward algorithm.

Given a trellis section $T_k \subseteq \Sigma_k \times A_k \times \Sigma_{k+1}$ and branch weights $\gamma_k = \{\gamma_k(a_k), a_k \in A_k\}$, the forward recursion of this algorithm computes $\alpha_k = \{\alpha_k(s_k), s_k \in \Sigma_k\}$ by:

$$\alpha_{k+1}(s_{k+1}) = \sum \alpha_k(s_k)\gamma_k(a_k)\phi_k(s_k, a_k, s_{k+1}),$$

for each $s_{k+1} \in \Sigma_{k+1}$, where the sum is implicitly over $\Sigma_k \times A_k$, and $\alpha_0 = (1)$. This may be viewed as a matrix multiplication

$$\alpha_{k+1} = \alpha_k \Gamma_k,$$

where α_k and α_{k+1} are regarded row vectors, and Γ_k is the matrix with entries:

(3.1) $$\Gamma_k(s_k, s_{k+1}) = \sum \gamma_k(a_k)\phi_k(s_k, a_k, s_{k+1}),$$

where the sum is implicitly over A_k. Thus

$$\alpha_k = \alpha_0 \Gamma_0 \Gamma_1 \cdots \Gamma_{k-1}.$$

It is straightforward to verify by induction that

$$\alpha_k(s_k) = \sum \gamma_{[0,k)}(a_{[0,k)}) \phi_{[0,k]}(s_{[0,k)}, a_{[0,k)}, s_k),$$

where

$$\gamma_{[0,k)}(a_{[0,k)}) = \prod_{0 \le j < k} \gamma_j(a_j),$$

and

$$\phi_{[0,k]}(s_{[0,k)}, a_{[0,k)}, s_k) = \prod_{0 \le j < k} \phi_j(s_j, a_j, s_{j+1}).$$

In other words, $\alpha_k(s_k)$ is the sum of the product $\gamma_{[0,k)}(a_{[0,k)})$ of weights $\gamma_j(a_j)$ of all valid trellis paths that start at the unique time-0 state and reach state s_k at time k.

If the weight $\gamma_k(a_k) = p(r_k \mid a_k)$, the likelihood of the received symbol r_k given the transmitted symbol a_k, and the channel is memoryless, then

$$\begin{aligned} \gamma_{[0,k)}(a_{[0,k)}) &= p(r_{[0,k)} \mid a_{[0,k)}); \\ \alpha_k(s_k) &= p(r_{[0,k)} \mid s_k). \end{aligned}$$

If moreover all trellis paths are *a priori* equiprobable, then the vector α_k is equal, up to scale, to the vector of *a posteriori* probabilities (APPs) $p(s_k \mid r_{[0,k)})$, and so

$$\alpha_k = \{p(r_{[0,k)} \mid s_k)\} \sim \{p(s_k \mid r_{[0,k)})\},$$

(where the symbol $\sim$ denotes "equal up to scale"). Thus the sum-product algorithm is sometimes called the "APP algorithm." (It is sometimes also called the "MAP algorithm," but since no maximization occurs, we prefer the term "APP algorithm".)

Similarly, the backward recursion computes

$$\beta_k(s_{k+1}) = \sum \beta_{k+1}(s_{k+1}) \gamma_k(a_k) \phi_k(s_k, a_k, s_{k+1}), \quad s_k \in \Sigma_k,$$

which may be viewed as a matrix multiplication

$$\beta_k = \Gamma_k \beta_{k+1},$$

where β_k and β_{k+1} are regarded as column vectors, and Γ_k is the same matrix as in the forward recursion. Thus

$$\beta_k = \Gamma_k \Gamma_{k+1} \cdots \Gamma_{n-1},$$

where $\beta_n = (1)$. So

$$\beta_k(s_k) = \sum \gamma_{[k,n)}(a_{[k,n)})\phi_{[k,n]}(s_{[k,n)}, a_{[k,n)}, s_{[k,n)}),$$

the sum of the product $\gamma_{[k,n)}(a_{[k,n)})$ of weights $\gamma_j(a_j)$ of all valid trellis paths that start at state s_k at time k and end at the unique time-n state.

If $\gamma_k(a_k) = p(r_k \mid a_k)$, the channel is memoryless and all trellis paths are *a priori* equiprobable, then

$$\beta_k \sim \{p(s_k \mid r_{[k,n)})\}.$$

The final outputs of the algorithm are as follows. The final state weight vector is the componentwise product λ_k of the α_k and β_k vectors, denoted

$$\lambda_k = \alpha_k \star \beta_k;$$

therefore

$$\lambda_k(s_k) = \alpha_k(s_k)\beta_k(s_k) = \sum \gamma_{[0,n)}(a_{[0,n)})\phi(s, a),$$

where the sum is over trellis paths (s, a) that pass through state s_k at time k. Thus, with the same assumptions as above, λ_k is equal up to scaling to the APP vector

$$\lambda_k \sim \{p(s_k \mid r_{[0,n)})\}.$$

Similarly, the final output weights are the products

$$\sigma_k(a_k) = \gamma_k(a_k) \sum \alpha_k(s_k)\beta_{k+1}(s_{k+1})\phi_k(s_k, a_k, s_{k+1}),$$

where the sum is implicitly over $\Sigma_k \times \Sigma_{k+1}$, which yields

$$\sigma_k(a_k) = \sum \gamma_{[0,n)}(a_{[0,n)})\phi(s, a),$$

where the sum is over trellis paths (s, a) that contain symbol a_k at time k. Thus

$$\sigma_k \sim \{p(a_k \mid r_{[0,n)})\}.$$

4. Matrices, graphs, and tail-biting trellises. In preparation for discussing the convergence of the iterative decoding algorithms on tail-biting trellises, we review a bit of the Perron-Frobenius theory of nonnegative matrices as well as notions of weighted graphs. We also show how TB trellises are related to weighted graphs and nonnegative matrices.

4.1. Perron-Frobenius theory. A real matrix or vector is said to be positive (resp. nonnegative) if all of its entries are positive (resp. nonnegative). A nonnegative square matrix Γ is **irreducible** if for any pair of indices (s, s'), for some integer $n = n(s, s')$ the corresponding entry of the nth power is positive: $\Gamma^n(s, s') > 0$. A stronger condition than irreducibility is primitivity: a nonnegative square matrix Γ is **primitive** if for some integer n, $\Gamma^n > 0$. Note that any positive matrix is automatically primitive.

THEOREM 4.1. *(Perron-Frobenius theorem (see Seneta [16, Chapter 1] or [12, Chapter 4]): Let Γ be an irreducible (nonnegative, square) matrix. Then*

*(a) Γ has a positive eigenvalue μ, called the **Perron eigenvalue**, which dominates all other eigenvalues: i.e., every eigenvalue ν of Γ has magnitude $|\nu| \leq \mu$;*

*(b) Γ has a positive (left, resp. right) eigenvector corresponding to the Perron eigenvalue μ (called the **left Perron eigenvector** v, resp. **right Perron eigenvector** w).*

(c) The Perron eigenvalue μ is algebraically simple (i.e., it is a simple root of the characteristic polynomial, $\chi(x) = det(xI - \Gamma)$ of Γ); in particular, it is geometrically simple (i.e., the space of corresponding eigenvectors has dimension 1 and is simply the space generated by v);

(d) Every positive eigenvector of Γ is a positive multiple of v.

Moreover, if Γ is primitive, then

(e) the inequality in (a) above is strict, and

(f) for any positive vector α, $\alpha \Gamma^m / \mu^m$ converges (up to scale) to v: i.e., in the limit as $m \to \infty$, $\alpha \Gamma^m \sim v$. In particular, the sequence of matrices Γ^m / μ^m converges.

A nonnegative square matrix Γ which is not irreducible is called **reducible**. A reducible (nonnegative) matrix is usually studied via its irreducible pieces: an **irreducible component** of a nonnegative matrix is a maximal irreducible submatrix, i.e., a square submatrix which is irreducible and is not contained in any larger irreducible submatrix. A reducible matrix breaks down into irreducible components with possibly some "transient" connections between the components. This can be expressed by saying that given a reducible matrix Γ (other than the 1×1 zero matrix), for some ordering of the indices, Γ is block lower triangular with more than one block.

A **principal component** is an irreducible component whose Perron eigenvalue is maximal among all irreducible components. Equivalently, a principal component of Γ is an irreducible component whose Perron eigenvalue is the largest (in magnitude) eigenvalue of Γ. A **sink** is an irreducible component from which there is "no escape"— i.e., an irreducible component C such that for any index s of C, if $\Gamma(s, s') > 0$, then s' must also be an index of C. Note that any irreducible matrix automatically has only one principal component, and that component is automatically a sink.

Part (f) of the Perron-Frobenius Theorem extends to a fairly general class of matrices, including irreducible matrices, as follows. By the **period** of an irreducible matrix Γ, we mean the g.c.d. of positive integers k such that $\mathrm{Tr}(\Gamma^k) > 0$; by the period of a nonnegative matrix we mean the l.c.m. of the periods of its irreducible components.

THEOREM 4.2. *(see Seneta [16, Chapter 1]) Let Γ be a nonnegative square matrix. Let p denote the period of Γ. Let μ denote the largest (in magnitude) eigenvalue of Γ. If all principal components of Γ are sinks (in particular, if Γ is irreducible), then the sequence of matrices Γ^{mp}/μ^{mp} converges as $m \to \infty$.*

In principle, a direct way of computing the Perron eigenvectors is given as follows (for more detail, see Seneta [16, pp. 7–8] or [12, p. 112]). Given a square matrix M, its **adjugate**, $\mathrm{adj}(M)$, is defined as the matrix of cofactors of M. Define

$$A(x, \Gamma) = \mathrm{adj}(xI - \Gamma),$$

where x is an indeterminate. The rows (resp. columns) of $A(\mu, \Gamma)$ turn out to be positive multiples of the left (resp. right) Perron eigenvectors of Γ; in particular, $A(\mu, \Gamma)$ is a rank-1 matrix formed as the outer product of left and right Perron eigenvectors of Γ.

4.2. Weighted graphs. A **finite directed graph** (or simply **graph**) G is defined by a set Σ of states and a set of (directed) edges, with each edge having an initial state and a terminal state. Such a graph is essentially a trellis section with the same starting and ending state set Σ. A **path** in a graph is simply a concatenation of edges, where the terminal state of each edge coincides with the initial edge of the succeeding edge. A **cycle** is a path whose initial and terminal states agree.

A **finite directed weighted graph** (or simply **weighted graph**) is a finite directed graph with weights assigned to its edges. An **unweighted graph** is a weighted graph where the weight of every edge is 1.

There is a natural one-to-one correspondence between nonnegative square matrices and finite directed weighted graphs with positive weights: namely, given such a graph G with state set Σ, we associate the square matrix Γ with index set Σ that is obtained by defining $\Gamma(s, s')$ to be the sum of weights of all edges from s to s'.

Properties of nonnegative matrices may be interpreted in terms of the corresponding graphs. For example, a matrix is irreducible if and only if for any pair of states (s, s') in the corresponding graph there is a path from s to s'. Also, the period of an irreducible matrix is the g.c.d. of lengths of cycles in the corresponding graph.

4.3. Relationship to TB trellises. A TB trellis T defines a graph $G = G(T)$; here, the state set is Σ_0 and the edges of G are trellis paths of length n from Σ_0 to Σ_0 (but the terminal state s_n need <u>not</u> agree with the initial state s_0). If T is endowed with branch weights, then paths of

T inherit weights by multiplying weights along the branches (here, what is meant by "multiplying" may depend on a semiring structure chosen for the branch weights of T). Then G inherits these weights and thus becomes a weighted graph. The nonnegative matrix corresponding to G is the composite matrix multiplication around one trellis length:

$$(4.1) \qquad \Gamma = \Gamma_0 \Gamma_1 \cdots \Gamma_{n-1},$$

where Γ_k is defined as in (3.1).

5. The sum-product algorithm on tail-biting trellises. The sum-product algorithm may be applied to TB trellises by assigning an arbitrary initial positive weight vector α_0 (usually $\alpha_0 = 1$, the all-one vector) to the state space Σ_0 and then applying the algorithm to compute

$$\alpha_k = \Gamma_0 \Gamma_1 \cdots \Gamma_{k-1}$$

for arbitrarily large $k \in Z$ (see [2]).

Since $\Gamma_k = \Gamma_{k-n}$ for $k > n$, we have

$$\alpha_{mn} = \alpha_0 \Gamma^m,$$

where Γ is as defined in (4.1) above.

For the remainder of this paper, we assume:

Standing assumption: Γ is primitive.

Note that the primitivity condition is much weaker than requiring Γ to be positive, which would be equivalent to the condition that for every pair of states (s, s') of Σ_0, there is a path in the TB trellis of length n from s to s'; this latter condition holds for most good codes.

If v is the left Perron eigenvector of $\Gamma = \Gamma_0 \Gamma_1 \cdots \Gamma_{n-1}$ with eigenvalue μ, then $v\Gamma_0$ is the left Perron eigenvector of $\Gamma' = \Gamma_1 \cdots \Gamma_{n-1}\Gamma_0$ with eigenvalue μ, since

$$(v\Gamma_0)\Gamma_1 \cdots \Gamma_{n-1}\Gamma_0 = \mu v\Gamma_0.$$

Thus by part (f) of the Perron-Frobenius Theorem, for each $0 \le j \le n-1$, as $m \to \infty$, α_{mn+j} converges, up to scale, to the left Perron eigenvector $v\Gamma_0\Gamma_1 \ldots \Gamma_{j-1}$. In other words, the sum-product algorithm applied to $v_0 = v$ produces a cycle of compatible vectors $v_1 = v\Gamma_0, v_2 = v\Gamma_0\Gamma_1, \ldots$, that comes around to $v_n = v\Gamma = \mu v \sim v$. In principle, the rate of convergence is determined to first order by the ratio of the magnitudes of the two largest eigenvalues of Γ. In practice, the memory of the forward recursion is of the order of a few times the code constraint length (the controller memory of the trellis), and convergence often takes only one trellis length plus a few constraint lengths [2].

Similarly, in the reverse direction, β_{mn} converges in direction to the right Perron eigenvector w of Γ; i.e., in the limit $\beta_{mn} \sim w$. More generally, for $0 \leq j \leq n - 1$, in the limit as $m \to \infty$, $\beta_{mn-j} \sim w_j \equiv \Gamma_{n-j} \ldots \Gamma_{n-1} w$. Since the eigenvalues of Γ and its transpose are the same, the rate of convergence is the same in the reverse direction.

The final state weight vectors λ_k and final output weights σ_k are computed from the α's and β's as in Section 3. Thus, we have:

PROPOSITION 5.1. *For* $0 \leq j \leq n - 1$, *in the limit as* $k \to \infty$, *with* $k \equiv j \pmod{n}$,

$$\lambda_k \sim v_j \star w_{n-j}.$$

In particular, setting $j = 0$, *we have: in the limit as* $k \to \infty$, *with* $k \equiv 0 \pmod{n}$,

$$\lambda_k \sim v_0 \star w_n \sim v \star w.$$

It follows from Proposition 5.1 and the discussion of the adjugate matrix in Section 4.1 that the limiting (over multiples k of n) final state weight vector λ_k is, up to scale, the componentwise product of any row and any column of the adjugate matrix $A(\mu, \Gamma)$; in particular, λ_k may be taken to be proportional to the diagonal entries of $A(\mu, \Gamma)$.

6. The min-sum algorithm on tail-biting trellises. The sum-product algorithm may be generalized to any commutative semiring (S, $+$, $\times$), whose addition and multiplication operations ($+$, $\times$) satisfy the associative, commutative and distributive rules of ordinary arithmetic.

The **min-sum algorithm** is the sum-product algorithm over the "min-sum" semiring ($\mathbf{R}$, min, sum); i.e., "sum" is replaced by "min" and "product" is replaced by "sum," and the branch weights $\gamma_k(a_k) = p(r_k \mid a_k)$ are replaced by their negative logarithms $-\log \gamma_k(a_k) = -\log p(r_k \mid a_k)$. Matrix multiplications over this semiring become the "add-compare-select" operations of the Viterbi algorithm, and the min-sum algorithm becomes a bidirectional Viterbi algorithm.

By a **maximum-likelihood (ML) path**, we mean a path with minimum weight. For instance, a ML past trellis path to state s_k is a path of length k ending at state s_k with minimum weight. It follows from the definitions of the forward and backward recursions that a state weight $-\log \alpha_k(s_k)$ (resp. $-\log \beta_k(s_k)$) is equal to the negative log likelihood of the ML past (resp. future) trellis path to state s_k, i.e., the path of length k ending at state s_k with minimum weight. The final state weight

$$-\log \lambda_k(s_k) = -\log \alpha_k(s_k) - \log \beta_k(s_k)$$

is the negative log likelihood of the ML trellis path passing through state s_k. Similarly, the final output weight $-\log \sigma_k(a_k)$ of a_k becomes the negative log likelihood of the ML trellis path that includes a_k.

Any TB trellis T yields a corresponding bi-infinite trellis T_∞ obtained by declaring the received weights to be repeated periodically with period n; i.e., $\gamma_k(a_k) = \gamma_{k-n}(a_k)$, for $k > n$. (T_∞ is sometimes referred to as the "unwrapped" or "periodic" version of the TB trellis; this notion of "unwrapping" a graph with cycles into a cycle-free "computation tree" was first introduced by Gallager [8], and has been substantially developed in [19, 20, 7].

We define a **pseudocodeword** (after [7]) as a length-mn path $(s_{[0,mn]}, a_{[0,mn)})$ on T_∞ that starts and ends in the same state: $s_0 = s_{mn} \in \Sigma_0$. The valid TB trellis paths (**codewords**) are precisely the length-n pseudocodewords. A pseudocodeword $(s_{[0,mn]}, a_{[0,mn)})$ is called **simple** if it never passes through the same state twice at times $\{0, n, 2n, \ldots, (m-1)n\}$. Every pseudocodeword can, in some sense, be "decomposed" into simple pseudocodewords, although not necessarily as a simple concatenation.

The **weight-per-symbol** (**wps**) of a pseudocodeword $(s_{[0,mn]}, a_{[0,mn)})$ is defined to be:

$$-\frac{1}{m} \sum_{0 \le k < mn} \log \gamma_k(a_k) = -\frac{1}{m} \log p(r_{[0,mn)} \mid a_{[0,mn)}).$$

We define the **dominant pseudocodeword(s)** to be the pseudocodeword(s) with the **minimum wps**, which we denote by w_0.

Recall that $G = G(T)$ denotes the weighted graph corresponding to a TB trellis T (as described in Section 4.3). Using the min-sum semiring for weights in T, the weight of an edge in G is defined to be the sum of the weights (the negative logarithms) of the branches that define the edge. Note that paths in G correspond to paths in T_∞ which begin and end in Σ_0. In particular, a pseudocodeword $(s_{[0,mn]}, a_{[0,mn)})$ corresponds to a cycle in G of length m.

Now, define the **minwps graph** $G_0 = G_0(T)$ to be the subgraph of G consisting of all edges in G that belong to a dominant pseudocodeword. It can be shown that the dominant pseudocodewords correspond precisely to the cycles in G_0 (in particular, any cycle formed from edges in G_0 turns out to be a dominant pseudocodeword) and that G_0 is the disjoint union of its irreducible components, i.e., there are no "transient" connections between the components (see [14, Proposition 3.9]). If there is a unique dominant pseudocodeword, then G_0 will consist of a single simple cycle. If not, then G_0 can be much more complicated.

The following result says that for large $k \equiv 0 \bmod n$, the past and future ML paths spend all but a vanishingly small fraction of time in the minwps graph. Since it is clear that any ML path that passes through a state s_k must be the concatenation of a past ML path to s_k and a future ML path from s_k, this gives a description of the result of the min-sum algorithm on TB trellises.

PROPOSITION 6.1. *Given a TB trellis T, there is a constant K_0 such that for any $k \equiv 0 \pmod{n}$, any state $s_k \in \Sigma_0$, and any past (resp. future) ML path to (resp. from) s_k, the number of edges in which the path does not lie in the minwps graph $G_0(T)$ is at most K_0.*

Proof. Any finite path in the graph G can be "decomposed" (again, not necessarily as a simple concatenation) into simple cycles of G together with a path of bounded length at most $|\Sigma_0|$. In particular, given $k \equiv 0 \pmod{n}$ and state $s_k \in \Sigma_0$, this holds for a path π in G corresponding to an ML path in the trellis which ends at s_k.

Now, there is a path π' of length k to state s_k that lies in the minwps graph G_0 except for perhaps a suffix of bounded length of length at most $|\Sigma_0|$. Such a path can be "decomposed" into simple cycles of G_0 together with a path of length at most $2|\Sigma_0|$. Since π corresponds to an ML path, the weight of π' is at most equal to that of π. But since the simple cycles of G_0 have minimum wps (among all possible cycles in G), it follows that only a bounded number of the simple cycles in the decomposition of π can fail to be contained in G_0. With the exception of these cycles and a path of bounded length, the path π lies within G_0. $\square$

7. The generating-function sum-product algorithm.

The sum-product algorithm computes a sum over paths of likelihood weights. The largest term in this sum is contributed by the ML path, which is computed by the min-sum algorithm. We may therefore expect some relation between the convergence behavior of the sum-product and min-sum algorithms. More precisely, we expect that the min-sum algorithm should in some sense be a limiting case of a sequence of sum-product algorithms, which in the limit detects ML paths.

To this end, we introduce a generating-function sum-product algorithm as follows.

By a **monomial** in an indeterminate x we mean a function of the form x^r for some real number r. By a **generating function** we mean a linear combination of monomials with positive coefficients over $\mathbf{R}$. Generating functions form a semiring under the usual definition of sum and product. The **generating-function sum-product algorithm** is defined as the sum-product algorithm applied over this semiring to a TB trellis, where the weight of a branch in the kth section with label a_k is taken to be the monomial $x^{-\log \gamma_k(a_k)}$.

The forward recursion of this algorithm yields forward generating-function vectors

$$\alpha_k(x) = \alpha_0 \Gamma_0(x) \cdots \Gamma_{k-1}(x),$$

where $\Gamma_k(x)$ is the **generating-function matrix** with components defined by:

$$\Gamma_k(x, s_k, s_{k+1}) = x^{-\log \gamma_k(a_k)} \phi_k(s_k, a_k, s_{k+1}),$$

and we again choose $\alpha_0 = 1$. Observe that

$$\alpha_k(x, s_k) = \sum x^{-\log \gamma_{[0,k)}(a_{[0,k)})} \phi_{[0,k]}(s_{[0,k)}, a_{[0,k)}, s_k).$$

Analogous expressions are obtained for the generating function vectors $\beta_k(x)$ obtained from the backward recursion of this algorithm, and similarly for the final state weights $\lambda_k(x)$ and final output weights $\sigma_k(x)$.

For $x > 0$, let

(7.1) $$\Gamma(x) = \Gamma_0(x) \cdots \Gamma_{n-1}(x),$$

and let $\mu(x)$, $v(x)$, $w(x)$ denote the Perron eigenvalue, left Perron eigenvector and right Perron eigenvector, respectively of $\Gamma(x)$ (with $v(x, s)$, $w(x, s)$, $s \in \Sigma_0$ denoting components of these vectors). Since we have assumed $\Gamma = \Gamma(e^{-1})$ to be primitive, we can apply the Perron-Frobenius Theorem to see that for each $x > 0$, these quantities are well defined, and as $m \to \infty$, up to scale, $\alpha_{mn}(x)/\mu(x)^m$ converges to $v(x)$ and $\beta_{mn}(x)/\mu(x)^m$ converges to $w(x)$. Applying Proposition 5.1, we immediately obtain:

PROPOSITION 7.1. *For each $x > 0$, in the limit as $k \to \infty$, with $k \equiv 0 \pmod{n}$,*

$$\lambda_k(x) \sim v(x) \star w(x).$$

Note that the generating-function sum-product algorithm reduces to the ordinary sum-product algorithm if we substitute $x = e^{-1}$. As $x \to 0$, more emphasis is given to the branches with small $-\log \gamma_k(a_k)$ (equivalently, large $\gamma_k(a_k)$), and this is why one might expect the convergence behavior of this algorithm as $x \to 0$ to depend heavily on the ML paths.

Generating function matrices corresponding to finite directed weighted graphs have been used in the study of Markov chains in symbolic dynamics (actually, where the matrix entries are polynomials in several variables). The states of the graphs are the states of the Markov chain and the weights are representations of the transition probabilities. A collection of induced subgraphs (including the minwps graph) is used as an invariant for problems of isomorphism between Markov chains (see, e.g., [14, 17, 15, 4]).

8. Behavior of the generating-function sum-product algorithm as x → 0. We are interested in the behavior of $\lambda_k(x)$ as $x \to 0$. For this purpose, we would like to express each $v(x, s)$ and each $w(x, s)$ as a power series expansion in a neighborhood of $x = 0$, and then pull out the leading terms of these power series. But as x varies, the scaling factors could vary wildly, and so $v(x)$ and $w(x)$ could behave poorly.

We may tame the variation of scaling factors by use of the adjugate matrix:

$$A(\mu(x), \Gamma(x)) = \text{adj}(\mu(x)I - \Gamma(x))$$

(see the end of section 4.1). So, we fix $v(x)$ and $w(x)$ to be an arbitrary row and column of $A(\mu(x), \Gamma(x))$.

Now, since $v(x, s)$ and $w(x, s)$ are cofactors of the matrix $\mu(x)I - \Gamma(x)$, each is expressible as a polynomial in x and $\mu(x)$. So it suffices to express $\mu(x)$ as a power series expansion.

Let $\chi_\Gamma(t)$ denote the characteristic polynomial of the matrix $\Gamma(x)$. This is a polynomial in one variable t whose coefficients lie in the polynomial ring $Z[x]$ (i.e, the coefficients are polynomials in x). Now, by definition, $\mu(x)$ is a root of the characteristic polynomial (i.e., $\chi_\Gamma(\mu(x)) = 0$). The ring $Z[x][t]$ is a unique factorization domain (UFD) (i.e., elements factor uniquely into irreducible factors), since Z is a UFD and the ring obtained by adjoining an indeterminate to a UFD is still a UFD [6, Theorem 32.3]. So $\chi_\Gamma(t)$ can factored uniquely into irreducible factors. One of these factors is the **minimal polynomial** of $\mu(x)$. This particular factor can be identified readily as the factor that contains the (unique, by virtue of part (e) of the Perron-Frobenius Theorem) dominant root $\mu(x)$ for any evaluation of $x > 0$. Let d denote the degree of this minimal polynomial; then $\mu(x)$ is said to be an **algebraic function of degree** d.

As an example, if $d = 2$, then the minimal polynomial would be of the form:

$$(8.1) \qquad\qquad t^2 + a(x)t + b(x)$$

where $a(x)$ and $b(x)$ are polynomials in x. In this case, we could use the quadratic formula to solve for $\mu(x)$. Because of the square root in the quadratic formula, $\mu(x)$ may fail to be differentiable, in particular at $x = 0$, and this might dash our hopes for a power series expansion valid in a neighborhood of $x = 0$. However, it would still remain plausible that $\mu(x)$ has a power series expansion in $\sqrt{x}$ instead of x. Indeed, it does! In fact, the general theory of algebraic functions (see [9, Chapter 12]) shows that any algebraic function of degree d always has a power series expansion in powers of $x^{1/u}$ for some positive integer $u \le d$.

EXAMPLE 1. As an example, consider

$$\Gamma(x) = \begin{bmatrix} 1 & x + x^2 \\ 1 & 1 + x^3 \end{bmatrix}.$$

The root $\mu(x)$ satisfies

$$\mu(x)^2 - (2 + x^3)\mu(x) + 1 - x - x^2 + x^3 = 0.$$

Applying the quadratic formula, we obtain

$$\begin{aligned} \mu(x) &= \left((2 + x^3) + \sqrt{(2 + x^3)^2 - 4(1 - x - x^2 + x^3)} \right) / 2 \\ &= 1 + x^3/2 + (\sqrt{4x + 4x^2 + x^6})/2 \\ &= 1 + x^3/2 + \sqrt{x}\sqrt{1 + O(x)} \\ &= 1 + \sqrt{x} + O(x). \end{aligned}$$

So, the initial part of the expansion of $\mu(x)$ is $1 + x^{1/2}$. Note that we chose the "+" root for the solution to the quadratic because $\mu(x)$ is, by definition, the largest eigenvalue.

EXAMPLE 2. As another example, consider

$$\Gamma(x) = \left[\begin{array}{cc} 1 + x & x^2 + x^3 \\ x + x^2 & 1 + x^3 \end{array} \right].$$

The root $\mu(x)$ satisfies

$$\mu(x)^2 - (2 + x + x^3)\mu(x) + (1 + x)(1 + x^3) - (x^2 + x^3)(x + x^2) = 0.$$

Applying the quadratic formula, we obtain

$$\begin{aligned} \mu(x) &= \frac{1}{2}\left((2 + x + x^3) \right. \\ &\quad \left. +\sqrt{(2 + x + x^3)^2 - 4((1 + x)(1 + x^3) - (x^2 + x^3)(x + x^2))}\right) \\ &= \frac{1}{2}\left((2 + x + x^3) + \sqrt{x^2 + O(x^3)}\right) \\ &= 1 + x/2 + +x^3/2 + (x/2)\sqrt{1 + O(x)} \\ &= 1 + x + O(x^2). \end{aligned}$$

So, the initial part of the expansion of $\mu(x)$ is $1 + x$ (in this case, the expansion turns out to be in powers of x, not $x^{1/2}$).

Of course, in general the quadratic formula is not available. But the terms of the power series expansions can be determined inductively from the minimal polynomial using either implicit differentiation or by setting the minimal polynomial to 0, substituting into the minimal polynomial an expansion of $t = \mu(x)$ with unknown coefficients, and solving for the coefficients [9, Chapter 12].

For an algebraic function $f = f(x)$ (such as a polynomial or the function $\mu(x)$), let $f(x)^{(0)}$ denote the leading term in the power series expansion (in some fractional power of x) of f at $x = 0$:

$$f(x) = f(x)^{(0)} + \text{ higher order terms,}$$

where

$$f(x)^{(0)} = c(f)x^{\delta(f)};$$

here $c = c(f) \neq 0$ denotes the coefficient of the leading term, and $\delta = \delta(f)$ denotes the degree of this term. For a vector or matrix $f(x)$ of algebraic functions define $f(x)^{(0)}$ (resp. $c(f)$, $\delta(f)$) to be the vector or matrix of leading terms (resp. corresponding coefficients, degrees).

The following result, which follows from [17, p. 411], characterizes the leading term in the expansion of $\mu(x)$.

PROPOSITION 8.1. $\mu(x)^{(0)} = \mu_0 x^{w_0}$, *where* w_0 *is the minimum wps and* μ_0 *is the largest eigenvalue of the matrix corresponding to the unweighted (i.e., weights of all edges are 1) minwps graph.*

Since every pseudocodeword can be decomposed into simple pseudocodewords, the minimum wps occurs as the wps of a cycle of length at most $|\Sigma_0|$. Thus, the minwps w_0 can be determined from the traces of the matrix powers $\Gamma(x)^i, i = 1, \ldots, |\Sigma_0|$.

For each state $s \in \Sigma_0$, define the **order** of s to be $\delta(v(x,s)) + \delta(w(x,s))$. Let S_0 denote the set of states with minimal order (over all states $s \in \Sigma_0$), and let ω be the minimal order itself. The following result, together with Proposition 7.1, gives a sense in which the "limiting version" (as $x \to 0$) of the generating-function sum-product algorithm picks out the states of minimal order. A complete proof of a much more general result is given in [4, Theorem on p. 117].

PROPOSITION 8.2. *Let* $\Gamma(x)$ *be as in (7.1) above with corresponding left and right Perron eigenvectors* $v(x)$ *and* $w(x)$. *Then*

1. *(a) For each state* $s \in S_0$,

$$\lim_{x \to 0} v(x,s)w(x,s)/x^\omega = c(v(x,s))c(w(x,s)) \neq 0.$$

(b) For each state $s \in \Sigma_0 \setminus S_0$,

$$\lim_{x \to 0} v(x,s)w(x,s)/x^\omega = 0.$$

2. *The set* S_0 *of minimal-order states is the union of the sets of states of a collection of principal components of the minwps graph.*

For each phase $j = 0, \ldots, n-1$, let $v_j(x)$ and $w_j(x)$ denote the corresponding left and right Perron eigenvectors for $\Gamma_j(x) \cdots \Gamma_{n-1}(x)\Gamma_0(x) \cdots \Gamma_{j-1}(x)$. Just as in Section 5, these vectors can be obtained inductively from $v_0(x) = v(x)$ and $w_n(x) = w(x)$ by

$$v_j(x) = v_{j-1}(x)\Gamma_{j-1}(x)$$

and

$$w_j(x) = \Gamma_j(x)w_{j+1}(x)$$

(for $j \geq n$, we find it convenient to also define $v_j(x) = v_{j-1}(x)\Gamma_{j-1}(x) = \mu(x)v_{j-n}(x)$). The corresponding vectors of leading terms, $v_j(x)^{(0)}$ and $w_j(x)^{(0)}$, can then be readily determined. As a result, we obtain a set S_j of minimal-order states for each phase.

The sets S_j can be used to trace out dominant pseudocodewords. For this, we need a criterion to tell which branches actually participate in dominant pseudocodewords.

For $0 \leq j < n-1$, a branch (s_j, a_j, s_{j+1}) is called a **minimal-weight branch** if

$$\delta(v_j(x, s_j)) - \log \gamma_j(a_j) = \delta(v_{j+1}(x, s_{j+1})).$$

Here, for $j = n - 1$, we take

$$v_{j+1}(x) = v_n(x) = v_{n-1}(x)\Gamma_{n-1}(x) = \mu(x)v_0(x),$$

not $v_0(x)$.

PROPOSITION 8.3. *A path is a dominant pseudocodeword if and only if it is a cycle consisting of a concatenation of minimal-weight branches.*

Proof. This follows from the discussion in [17] and [4]. For completeness, we sketch the rough idea. Since each $v_j(x) > 0$ whenever $x > 0$, it follows that the leading terms $v_j(x, s_j)^{(0)}$ have positive coefficients. Also, by definition, the entries of $\Gamma_j(x)$ have positive coefficients. It follows that there can be no cancellation in the branches involved in the part of the sum

$$\sum_{\{(s_j, a_{j+1}, s_{j+1})\}} v_j(x, s_j)\Gamma(x, s_j, s_{j+1})$$

that contributes to $v_{j+1}(x, s_{j+1})^{(0)}$, and these are precisely the minimal-weight branches. For the same reason, no branch can contribute to a term of degree lower than $\delta(v_{j+1}(x, s_{j+1}))$, and so for all branches other than minimal-weight branches we have

$$\delta(v_j(x, s_j)) - \log \gamma_j(a_j) > \delta(v_{j+1}(x, s_{j+1})).$$

By following the progression of eigenvectors $v_0(x), v_1(x), \ldots$, and using the fact that they are indeed eigenvectors for the eigenvalue $\mu(x) = \mu_0 x^{w_0}$ (+ higher order terms), we see that the only branch that can contribute to a dominant pseudocodeword is one of minimal weight and that any cycle of minimal-weight branches is indeed a dominant pseudocodeword. □

By a **minimal-weight edge**, we mean an edge of the graph $G(T)$ obtained as a concatenation of minimal-weight branches in T. From [4, discussion on p. 118], it follows that the principal components of the min-wps graph are sinks of the subgraph defined by the minimal-weight edges. By Proposition 8.2 (part 2), any minimal order state s_0 belongs to a principal component, and so any minimal-weight branch that emanates from s_0 must terminate at a minimal-order state in the same principal component. Thus, if at each minimal-order state we recursively pick out minimal-weight branches, then we will trace out entire principal components. Any cycle in such a component will be a dominant pseudocodeword (by Proposition 8.3). So, in this sense, the "limiting version" (as $x \to 0$) of the generating-function sum-product algorithm detects at least some of the dominant pseudocodewords. However, Example 4 below shows that it may *not* detect all dominant pseudocodewords.

9. Examples.

EXAMPLE 3. Figure 2 shows a trellis section for the standard rate-1/2 4-state binary linear convolutional code. Anderson and Hladik [2] consider the TB trellis T with length $n = 5$ that is based on this code. The channel is a binary symmetric channel with error probability $p < 1/2$. (For our purposes, the value of p is immaterial.) The branch weights are set equal to the Hamming distance $d_H(a_k, r_k)$ between the transmitted and received binary 2-tuples (since $-\log p(r \mid a) = d_H(a_k, r_k) \log(\frac{p}{1-p})$ plus a constant). Branches therefore have weights 0, 1 or 2, which will be indicated in some of the figures by solid, dashed, or invisible lines, respectively.

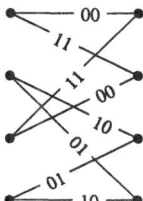

FIG. 2. *Trellis section for rate-1/2 4-state binary linear convolutional code.*

Assume that the received sequence is (00 10 10 00 00). This defines branch weights on T, and therefore also the matrices $\Gamma_j(x)$, $j = 0, 1, \ldots, 4$ as well as the matrix $\Gamma(x)$, defined in (7.1), computed over one trellis length of T. A straightforward computation shows that the leading terms of the entries of $\Gamma(x)$ are:

$$\Gamma(x)^{(0)} = \begin{bmatrix} x^2 & x^3 & x^2 & x^2 \\ x^3 & x^2 & 2x^3 & 2x^3 \\ x & x^2 & x^3 & x^3 \\ x^3 & x^2 & 2x^3 & 2x^3 \end{bmatrix}$$

Examination of the first four powers of this matrix reveals that the minimum wps w_0 is $3/2$, and the only cycle that achieves the minwps is the cycle of length 2 which alternates between states 1 and 3 with weights x and x^2 (so the minwps graph $G_0(T)$ consists of a single cycle of length 2). This cycle corresponds to the following pseudocodeword at state 1

$$\text{(00 11 10 00 10 00 10 11 00 00)},$$

which is shown in darkened lines in Figure 3. The two phases of this pseudocodeword are then the only dominant pseudocodewords.

According to Proposition 8.1, we thus have $\mu(x)^{(0)} = x^{3/2}$. In this case, the degree of $\mu(x)$ turns out to be 3, so $v(x)$ and $w(x)$ all have power series expansions in some fractional power $1/u$ of x with $u \leq 3$, valid in a neighborhood of $x = 0$. (It turns out that the expansion is actually in powers of $x^{1/2}$.)

The vectors of leading terms of $v(x), w(x)$ are entries of the adjugate matrix $A(\mu(x), \Gamma(x))^{(0)}$. We would like to compute this matrix using only

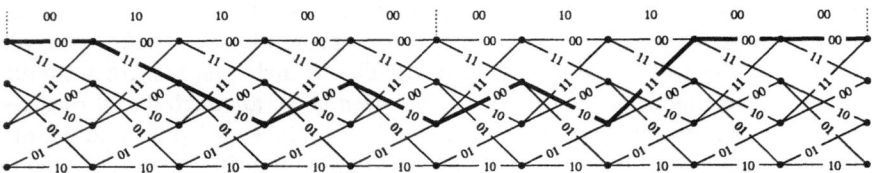

FIG. 3. *The dominant pseudocodeword.*

$\mu(x)^{(0)}$ and $\Gamma(x)^{(0)}$. However, this may not be enough information. For instance, consider the (2,2) entry of $A(\mu(x)^{(0)}, \Gamma(x)^{(0)})$; this entry is $\det(S)$ where

$$S = \begin{bmatrix} x^{3/2} - x^2 & -x^2 & -x^2 \\ -x & x^{3/2} - x^3 & -x^3 \\ -x^3 & -2x^3 & x^{3/2} - 2x^3 \end{bmatrix}.$$

Expanding this determinant, we see that the lowest-degree term in $\det(S)$ is $x^{9/2} - x^{9/2} = 0$; the next-lowest-degree term is $-x^5$. Since this term is negative and $A(\mu(x) - \Gamma(x)) > 0$ for any $x > 0$, it follows that $-x^5$ cannot possibly be the leading term of the (2, 2) entry of $A(\mu(x), \Gamma(x))$. Thus the leading term can be resolved only by plugging in more of the initial part of the expansion of $\mu(x)$.

On the other hand, consider the (1,2) entry, which is $-\det(S)$ where

$$S = \begin{bmatrix} -x^3 & -x^2 & -x^2 \\ -x^2 & x^{3/2} - x^3 & -x^3 \\ -x^2 & 2x^3 & x^{3/2} - 2x^3 \end{bmatrix}.$$

Expanding this determinant, we see that the lowest-degree term has degree $11/2$ and that there is no cancellation, so we are left with $2x^{11/2}$. Any higher-degree term in the expansion of $\mu(x)$ or any higher-degree terms in the entries of $\Gamma(x)$ must therefore contribute to terms of degree $> 11/2$. So in this case we can reliably say that $2x^{11/2}$ is the leading term of the $(1, 2)$-entry of $A(\mu(x), \Gamma(x))$.

If we do this for each entry, we obtain the following (where $0x^\kappa$ means that cancellation occurred and we were unable to determine the lowest-degree term of the entry):

$$\begin{bmatrix} x^{9/2} & 2x^{11/2} & x^5 & x^5 \\ 2x^{11/2} & 0x^{9/2} & 2x^6 & 2x^6 \\ x^4 & 2x^5 & x^{9/2} & x^{9/2} \\ 2x^{11/2} & 0x^5 & 2x^6 & 0x^{9/2} \end{bmatrix}.$$

Since there are no zero entries in the first row and column, we can read off the leading terms of $v(x)$ and $w(x)$ as:

$$v^{(0)}(x) = [x^{9/2}, 2x^{11/2}, x^5, x^5] \sim [1, 2x, x^{1/2}, x^{1/2}]$$

and

$$w^{(0)}(x) = [x^{9/2}, 2x^{11/2}, x^4, 2x^{11/2}] \sim [x^{1/2}, 2x^{3/2}, 1, 2x^{3/2}].$$

Figures 4 and 5 show the sequences of vectors $v_0^{(0)}(x), v_1^{(0)}(x), \ldots, v_4^{(0)}(x)$ and $w_0^{(0)}(x), w_1^{(0)}(x), \ldots, w_4^{(0)}(x)$. From this, we see that

$$v^{(0)} \star w^{(0)} = [x^{1/2}, 4x^{5/2}, x^{1/2}, 2x^2] \sim [1, 4x^2, 1, 2x^{3/2}],$$

and so the minimal-order states in phase 0 are states 1 and 3.

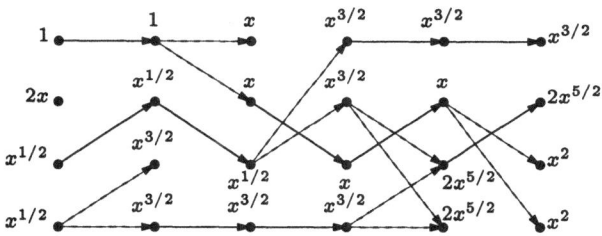

FIG. 4. *The forward recursion.*

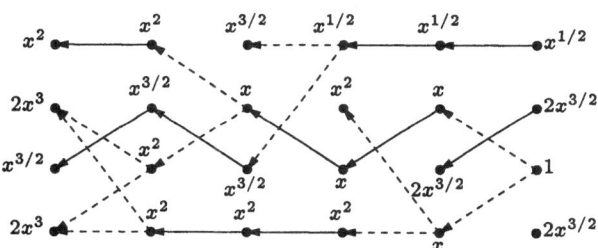

FIG. 5. *The backward recursion.*

In Figure 6, the minimal-order states are shown as darkened dots and the minimal-weight branches passing through these states are darkened. Note that these branches trace out the two halves of the (unique) dominant pseudocodeword in Figure 3.

Notice that the best candidates among the actual codewords are the all-zero codeword (00 00 00 00 00) and the nonzero codeword (01 10 10 01 00); each has Hamming distance 2 per trellis length from the received sequence. However, the dominant pseudocodeword (00 11 10 00 10 00 10 11 00 00) has Hamming distance 3 per two trellis lengths; therefore it has the minimum wps and is detected instead.

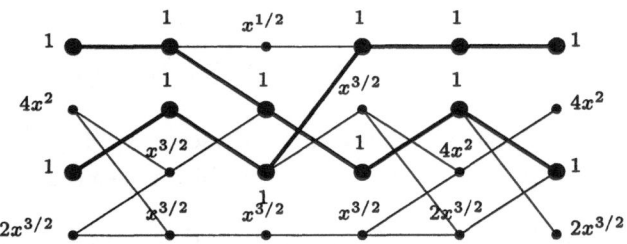

FIG. 6. *Minimal-order states and minimal-weight branches.*

EXAMPLE 4. Consider the (rather poor) single-error-correcting, rate-2/7 convolutional code defined by the generator matrix:

$$\begin{bmatrix} 1 & 1 & 1 & 0 & 0 & 0 & 0 \\ 0 & D & 1 & 1+D & D & D & 1 \end{bmatrix}.$$

This code is also defined by the trellis section in Figure 7.

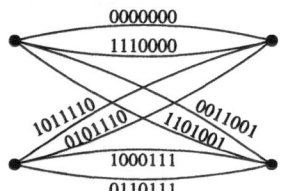

FIG. 7. *Trellis section for a rate-2/7 2-state convolutional code.*

Consider the TB trellis T with trellis length $n = 1$ defined by this code. Assume that the received sequence is

$$(1\ 0\ 0\ 0\ 1\ 0\ 0).$$

Then we have:

$$\Gamma(x) = \begin{bmatrix} x^2 + x^3 & x^4 + x^5 \\ x^3 + x^4 & x^2 + x^5 \end{bmatrix}.$$

Clearly, the minwps graph $G_0(T)$ consists of the union of the two self-loops at states 1 and 2 with weight x^2; thus $w_0 = 2$, and the dominant pseudocodewords are actually the codewords

$$(0\ 0\ 0\ 0\ 0\ 0\ 0)\ \text{and}\ (1\ 0\ 0\ 0\ 1\ 1\ 1)$$

(each with distance 2 from the received word). So, by Proposition 8.1, the leading term in the power series expansion for $\mu(x)$ is x^2. Substituting this for $\mu(x)$ in the adjugate $A(\mu(x), \Gamma(x))$ yields:

$$\begin{bmatrix} 0x^2 & x^4 \\ x^3 & 0x^2 \end{bmatrix};$$

the zero entries occur because there may be a degree 3 term in the expansion of $\mu(x)$. Indeed, the second-to-leading term in the expansion is x^3; in Example 2, we computed $\mu(x)$ for the matrix $\Gamma(x)/x^2$, and found the initial part of that expansion to be $1 + x$. Now, plugging $x^2 + x^3$ in for $\mu(x)$ yields:

$$\begin{bmatrix} x^3 & x^4 \\ x^3 & 0x^2 \end{bmatrix}.$$

The first row and column of this matrix are nonzero, and so we obtain:

$$v(x)^{(0)} = [x^3, x^4], \quad w(x)^{(0)} = [x^3, x^3].$$

From this, we see that state 1 has order $3 + 3 = 6$, while state 2 has order $3 + 4 = 7$. Thus, only state 1 is a minimal-order state, so the limiting version of the generating-function sum-product algorithm detects only one (of the two) dominant pseudocodewords (in this case, these are actually codewords). In other words, the algorithm would conclude that $(0\ 0\ 0\ 0\ 0\ 0\ 0)$ was sent, whereas it is equally likely that $(1\ 0\ 0\ 0\ 1\ 1\ 1)$ was sent.

10. The generating-function min-sum algorithm. In the preceding section, we defined the leading term of an algebraic function (in particular a polynomial) or a vector or matrix of such functions, and we explained the significance of the leading terms of the Perron eigenvector function. This motivates us to define a sum-product algorithm over single-term generating functions, i.e., scalar multiples of monomials. The set of such functions carries a natural semiring structure, called the **leading-term semiring**. Let "sum" be defined by:

$$ax^r + bx^s = \begin{cases} ax^r, & \text{if } r < s; \\ bx^s, & \text{if } r > s; \\ (a+b)x^r, & \text{if } r = s, \end{cases}$$

and let "product" be the usual notion of product. With this semiring structure, we can define matrix multiplication and the sum-product algorithm, which we call the **generating-function min-sum algorithm**. Its forward recursion computes

$$\alpha_k(x)^{(0)} = \alpha_0 \Gamma_0(x)^{(0)} \dots \Gamma_{k-1}(x)^{(0)},$$

where the matrix multiplications are over this semiring (i.e., do ordinary matrix multiplication over the polynomial ring $Z[x]$, but keep only the lowest-degree term in each component).

Thus the components of the vector generating function $\alpha_k(x)^{(0)}$ are:

$$\alpha_k(x, s_k)^{(0)} = N_{\max} \, x^{\min \{-\log \gamma_{[0,k)}(a_{[0,k)})\}},$$

where the minimum is taken over all past paths $(s_{[0,k)}, a_{[0,k)}, s_k)$ that terminate at s_k, and $N_{\max}$ is the number of past paths achieving this minimum. Thus this algorithm computes both the ML past path weight, $\max\{\gamma_{[0,k)}(a_{[0,k)})\}$, and also the number $N_{\max}$ of past paths with this weight, so it gives slightly more information than the ordinary min-sum algorithm.

The backward recursion is defined similarly, as well as the computation of final state weights and final output weights.

If Perron-Frobenius theory extended nicely to this semiring, then we might expect the forward recursion of this algorithm to converge to some kind of eigenvector (i.e., an eigenvector for the matrix $\Gamma(x)^{(0)}$ over the leading-term semiring). Proposition 10.1 below shows that we can find meaningful eigenvectors over this semiring, and Proposition 10.2 below gives a version of convergence of the generating-function min-sum algorithm. However, we will see in Example 5 below that we do not always obtain convergence to these eigenvectors.

PROPOSITION 10.1. *Let $\Gamma(x)$ be as in (7.1) above with corresponding left, right Perron eigenvectors $v(x), w(x)$. Then the leading-term vectors $v(x)^{(0)}, w(x)^{(0)}$ are eigenvectors corresponding to the eigenvalue $\mu^{(0)}(x)$ for the leading-term matrix $\Gamma(x)^{(0)}$ over the leading-term semiring:*

$$(10.1) \qquad\qquad v(x)^{(0)} \Gamma(x)^{(0)} = \mu(x)^{(0)} v(x)^{(0)}$$

and

$$(10.2) \qquad\qquad \Gamma(x)^{(0)} w(x)^{(0)} = \mu(x)^{(0)} w(x)^{(0)}.$$

Proof. The following proof is essentially contained in [17, Theorem 20(d)]. By definition, the entries of $\Gamma(x)$ have positive coefficients. Since $\mu(x), v(x)$ and $w(x)$ are positive for all positive x, it follows that the the leading terms in the power series expansion of $\mu(x)$ and each component of $v(x), w(x)$ have positive coefficient. Thus, there can be no cancellation in determining the lowest-degree terms in the equations $v(x)\Gamma(x) = \mu(x)v(x)$ and $\Gamma(x)w(x) = \mu(x)w(x)$. It then follows that equations (10.1) and (10.2) hold. $\qquad\qquad\square$

PROPOSITION 10.2. *Let $\Gamma(x)$ be as in (7.1) above. Let n be the trellis length and p be the period of the minwps graph G_0. Then the forward recursion of the generating-function min-sum algorithm converges on the subsequence of multiples of np.*

Proof. Let $D(x)$ denote the diagonal matrix defined by $D(x, s, s) = x^{d_s}$, where $d_s = \delta(v(x, s))$. Let

$$B = (D(x)\Gamma(x)D^{-1}(x)/x^{w_0})^{(0)}.$$

It turns out (see [17, pp. 411–412] and [4, pp. 117–118]) that B is a nonnegative matrix, all principal components of B are sinks and for every state s, we have $\Gamma^n(s, s') \neq 0$ for some n and some state s' belonging to a sink. Note that we can write:

$$(10.3) \qquad \Gamma(x) = x^{w_0} D^{-1}(x)(B + H(x))D(x),$$

where $H(x)$ is a matrix whose nonzero entries are polynomials with positive integer coefficients in some fractional power of x and have no constant terms.

Now, $\Gamma^k(x)^{(0)}$ is the matrix obtained from the kth power of $\Gamma(x)$ obtained by keeping only the lowest-degree term in each entry. From equation (10.3) above, we have:

$$\Gamma^k(x, s, t)^{(0)} = z^{d_t - d_s + k w_0} \mathbf{x}_{s,t,k}$$

where $\mathbf{x}_{s,t,k}$ is the lowest-degree term (with coefficient) in the (s, t)-entry of $(B + H(x))^k$. Now, $\mathbf{x}_{s,t,k}$ involves only those terms in the expansion of $(B + H(x))^k$ that contain a bounded number of appearances of $H(x)$. Suppose for simplicity that at most one appearance of $H(x)$ contributes to $\mathbf{x}_{s,t,k}$ (the general case will follow from similar arguments). Then, only terms of the form B^k or $B^i H(x)B^{k-i-1}, i = 1, \ldots, k-1$, can contribute to $\mathbf{x}_{s,t,k}$.

Now, since all principal components of B are sinks, it follows from Theorem 4.2 that B^{mp}/μ_0^{mp} converges as $m \to \infty$ to some limit matrix L. If $B^{mp}(s, t)$ contributes to $\mathbf{x}_{s,t,mp}$ for some m, then for all sufficiently large m, we have $\mathbf{x}_{s,t,mp} = B^{mp}(s, t)$, and so we have convergence, up to scale, to $L(s, t)$. For the remainder of the proof, we may assume that for all m, $B^{mp}(s, t)$ does not contribute to $\mathbf{x}_{s,t,mp}$.

For a given pair of states s, t, let $\mathcal{C}$ denote the set of congruence classes c mod p such that there exist an m and an $i \equiv c$ mod p for which $(B^i H(x)B^{mp-i-1})(s, t)$ contributes to $\mathbf{x}_{s,t,mp}$. There is a constant N_0 such that for all sufficiently large m, all $c \in \mathcal{C}$ and all $i \equiv c$ mod p, with $i > N_0$ and $i < mp - N_0$, $(B^i H(x)B^{mp-i-1})(s, t)$ contributes to $\mathbf{x}_{s,t,mp}$ (this follows from the fact that for every state s, we have $\Gamma^n(s, s') \neq 0$ for some n and some state s' belonging to a sink).

Fix $\nu \in (0, 1/2)$. Then for sufficiently large m, all $c \in \mathcal{C}$, and all $i \equiv c$ mod p, with $i > \nu mp$ and $i < (1 - \nu)mp$, we have that $(B^i H(x)B^{mp-i-1})$ $(s, t)/\mu_0^{mp}$ is close to $(LB^c H(x)LB^{c-1})(s, t)/\mu_0^{2c}$. We may assume that m is so large that $\nu mp > N_0$.

Decompose the set of all $i = 0, \ldots, mp-1$ such that $(B^i H(x)B^{mp-i-1})$ (s, t) contributes to $\mathbf{x}_{s,t,mp}$ into two subsets: $\mathcal{G}(s, t, m)$ is the set of all such i with $\nu mp \leq i \leq (1 - \nu)mp$, and $\mathcal{B}(s, t, m)$ is the set of all such i with $i < \nu mp$ or $i > (1 - \nu)mp$. Now, write

$$\mathbf{x}_{s,t,mp}/(mp\mu_0^{mp}) = \sum_{i \in \mathcal{G}(s,t,mp)} (B^i H(x)B^{mp-i-1})(s, t)/(mp\mu_0^{mp})$$
$$+ \sum_{i \in \mathcal{B}(s,t,mp)} (B^i H(x)B^{mp-i-1})(s, t)/(mp\mu_0^{mp}).$$

Since $\nu m p > N_0$, the first term in the sum is close to:

$$(1 - 2\nu) \sum_{c \in C} (LB^c H(x) LB^{c-1})(s, t)/(\mu_0)^{2c}.$$

The second term in this sum is bounded by a function of ν that tends to 0 as ν tends to 0 (uniformly in m). Since ν is an arbitrary number in $(0, 1/2)$, it follows that the second term can be made arbitrarily small. Thus, by taking ν smaller and smaller and then n larger and larger, depending on ν, we obtain convergence to

$$\sum_{c \in C} (LB^c H(x) LB^{c-1})(s, t)/(\mu_0)^{2c}.$$

$\square$

EXAMPLE 5. We continue with Example 3. Recall that the trellis length is $n = 5$. The minwps graph G_0 consists of one cycle of length 2, and thus the period of G_0 is $p = 2$.

Figure 8 illustrates the behavior of the forward recursion of the generating-function min-sum algorithm (normalized by division by x^3 after every two trellis lengths). Here, the forward recursion over $2k$ trellis lengths yields:

$$\alpha_{2k} = [(k + 1)x^3, (2k + 3)x^4, x^3, x^3].$$

Thus, over the subsequence of multiples of $10 = np$, the forward recursion converges to

$$[1, 2x, 0, 0].$$

Note that this is different from

$$v(x)^{(0)} \sim [1, 2x, x^{1/2}, x^{1/2}].$$

The backward recursion over $2k$ trellis lengths gives $[1, 2x, k + 2, 2x]$. So, over the subsequence of multiples of 10, the backward recursion converges to

$$(0, 0, 1, 0).$$

Note that this is different from

$$w(x)^{(0)} \sim [x^{1/2}, 2x^{3/2}, 1, 2x^{3/2}].$$

The final weights (after $2k$ trellis lengths) are

$$\lambda_k = [(k + 1)x^3, (4k + 6)x^5, (k + 2)x^3, 2x^4],$$

which in the limit is (up to scale)

$$[1, 4x^2, 1, 0].$$

Thus the algorithm detects the states of the dominant pseudocodeword (the states 1 and 3 with minimal degree in the final weight vector), but the final weights are different from the final weights of the limiting version of the generating-function sum-product algorithm:

$$[1, 4x^2, 1, 2x^{3/2}].$$

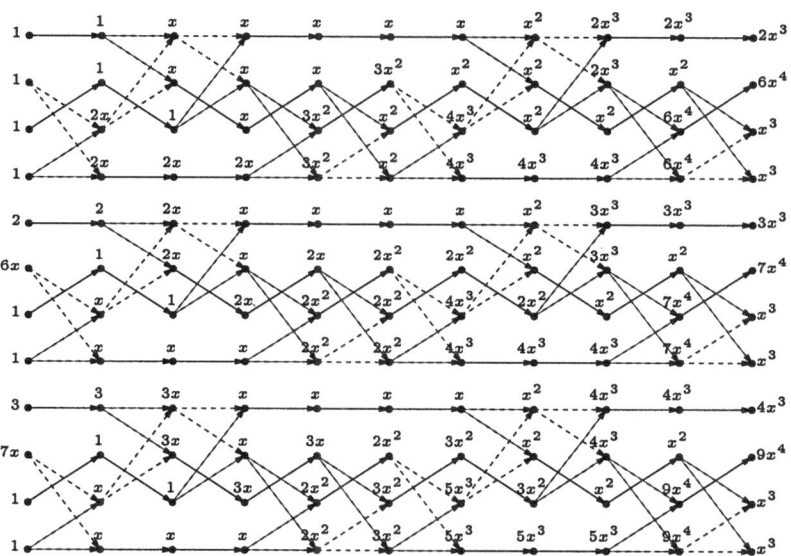

FIG. 8. *Behavior of the generating-function min-sum algorithm.*

Note that in these computations, the result of the generating-function min-sum algorithm agrees with the limiting (as $x \to 0$) version of the generating-function sum-product algorithm, *except* for the components that have fractional exponents. Is this an accident?

Finally, note that the forward recursion over an odd number $2k + 1$ of trellis lengths gives:

$$\alpha_{2k+1} = [1, 2x, (k + 1)x, (k + 1)x].$$

which converges to

$$[0, 0, 1, 1].$$

Thus, the result of Proposition 10.2 cannot be extended to convergence over the entire sequence of multiples of $n = 5$.

Acknowledgments. The results in this paper are closely related to work of others, including D. Agrawal, S. Aji, J. Anderson, B. Frey, S. Hladik, R. Koetter, R. McEliece, H.-A. Loeliger, A. Reznik, A. Vardy,

N. Wiberg and Y. Weiss (cf. [1, 18, 11]), whose help is gratefully acknowledged. The authors wish to thank the IMA for hosting the 1999 Summer Workshop on Codes, Systems and Graphical Models, where some of this work was done. The second author also wishes to acknowledge the hospitality of M.I.T. and Motorola.

REFERENCES

[1] S.M. AJI AND R.J. McELIECE, The generalized distributive law, IEEE Trans. Inform. Theory, 46 (2000), pp. 325–343.

[2] J.B. ANDERSON AND S.M. HLADIK, Tailbiting MAP decoders, IEEE J. Select. Areas Commun., 16 (1998), pp. 297–302.

[3] A.R. CALDERBANK, G.D. FORNEY, JR., AND A. VARDY, Minimal tail-biting trellises: The Golay code and more, IEEE Trans. Inform. Theory, 45 (1999), pp. 1435–1455.

[4] E. CAWLEY, B. MARCUS, AND S. TUNCEL, Boundary measures of Markov chains, Israel J. Math., 94 (1996), pp. 111–123.

[5] G.D. FORNEY, JR., On iterative decoding and the two-way algorithm, in Proc. Intl. Symp. Turbo Codes and Related Topics, Brest, France, Sept. 1997.

[6] J.B. FRALEIGH, A First Course in Abstract Algebra, Addison-Wesley, 1977.

[7] B.J. FREY, R. KOETTER, AND A. VARDY, Skewness and pseudocodewords in iterative decoding, in Proc. IEEE Int. Symp. Inform. Theory, Cambridge, MA, Aug. 1998, p. 148.

[8] R.G. GALLAGER, Low-Density Parity-Check Codes, MIT Press, Cambridge, MA, 1963.

[9] E. HILLE, Analytic Function Theory, Chelsea, 1962.

[10] F.R. KSCHISCHANG AND B.J. FREY, Iterative decoding of compound codes by probability propagation in graphical models, IEEE J. Select. Areas Commun. (1998), pp. 219–230.

[11] F.R. KSCHISCHANG, B.J. FREY, AND H.-A. LOELIGER, Factor graphs and the sum-product algorithm. submitted to IEEE Trans. Info. Theory, 1998.

[12] D. LIND AND B. MARCUS, An Introduction to Symbolic Dynamics and Coding, Cambridge U. Press, Cambridge, UK, 1995.

[13] D.J.C. MACKAY, R.J. McELIECE, AND J.-F. CHENG, Turbo decoding as an instance of Pearl's 'belief propagation' algorithm, IEEE J. Select. Areas Commun. (1998), pp. 140–152.

[14] B. MARCUS AND S. TUNCEL, The weight-per-symbol polytope and scaffolds of invariants associated with Markov chains, Ergod. Thy. and Dynam. Sys., 11 (1991), pp. 129–180.

[15] ———, Matrices of polynomials, positivity, and finite equivalence of Markov chains, J. Am. Math. Soc., 6 (1993), pp. 131–147.

[16] E. SENETA, Nonnegative Matrices and Markov Chains, Springer, 1981.

[17] S. TUNCEL, Faces of Markov chains and matrices of polynomials, Contemp. Math., 135 (1992), pp. 391–422.

[18] Y. WEISS, Correctness of local probability propagation in graphical models with loops, Neural Comp., 12 (2000), pp. 1–41.

[19] N. WIBERG, Codes and decoding on general graphs, PhD thesis, Dept. Elec. Engg., U. Linköping, Sweden, Apr. 1996.

[20] N. WIBERG, H.-A. LOELIGER, AND R. KOETTER, Codes and iterative decoding on general graphs, Euro. Trans. Telecomm., 6 (1995), pp. 513–525.

ALGORITHMS FOR DECODING AND INTERPOLATION

MARGREET KUIJPER*

Abstract. In this paper we consider various algorithms for decoding BCH/RS/ Goppa codes, in particular the euclidean algorithm, the Berlekamp-Massey algorithm and the Welch-Berlekamp algorithm. We focus on relationships of these algorithms with interpolation methods in system theory. We note that the problem statements in the two areas can be different: from a system theoretic point of view, rational interpolating functions with common factors between numerator and denominator are undesirable whereas common factors can be required in a decoding context.

The behavioral approach was introduced by Jan C. Willems into system theory in the eighties. It proposes the family of trajectories of a system as its central focus. This makes the approach attractive for coding theorists (most naturally in the context of convolutional codes where the family of trajectories corresponds to the code). In this paper we focus on a connection between behavioral modeling and the decoding of BCH/RS/Goppa codes. In this context, the behavioral modeling approach is attractive because it naturally generates solutions with common factors.

We present slight modifications of both the Berlekamp-Massey and the Welch-Berlekamp algorithm and give a derivation in terms of behavioral modeling. In particular, we derive the latter algorithm directly from Reed & Solomon's original approach. We demonstrate the similarity of the two algorithms and show that they are special instances of one general iterative behavioral modeling procedure.

Key words. Reed-Solomon codes, Berlekamp-Massey algorithm, Welch-Berlekamp algorithm, behaviors, exact modeling.

1. Introduction. Interpolation techniques have proved useful tools for various system-theoretic problems, see [2] and references therein. In a system-theoretic context the interpolant is usually required to be a rational function, say N/D, where N and D are polynomials. Thus any common factors between N and D are cancelled. Then, if the objective is to obtain an interpolant of minimal McMillan degree (i.e. max { deg D, deg N } minimal), the occurrence of common factors can interfere with the minimality requirement, see also [7]. Let us illustrate this with a small example.

EXAMPLE 1. *Consider an interpolation problem over* $\mathbb{R}$. *The interpolation data are given by* $(x_1, y_1) = (0, 1)$, $(x_2, y_2) = (1, 0)$ *and* $(x_3, y_3) = (-1, 1)$. *Then a unique minimal polynomial solution is given by*

$$(D(s), N(s)) = (s - 1, s - 1).$$

Indeed, for $i = 1, 2, 3$, $D(x_i)y_i = N(x_i)$. *However, rational interpolants are all of degree* ≥ 2. *(For example* $Y(s) = \frac{s^2 - 2s + 1}{-3s + 1}$ *is a minimal rational interpolant; indeed* $Y(x_i) = y_i$ *for* $i = 1, 2, 3$.)

However, as we will see in this paper, in a coding-theoretic context, the minimal interpolation problem is most naturally formulated as a *polyno-*

*Department of EE Engineering, University of Melbourne, VIC 3010, Australia; m.kuijper@ee.mu.oz.au.

mial interpolation problem, in which common factors play a crucial role in retrieving the transmitted message.

In recent years, Reed-Solomon decoding methods based on interpolation have captured a renewed interest as they provide a natural setting for successive erasure decoding ([5]) and list decoding [25, 26, 12]. In the latter references it was shown how Reed-Solomon error correction beyond half the minimum distance can be achieved by a list decoding technique based on interpolation.

In this paper we show how a system-theoretic modeling procedure, based on a behavioral approach, gives rise to interpolating solutions with common factors in a natural way, thus making it relevant to coding theoretic applications.

The behavioral approach to system theory has been introduced in [28]–[30]. Using this approach, ideas concerning the modeling of data have been developed in [29, 30, 3, 1, 31]. More specifically, [30, p. 289] gives a modeling procedure for the construction of exact linear models. In [16] it was shown how the Berlekamp key equation of [4] can be reformulated as minimal modeling of a certain finite dimensional behavior. It was further shown in [16] how the Berlekamp-Massey algorithm can be interpreted as a special instance of the modeling procedure of [30, p. 289], involving a clever choice of update matrix at each step. This work formed the basis for the multivariable algorithm of [17] which was then used to achieve improved decoding of BCH codes in [19, 14, 15].

An alternative decoding method ([24]) that is noniterative is the euclidean algorithm. It was shown in [18] how the euclidean algorithm fits into this approach. This extends (see subsection 4.3 of this paper) to interpolation-based decoding, connecting with results in [8].

In this paper we show how the Welch-Berlekamp key equation (see e.g. [5]) can be reformulated as minimal modeling of a finite dimensional behavior. The Welch-Berlekamp algorithm [27] exhibits similarities with the Berlekamp-Massey algorithm, see e.g. the presentation in [9]. However, unlike the Berlekamp-Massey algorithm, it makes no direct use of the solution's degree ℓ at each step to determine which type of update matrix is used. In this paper we present a modification of the Welch-Berlekamp algorithm that does use this parameter ℓ. We believe that its workings are then particularly transparent in showing that this algorithm is, like the Berlekamp-Massey algorithm, another special instance of the modeling procedure of [30, p. 289].

2. The behavioral approach and exact modeling of data. In this section we first present some preliminaries on the behavioral approach and then proceed with a brief outline of the relevant part of the behavioral theory of exact modeling. In particular, we present the general iterative modeling procedure of [30] in Procedure 2.1. In the sequel, trajectories will take values in a field, denoted as $\mathbb{F}$. For system-theoretic applications, $\mathbb{F}$ is usually infinite ($\mathbb{R}$ or $\mathbb{C}$); however, in the coding-theoretic context of the sequel, $\mathbb{F}$ will be a finite field.

In the behavioral approach [22], [28]–[30], a system is essentially defined as a set of trajectories. We will be concerned with linear shift-invariant behaviors on the time-set $\mathbb{Z}_+$ of the form $\mathcal{B} = \ker R(\sigma)$, where R is a polynomial matrix of, say, size $p \times m$ and σ is the backward shift operator:

$$\sigma(w_0, w_1, w_2, \ldots) := (w_1, w_2, \ldots).$$

The behavior $\mathcal{B}$ consists of trajectories $w : \mathbb{Z}_+ \mapsto \mathbb{F}^m$, for which

(2.1) $R(\sigma)w = 0.$

The representation (2.1) is called a *kernel representation* of $\mathcal{B}$.

Let us repeat some notions from [28], and start with the following lemma (see e.g. [13, Th. 3.9] for a detailed proof).

LEMMA 2.1. *Let $R_1 \in \mathbb{F}^{p_1 \times m}[s]$ and $R_2 \in \mathbb{F}^{p_2 \times m}[s]$. Then*

$$ker\ R_1(\sigma) \subset ker\ R_2(\sigma)$$

if and only if there exists a polynomial matrix $F \in \mathbb{F}^{p_2 \times p_1}[s]$ such that

$$R_2 = FR_1.$$

It is a corollary of the above lemma that polynomial matrices R_1 and R_2 of full row rank represent the same behavior if and only if there exists a unimodular matrix U (i.e., a polynomial matrix with constant nonzero determinant) such that $R_2 = UR_1$.

As a measure of complexity of a model we introduce the *order* $n(\mathcal{B})$ of a behavior. It is defined as the minimum value of the sum of the row degrees of R, where the minimum is taken over all possible full row rank kernel representations (2.1) of $\mathcal{B}$. This minimum is attained exactly when R is "row reduced":

DEFINITION 2.1. *Let $R \in \mathbb{F}^{p \times m}[s]$ have full row rank. Define $R_d \in \mathbb{R}^{p \times m}$ as the leading row coefficient matrix of R, i.e., the constant matrix that consists of the coefficients of the highest degree terms in each row of R. Define R to be row reduced if R_d has full row rank.*

When a matrix R is not row reduced, a unimodular matrix U can be found such that UR is row reduced. A procedure is given in [32, p. 27], see also [13, p. 24] where it is shown that not only the sum of the minimal row degrees is an invariant of a behavior, but also the minimal row degrees themselves are invariants of a behavior.

Let us next repeat some standard notions and terminology from [29] (see also [30, 3]) and assume that we have a *data set* $\mathbf{D} = \{\boldsymbol{b_1}, \ldots, \boldsymbol{b_\nu}\}$ where $\boldsymbol{b_i} \in (\mathbb{R}^m)^{\mathbb{Z}_+}$ are observed trajectories $(i = 1, \ldots, \nu)$. A behavior $\mathcal{B}$ is called an *unfalsified model* for $\mathbf{D}$ if $\mathbf{D} \subseteq \mathcal{B}$. A model $\mathcal{B}_1$ is called *more powerful* than a model $\mathcal{B}_2$ if $\mathcal{B}_1 \subseteq \mathcal{B}_2$. A model $\mathcal{B}^*$ is called the *most powerful unfalsified model (MPUM)* for $\mathbf{D}$, if $\mathcal{B}^*$ is unfalsified for $\mathbf{D}$ and $\mathbf{D} \subseteq \mathcal{B} \implies \mathcal{B}^* \subseteq \mathcal{B}$. It has been shown in [29] that a unique MPUM $\mathcal{B}^*$ exists for $\mathbf{D}$. However, note that a kernel representation (2.1) of $\mathcal{B}^*$ is far from unique.

We are now ready to present the procedure of [30, p. 289], which provides a framework for the iterative construction of a kernel representation of the MPUM for $\mathbf{D} = \{\boldsymbol{b_1}, \ldots, \boldsymbol{b_\nu}\}$. It can be easily understood from Lemma 2.1.

PROCEDURE 2.1. *([30]) Initially define*

$$R_0 := I \quad \text{(where } I \text{ is the identity matrix)}.$$

Proceed iteratively as follows for $k = 1, \ldots, \nu$. *Define, after receiving* $\{\boldsymbol{b_1}, \ldots, \boldsymbol{b_k}\}$, *the* k-th *error trajectory* $\boldsymbol{e_k}$ *as*

$$\boldsymbol{e_k} := R_{k-1}(\sigma)\boldsymbol{b_k}.$$

Compute a kernel representation $V_k(\sigma)\boldsymbol{w} = 0$ *of the MPUM for* $\{\boldsymbol{e_k}\}$. *Then define*

$$R_k := V_k R_{k-1}.$$

THEOREM 2.1. *([30]) For* $k = 1, \ldots, \nu$, *the kernel representation*

$$R_k(\sigma)\boldsymbol{w} = 0,$$

with R_k *defined in Procedure 2.1, represents the MPUM for* $\{\boldsymbol{b_1}, \ldots, \boldsymbol{b_k}\}$. With the above procedure we only need to be able to compute a MPUM representation for a *single* trajectory (the error trajectory) in order to derive an MPUM representation for a *finite set* of trajectories. Furthermore, row reducedness (Definition 2.1) of the representation $R_k(\sigma)\boldsymbol{w} = 0$ can be achieved by choosing V_k in such a way that $V_k R_{k-1}$ remains row reduced at each step k $(k = 1, \ldots, \nu)$.

3. Decoding and interpolation.

There are various ways of defining Reed-Solomon codes. In their original presentation [23], Reed & Solomon defined their code as a set of polynomial evaluations at all points of a Galois field. More specifically, writing

$$\mathbb{F} = GF(q) = \{0, \alpha, \ldots, \alpha^{q-1} = 1\}$$

with α a primitive element, the code consists of codewords $\boldsymbol{c}$, given by

$$(3.1) \qquad \boldsymbol{c} = \big(M(0), M(1), M(\alpha), \ldots, M(\alpha^{q-2})\big)$$

where M is a polynomial of degree $< \kappa$. Decoding then amounts to curve fitting:

given a received word $r = (r_0, r_1, \ldots, r_{q-1})$, find a polynomial M of degree $< \kappa$ such that the vectors

$$\begin{bmatrix} r_0 \\ r_1 \\ \vdots \\ r_{q-1} \end{bmatrix} \quad \text{and} \quad \begin{bmatrix} M(0) \\ M(1) \\ M(\alpha) \\ \vdots \\ M(\alpha^{q-2}) \end{bmatrix}$$

agree in $q - e$ places, with e minimal.

In Reed & Solomon's original paper [23] this was simply solved by repeated Lagrange interpolation followed by majority voting. The inefficiency of this method caused attention to shift to the generator polynomial approach ([11]), which then became common in textbooks. In this approach it was recognized that Reed & Solomon's original code is an extension of a $(q-1, \kappa)$ cyclic code C, consisting of codewords of the form

$$(3.2) \qquad c = \big(M(1), M(\alpha), \ldots, M(\alpha^{q-2}) \big)$$

where M is a polynomial of degree $< \kappa$. The code C is cyclic since each codeword can be shown to have zeros at $\alpha, \alpha^2, \ldots, \alpha^{q-1-\kappa}$. The code C was subsequently called a $(q - 1, \kappa)$ Reed-Solomon code and attention focused on the derivation of efficient decoding methods based on the syndrome sequence $(r(\alpha), \ldots, r(\alpha^{q-1-\kappa}))$, where r denotes a received polynomial. Methods that were derived include the Berlekamp-Massey algorithm [4, 20] and the euclidean algorithm [24]. Both these algorithms compute a shortest linear recurrence relation for the above syndrome sequence. Let us consider this in somewhat more detail.

3.1. The first key equation. Early work by Berlekamp recognized that the crucial step in syndrome decoding of Reed-Solomon/BCH codes amounts to solving the following (here $n := q - 1$ and $\nu := n - \kappa$):

Given a received polynomial r, compute syndromes

$$\begin{bmatrix} a_1 \\ a_2 \\ \vdots \\ a_\nu \end{bmatrix} = \begin{bmatrix} r(\alpha) \\ r(\alpha^2) \\ \vdots \\ r(\alpha^\nu) \end{bmatrix}$$

and solve

$$C(s)(a_1 s + a_2 s^2 + \cdots + a_\nu s^\nu) = P(s) \mod s^{\nu+1} \qquad \textbf{(key equation 1)}$$

such that $C(0) = 1$ and $\max \{ \deg C, \deg P\}$ is minimal.

We can readily recognize this as minimal rational interpolation at multiple points $s = 0$. Of course, another constraint is that C has distinct zeros whose reciprocals are code locations. (In the case of BCH codes they should in addition give rise to errors of value 1, see [14, 15]). For the sake of brevity, we will not repeat these constraints in the sequel.

The polynomial C gives error locations $\alpha_1, \ldots, \alpha_\ell$ through

$$(3.3) \qquad\qquad C(s) = \prod_{i=1}^{\ell} (1 - \alpha_i s).$$

Error values are obtained from

$$e_j = \frac{P(\alpha_j^{-1})}{\prod_{i \neq j}(1 - \alpha_i \alpha_j^{-1})} = -\frac{\alpha_j P(\alpha_j^{-1})}{C'(\alpha_j^{-1})} \qquad \text{for } j = 1, \ldots, \ell.$$

The above key equation was originally put forward by Berlekamp in [4], together with an iterative algorithm for solving it, that later resulted in the Berlekamp-Massey algorithm, incorporating ideas from [20]. Let us here reformulate this key equation plus accompanying constraints in a behavioral framework, recalling [16].

With the syndrome sequence $a_1, a_2, \ldots, a_\nu$ we first associate a trajectory $\boldsymbol{b} : \mathbb{Z}_+ \mapsto \mathbb{F}^2$ given by

$$\boldsymbol{b} := \left(\begin{bmatrix} a_\nu \\ 0 \end{bmatrix}, \ldots, \begin{bmatrix} a_1 \\ 0 \end{bmatrix}, \begin{bmatrix} 0 \\ 1 \end{bmatrix}, \begin{bmatrix} 0 \\ 0 \end{bmatrix}, \ldots \right).$$

The MPUM $\mathcal{B}^\star$ for $\boldsymbol{b}$ is given by

$$(3.4) \qquad \begin{bmatrix} 1 & -(a_1\sigma + a_2\sigma^2 + \cdots + a_\nu\sigma^\nu) \\ 0 & \sigma^{\nu+1} \end{bmatrix} \boldsymbol{w} = 0.$$

Solving Key equation 1 with the accompanying constraints is now equivalent to the construction of a minimal representation of $\mathcal{B}^\star$, i.e. a representation $R(\sigma)\boldsymbol{w} = 0$ where the 2×2 polynomial matrix R has minimal row degrees. The row of R that does not vanish at $s = 0$ constitutes the solution $[C \quad -P]$. Note that there always is such a row since in a minimal representation of $\mathcal{B}^\star$ not both rows can vanish at $s = 0$.

A slightly different key equation which involves the reversed syndrome sequence, is detailed in the next subsection.

3.2. The second key equation. It is easily verified that an alternative key equation that involves the reciprocal of the polynomial of (3.3), namely the polynomial $D(s) = \prod_{i=1}^{\ell}(s - \alpha_i)$, is given by

$$D(s)(a_\nu s + a_{\nu-1}s^2 + \cdots + a_1 s^\nu) = H(s) \mod s^{\nu+1}, \qquad \textbf{(key equation 2)}$$

such that $\deg H \leq \deg D$ and $\deg D$ is minimal. (In this paper we will call $D(s)$ the "error locator polynomial"; note though that this terminology is sometimes used for the polynomial of (3.3)). This amounts to minimal polynomial interpolation at $s = 0$ with the additional constraint that $\deg H \leq \deg D$. Error values are now obtained from

$$e_j = \frac{-H(\alpha_j)}{\alpha_j^{\nu+2} \prod_{i \neq j}(\alpha_i - \alpha_j)} = \frac{-H(\alpha_j)}{\alpha_j^{\nu+2} D'(\alpha_j)} \qquad \text{for } j = 1, \ldots, \ell.$$

In [24] a slightly different key equation (requiring $\deg H < \deg D$) was presented and it was then shown how the euclidean algorithm acts as a solution method. As we will see in subsection 4.3, for key equation 2 above, the euclidean algorithm can also be applied, namely to the polynomials $s^{\nu+1}$ and $a_\nu s + a_{\nu-1}s^2 + \cdots + a_1 s^\nu$ to produce a solution that satisfies the accompanying constraints.

Adopting a behavioral approach, we associate a trajectory $\bar{b} : \mathbb{Z}_+ \mapsto \mathbb{F}^2$ with the syndrome sequence $a_1, a_2, \ldots, a_\nu$ as follows

$$(3.5) \qquad \bar{b} := \left(\begin{bmatrix} a_1 \\ 0 \end{bmatrix}, \ldots, \begin{bmatrix} a_\nu \\ 0 \end{bmatrix}, \begin{bmatrix} 0 \\ 1 \end{bmatrix}, \begin{bmatrix} 0 \\ 0 \end{bmatrix}, \ldots \right).$$

This time the MPUM for $\bar{b}$ consists of all solutions of

$$(3.6) \qquad \begin{bmatrix} 1 & -(a_\nu \sigma + a_{\nu-1}\sigma^2 + \cdots + a_1 \sigma^\nu) \\ 0 & \sigma^{\nu+1} \end{bmatrix} w = 0.$$

key equation 2 with the accompanying constraints can now be solved by constructing a minimal representation of the above MPUM, i.e. a representation $R(\sigma)w = 0$ where the polynomial matrix R has minimal row degrees. The row $[D \quad -H]$ of R that has minimal row degree and obeys the constraint $\deg H \leq \deg D$ then constitutes the solution. An important observation is that there always is such a row because of row reducedness of R.

3.3. The third Key equation. Next, let $F(s) = F_0 + \cdots + F_\nu s^\nu$ be a polynomial with $F_\nu \neq 0$ and let $\beta \in \mathbb{F}$. define $\bar{F}(s) := (s - \beta)F(s)$ and define transformed syndromes by

$$\begin{bmatrix} \bar{a}_1 \\ \vdots \\ \bar{a}_N \end{bmatrix} = \begin{bmatrix} F_N & 0 & 0 \\ \vdots & \ddots & \vdots \\ F_1 & \cdots & F_N \end{bmatrix} \begin{bmatrix} a_1 \\ \vdots \\ a_N \end{bmatrix}.$$

Next, define a polynomial $\bar{A}$ by $\bar{A}(s) := (s - \beta)(\bar{a}_\nu + \bar{a}_{\nu-1}s + \cdots + \bar{a}_1 s^{\nu-1})$. Then it can be proven that solving key equation 2 under accompanying constraints is equivalent to solving the following equation

$$D(s)\bar{A}(s) = N(s) \mod \bar{F}(s) \qquad \text{(key equation 3)}$$

such that $\deg N \le \deg D$ and $\deg D$ is minimal. Indeed, with $D(s) = \prod_{i=1}^{\ell}(s - \alpha_i)$ and

$$\begin{bmatrix} a_1 \\ \vdots \\ a_\nu \end{bmatrix} = e_1 \begin{bmatrix} \alpha_1 \\ \vdots \\ \alpha_1^\nu \end{bmatrix} + \cdots + e_\ell \begin{bmatrix} \alpha_\ell \\ \vdots \\ \alpha_\ell^\nu \end{bmatrix},$$

we have that

$$\begin{aligned} D(s)\tilde{A}(s) &= D(s)(s - \beta)(\tilde{a}_\nu + \tilde{a}_{\nu-1}s + \cdots + \tilde{a}_1 s^{\nu-1}) \\ &= (s - \beta)\Sigma_{j=1}^\ell (\prod_{i \ne j}(s - \alpha_i)\alpha_j e_j (F(s) - F(\alpha_j))) \\ &= -(s - \beta)\Sigma_{j=1}^\ell (\prod_{i \ne j}(s - \alpha_i)\alpha_j e_j F(\alpha_j)) \mod \tilde{F}(s) \end{aligned}$$

has degree $\le \ell$.

A solution can be derived by applying the euclidean algorithm to the polynomials $\tilde{A}$ and $\tilde{F}$. In the case that the polynomial F is a Goppa polynomial, having no zeros at code locations, this accomplishes Goppa decoding, see [24]. Error values are calculated as

$$e_j = \frac{-N(\alpha_j)}{\alpha_j \tilde{F}(\alpha_j)D'(\alpha_j)} \qquad \text{for } j = 1, \ldots, \ell.$$

Let us now step away from the requirement that F has no zeros at code locations and consider the case $\beta = \alpha^\nu$ and

$$F(s) = (s - 1)(s - \alpha) \cdots (s - \alpha^{\nu-1}).$$

Define $\tilde{a}_1, \ldots, \tilde{a}_\nu$ and $\tilde{A}$ as before. Now key equation 3 can be reformulated as the interpolation requirement

$$(3.7) \qquad D(x_i)y_i = N(x_i) \quad \text{for } x_i = \alpha^i \quad (i = 0, \ldots, \nu),$$

where $y_i = \tilde{A}(x_i)$ for $i = 0, \ldots, \nu$. We will see in subsection 3.4 that this equation constitutes the key equation for our modified Welch-Berlekamp algorithm.

Let us now reformulate the problem of solving (3.7) such that $\deg D$ is minimal and $\deg N \le \deg D$ in a behavioral framework. With the data (x_i, y_i) $(i = 0, \ldots, \nu)$ we associate trajectories $b_i : \mathbb{Z}_+ \mapsto \mathbb{F}^2$ given by

$$b_i = \left(\begin{bmatrix} y_i \\ 1 \end{bmatrix}, \begin{bmatrix} y_i x_i \\ x_i \end{bmatrix}, \begin{bmatrix} y_i x_i^2 \\ x_i^2 \end{bmatrix}, \cdots \right).$$

Let L be the Lagrange interpolating polynomial that maps x_i to y_i $(i = 0, \ldots, \nu)$. Then the MPUM $\mathcal{B}$ for $b_0, \ldots, b_\nu$ is given by

$$\begin{bmatrix} 1 & -L(\sigma) \\ 0 & (\sigma - \alpha^\nu)F(\sigma) \end{bmatrix} w = 0.$$

Solving (3.7) under the above constraints is now equivalent to the construction of a minimal representation of $\mathcal{B}$, i.e. a representation $R(\sigma)w = 0$ where the 2×2 polynomial matrix R has minimal row degrees. The row $[D \quad -N]$ of R that has minimal row degree and obeys the constraint deg $N \leq$ deg D then constitutes the solution. Again, such a row always exists because of row reducedness of R.

3.4. The fourth key equation. Let us now move away from the generator approach and go back to Reed & Solomon's original curve fitting formulation outlined in the beginning of this section. This is readily reformulated as an interpolation problem involving the error locator polynomial D:

> given a received word $r = (r_0, r_1, \ldots, r_{n-1})$, find polynomials E and D such that
>
> $$D(x_i)r_i = E(x_i)$$
>
> for $x_i = \alpha^i$ and $i = 0, \ldots, n-1$ with
> - deg D minimal
> - $E(s) = D(s)M(s)$ with deg $M < \kappa$.

Note that the common factor M between D and E is of crucial importance. In a behavioral approach we again can associate trajectories

$$\left(\begin{bmatrix} r_i \\ 1 \end{bmatrix}, \begin{bmatrix} r_i x_i \\ x_i \end{bmatrix}, \begin{bmatrix} r_i x_i^2 \\ x_i^2 \end{bmatrix}, \ldots \right).$$

with the data (x_i, r_i) ($i = 0, \ldots, n-1$). It is however important to note that the above problem statement is **not** solved by constructing a minimal representation for the corresponding MPUM. The reason for this is that the degree requirements on D and E are not equivalent to a row degree requirement.

In order to reformulate the problem in such a way that we can use the theory of behavioral modeling, let us now re-encode the last κ entries of r, resulting in a codeword $c = (c_0, \ldots, c_{n-1})$ such that (recall that $\nu = n - \kappa$)

$$c_i = r_i \qquad \text{for } i = \nu, \ldots, n-1.$$

(Denoting the code's generator polynomial by g, we then have $c(s) = r(s) \mod g(s) + r(s)$, which is why this approach is sometimes called "remainder decoding", see [10]).

Since r and $\tilde{r} := r - c$ are necessarily disturbed by the same error pattern, we might just as well decode $\tilde{r}$. In the following we use the fact that the last κ entries of $\tilde{r}$ are zero to reformulate the problem statement at the beginning of this subsection as a case of behavioral minimal modeling.

For $i = 0, \ldots, n-1$, we first introduce trajectories

$$\tilde{b}_i := \left(\begin{bmatrix} \tilde{r}_i \\ 1 \end{bmatrix}, \begin{bmatrix} \tilde{r}_i x_i \\ x_i \end{bmatrix}, \begin{bmatrix} \tilde{r}_i x_i^2 \\ x_i^2 \end{bmatrix}, \ldots \right).$$

Next we define the polynomial G as $G(s) := (s - \alpha^{\nu+1}) \cdots (s - \alpha^{n-1})$ and we define $\mathcal{B}$ as the behavior spanned by trajectories

$$\bar{\bar{b}}_i := \begin{bmatrix} 1 & 0 \\ 0 & G(\sigma) \end{bmatrix} \bar{b}_i \qquad (i = 0, \ldots, n-1).$$

Then, since $G(x_i) = 0$, we have that $\bar{\bar{b}}_i = 0$ for $i = \nu+1, \ldots, n-1$, so that $\mathcal{B}$ is of dimension $\nu + 1$. Furthermore, for $i = 0, \ldots, \nu$, we have

$$\bar{\bar{b}}_i := \left(\begin{bmatrix} \tilde{r}_i \\ G(x_i) \end{bmatrix}, \begin{bmatrix} \tilde{r}_i x_i \\ x_i G(x_i) \end{bmatrix}, \begin{bmatrix} \tilde{r}_i x_i^2 \\ x_i^2 G(x_i) \end{bmatrix}, \cdots \right).$$

Let us now define, for $i = 0, \ldots, \nu$,

$$y_i := \frac{\tilde{r}_i}{G(x_i)} \quad \text{and} \quad \boldsymbol{b}_i := \frac{1}{G(x_i)} \bar{\bar{b}}_i.$$

It can then be shown that the interpolation data (x_i, y_i) coincide with the data from subsection 3.3, that were obtained from the conventional syndrome sequence, see also [10]. In particular, the trajectories $\boldsymbol{b}_0, \ldots, \boldsymbol{b}_\nu$ are as in subsection 3.3 (note that $y_\nu = 0$). As outlined in subsection 3.3, the decoding problem is now readily formulated as the problem of finding a minimal representation

$$R(\sigma)\boldsymbol{w} = 0$$

for $\mathcal{B} = \{\boldsymbol{b}_0, \boldsymbol{b}_1, \ldots, \boldsymbol{b}_\nu\}$. Indeed, the row $[D \quad -N]$ of the polynomial matrix R that has minimal row degree and for which $\deg N \leq \deg D$ gives rise to error locations and values as follows: the error locations $\alpha_1, \ldots, \alpha_\ell$ are the zeros of D, whereas the error values are obtained from the polynomial $\tilde{N} := NG$, namely by writing $\tilde{N} = \tilde{M}D$ with $\deg \tilde{M} < \kappa$ and calculating $e_j := \tilde{r}_j - \tilde{M}(\alpha_j)$ $(j = 1, \ldots, \ell)$. In other words, the decoding problem can be reformulated as follows:

> Given a received word $\boldsymbol{r} = (r_0, r_1, \ldots, r_{n-1})$, compute y_i as above for $i = 0, \ldots, \nu$. Now find polynomials D and N such that

$$D(x_i)y_i = N(x_i) \qquad \text{(key equation 4)}$$

> for $i = 0, \ldots, \nu$ with $\deg D$ minimal and $\deg N \leq \deg D$.

Observe that this is a problem of minimal polynomial interpolation at distinct points $x_0, \ldots, x_\nu$ in which common factors between D and N are of importance.

EXAMPLE 2. *Consider a $(7, 3)$ Reed-Solomon code over GF(8) given by (3.2). Suppose that $c = (\alpha^3, \alpha, 1, \alpha^3, 1, 0, 0)$ is sent but*

$$\boldsymbol{r} = (\alpha, \alpha, 1, \alpha^3, 1, 0, \alpha)$$

is received. Reencoding the last 3 *entries of* r*, we get*

$$\tilde{r} = r - (\alpha^4, \alpha^6, \alpha^3, \alpha^2, 1, 0, \alpha)$$
$$= (\alpha^2, \alpha^5, \alpha, \alpha^5, 0, 0, 0).$$

With $G(s) = (s - \alpha^5)(s - \alpha^6)$*, the new interpolation data are given by* $x_0 = 1, x_1 = \alpha, x_2 = \alpha^2, x_3 = \alpha^3, x_4 = \alpha^4$ *and*

$$y_0 = \frac{\alpha^2}{G(1)} = \alpha^3$$

$$y_1 = \frac{\alpha^5}{G(\alpha)} = \alpha$$

$$y_2 = \frac{\alpha}{G(\alpha^2)} = \alpha^5$$

$$y_3 = \frac{\alpha^5}{G(\alpha^3)} = \alpha^6$$

$$y_4 = 0.$$

As we will see in the next section, a solution of key equation 4 under accompanying constraints is obtained as $D(s) = (s - 1)(s - \alpha^6)$ *and* $N(s) = \alpha^4(s - 1)(s - \alpha^4)$*. From this we obtain* $\tilde{N}(s) = N(s)G(s) = \alpha^4(s - 1)(s - \alpha^4)(s - \alpha^5)(s - \alpha^6)$*, so that* $\tilde{M}(s) = \alpha^4(s - \alpha^4)(s - \alpha^5)$*. This leads to* $e_1 = \alpha^2 - \alpha^6 = 1$ *and* $e_2 = 0 - \alpha = \alpha$*.*

4. Algorithms. In this section we present algorithms for solving the various key equations of the previous section. Our starting point will be the corresponding behavioral formulations.

4.1. Iterative algorithm for key equation 1. Let us start with solving Key equation 1, i.e. construct a representation for $\mathcal{B}^*$, given by (3.4), of the form $R(\sigma)w = 0$, where R has minimal row degrees (i.e. is row reduced). An iterative algorithm would, at each step k, aim to construct a row reduced representation $R_k(\sigma)w = 0$ of the MPUM corresponding to $a_1, a_2, \ldots, a_k$ $(k = 1, \ldots, \nu)$. Such an algorithm can be designed by using the iterative modeling procedure outlined in Procedure 2.1. It has been shown in [16] how the Berlekamp-Massey algorithm ([4, 20, 6]) chooses the update matrix V_k such that R_k is not only row reduced but also has its second row vanish at $s = 0$, so that the solution can be read off from the first row of R_k. We here present a slightly modified version of the Berlekamp-Massey algorithm which involves a different initialization and replaces the condition $\ell_{k-1} \leq k/2$ by $\ell_{k-1} < k/2$. Neither of these modifications affects the outcome of the algorithm—they are here introduced to increase similarity with the algorithms of subsection 4.2 and subsection 4.4.

ALGORITHM 4.1.

For $k = 0, \ldots, \nu$ denote $C_k := \begin{bmatrix} 1 & 0 \end{bmatrix} R_k \begin{bmatrix} 1 \\ 0 \end{bmatrix}$. Initially define

$$R_0 := \begin{bmatrix} 1 & 0 \\ 0 & s \end{bmatrix}, \text{ and } \ell_0 := 0.$$

Proceed iteratively as follows for $k = 1, \ldots, \nu$. Define, after receiving $a_1, a_2, \ldots, a_k$, the number Δ_k as the coefficient of s^k in $C_{k-1}(s)(a_1 s + \cdots + a_k s^k)$.

Compute the matrix R_k and the integer ℓ_k as follows:

$$R_k := V_k R_{k-1},$$

where, if $\Delta_k \neq 0$ and $\ell_{k-1} < k/2$,

$$V_k(s) := \begin{bmatrix} 1 & -\Delta_k \\ s/\Delta_k & 0 \end{bmatrix}; \quad \ell_k := k - \ell_{k-1}.$$

and, if otherwise,

$$V_k(s) := \begin{bmatrix} 1 & -\Delta_k \\ 0 & s \end{bmatrix}; \quad \ell_k := \ell_{k-1}.$$

In the algorithm the integer ℓ_k denotes the first row degree of R_k. Denoting the second row degree of R_k by $\bar{\ell}_k$, we have $\ell_k + \bar{\ell}_k = k + 1$, so that the algorithm's condition $\ell_{k-1} < k/2$ is nothing else than $\ell_{k-1} < \bar{\ell}_{k-1}$, whereas $\ell_k := k - \ell_{k-1}$ is equivalent to $\ell_k := \bar{\ell}_{k-1}$.

It has been shown in [16] that the first row $[C \quad - P]$ of R_ν constitutes a solution of key equation 1 such that max $\{$ deg C, deg $P\}$ is minimal and $C(0) \neq 0$. In fact, the matrix R_ν contains all information necessary to derive solutions of any degree (not necessarily minimal). Indeed, we need only apply Lemma 2.1 to get a parametrization in terms of the entries of R_ν, see also [14, 15]. This could, in principle, be used for the purposes of list decoding. However, the efficiency of this method is low since checking for the validity of zeros of candidate error locator polynomials is a computationally intensive operation.

4.2. Iterative algorithm for key equation 2. The algorithm that we will present in this subsection, follows the main outline of Algorithm 4.1, except for the fact that this time the algorithm is tailored to produce, at each step k, a row reduced matrix

$$R_k(s) := \begin{bmatrix} D_k & -H_k \\ K_k & -Q_k \end{bmatrix}$$

for which deg $H_k \leq$ deg D_k. The resulting algorithm bears close resemblance to the Berlekamp-Massey algorithm but does not exhibit any jumps in the integer parameter ℓ_k.

ALGORITHM 4.2.

For $k = 0, \ldots, \nu$ denote $R_k := \begin{bmatrix} D_k & -H_k \\ K_k & -Q_k \end{bmatrix}$. Initially define

$$R_0 := \begin{bmatrix} 1 & 0 \\ 0 & s \end{bmatrix}, \text{ and } \ell_0 := 0.$$

Proceed iteratively as follows for $k = 1, \ldots, \nu$. Compute, after receiving $a_\nu, \ldots, a_{\nu-k+1}$, the numbers Δ_k and Γ_k as follows:

$\Delta_k :=$ *the coefficient of s^k in $D_{k-1}(s)(a_\nu s + \cdots + a_{\nu-k+1}s^k)$*

$\Gamma_k :=$ *the coefficient of s^k in $K_{k-1}(s)(a_\nu s + \cdots + a_{\nu-k+1}s^k) - Q_{k-1}(s)$.*

Compute the matrix R_k and the integer ℓ_k as follows:

$$R_k := V_k R_{k-1},$$

where, if $\Delta_k \neq 0$ and $(\ell_{k-1} < k/2 \quad$ or $\Gamma_k = 0)$,

$$V_k(s) := \begin{bmatrix} s & 0 \\ -\frac{\Gamma_k}{\Delta_k} & 1 \end{bmatrix}; \quad \ell_k := \ell_{k-1} + 1,$$

and, if otherwise,

$$V_k(s) := \begin{bmatrix} 1 & -\frac{\Delta_k}{\Gamma_k} \\ 0 & s \end{bmatrix}; \quad \ell_k := \ell_{k-1}.$$

The next theorem shows that the above algorithm produces a solution to key equation 2 under accompanying constraints.

THEOREM 4.1. *Let Algorithm 4.2 operate on a sequence $a_\nu, a_{\nu-1}, \ldots, a_1$. Then for $k = 1, \ldots, \nu$ the polynomials D_k and H_k satisfy*

$$D_k(s)(a_\nu s + a_{\nu-1} + \cdots + a_{\nu-k+1}s^k) = H_k(s) \mod s^{k+1}$$

with $\ell_k = \deg D_k$ minimal and $\deg H_k \leq \deg D_k$. In particular, D_ν and H_ν are a solution of key equation 2 with $\ell_\nu = \deg D_\nu$ minimal and $\deg H_\nu \leq \deg D_\nu$.

Proof. Let the trajectory $\bar{b}$ be defined as in (3.5). In the following we show that the above algorithm is a special instance of Procedure 2.1 applied to the data set $\{\sigma^\nu \bar{b}, \sigma^{\nu-1}\bar{b}, \cdots, \bar{b}\}$. At each step k ($k = 1, \ldots, \nu$), the error trajectory $e_k = R_{k-1}(\sigma)\sigma^{\nu-k}\bar{b}$ satisfies $\sigma e_k = \sigma R_{k-1}(\sigma)\sigma^{\nu-k}\bar{b} = R_{k-1}(\sigma)\sigma^{\nu-k+1}\bar{b} = 0$, so that e_k is of the form

$$e_k = \left(\begin{bmatrix} \tilde{\Delta}_k \\ \Gamma_k \end{bmatrix}, \begin{bmatrix} 0 \\ 0 \end{bmatrix}, \begin{bmatrix} 0 \\ 0 \end{bmatrix}, \cdots \right).$$

Here $\tilde{\Delta}_k$ is the coefficient of s^k in $D_{k-1}(s)(a_\nu s + \cdots + a_{\nu-k+1}s^k) - H_{k-1}(s)$ and Γ_k is as defined in Algorithm 4.2. Again, the integer ℓ_k in Algorithm 4.2

should be interpreted as the first row degree of R_k. Denoting the second row degree of R_k by $\bar{\ell}_k$, we have $\ell_k + \bar{\ell}_k = k + 1$. Because of this, it is now easily proven that $\ell_k \leq k$ for $k = 0, \ldots, \nu$, so that deg $H_k \leq k$. Because of this, we can reformulate the expression for $\tilde{\Delta}_k$ as Δ_k, defined in Algorithm 4.2. Note also that the condition $\ell_{k-1} < k/2$ translates into $\ell_{k-1} < \bar{\ell}_{k-1}$.

The update matrix V_k clearly represents the MPUM for $\{e_k\}$, so that for $k = 1, \ldots, \nu$, $R_k(\sigma)w = 0$ represents the MPUM for $\{\sigma^\nu \bar{b}, \ldots, \sigma^{\nu-k}\bar{b}\}$. It follows by induction that the choice of V_k's ensures that each R_k has minimal row degrees. Indeed, this holds trivially for $k = 0$ and the assumption that R_{k-1} has minimal row degrees implies that R_k has minimal row degrees because of the fact that V_k increases the degree of only one row of R_{k-1} by one.

It also follows by induction that deg $Q_k > $ deg K_k, so that, because of row reducedness of R_k,

$$\deg H_k \leq \deg D_k = \ell_k \qquad k = 0, \ldots, \nu.$$

Now any solution of key equation 2 of smaller degree that satisfies the accompanying constraints leads to a row reduced MPUM representation with smaller sum of row degrees. This contradicts the minimality of the row degrees of the matrix R_k and proves the theorem. □

4.3. The euclidean algorithm. A noniterative method to solve key equations 1–4 under accompanying constraints is the euclidean algorithm. Its usefulness for key equations 2 and 3 was first detailed in [24]. Let us illustrate the workings of this algorithm for key equation 4, compare [8]. The reasoning is immediately extendable to key equations 1–3 ([18]). Recall that we need to construct a minimal representation for the behavior $\mathcal{B}$ associated with the interpolation data (x_i, y_i), as specified in subsection 3.3. One way of constructing such a representation is to left multiply

$$\begin{bmatrix} 1 & -L(s) \\ 0 & -(s - x_\nu)F(s) \end{bmatrix}$$

by a unimodular polynomial matrix U such that the resulting matrix is row reduced. This can be achieved by applying the euclidean algorithm to $L(s)$ and $(s - x_\nu)F(s)$ as follows. Initializing $r_0(s) := (s - x_\nu)F(s)$ and $r_1(s) := L(s)$, we compute, for $k = 0, 1, \ldots$, polynomials r_k and q_{k+1} such that

$$r_k = q_{k+1}r_{k+1} + r_{k+2}$$

with deg $r_{k+2} < $ deg r_{k+1}. Next define $t_0(s) := 0$, $t_1(s) := 1$ and

$$t_k = q_{k+1}t_{k+1} + t_{k+2}.$$

Let k^* be the smallest integer such that $\deg r_{k^*} \leq \deg t_{k^*}$. Then, with

$$U = \begin{bmatrix} -q_{k^*-1} & 1 \\ 1 & 0 \end{bmatrix} \begin{bmatrix} -q_{k^*-2} & 1 \\ 1 & 0 \end{bmatrix} \cdots \begin{bmatrix} -q_1 & 1 \\ 1 & 0 \end{bmatrix}$$

it follows that

$$UR = \begin{bmatrix} t_{k^*} & -r_{k^*} \\ t_{k^*-1} & -r_{k^*-1} \end{bmatrix}$$

is row reduced. As a result, $D := t_{k^*}$ and $N := r_{k^*}$ solve key equation 4 under the accompanying constraints.

4.4. Iterative algorithm for key equation 4. In this subsection we present an algorithm that is a direct generalization of the algorithm of subsection 4.2. The algorithm produces a solution to key equation 4 that obeys the accompanying constraints. At each step it produces a row reduced matrix R_k whose first row contains the solution corresponding to the interpolation data processed so far.

ALGORITHM 4.3.

For $k = -1, \ldots, \nu - 1$ denote $R_k := \begin{bmatrix} D_k & -N_k \\ K_k & -Q_k \end{bmatrix}$. Initially define

$$R_{-1} := \begin{bmatrix} 1 & 0 \\ 0 & s - x_\nu \end{bmatrix}, \text{ and } \ell_{-1} := 0.$$

Proceed iteratively as follows for $k = 0, \ldots, \nu - 1$. Compute, after processing (x_i, y_i) for $i = 0, \ldots, k$, the numbers Δ_k and Γ_k as follows:

$$\Delta_k := D_{k-1}(x_k)y_k - N_{k-1}(x_k)$$
$$\Gamma_k := K_{k-1}(x_k)y_k - Q_{k-1}(x_k).$$

Compute the matrix R_k and the integer ℓ_k as follows:

$$R_k := V_k R_{k-1},$$

where, if $\Delta_k \neq 0$ and $(\ell_{k-1} < (k+1)/2 \quad$ or $\Gamma_k = 0)$,

$$V_k(s) := \begin{bmatrix} s - x_k & 0 \\ -\frac{\Gamma_k}{\Delta_k} & 1 \end{bmatrix}; \quad \ell_k := \ell_{k-1} + 1,$$

and, if otherwise,

$$V_k(s) := \begin{bmatrix} 1 & -\frac{\Delta_k}{\Gamma_k} \\ 0 & s - x_k \end{bmatrix}; \quad \ell_k := \ell_{k-1}.$$

The next theorem shows that the above algorithm produces a solution to key equation 4 under accompanying constraints.

THEOREM 4.2. *Let interpolation data* (x_i, y_i) $(i = 0, \ldots, \nu)$ *be given as in subsection 3.4. Let the above algorithm operate on these data. Then for* $k = 0, \ldots, \nu - 1$ *the polynomials* D_k *and* N_k *satisfy*

$$D_k(x_i)y_i = N_k(x_i)$$

for $i = \nu, 0, 1, \ldots, k$ *with* $\ell_k = \deg D_k$ *minimal and* $\deg N_k \leq \deg D_k$. *In particular,* $D_{\nu-1}$ *and* $N_{\nu-1}$ *are a solution of key equation 4 with* $\ell_{\nu-1} = \deg D_{\nu-1}$ *minimal and* $\deg N_{\nu-1} \leq \deg D_{\nu-1}$.

Proof. Define trajectories $\boldsymbol{b}_i : \mathbb{Z}_+ \mapsto \mathbb{F}^2$ as in subsection 3.4 $(i = 0, \ldots, \nu)$:

$$\boldsymbol{b}_i = \left(\begin{bmatrix} y_i \\ 1 \end{bmatrix}, \begin{bmatrix} y_i x_i \\ x_i \end{bmatrix}, \begin{bmatrix} y_i x_i^2 \\ x_i^2 \end{bmatrix}, \ldots \right).$$

In the following we show that Algorithm 4.3 is a special instance of Procedure 2.1 applied to the data set $\{\boldsymbol{b}_\nu, \boldsymbol{b}_0, \boldsymbol{b}_1, \ldots, \boldsymbol{b}_{\nu-1}\}$, starting with $R_{-2} := \begin{bmatrix} 1 & 0 \\ 0 & 1 \end{bmatrix}$. Then the error trajectory $\boldsymbol{e}_{-1} = \boldsymbol{b}_\nu$, whereas for $k = 0, \ldots, \nu - 1$, $\boldsymbol{e}_k = R_{k-1}(\sigma)\boldsymbol{b}_k$ is given by

$$\boldsymbol{e}_k = \left(\begin{bmatrix} \Delta_k \\ \Gamma_k \end{bmatrix}, \begin{bmatrix} \Delta_k x_k \\ \Gamma_k x_k \end{bmatrix}, \begin{bmatrix} \Delta_k x_k^2 \\ \Gamma_k x_k^2 \end{bmatrix}, \ldots \right).$$

Here Δ_k and Γ_k are given as in Algorithm 4.3. In particular, $\Delta_{-1} = 0$, $\Gamma_{-1} = 1$ and $R_{-1}(\sigma)\boldsymbol{w} = 0$ is a minimal representation of the MPUM of $\{\boldsymbol{e}_{-1}\}$. Further, the update matrices V_k, defined in Algorithm 4.3, represent the MPUM for $\{\boldsymbol{e}_k\}$ $(k = 0, \ldots, \nu - 1)$, so that $R_k(\sigma)\boldsymbol{w} = 0$ represents the MPUM for $\{\boldsymbol{b}_\nu, \boldsymbol{b}_0, \boldsymbol{b}_1, \ldots, \boldsymbol{b}_k\}$. The rest of the proof is completely analogous to the proof of Theorem 4.1. $\square$

5. Conclusions. In approaching the decoding of Reed-Solomon codes we adopted a behavioral view. In contrast to transfer function oriented approaches, such a view enables the study of finite dimensional sets of trajectories which turns out to be of crucial importance in this context. In particular, the simple iterative procedure of [30] for modeling such sets turns out to be a key ingredient. In the main results of this paper two algorithms were formulated as special instances of this modeling procedure. The first algorithm (Algorithm 4.2) solves the Berlekamp-Massey key equation, whereas the second algorithm (Algorithm 4.3) solves the Welch-Berlekamp key equation. Both algorithms are similar to the original Berlekamp-Massey algorithm in that they make use of the solution's degree ℓ at each step to determine which type of update matrix is used.

In addition, a simple derivation in terms of behavioral modeling was presented to convert Reed & Solomon's original decoding formulation into the Welch-Berlekamp key equation. It was also shown how the Welch-Berlekamp key equation can be derived from the Berlekamp-Massey key equation.

It is a topic of further research to extend this approach towards list decoding based on interpolation [12, 21, 25, 26] and to derive efficient algorithms for improved Reed-Solomon decoding.

REFERENCES

[1] ANTOULAS, A.C., *Recursive modeling of discrete-time time series*, in "Linear Algebra for Control Theory", P. Van Dooren and B. Wyman eds., Springer-Verlag, IMA, **62**, 1994, 1–20.

[2] ANTOULAS, A.C., J.A. BALL, J. KANG AND J.C. WILLEMS, *On the solution of the minimal rational interpolation problem*, Linear Alg. Appl., **137**, 1990, 511–573.

[3] ANTOULAS, A.C. AND J.C. WILLEMS, *A behavioral approach to linear exact modeling*, IEEE Trans. Aut. Control, **38**, 1993, 1776–1802.

[4] BERLEKAMP, E.R., *Algebraic Coding Theory*, New York, McGraw-Hill, 1968.

[5] BERLEKAMP, E.R., *Bounded distance + 1 soft-decision Reed-Solomon decoding*, IEEE Trans. Inform. Theory, **42**, 1996, 704–720.

[6] BLAHUT, R.E., *Theory and Practice of Error Control Codes*, Addison-Wesley, 1983.

[7] BLACKBURN, S.R., *A generalized rational interpolation problem and the solution of the Welch-Berlekamp algorithm*, Designs, Codes and Cryptography, **11**, 1997, 223–234.

[8] CHAMBERS, W.G., *Solution of Welch-Berlekamp key equation by Euclidean algorithm*, Electronics Letters, **29**, 1993, p. 1031.

[9] CHAMBERS, W.G., R.E. PEILE, K.Y. TSIE AND N. ZEIN, *Algorithm for solving the Welch-Berlekamp key-equation, with a simplified proof*, Electronics Letters, **29**, 1993, 1620–1621.

[10] DABIRI, D. AND I.F. BLAKE, *Fast parallel algorithms for decoding Reed-Solomon codes based on remainder polynomials*, IEEE Trans. Info. Theory, **41**, 1995, 873–885.

[11] GORENSTEIN, D. AND N. ZIERLER, *A class of error correcting codes in p^m symbols*, Journal of the Society of Industrial and Applied Mathematics, **9**, 1961, 207–214.

[12] GURUSWAMI, V. AND M. SUDAN, *Improved decoding of Reed-Solomon and algebraic-geometric codes*, IEEE Trans. Info. Theory, **45**(6), 1999, 1757–1768.

[13] KUIJPER, M., *First-Order Representations of Linear Systems*, Series on "Systems and Control: Foundations and Applications", Birkhäuser, Boston, 1994.

[14] KUIJPER, M., *Parametrizations and finite options*, in "The Mathematics of Systems and Control: from Intelligent Control to Behavioral Systems" (Festschrift on the occasion of the 60th birthday of Jan C. Willems), H.L. Trentelman, J.W. Polderman eds., ISBN 90-367-1112-6, 1999, 59–72.

[15] KUIJPER, M., *Further results on the use of a generalized B-M algorithm for BCH decoding beyond the designed error-correcting capability*, in "Proceedings of the 13th Symposium on Applied Algebra, Algebraic Algorithms, and Error-Correcting Codes (AAECC)", Hawaii, USA, 1999, 98–99.

[16] KUIJPER, M. AND J.C. WILLEMS, *On constructing a shortest linear recurrence relation*, IEEE Trans. Aut. Control, **42**, 1997, 1554–1558.

[17] KUIJPER, M., *An algorithm for constructing a minimal partial realization in the multivariable case*, Systems & Control Letters, **31**, 1997, 225–233.

[18] KUIJPER, M., *Partial realization and the Euclidean algorithm*, IEEE Trans. Aut. Control, **44**(5), 1999, 1013–1016.

[19] KUIJPER, M., *The Berlekamp-Massey algorithm, error-correction, keystreams and modeling*, in "Dynamical Systems, Control, Coding, Computer Vision: New trends, Interfaces, and Interplay", G. Picci, D.S. Gilliam (eds.), Birkhäuser's series "Progress in Systems and Control Theory", 1999, 321–341.

[20] MASSEY, J.L., *Shift-register synthesis and BCH decoding*, IEEE Trans. Info. Theory, **15**, 1969, 122–127.

[21] NIELSEN, R.R. AND T. HOEHOLDT, *Decoding Reed-Solomon codes beyond half the minimum distance*, Draft manuscript, 1999.

[22] POLDERMAN, J.W. AND J.C. WILLEMS, *Introduction to Mathematical Systems Theory—a behavioral approach*, Springer Verlag, New York, 1998.

[23] REED, I.S. AND G. SOLOMON, *Polynomial codes over certain finite fields*, SIAM Journal on Applied Mathematics, **8**, 1960, 300–304.

[24] SUGIYAMA, Y., M. KASAHARA, S. HIRASAWA AND T. NAMEKAWA, *A method for solving key equation for decoding Goppa codes*, Information and Control, **27**, 1975, 87–99.

[25] SUDAN, M., *Decoding of Reed-Solomon codes beyond the error correction bound*, Journal of Complexity, **13**, 1997, 180–193.

[26] SUDAN, M., *Decoding of Reed-Solomon codes beyond the error correction diameter*, in "Proceedings of the 35th Allerton Conference on Communication, Control and Computing", 1997, http://theory.lcs.mit.edu/~madhu/papers.html.

[27] WELCH, L.R. AND E.R. BERLEKAMP, *Error correction for algebraic block codes*, U.S. Patent 4 633 470, issued Dec. 30, 1986.

[28] WILLEMS, J.C., *From time series to linear system. Part I: Finite-dimensional linear time invariant systems*, Automatica, **22**, 1986, 561–580.

[29] WILLEMS, J.C., *From time series to linear system. Part II: Exact modeling*, Automatica, **22**, 1986, 675–694.

[30] WILLEMS, J.C., *Paradigms and puzzles in the theory of dynamical systems*, IEEE Trans. Aut. Control, **36**, 1991, 259–294.

[31] WILLEMS, J.C., *Fitting data sequences to linear systems*, in "Systems and Control in the Twenty-First Century", C.I. Byrnes, B.N. Datta, C.F. Martin and D.S. Gilliam, eds., Birkhäuser, Boston, 1997, 405–416.

[32] WOLOVICH, W.A., *Linear Multivariable Systems*, Springer Verlag, New York, 1974.

AN ALGEBRAIC DESCRIPTION OF ITERATIVE DECODING SCHEMES*

ELKE OFFER[†] AND EMINA SOLJANIN[‡]

Abstract. Several popular, suboptimal algorithms for bit decoding of binary block codes such as turbo decoding, threshold decoding, and message passing for LDPC, were developed almost as a *common sense* approach to decoding of some specially designed codes. After their introduction, these algorithms have been studied by mathematical tools pertinent more to computer science than the conventional algebraic coding theory. We give an algebraic description of the optimal and suboptimal bit decoders and of the optimal and suboptimal message passing. We explain exactly how suboptimal algorithms approximate the optimal, and show how good these approximations are in some special cases.

Key words. Iterative decoding, soft-output decoding, suboptimal decoding, bit-optimal decoding.

AMS(MOS) subject classifications. 94B05, 94B25, 94B35, 94B60.

1. Introduction. We propose an entirely new approach to the problem of iterative decoding, which is algebraic in nature and *derives* the well known suboptimal algorithms from the bit-optimal as a starting point. The approach gives new insights into the issues of iterative decoding from the algebraic coding theorist's point of view.

We are concerned with a binary block code $\mathcal{C}$ defined by its parity-check matrix $H = \{h_{ij}\}_{(n-k)\times n}$, *i.e.*, by the group generators $h_i = \{h_{ij}\}_{1\times n}$, $i \in \mathcal{I}$, of the dual code $\mathcal{C}'$, where $\mathcal{I}$ is used to denote the index set $\mathcal{I} = \{0, 1, \ldots, n - k - 1\}$. The channel is assumed to be memoryless.

The bit-optimal decoder computes the probability that the bit at position m is equal to $b \in \{0, 1\}$ given the received word r. To do this, it computes and adds probabilities (given r) of each codeword in the code $\mathcal{C}$ with b at position m. It can equivalently compute the Fourier Transform of the probabilities and add them up over the dual code $\mathcal{C}'$. This equivalence was first shown by Hartmann and Rudolph in [1], and Battail et al. in [2]. We point out that it is merely based on the Poisson summation formula. Working with the the dual code is in practice preferred when $\mathcal{C}'$ has fewer codewords than $\mathcal{C}$, *i.e.*, when dealing with high rate codes.

Since the suboptimal algorithms we are concerned with (turbo decoding, threshold decoding, and message passing for LDPC) operate with the generators h_i of the dual code, we derive an expression for optimal decod-

*This work was supported by the 1999 German-American Networking Research Grant given by the national academies of engineering of Germany and the USA.

[†]Institute for Communications Engineering, Munich University of Technology, D-80290 Munich, Germany; elke@lnt.e-technik.tu-muenchen.de.

[‡]Mathematical Sciences Research Center, Bell Labs, Lucent Technologies, Murray Hill, NJ 07974, USA; emina@lucent.com.

ing based on the dual code and then rewrite it so that it explicitly involves only h_i. Since sub optimal algorithms operate on subcodes of the dual code, we define L subcodes of C' by partitioning its set of generators h_i, $i \in \mathcal{I}$. Thus each $C'_l \subset C'$ is defined by its set of generators h_{i_l}, $i_l \in \mathcal{I}_l$, $\mathcal{I}_l \subset \mathcal{I}$, where the index sets $\mathcal{I}_l$ are disjoint and $\cup_{l=1}^{L} \mathcal{I}_l = \mathcal{I}$. We then expand our expression for optimal decoding to show the participation of each subcode C'_l. We use this expression to point out the approximation each of the suboptimal schemes performs.

2. Bit optimal decoding. Let block code C be a subgroup of the additive group of $\mathbb{F}_2^n$. The bit optimal decoding rule maximizes $P(c_m = b|r)$, the probability that c_m equals $b \in \{0, 1\}$ given the received word r. This probability, given by

$$P(c_m = b|r) = \sum_{c \in C, c_m = b} P(c|r),$$

can be expressed as

$$P(c_m = b|r) = \sum_{c \in C} \frac{P(c)}{p(r)} p(r|c) \delta_{b, c \cdot e_m}.$$

Therefore the log-likelihood of bit m over code C, L_m^C, is given by

$$(2.1) \qquad L_m^C = \log \frac{P(c_m = 0|r)}{P(c_m = 1|r)} = \log \frac{\sum_{c \in C} p(r|c) \delta_{0, c \cdot e_m}}{\sum_{c \in C} p(r|c) \delta_{1, c \cdot e_m}}$$

when the codewords are equiprobable. In the above equation e_m denotes the unit vector whose all components are 0 except the m-th which is 1, and $\delta_{\cdot, \cdot}$ denotes the Kronecker delta function.

Let $f : \mathbb{F}_2^n \to \mathbb{R}$ be a real valued function defined on $\mathbb{F}_2^n$ and $\hat{f}$ its DFT on the additive group $\mathbb{F}_2^n$, *i.e.*,

$$\hat{f}(v) = \sum_{u \in \mathbb{F}_2^n} f(u)(-1)^{u \cdot v}.$$

Then, by the Poisson summation formula

$$(2.2) \qquad \frac{1}{|C|} \sum_{c \in C} f(c) = \frac{1}{|\mathbb{F}_2^n|} \sum_{c' \in C'} \hat{f}(c'),$$

where C' is the dual of C in $\mathbb{F}_2^n$. For $f(u)$, $u \in \mathbb{F}_2^n$, defined as

$$f(u) = p(r|u) \delta_{b, u \cdot e_m}$$

and $p(\boldsymbol{r}|\boldsymbol{u}) = \prod_{j=1}^{n} p(r_j|u_j)$, it follows that $\hat{f}(v)$, $v \in \mathbb{F}_2^n$, is given by

$$
(2.3) \quad
\begin{aligned}
\hat{f}(v) &= \sum_{u \in \mathbb{F}_2^n} \delta_{b,u_m} \prod_{j=0}^{n-1} p(r_j|u_j)(-1)^{u_j v_j} \\
&= p(r_m|b)(-1)^{bv_m} \prod_{\substack{j=0 \\ j \neq m}}^{n-1} \left[p(r_j|0) + p(r_j|1)(-1)^{v_j} \right].
\end{aligned}
$$

By using (2.2) with $\hat{f}$ defined by (2.3), equation (2.1) can be expressed as

$$
(2.4) \quad L_m^{\mathcal{C}} = \log \frac{p(r_m|0)}{p(r_m|1)} + \log \frac{\displaystyle\sum_{i=0}^{2^{n-k}-1} \prod_{\substack{j=0 \\ j \neq m}}^{n-1} \left(\frac{p(r_j|0) - p(r_j|1)}{p(r_j|0) + p(r_j|1)} \right)^{c'_{ij}}}{\displaystyle\sum_{i=0}^{2^{n-k}-1} (-1)^{c'_{im}} \prod_{\substack{j=0 \\ j \neq m}}^{n-1} \left(\frac{p(r_j|0) - p(r_j|1)}{p(r_i|0) + p(r_j|1)} \right)^{c'_{ij}}},
$$

which is basically the result of Hartmann and Rudolph derived and presented somewhat differently in [1], and of Battail et al. in [2].

Note that expression (2.4) explicitly involves all codewords of the dual code $\mathcal{C}'$. Recall that suboptimal algorithms we are concerned with (turbo decoding, threshold decoding, and message passing for LDPC) operate with the parity-check matrix $H = \{h_{ij}\}_{(n-k)\times n}$, i.e., with the generators (independent parity checks) of the dual code, $\boldsymbol{h}_i = \{h_{ij}\}_{1\times n}$, $0 \leq i < n-k$. Because of that, we derive an expression for optimal decoding (2.4) which explicitly involves only $\boldsymbol{h}_i$. We denote

$$
(2.5) \quad L_j = \log \frac{p(r_j|0)}{p(r_j|1)} \text{ and } \lambda_j = \frac{p(r_j|0) - p(r_j|1)}{p(r_j|0) + p(r_j|1)} = \tanh(L_j/2),
$$

which when substituted in (2.4) gives

$$
(2.6) \quad L_m^{\mathcal{C}} = \log \frac{1 + \lambda_m}{1 - \lambda_m} + \log \frac{\displaystyle\sum_{i=0}^{2^{n-k}-1} \prod_{\substack{j=0 \\ j \neq m}}^{n-1} \lambda_j^{c'_{ij}}}{\displaystyle\sum_{i=0}^{2^{n-k}-1} (-1)^{c'_{im}} \prod_{\substack{j=0 \\ j \neq m}}^{n-1} \lambda_j^{c'_{ij}}}.
$$

We introduce *special multiplication* as a commutative and associative binary operation $\otimes$:

$$
\lambda_i \otimes \lambda_j = (\lambda_i \cdot \lambda_j)^{1-\delta_{i,j}} = \begin{cases} \lambda_i \cdot \lambda_j, & i \neq j, \\ 1, & i = j. \end{cases}
$$

For $a \in \mathbf{R}$, $a \otimes \lambda_j = a\lambda_j$. For $a_j, b_j \in \mathbb{F}_2$,

$$\left(\prod_j \lambda_j^{a_j}\right) \otimes \left(\prod_j \lambda_j^{b_j}\right) = \prod_j \lambda_j^{a_j \oplus b_j},$$

where $\oplus$ is the addition in $\mathbb{F}_2$

It is easy to see that equation (2.4) can be formally written as

$$(2.7) \qquad L_m^{\mathcal{C}} = \log \frac{1 + \lambda_m}{1 - \lambda_m} + \log \frac{\displaystyle\prod_{i \in \mathcal{I}}{}_{\otimes} \left[1 + \prod_{\substack{j=0 \\ j \neq m}}^{n-1} \lambda_j^{h_{ij}}\right]}{\displaystyle\prod_{i \in \mathcal{I}}{}_{\otimes} \left[1 + (-1)^{h_{im}} \prod_{\substack{j=0 \\ j \neq m}}^{n-1} \lambda_j^{h_{ij}}\right]},$$

where $\mathcal{I}$ is used to denote the index set $\mathcal{I} = \{0, 1, \ldots, n-k-1\}$. We now derive a formal companion expression for $\lambda_m^{\mathcal{C}} = \tanh(L_m^{\mathcal{C}}/2)$.

THEOREM 2.1. *Let λ_j defined by (2.5) be the j-th soft input to the bit optimal decoder for code $\mathcal{C}$. Then its soft output for bit m is given by*

$$(2.8) \qquad \lambda_m^{\mathcal{C}} = \tanh(L_m^{\mathcal{C}}/2) = \frac{\lambda_m \otimes \displaystyle\prod_{i \in \mathcal{I}}{}_{\otimes} \left[1 + \prod_{j=0}^{n-1} \lambda_j^{h_{ij}}\right]}{\displaystyle\prod_{i \in \mathcal{I}}{}_{\otimes} \left[1 + \prod_{j=0}^{n-1} \lambda_j^{h_{ij}}\right]}.$$

Proof. Let

$$A_m^+ = \prod_{i \in \mathcal{I}}{}_{\otimes} \left[1 + \prod_{\substack{j=0 \\ j \neq m}}^{n-1} \lambda_j^{h_{ij}}\right] \quad \text{and} \quad A_m^- = \prod_{i \in \mathcal{I}}{}_{\otimes} \left[1 + (-1)^{h_{im}} \prod_{\substack{j=0 \\ j \neq m}}^{n-1} \lambda_j^{h_{ij}}\right].$$

Then

$$L_m^{\mathcal{C}} = \log \frac{(1 + \lambda_m) A_m^+}{(1 - \lambda_m) A_m^-}.$$

By using the definition $\tanh(x/2) = (e^x - 1)/(e^x + 1)$, we get

$$\lambda_m^{\mathcal{C}} = \frac{(A_m^+ - A_m^-) + \lambda_m(A_m^+ + A_m^-)}{(A_m^+ + A_m^-) + \lambda_m(A_m^+ - A_m^-)} = \frac{\displaystyle\sum_{i=0}^{2^{n-k}-1} \lambda_m^{1-c_{im}'} \prod_{\substack{j=0 \\ j \neq m}}^{n-1} \lambda_j^{c_{ij}'}}{\displaystyle\sum_{i=0}^{2^{n-k}-1} \prod_{j=0}^{n-1} \lambda_j^{c_{ij}'}}.$$

Note that $(A_m^+ + A_m^-)/2$ represents the summation over all dual codewords with a '0' at position m, where $\lambda_m(A_m^+ - A_m^-)/2$ is the summation over all dual codewords with a '1' at position m. Therefore,

$$\lambda_m^{\mathcal{C}} = \frac{\lambda_m \otimes [(A_m^+ + A_m^-) + \lambda_m(A_m^+ - A_m^-)]}{(A_m^+ + A_m^-) + \lambda_m(A_m^+ - A_m^-)} = \frac{\lambda_m \otimes \displaystyle\sum_{i=0}^{2^{n-k}-1} \prod_{j=0}^{n-1} \lambda_j^{c'_{ij}}}{\displaystyle\sum_{i=0}^{2^{n-k}-1} \prod_{j=0}^{n-1} \lambda_j^{c'_{ij}}}.$$

which can formally be written as (2.8). □

3. Optimal message passing bit decoders. We define L subcodes of the dual code $\mathcal{C}'$ by partitioning its set of generators h_i, $i \in \mathcal{I}$. Thus each $\mathcal{C}'_l \subset \mathcal{C}'$ is defined by its set of generators h_{i_l}, $i_l \in \mathcal{I}_l$, $\mathcal{I}_l \subset \mathcal{I}$, where index sets $\mathcal{I}_l$ are disjoint and $\cup_{l=1}^{L} \mathcal{I}_l = \mathcal{I}$. Note that in this scenario equation (2.8) becomes

$$(3.1) \qquad \lambda_m^{\mathcal{C}} = \frac{\lambda_m \otimes \prod_{i \in \mathcal{I}_1 \otimes} \left[1 + \prod_{j=0}^{n-1} \lambda_j^{h_{ij}}\right] \otimes \cdots \otimes \prod_{i \in \mathcal{I}_L \otimes} \left[1 + \prod_{j=0}^{n-1} \lambda_j^{h_{ij}}\right]}{\prod_{i \in \mathcal{I}_1 \otimes} \left[1 + \prod_{j=0}^{n-1} \lambda_j^{h_{ij}}\right] \otimes \cdots \otimes \prod_{i \in \mathcal{I}_L \otimes} \left[1 + \prod_{j=0}^{n-1} \lambda_j^{h_{ij}}\right]}.$$

THEOREM 3.1. *Let $\mathcal{C}'^M_1$ be the subcode of $\mathcal{C}'$ defined by its index set $\mathcal{I}_1^M = \cup_{l=1}^{M} \mathcal{I}_l$. Then*

$$(3.2) \qquad \lambda_m^{\mathcal{C}_1^{M+1}} = \frac{\lambda_m^{\mathcal{C}_1^M} \otimes \prod_{i \in \mathcal{I}_{M+1} \otimes} \left[1 + \prod_{j=0}^{n-1} (\lambda_j^{\mathcal{C}_1^M})^{h_{ij}}\right]}{\prod_{i \in \mathcal{I}_{M+1} \otimes} \left[1 + \prod_{j=0}^{n-1} (\lambda_j^{\mathcal{C}_1^M})^{h_{ij}}\right]}.$$

Proof. Without loss of generality, we assume $M = 2$. Thus $\mathcal{C}_1^1 = \mathcal{C}_1$ and $\mathcal{C}_1^2 = \mathcal{C}$. From equation (2.8), for $\mathcal{C}_1$ and $\mathcal{C}_2$ we have

$$(3.3) \qquad \lambda_j^{\mathcal{C}_1} = \frac{\lambda_j \otimes \prod_{i \in \mathcal{I}_1 \otimes} \left[1 + \prod_{l=0}^{n-1} \lambda_l^{h_{il}}\right]}{\prod_{i \in \mathcal{I}_1 \otimes} \left[1 + \prod_{l=0}^{n-1} \lambda_l^{h_{il}}\right]} = \frac{\lambda_j \otimes A_{\mathcal{C}_1}}{A_{\mathcal{C}_1}}$$

and

$$(3.4) \qquad \lambda_m^{\mathcal{C}_2} = \frac{\lambda_m \otimes \prod_{\substack{\otimes \\ i \in \mathcal{I}_2}} \left[1 + \prod_{j=0}^{n-1} \lambda_j^{h_{ij}}\right]}{\prod_{\substack{\otimes \\ i \in \mathcal{I}_2}} \left[1 + \prod_{j=0}^{n-1} \lambda_j^{h_{ij}}\right]}.$$

where $A_{\mathcal{C}_1} = \prod_{\otimes i \in \mathcal{I}_1} \left[1 + \prod_{l=0}^{n-1} \lambda_l^{h_{il}}\right]$. We replace (update) each λ_j in equation (3.4) by $\lambda_j^{\mathcal{C}_1}$ given by (3.3), and change the real product in equation (3.4) by the special multiplication product, thus obtaining

$$(3.5) \qquad [\lambda_m^{\mathcal{C}_2}]_2 = \frac{\frac{\lambda_m \otimes A_{\mathcal{C}_1}}{A_{\mathcal{C}_1}} \otimes \prod_{\substack{\otimes \\ i \in \mathcal{I}_2}} \left[1 + \prod_{j=0}^{n-1} \left(\frac{\lambda_j \otimes A_{\mathcal{C}_1}}{A_{\mathcal{C}_1}}\right)^{h_{ij}}\right]}{\prod_{\substack{\otimes \\ i \in \mathcal{I}_2}} \left[1 + \prod_{j=0}^{n-1} \left(\frac{\lambda_j \otimes A_{\mathcal{C}_1}}{A_{\mathcal{C}_1}}\right)^{h_{ij}}\right]},$$

where $[\lambda_m^{\mathcal{C}_2}]_2$ denotes the result for bit m of the *second iteration* in decoding of $\mathcal{C}_2$. We can further transform the above equation by taking into account that

$$(3.6) \qquad \underbrace{A_{\mathcal{C}_1} \otimes A_{\mathcal{C}_1} \otimes \cdots \otimes A_{\mathcal{C}_1}}_{\nu} = 2^{(\nu-1)\cdot|\mathcal{I}_1|} \cdot A_{\mathcal{C}_1}$$

and obtain

$$(3.7) \qquad [\lambda_m^{\mathcal{C}_2}]_2 = \frac{\lambda_m \otimes \prod_{\substack{\otimes \\ i \in \mathcal{I}_1}} \left[1 + \prod_{j=0}^{n-1} \lambda_j^{h_{ij}}\right] \otimes \prod_{\substack{\otimes \\ i \in \mathcal{I}_2}} \left[1 + \prod_{j=0}^{n-1} \lambda_j^{h_{ij}}\right]}{\prod_{\substack{\otimes \\ i \in \mathcal{I}_1}} \left[1 + \prod_{j=0}^{n-1} \lambda_j^{h_{ij}}\right] \otimes \prod_{\substack{\otimes \\ i \in \mathcal{I}_2}} \left[1 + \prod_{j=0}^{n-1} \lambda_j^{h_{ij}}\right]}.$$

which is precisely (3.1) for $M = 2$. Therefore $\lambda_{\mathcal{C}} = [\lambda_m^{\mathcal{C}_2}]_2$, and the optimal solution is reached in 2 steps.   □

With optimal message passing in the L subcode case, we obtain $\lambda_m^{\mathcal{C}}$ in L steps. We start with computing $\lambda_m^{\mathcal{C}_1}$ by using (3.3), and continue with computing $\lambda_m^{\mathcal{C}_1^{M+1}}$ by optimal message passing (3.2) for $1 \leq M \leq L$.

REMARK 3.1. *The optimal message passing (3.5) from decoder of $\mathcal{C}_1$ to decoder of $\mathcal{C}_2$ doesn't simplify the computations required for the optimal solution (3.7). It only shows how the information from decoder of $\mathcal{C}_1$ should be used by decoder of $\mathcal{C}_2$ to obtain the optimal solution, which we need as a reference for comparison with suboptimal algorithms.*

REMARK 3.2. *Once λ_m^C is computed, further iterations will leave it unchanged. Namely, λ_j^C used as new information to compute $[\lambda_m^{C_1}]_2$ will not change its value, starting with C_1 and so on. This is again easily shown by using equalities of type (3.6).*

In what follows, we consider several popular, suboptimal bit detection algorithms. For each algorithm, we give an expression of (2.4) or (2.8) which is the most suitable for the analysis of its relation to the optimal algorithm.

4. Turbo codes. We consider a turbo coding scheme with two component codes C_1 and C_2 as introduced in [3], [4]. The bit-optimal decoder computes the log-likelihood for bit m by (2.7), which can be expressed as

$$
L_m^C = \log \frac{1 + \lambda_m}{1 - \lambda_m} +
$$

$$
\log \frac{\displaystyle\prod_{i \in \mathcal{I}_1}{}_{\otimes}\Big[1 + \prod_{\substack{j=0 \\ j \neq m}}^{n-1} \lambda_j^{h_{ij}}\Big] \otimes \prod_{i, \mathcal{I}_2}{}_{\otimes}\Big[1 + \prod_{\substack{j=0 \\ j \neq m}}^{n-1} \lambda_j^{h_{ij}}\Big]}{\displaystyle\prod_{i, \in \mathcal{I}_1}{}_{\otimes}\Big[1 + (-1)^{h_{im}}\prod_{\substack{j=0 \\ j \neq m}}^{n-1} \lambda_j^{h_{ij}}\Big] \otimes \prod_{i \in \mathcal{I}_2}{}_{\otimes}\Big[1 + (-1)^{h_{im}}\prod_{\substack{j=0 \\ j \neq m}}^{n-1} \lambda_j^{h_{ij}}\Big]}.
$$

The corresponding soft output λ_m^C can be written as

$$
(4.1) \qquad \lambda_m^C = \frac{\displaystyle\lambda_m \otimes \prod_{i \in \mathcal{I}_1}{}_{\otimes}\Big[1 + \prod_{j=0}^{n-1} \lambda_j^{h_{ij}}\Big] \otimes \prod_{i \in \mathcal{I}_2}{}_{\otimes}\Big[1 + \prod_{j=0}^{n-1} \lambda_j^{h_{ij}}\Big]}{\displaystyle\prod_{i \in \mathcal{I}_1}{}_{\otimes}\Big[1 + \prod_{j=0}^{n-1} \lambda_j^{h_{ij}}\Big] \otimes \prod_{i \in \mathcal{I}_2}{}_{\otimes}\Big[1 + \prod_{j=0}^{n-1} \lambda_j^{h_{ij}}\Big]}.
$$

As described in the previous section, the optimal decoding (4.1) can be performed sequentially. First, the optimal decoder of C_1 computes soft information (3.3). Then, decoder of C_2 uses the soft information in the optimal message passing equation (3.5) to obtain the optimal solution (4.1).

Turbo decoding also starts with the optimal decoding of C_1 by computing the *first iteration* log-likelihood for bit m as

$$
(4.2) \qquad [L_m^{C_1}]_1 = \log \frac{1 + \lambda_m}{1 - \lambda_m} + \underbrace{\log \frac{\displaystyle\prod_{i \in \mathcal{I}_1}{}_{\otimes}\Big[1 + \prod_{\substack{j=0 \\ j \neq m}}^{n-1} \lambda_j^{h_{ij}}\Big]}{\displaystyle\prod_{i \in \mathcal{I}_1}{}_{\otimes}\Big[1 + (-1)^{h_{im}}\prod_{\substack{j=0 \\ j \neq m}}^{n-1} \lambda_j^{h_{ij}}\Big]}}_{[L_{m,e}^{C_1}]_1}
$$

with corresponding

$$[\lambda_m^{\mathcal{C}_1}]_1 = \tanh([L_m^{\mathcal{C}_1}]_1/2) = \frac{\lambda_m \otimes \prod_{\substack{i \in \mathcal{I}_1}}^{\otimes} \left[1 + \prod_{j=0}^{n-1} \lambda_j^{h_{ij}}\right]}{\prod_{\substack{i \in \mathcal{I}_1}}^{\otimes} \left[1 + \prod_{j=0}^{n-1} \lambda_j^{h_{ij}}\right]}.$$

The second term in equation (4.2) is known as (first iteration) *extrinsic* information of decoder $\mathcal{C}_1$ for bit m. It is denoted by $[L_{m,e}^{\mathcal{C}_1}]_1$ with corresponding $[\lambda_{m,e}^{\mathcal{C}_1}]_1 = \tanh([L_{m,e}^{\mathcal{C}_1}]_1/2)$, and transfered to the decoder of $\mathcal{C}_2$ for all bit positions m checked by both codes. Note that

$$[\lambda_m^{\mathcal{C}_1}]_1 = \frac{\lambda_m + [\lambda_{m,e}^{\mathcal{C}_1}]_1}{1 + \lambda_m [\lambda_{m,e}^{\mathcal{C}_1}]_1}.$$

Turbo decoding continues with the decoding of $\mathcal{C}_2$ by computing

$$(4.3) \qquad [L_m^{\mathcal{C}_2}]_2 = \log\frac{1+\lambda_m}{1-\lambda_m} + \log \underbrace{\frac{\prod_{\substack{i \in \mathcal{I}_2}}^{\otimes}\left[1 + \prod_{\substack{j=0 \\ j \neq m}}^{n-1}[\lambda_j^{\mathcal{C}_1}]_1^{h_{ij}}\right]}{\prod_{\substack{i \in \mathcal{I}_2}}^{\otimes}\left[1 + (-1)^{h_{im}}\prod_{\substack{j=0 \\ j \neq m}}^{n-1}[\lambda_j^{\mathcal{C}_1}]_1^{h_{ij}}\right]}}_{[L_{m,e}^{\mathcal{C}_2}]_2},$$

where $[\lambda_j^{\mathcal{C}_1}]_1 = \lambda_j$ if position j is checked only by $\mathcal{C}_2$. Now, the extrinsic information $[L_{m,e}^{\mathcal{C}_2}]_2$ is transfered to the decoder of $\mathcal{C}_1$ for the bit positions m checked by both codes, and the process continues iteratively, as illustrated in Fig. 1. In the iteration ν, when component decoder $\mathcal{C}_1$ is active, we have the following scenario: The input information are the channel values λ_r for the bit positions r checked only by code $\mathcal{C}_1$ and channel values λ_m together with the extrinsic information $[\lambda_{m,e}^{\mathcal{C}_2}]_{\nu-1}$ of code $\mathcal{C}_2$ from the previous iteration for bit positions m checked by both component codes. As described above for the two first iterations, the soft output of decoder $\mathcal{C}_1$ is calculated as

$$[L_m^{\mathcal{C}_1}]_\nu = \log\frac{1+\lambda_m}{1-\lambda_m} + \log \underbrace{\frac{\prod_{\substack{i \in \mathcal{I}_1}}^{\otimes}\left[1 + \prod_{\substack{j=0 \\ j \neq m}}^{n-1}[\lambda_j^{\mathcal{C}_2}]_{\nu-1}^{h_{ij}}\right]}{\prod_{\substack{i \in \mathcal{I}_1}}^{\otimes}\left[1 + (-1)^{h_{im}}\prod_{\substack{j=0 \\ j \neq m}}^{n-1}[\lambda_j^{\mathcal{C}_2}]_{\nu-1}^{h_{ij}}\right]}}_{[L_{m,e}^{\mathcal{C}_1}]_\nu}$$

from the channel-detector

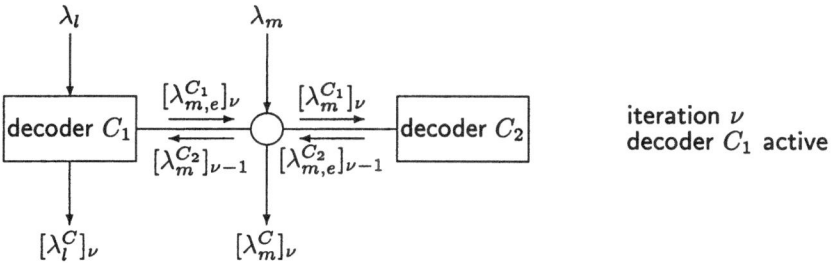

from the channel-detector

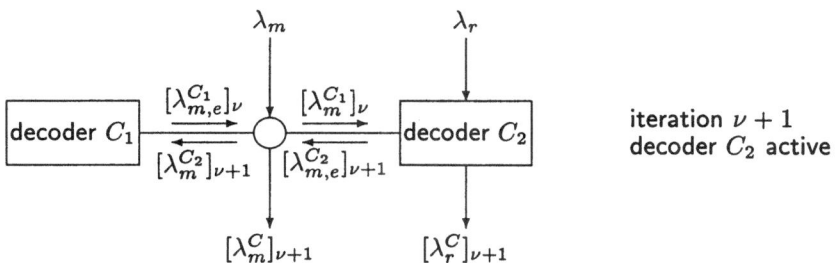

FIG. 1. *Message passing procedure and the corresponding soft-output in a serial turbo decoding procedure. Here the symbol positions checked by both component codes C_1 and C_2 are denoted by m. The symbol positions involved only in the parity check equations of code C_1, respectively C_2, are denoted by r, respectively l.*

giving

$$[\lambda_m^{C_1}]_\nu = \frac{\lambda_m \otimes \prod_{\substack{i \in \mathcal{I}_1}}^{\otimes} \left[1 + \lambda_m^{h_{im}} \prod_{\substack{j=1 \\ j \neq m}}^{n-1} [\lambda_j^{C_2}]_{\nu-1}^{h_{ij}}\right]}{\prod_{\substack{i \in \mathcal{I}_1}}^{\otimes} \left[1 + \lambda_m^{h_{im}} \prod_{\substack{j=1 \\ j \neq m}}^{n-1} [\lambda_j^{C_2}]_{\nu-1}^{h_{ij}}\right]}.$$

The overall soft-output of the decoding scheme after iteration ν is computed by

$$\begin{aligned}
[L_m^C]_\nu &= \log \frac{1 + \lambda_m}{1 - \lambda_m} + [L_{m,e}^{C_2}]_{\nu-1} + [L_{m,e}^{C_1}]_\nu \\
&= [L_m^{C_2}]_{\nu-1} + [L_{m,e}^{C_1}]_\nu \\
&= [L_{m,e}^{C_2}]_{\nu-1} + [L_m^{C_1}]_\nu.
\end{aligned}$$

Therefore the corresponding $[\lambda_m^{\mathcal{C}}]_\nu$ can be expressed as

$$[\lambda_m^{\mathcal{C}}]_\nu = \frac{[\lambda_m^{\mathcal{C}_2}]_{\nu-1} \otimes \prod\limits_{i \in \mathcal{I}_1 \otimes} \left[1 + \prod\limits_{j=0}^{n-1} [\lambda_j^{\mathcal{C}_2}]_{\nu-1}^{h_{ij}}\right]}{\prod\limits_{i \in \mathcal{I}_1 \otimes} \left[1 + \prod\limits_{j=0}^{n-1} [\lambda_j^{\mathcal{C}_2}]_{\nu-1}^{h_{ij}}\right]}$$

and, similarly,

$$[\lambda_m^{\mathcal{C}}]_{\nu+1} = \frac{[\lambda_m^{\mathcal{C}_1}]_\nu \otimes \prod\limits_{i \in \mathcal{I}_2 \otimes} \left[1 + \prod\limits_{j=0}^{n-1} [\lambda_j^{\mathcal{C}_1}]_{\nu}^{h_{ij}}\right]}{\prod\limits_{i \in \mathcal{I}_2 \otimes} \left[1 + \prod\limits_{j=0}^{n-1} [\lambda_j^{\mathcal{C}_1}]_{\nu}^{h_{ij}}\right]}.$$

We have thus proved the following:

THEOREM 4.1. *Two-iterations turbo decoding algorithm of [3], [4] approximates the optimal by substituting the term*

$$(4.4) \qquad \prod\limits_{i \in \mathcal{I}_2 \otimes} \left[1 + \prod\limits_{j=0}^{n-1} \left(\lambda_j^{\mathcal{C}_1}\right)^{h_{ij}}\right]$$

for

$$(4.5) \qquad \prod\limits_{i \in \mathcal{I}_2 \otimes} \left[1 + \prod\limits_{j=0}^{n-1 \otimes} \left(\lambda_j^{\mathcal{C}_1}\right)^{h_{ij}}\right]$$

in the optimal message passing equation (3.5). An example is given in the Appendix.

Expression (4.5) evaluates to

$$(4.6) \qquad a \sum\limits_{i=0}^{2^{n-k}-1} \prod\limits_{j=0}^{n-1} \lambda_j^{c'_{ij}},$$

whereas expression (4.4) evaluates to

$$(4.7) \qquad \sum\limits_{i=0}^{2^{n-k}-1} a_i \prod\limits_{j=0}^{n-1} \lambda_j^{c'_{ij}}, \quad a_i \geq 0.$$

Both expressions are sums over all dual codewords c' of the turbo code $\mathcal{C}$ of the same weighted terms. The weights in (4.5) are constant; in (4.4) they are in general not and each is equal to a_i. These factors a_i originate from the difference between the special multiplication $\otimes$ and the real number multiplication. The convergence of the suboptimal decoder is thus related to the asymptotic properties of the factors a_i.

5. Low-density parity-check codes. Equation (2.7) can be expressed as

$$
(5.1) \qquad L_m^{\mathcal{C}} = L_m + \log \frac{\displaystyle\prod_{\substack{i=0 \\ h_{im}=0}}^{n-k-1} \left[1 + \prod_{j=0}^{n-1} \lambda_j^{h_{ij}}\right] \otimes \prod_{\substack{i=0 \\ h_{im}=1}}^{n-k-1} \left[1 + \prod_{\substack{j=0 \\ j\neq m}}^{n-1} \lambda_j^{h_{ij}}\right]}{\displaystyle\prod_{\substack{i=0 \\ h_{im}=0}}^{n-k-1} \left[1 + \prod_{j=0}^{n-1} \lambda_j^{h_{ij}}\right] \otimes \prod_{\substack{i=0 \\ h_{im}=1}}^{n-k-1} \left[1 - \prod_{\substack{j=0 \\ j\neq m}}^{n-1} \lambda_j^{h_{ij}}\right]}.
$$

Let $\mathcal{C}$ be an LDPC code such that, for all pairs of bits m and l, there is at most one index $i \in \mathcal{I}$ for which $h_{im} = 1$ and $h_{il} = 1$. In the bipartite graph terminology, we say that, for any two variable nodes m and l, there is at most one common check node $c(m,l)$ *i.e.*, there are no loops of length 4. We first illustrate Gallager's belief propagation algorithm [5] in our terminology on a special one-step case, and then consider a general case in terms of the results presented earlier in the paper.

The optimal soft output for bit m based solely on the channel information L_m and the knowledge of the parity check equations in which it participates is given by

$$
(5.2) \qquad L_m + \log \frac{\displaystyle\prod_{\substack{i=0 \\ h_{im}=1}}^{n-k-1} \left(1 + \prod_{\substack{j=0 \\ j\neq m}}^{n-1} \lambda_j^{h_{ij}}\right)}{\displaystyle\prod_{\substack{i=0 \\ h_{im}=1}}^{n-k-1} \left(1 - \prod_{\substack{j=0 \\ j\neq m}}^{n-1} \lambda_j^{h_{ij}}\right)}.
$$

Because of the low-density assumption the special multiplication above is equal to the real multiplication. Let λ_{ml} denote the message passed from variable node l to variable node m through their unique common check node $c(m,l)$, *i.e.*,

$$
\lambda_{ml} = \tanh(L_{ml}/2),
$$

where (as proposed by Gallager)

$$
L_{ml} = \log \frac{1+\lambda_l}{1-\lambda_l} + \log \frac{\displaystyle\prod_{\substack{i=0 \\ h_{il}=1 \\ i\neq c(m,l)}}^{n-k-1} \left(1 + \prod_{\substack{j=0 \\ j\neq l}}^{n-1} \lambda_j^{h_{ij}}\right)}{\displaystyle\prod_{\substack{i=0 \\ h_{il}=1 \\ i\neq c(m,l)}}^{n-k-1} \left(1 - \prod_{\substack{j=0 \\ j\neq l}}^{n-1} \lambda_j^{h_{ij}}\right)}.
$$

It can be easily shown (in the manner of proof of Theorem 2.1) by using the low-density assumption that

$$
\lambda_{ml} = \frac{\lambda_l \otimes \displaystyle\prod_{\substack{i=0 \\ h_{im}=0, h_{il}=1}}^{n-k-1} \left[1 + \prod_{j=0}^{n-1} \lambda_j^{h_{ij}}\right]}{\displaystyle\prod_{\substack{i=0 \\ h_{im}=0, h_{il}=1}}^{n-k-1} \left[1 + \prod_{j=0}^{n-1} \lambda_j^{h_{ij}}\right]} .
$$

Upon receiving this message, the update for bit m is computed by (5.2) with λ_{ml} in the place of λ_l, giving the following

$$
L_m + \log \frac{\displaystyle\prod_{\substack{i=0 \\ h_{im}=0, h_{il}=1}}^{n-k-1} \left[1 + \prod_{j=0}^{n-1} \lambda_j^{h_{ij}}\right] \otimes \prod_{\substack{i=0 \\ h_{im}=1}}^{n-k-1} \left[1 + \prod_{\substack{j=0 \\ j \neq m}}^{n-1} \lambda_j^{h_{ij}}\right]}{\displaystyle\prod_{\substack{i=0 \\ h_{im}=0, h_{il}=1}}^{n-k-1} \left[1 + \prod_{j=0}^{n-1} \lambda_j^{h_{ij}}\right] \otimes \prod_{\substack{i=0 \\ h_{im}=1}}^{n-k-1} \left[1 - \prod_{\substack{j=0 \\ j \neq m}}^{n-1} \lambda_j^{h_{ij}}\right]} .
$$

This brings us a step *closer* to the expression for optimal decoding of bit m given by (5.1).

We now consider a general case. Let $\mathcal{N}(m)$ denote the set of *neighbors* of m, namely

$$
\mathcal{N}(m) = \{l | 0 \leq l \leq n-1 \text{ and } \exists i \in \mathcal{I} \text{ such that } h_{im} = 1 \text{ and } h_{il} = 1\}.
$$

For our low-density assumption, we shall also require that no two members of $\mathcal{N}(m)$ are neighbors themselves, *i.e.*, there are no loops of length 6.

In connection with decoding of bit m, we look at the set of subcodes of $\mathcal{C}$ whose index sets are defined as follows:

$$
\mathcal{I}_m = \{i \in \mathcal{I} | h_{im} = 1\}
$$
$$
\mathcal{I}_{ml} = \{i \in \mathcal{I} | h_{il} = 1 \text{ and } h_{im} = 0\}
$$
$$
\mathcal{I}_{\mathcal{N}(m),m} = \left(\cup_{l \in \mathcal{N}(m)} \mathcal{I}_{ml}\right) \cup \mathcal{I}_m .
$$

Note that these index sets are disjoint.

THEOREM 5.1. *Let $\mathcal{C}$, λ_m, $\mathcal{I}_m$, $\mathcal{I}_{ml}$, and $\mathcal{I}_{\mathcal{N}(m),m}$ be as previously defined. Then Gallager's belief propagation algorithm [5] gives optimal decoding of bit m over the code $\mathcal{C}_{\mathcal{N}(m),m}$.*

Proof. By Theorem 2.1, the optimal decoder of $\mathcal{C}_{ml}$ outputs the following soft information for bit l:

$$\lambda_l^{C_{ml}} = \frac{\lambda_l \otimes \prod_{\substack{i \in \mathcal{I}_{ml}}}^{\otimes} \left[1 + \prod_{j=0}^{n-1} \lambda_j^{h_{ij}}\right]}{\prod_{\substack{i \in \mathcal{I}_{ml}}}^{\otimes} \left[1 + \prod_{j=0}^{n-1} \lambda_j^{h_{ij}}\right]}.$$

Gallager's algorithm replaces the special multiplication product $\prod_{\otimes}$ by the real product but, because of the low-density assumption (no loops of length 4), the result is the same.

By Theorem 3.1, the optimal soft output for $C_{\mathcal{N}(m),m}$ can be computed by the optimal message passing (3.2) as follows:

$$\lambda_m^{C_{\mathcal{N}(m),m}} = [\lambda_m^{C_m}]_2 = \frac{\dfrac{\lambda_m \otimes A_{C_{ml}}}{A_{C_{ml}}} \otimes \prod_{\substack{i \in \mathcal{I}_m}}^{\otimes} \left[1 + \prod_{\substack{l \in \mathcal{N}(m)}}^{\otimes} \left(\dfrac{\lambda_j \otimes A_{C_{ml}}}{A_{C_{ml}}}\right)^{h_{ij}}\right]}{\prod_{\substack{i \in \mathcal{I}_m}}^{\otimes} \left[1 + \prod_{\substack{l \in \mathcal{N}(m)}}^{\otimes} \left(\dfrac{\lambda_j \otimes A_{C_{ml}}}{A_{C_{ml}}}\right)^{h_{ij}}\right]},$$

Gallager's algorithm replaces both special multiplication products $\prod_{\otimes}$ by the real products but, because of the low-density assumption (no loops of length 4 for the *outer* product and no loops of length 6 for the *inner* product), the result is the same. □

Note that by the above proof, we showed that the LDPC codes are a special case of turbo codes for which the participating subcodes are also optimally decoded. Only the optimal decoding of subcodes of an LDPC code is performed by message passing (and can be because of the low-density assumption), rather than by a straight-forward optimal algorithm such as the one in [4].

6. Threshold decoding. Equation (2.7) can be expressed as

$$L_m^C = \log \frac{1 + \lambda_m}{1 - \lambda_m} +$$

$$\log \frac{\prod_{\substack{i=0 \\ h_{im}=0}}^{n-k-1} \left[1 + \prod_{\substack{j=0 \\ j \neq m}}^{n-1} \lambda_j^{h_{ij}}\right] \otimes \prod_{\substack{i=0 \\ h_{im}=1}}^{n-k-1} \left[1 + \prod_{\substack{j=0 \\ j \neq m}}^{n-1} \lambda_j^{h_{ij}}\right]}{\prod_{\substack{i=0 \\ h_{im}=0}}^{n-k-1} \left[1 + \prod_{\substack{j=0 \\ j \neq m}}^{n-1} \lambda_j^{h_{ij}}\right] \otimes \prod_{\substack{i=0 \\ h_{im}=1}}^{n-k-1} \left[1 - \prod_{\substack{j=0 \\ j \neq m}}^{n-1} \lambda_j^{h_{ij}}\right]}.$$

We follow the principle of threshold decoding introduced in [6]. Thus, for bit position m, the subcode over which we decode is determined by a subset of parity checks which have a 1 at position m and are *orthogonal* to each other masking out position m. With such parity checks, the special multiplication $\otimes$ in the above equation is the same as the real number multiplication, and we get for the sub-optimal threshold decoding algorithm of [6]:

$$L_m^{\mathcal{C}} = \log \frac{1+\lambda_m}{1-\lambda_m} + \log \frac{\prod\limits_{\substack{i=0 \\ h_{im}=1}}^{n-k-1} \left[1 + \prod\limits_{\substack{j=0 \\ j \neq m}}^{n-1} \lambda_j^{h_{ij}}\right]}{\prod\limits_{\substack{i=0 \\ h_{im}=1}}^{n-k-1} \left[1 - \prod\limits_{\substack{j=0 \\ j \neq m}}^{n-1} \lambda_j^{h_{ij}}\right]}$$

$$= L_m + \sum\limits_{\substack{i=0 \\ h_{im}=1}}^{n-k-1} \log \frac{\left[1 + \prod\limits_{\substack{j=0 \\ j \neq m}}^{n-1} \lambda_j^{h_{ij}}\right]}{\left[1 - \prod\limits_{\substack{j=0 \\ j \neq m}}^{n-1} \lambda_j^{h_{ij}}\right]}.$$

APPENDIX

A. Turbo decoding of an $(8,4)$ single-loop code. We consider a code $\mathcal{C}$ whose Tanner graph and the parity check matrix H are shown in Fig. 2. There are four parity check equations of code $\mathcal{C}$: h_0 and h_1 corre-

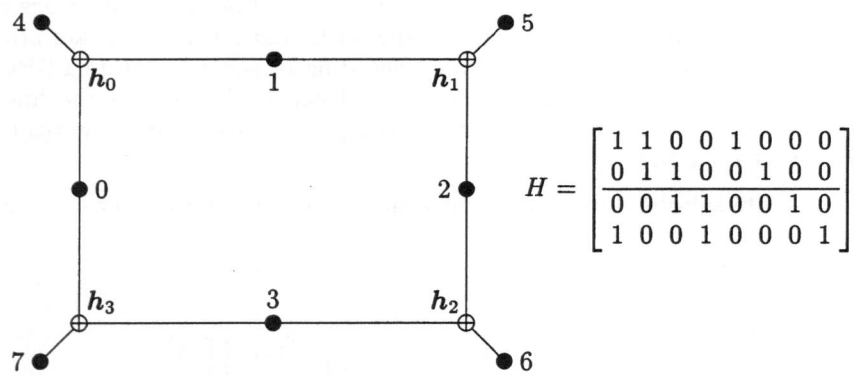

FIG. 2. *An $(8,4)$ single-loop code.*

sponding to code $\mathcal{C}_1$ and h_2 and h_3 corresponding to code $\mathcal{C}_2$, as indicated by the dividing line of H in the figure.

The optimal decoder of $\mathcal{C}$ computes

(A.1) $$\lambda_m^{\mathcal{C}} = \frac{\lambda_m \otimes A_{\mathcal{C}}}{A_{\mathcal{C}}},$$

where

$$A_{\mathcal{C}} = (1 + \lambda_0\lambda_1\lambda_4) \otimes (1 + \lambda_1\lambda_2\lambda_5) \otimes (1 + \lambda_2\lambda_3\lambda_6) \otimes (1 + \lambda_0\lambda_3\lambda_7).$$
$$= 1 + \lambda_0\lambda_1\lambda_4 + \lambda_1\lambda_2\lambda_5 + \lambda_0\lambda_2\lambda_4\lambda_5 +$$
$$\lambda_2\lambda_3\lambda_6 + \lambda_0\lambda_1\lambda_2\lambda_3\lambda_4\lambda_6 + \lambda_1\lambda_3\lambda_5\lambda_6 + \lambda_0\lambda_3\lambda_4\lambda_5\lambda_6 + \lambda_0\lambda_3\lambda_7 +$$
$$\lambda_0\lambda_1\lambda_2\lambda_3\lambda_5\lambda_7 + \lambda_2\lambda_3\lambda_4\lambda_5\lambda_7 + \lambda_0\lambda_1\lambda_5\lambda_6\lambda_7 + \lambda_1\lambda_3\lambda_4\lambda_7 +$$
$$\lambda_1\lambda_2\lambda_4\lambda_6\lambda_7 + \lambda_4\lambda_5\lambda_6\lambda_7 + \lambda_0\lambda_2\lambda_6\lambda_7.$$

The turbo decoding process starts with decoding of $\mathcal{C}_1$ by calculating the messages to be passed to the decoder of $\mathcal{C}_2$:

$$[\lambda_0^{\mathcal{C}_1}]_1 = \frac{\lambda_0 \otimes A_{\mathcal{C}_1}}{A_{\mathcal{C}_1}} \quad \text{and} \quad [\lambda_2^{\mathcal{C}_1}]_1 = \frac{\lambda_2 \otimes A_{\mathcal{C}_1}}{A_{\mathcal{C}_1}}.$$

Note that these are the only bit positions also checked by $\mathcal{C}_2$.

In the second iteration, the decoder of $\mathcal{C}_2$ is active. It uses the information received from the first component decoder, and calculates the messages to pass back to it in the next iteration:

$$[\lambda_0^{\mathcal{C}_2}]_2 = \frac{\lambda_0 \otimes (1 + \lambda_0\lambda_3\lambda_7) \otimes (1 + [\lambda_2^{\mathcal{C}_1}]_1\lambda_3\lambda_6)}{(1 + \lambda_0\lambda_3\lambda_7) \otimes (1 + [\lambda_2^{\mathcal{C}_1}]_1\lambda_3\lambda_6)}$$

and

$$[\lambda_2^{\mathcal{C}_2}]_2 = \frac{\lambda_2 \otimes (1 + [\lambda_0^{\mathcal{C}_1}]_1\lambda_3\lambda_7) \otimes (1 + \lambda_2\lambda_3\lambda_6)}{(1 + [\lambda_0^{\mathcal{C}_1}]_1\lambda_3\lambda_7) \otimes (1 + \lambda_2\lambda_3\lambda_6)}.$$

For the soft-output of the whole decoding system after the second iteration, we combine the results from both sub decoders:

$$(A.2) \qquad [\lambda_0^{\mathcal{C}}]_2 = \frac{[\lambda_0^{\mathcal{C}_1}]_1 \otimes (1 + [\lambda_0^{\mathcal{C}_1}]_1\lambda_3\lambda_7) \otimes (1 + [\lambda_2^{\mathcal{C}_1}]_1\lambda_3\lambda_6)}{(1 + [\lambda_0^{\mathcal{C}_1}]_1\lambda_3\lambda_7) \otimes (1 + [\lambda_2^{\mathcal{C}_1}]_1\lambda_3\lambda_6)}$$

and

$$[\lambda_2^{\mathcal{C}}]_2 = \frac{[\lambda_2^{\mathcal{C}_1}]_1 \otimes (1 + [\lambda_0^{\mathcal{C}_1}]_1\lambda_3\lambda_7) \otimes (1 + [\lambda_2^{\mathcal{C}_1}]_1\lambda_3\lambda_6)}{(1 + [\lambda_0^{\mathcal{C}_1}]_1\lambda_3\lambda_7) \otimes (1 + [\lambda_2^{\mathcal{C}_1}]_1\lambda_3\lambda_6)}$$

For bit position '0', one should compare the soft output $\lambda_0^{\mathcal{C}}$ of the optimal decoder given by (A.1) with the soft output of the turbo decoder $[\lambda_0^{\mathcal{C}}]_2$ given by (A.2).

Acknowledgement. The authors would like to thank J. Mazo, I. Sason, A. Shokrollahi, and R. Urbanke for their comments on an earlier version of this article.

REFERENCES

[1] C.R.P. HARTMANN AND L.D. RUDOLPH, "An optimum symbol by symbol decoding rule for linear codes," *IEEE Trans. Inform. Theory*, Vol. IT-22, pp. 514–517, Sept. 1976.

[2] G. BATTAIL, H.C. DECOUVELAERE, AND P. GODLEWSKI, "Replication decoding", *IEEE Trans. Inform. Theory*, Vol. IT-25, pp. 332–345, May 1979.

[3] C. BERROU, A. GLAVIEUX, AND P. THITIMAJSHIMA, "Near Shannon limit error correcting coding and decoding: Turbo-codes(1)," *Proc. 1993 IEEE Int. Conf. Commun. (ICC'93)*, Geneva, Switzerland, May 1993, pp. 1064–1070.

[4] J. HAGENAUER, E. OFFER, AND L. PAPKE, "Iterative decoding of binary block and convolutional codes," *IEEE Trans. Inform. Theory*, Vol. IT-42, pp. 429–445, March 1996.

[5] R. GALLAGER, "Low-density parity-check codes," *IRE Trans. Inform. Theory*, Vol. IT-8, pp. 21–28, Jan. 1962.

[6] J. MASSEY, "Threshold Decoding," Cambridge: The M.I.T. Press, 1963.

RECURSIVE CONSTRUCTION OF GRÖBNER BASES FOR THE SOLUTION OF POLYNOMIAL CONGRUENCES

HENRY O'KEEFFE* AND PATRICK FITZPATRICK*

Abstract. A number of questions in systems theory and coding theory can be formulated as the solution of a system of polynomial congruences in one or more variables. We present a new generalised algorithm, based on Gröbner basis techniques, that recursively solves a wide class of such problems.

Key words. Gröbner basis, systems theory, coding theory, partial realization, rational interpolation, discrete-time behaviours.

AMS(MOS) subject classifications. 93B, 93C, 94B.

1. Introduction. Let $A = F[x_1, \ldots, x_n]$ be a polynomial ring over a field F. Let $(b_1, \ldots, b_q) \in A^q$ be a solution of the system of congruences

$$(1.1) \qquad \sum_{i=1}^{q} b_i h_{ik} \equiv 0 \bmod I^{(k)}$$

where $h_{ik} \in A$, and $I^{(k)}$ is an ideal in A, for $1 \leq i \leq q$, $1 \leq k \leq p$. The set of solution vectors $(b_1, \ldots, b_q)$ forms a submodule M of A^q. Under suitable conditions on the ideals $I^{(k)}$, and with an appropriate term order, we shall construct a Gröbner basis of M, using an algorithm which recursively determines Gröbner bases of submodules in a descending sequence terminating in M. Since the method is independent of the term order chosen, it can be used to select solution vectors for a range of different criteria. In particular, the recursive nature of the construction and its flexibility with regard to the sequence in which the data are processed makes it suitable for systems with time-invariant constraints or where not all of the data are known at the outset. Although it is intended primarily as computational tool, the algorithm can produce closed theoretical solutions in certain cases. The algorithm presented here extends and generalises the results of [6, 7, 8, 9].

2. Definitions and notation. In this section we introduce some notation and terminology on Gröbner bases; detailed treatments can be found in ([1],[3],[4]). The standard basis vector with 1 in position i and 0 elsewhere (and length defined by the context) is denoted $\mathbf{e}_i$. A *term* in A^q is a vector of the type $\mathbf{X} = X\mathbf{e}_i$ where $X = x_1^{t_1} x_2^{t_2} \cdots x_n^{t_n}$ is a *term* in A. Thus a term in A^q is a vector all of whose components are zero except for one which is a term in A. A *term order* in A^q is a total order $<$ on terms satisfying
  (i) $\mathbf{X} < Z\mathbf{X}$ for each term $\mathbf{X}$ in A^q and each term $Z \neq 1$ in A,

---
*Department of Mathematics, National University of Ireland, Cork, Ireland; Email hok@ucc.ie, fitzpat@ucc.ie.

(ii) if $\mathbf{X} < \mathbf{Y}$ then $Z\mathbf{X} < Z\mathbf{Y}$ for all terms $\mathbf{X}, \mathbf{Y}$ in A^q and each term Z in A.

An example of such a term order is obtained as follows. Let $<_{\mathrm{lex}}$ denote lexicographic order in A and define $\mathbf{X} = X\mathbf{e_i} < Y\mathbf{e_j}$ if either $i < j$, or $i = j$ and $X <_{\mathrm{lex}} Y$. This is a *position-over-term* (or *POT*) order.

For non-zero $\mathbf{f} \in A^q$ we may write

$$(2.1) \qquad \mathbf{f} = a_1\mathbf{X_1} + \cdots + a_r\mathbf{X_r}$$

where the a_i are non-zero constants and the $\mathbf{X_i}$ are terms satisfying $\mathbf{X_1} > \mathbf{X_2} > \cdots > \mathbf{X_r}$. The *leading term* $\mathrm{lt}(\mathbf{f})$ of $\mathbf{f}$ is $\mathbf{X_1}$, and the *leading coefficient* $\mathrm{lc}(\mathbf{f})$ is a_1. These definitions are extended to all of A^q by setting $\mathrm{lt}(\mathbf{0}) = \mathbf{0}, \mathrm{lc}(\mathbf{0}) = 0$. For example, with $n = 2, q = 3$, $x_1 <_{\mathrm{lex}} x_2$, and $\mathbf{f} = (x_1^2 x_2 + 3x_1^3, x_2 + 1, 2x_1)$, and using the POT order defined above, we find that

$$f = 2x_1\mathbf{e_3} + x_2\mathbf{e_2} + \mathbf{e_2} + x_1^2 x_2\mathbf{e_1} + 3x_1^3\mathbf{e_1}$$

has leading term $x_1\mathbf{e_3}$ and leading coefficient 2.

If $\mathbf{X} = X\mathbf{e_i}$ and $\mathbf{Y} = Y\mathbf{e_j}$ are terms in A^q we say $\mathbf{X}$ *divides* $\mathbf{Y}$ provided $i = j$ and X divides Y in A, that is, if there is a term Z (the quotient) in A satisfying $Z\mathbf{X} = \mathbf{Y}$. A set of non-zero vectors $G = \{\mathbf{g_1}, \ldots, \mathbf{g_r}\}$ contained in the submodule M is called a *Gröbner basis* of M if for all $\mathbf{f} \in M$ there exists $i \in \{1, \ldots, r\}$ such that $\mathrm{lt}(\mathbf{g_i})$ divides $\mathrm{lt}(\mathbf{f})$. In particular G is a basis of M. A Gröbner basis is *minimal* if none of its elements has leading term a multiple of the leading term of another of its elements. Each $\mathbf{f} \in M$ has a *standard representation* with respect to a Gröbner basis $G = \{\mathbf{g_1}, \ldots, \mathbf{g_r}\}$ of the form

$$\mathbf{f} = \sum_{i=1}^{r} f_i\mathbf{g_i}$$

where $f_i \in A$ and $\mathrm{lt}(f_i\mathbf{g_i}) \leq \mathrm{lt}(\mathbf{f})$, $1 \leq i \leq r$. A *minimal* element in a submodule $M \subseteq A^q$ is one whose leading term is least among the elements of M, under the given term order $<$. It is unique up to a constant multiple and must appear in any Gröbner basis relative to $<$. If the leading terms of the elements of a Gröbner basis (resp. minimal Gröbner basis) are in non-decreasing (resp. increasing) order then we call it an *ordered* Gröbner basis.

3. The incremental step. The incremental step defined in the following theorem will be applied to ideals I_l, I_{l+1} in A with $I_l \supseteq I_{l+1}$ such that for each $s = 1, \ldots, n$ there exists $\beta_s \in F$ satisfying

$$(3.1) \qquad (x_s - \beta_s)I_l \subseteq I_{l+1}.$$

We also require an F-homomorphism

$$(3.2) \qquad \alpha : I_l \longrightarrow F$$

with $\ker(\alpha) = I_{l+1}$.

If $\mathcal{W}$ is an ordered set then $\mathcal{W}[j]$ denotes its jth element, and if $\mathcal{W}[j]$ is a vector then $\mathcal{W}[j]_i$ denotes its ith component.

THEOREM 3.1. *Let $I_l \supseteq I_{l+1}$ be ideals in A satisfying (3.1) and (3.2). Let S be the submodule of A^q of solution vectors $(b_1, \ldots, b_q)$ of*

$$(3.3) \qquad \sum_{i=1}^{q} b_i h_{ik} \equiv 0 \bmod I_l$$

and let $S' \subseteq S$ be the submodule of S of elements satisfying

$$(3.4) \qquad \sum_{i=1}^{q} b_i h_{ik} \equiv 0 \bmod I_{l+1}.$$

If $\mathcal{W}$ is an ordered minimal Gröbner basis of S relative to a given term order $<$ then a Gröbner basis $\mathcal{W}'$ of S' relative to $<$ can be constructed as follows.

Define $\alpha_j := \alpha \left(\sum_{i=1}^{q} \mathcal{W}[j]_i h_{ik} \right)$ for $1 \leq j \leq |\mathcal{W}|$.

If $\alpha_j = 0$ for all j then $\mathcal{W}' := \mathcal{W}$.

Otherwise

   $j^* :=$ *least j for which $\alpha_j \neq 0$*

   $\mathcal{W}_1 := \{\mathcal{W}[j] : j < j^*\}$

   $\mathcal{W}_2 := \{(x_s - \beta_s)\mathcal{W}[j^*] : 1 \leq s \leq n\}$

   $\mathcal{W}_3 := \{\mathcal{W}[j] - (\alpha_j/\alpha_{j^*})\mathcal{W}[j^*] : j > j^*\}$

   $\mathcal{W}' := \mathcal{W}_1 \bigcup \mathcal{W}_2 \bigcup \mathcal{W}_3$.

Proof. By definition, $\sum_{i=1}^{q} \mathcal{W}[j]_i h_{ik} \in I_l$, so if $\alpha_j = 0$, for all j, then $\mathcal{W} \subseteq S'$ so $S' = S$. Thus suppose some $\alpha_j \neq 0$ and let j^* be as defined. If $j < j^*$ then clearly $\mathcal{W}[j] \in S'$. Next, $\sum_{i=1}^{q} \mathcal{W}[j^*]_i h_{ik} \in I_l$, so $(x_s - \beta_s) \sum_{i=1}^{q} \mathcal{W}[j^*]_i h_{ik} \in I_{l+1}$, by (3.1), and hence $(x_s - \beta_s)\mathcal{W}[j^*] \in S'$. Finally, for $j > j^*$,

$$\alpha \left(\sum_{i=1}^{q} (\mathcal{W}[j]_i - (\alpha_j/\alpha_{j^*})\mathcal{W}[j^*]_i) h_{ik} \right) = \alpha_j - (\alpha_j/\alpha_{j^*})\alpha_{j^*} = 0$$

so $\mathcal{W}[j] - (\alpha_j/\alpha_{j^*})\mathcal{W}[j^*] \in S'$ by (3.2). We have now proved that $\mathcal{W}' \subseteq S'$.

We show that $\mathcal{W}'$ is a Gröbner basis as follows. By assumption, $\mathcal{W}$ is a minimal Gröbner basis of S so $\mathrm{lt}(\mathcal{W}[i])$ does not divide $\mathrm{lt}(\mathcal{W}[j]), i \neq j$. Now, $\mathrm{lt}((x_s - \beta_s)\mathcal{W}[j^*]) = x_s \mathrm{lt}(\mathcal{W}[j^*])$ and $\mathrm{lt}(\mathcal{W}[j] - (\alpha_j/\alpha_{j^*})\mathcal{W}[j^*])) = \mathrm{lt}(\mathcal{W}[j]), j > j^*$. Let $\mathbf{f} \in S' \subseteq S$. Then $\mathrm{lt}(\mathbf{f})$ is divisible by some $\mathrm{lt}(\mathcal{W}[j])$. If $j \neq j^*$ then $\mathrm{lt}(\mathbf{f})$ is divisible by the leading term of an element of $\mathcal{W}'$.

Thus, we may suppose that $\mathrm{lt}(\mathcal{W}[j^*])$ is the *only* leading term of the basis elements $\mathcal{W}[j]$ that divides $\mathrm{lt}(\mathbf{f})$. We show that $x_s \mathrm{lt}(\mathcal{W}[j^*])$ also divides $\mathrm{lt}(\mathbf{f})$ for some s. Consider the standard representation

$$\mathbf{f} = \sum_{j \in J} f_j \mathcal{W}[j]$$

with $f_j \neq 0$, and $J \subseteq \{1, \ldots, |\mathcal{W}|\}$. By definition of this representation, and by the assumption on $\mathrm{lt}(\mathcal{W}[j^*])$, it follows that $j^* \in J$, $\mathrm{lt}(\mathbf{f}) = \mathrm{lt}(f_{j^*}\mathcal{W}[j^*])$, and $\mathrm{lt}(f_j\mathcal{W}[j]) < \mathrm{lt}(\mathbf{f})$ for $j \neq j^*$. Let $X_j, j \in J$ be terms in A such that $\mathrm{lt}(f_j\mathcal{W}[j]) = X_j\mathrm{lt}(\mathcal{W}[j])$. Thus $X_j\mathrm{lt}(\mathcal{W}[j]) < X_{j^*}\mathrm{lt}(\mathcal{W}[j^*])$. Suppose that there is some $j \in J$ with $j > j^*$. If $X_{j^*} = 1$ then

$$X_j\mathrm{lt}(\mathcal{W}[j]) < \mathrm{lt}(\mathcal{W}[j^*]) \leq X_j\mathrm{lt}(\mathcal{W}[j^*])$$

which contradicts the increasing order of $\mathcal{W}$. Hence $X_{j^*} \neq 1$, so $x_s\mathrm{lt}(\mathcal{W}[j^*])$ divides $\mathrm{lt}(\mathbf{f})$ for some s. Otherwise, $J \subseteq \{1, \ldots, j^*\}$ and

$$\mathbf{f} - \sum_{j=1}^{j^*-1} f_j\mathcal{W}[j] = f_{j^*}\mathcal{W}[j^*]$$

lies in S'. Therefore $f_{j^*} \neq 1$ since $\mathcal{W}[j^*] \notin S'$. As a consequence, $X_{j^*} \neq 1$ and again $x_s\mathrm{lt}(\mathcal{W}[j^*])$ divides $\mathrm{lt}(\mathbf{f})$ for some s. This completes the proof. $\square$

REMARK 3.1.   Note that the position of the leading term of $(x_s - \beta)\mathbf{f}$ is the same as that of $\mathbf{f}$. Similarly, if $\mathrm{lt}(\mathbf{g}) < \mathrm{lt}(\mathbf{f})$ then $\mathbf{f} - c\mathbf{g}$, where c is a constant, also has leading term in the same position as $\mathbf{f}$.

3.1. The generalised algorithm. The title of this section indicates that the algorithm presented is a generalisation of that in [6, 7, 8, 9]. Moreover the proof is significantly simpler than the one given in [8] (for a special case — see Section 4.3).

We consider the problem of finding a Gröbner basis of the solution module M of (1.1), given a set of sequences of ideals $I_0^{(k)} \supseteq I_1^{(k)} \supseteq \cdots \supseteq I_{N_k}^{(k)} = I^{(k)}$, $k = 1, \ldots, p$. As before, all computations are carried out relative to a fixed term order $<$ in A^q. We begin with an ordered minimal Gröbner basis of the solution module M_0 of

$$(3.5) \qquad \sum_{i=1}^q b_i h_{ik} \equiv 0 \bmod I_0^{(k)} \quad k = 1, \ldots, p.$$

In particular, if $I_0^{(k)} = A$, for all k, then $M_0 = A^q$ and an ordered standard basis for A^q provides the required starting Gröbner basis. We also assume that there exist $\beta_{sk} \in F$ such that

$$(3.6) \qquad (x_s - \beta_{sk})I_{l_k}^{(k)} \subseteq I_{l_k+1}^{(k)}$$

and that we have a set of $F - linear$ functions

$$(3.7) \qquad \alpha_{l_k}^{(k)} : I_{l_k}^{(k)} \longrightarrow F$$

with $\ker(\alpha_{l_k}^{(k)}) = I_{l_k+1}^{(k)}$.

At each step we choose some k such that $l_k < N_k$ and use Theorem 3.1 to determine the Gröbner basis of the solution module for the system of congruences in which $I_{l_k}^{(k)}$ is replaced by $I_{l_k+1}^{(k)}$. The procedure ord(S) puts the elements of a list $S \subseteq A^q$ in non-decreasing order of leading term with respect to $<$ and removes any element whose leading term is a multiple of the leading term of another element. In the event of there being more than one element with the same leading term only one is retained. If S is a Gröbner basis then ord(S) an ordered minimal Gröbner basis. Verification of the algorithm is straightforward from the observation that it implements a finite sequence of incremental steps as defined in the previous section.

Algorithm

Input:

$h_{ik}, i = 1, \ldots, q, k = 1, \ldots, p$

$I_{l_k}^{(k)}, k = 1, \ldots, p, l_k = 0, \ldots, N_k$

$\beta_{sk}, s = 1, \ldots, n, k = 1, \ldots, p$

$\alpha_{l_k}^{(k)}$

$<$ a term order

M_0 the initial solution module

$\mathcal{W}_0$ an ordered minimal Gröbner basis of M_0

Output:

$\mathcal{W}$ an ordered minimal Gröbner basis of M

Main Routine:

$\mathcal{W} := \mathcal{W}_0$

WHILE $l_k < N_k$ for some k DO

    choose k such that $l_k < N_k$

    FOR j FROM 1 TO $|\mathcal{W}|$ DO

        $\alpha_j := \alpha_{l_k}^{(k)}(\sum_{i=1}^q \mathcal{W}[j]_i h_{ik})$

    IF $\alpha_j = 0$ for all j THEN

        $\mathcal{W}' = \mathcal{W}$

    ELSE

        $j^* :=$ least j for which $\alpha_j \neq 0$

        $\mathcal{W}' := \emptyset$

        FOR j FROM 1 TO $j^* - 1$ DO

            $\mathcal{W}' := \mathcal{W}' \bigcup \{\mathcal{W}[j]\}$

        FOR s FROM 1 TO n DO

            $\mathcal{W}' := \mathcal{W}' \bigcup \{(x_s - \beta_{sk})\mathcal{W}[j^*]\}$

        FOR j FROM $j^* + 1$ TO $|\mathcal{W}|$ DO

            $\mathcal{W}' := \mathcal{W}' \bigcup \{\mathcal{W}[j] - (\alpha_j/\alpha_{j^*})\mathcal{W}[j^*]\}$

    $\mathcal{W} := $ ord($\mathcal{W}'$)

3.2. A single indeterminate. When $A = F[x]$ several simplifications can be made.

LEMMA 3.1. *A Gröbner basis G for a module $M \subseteq F[x]^q$ is a minimal basis if and only if no two basis elements have leading terms in the same position.*

Proof. If the leading terms of the basis elements are all in different positions then none of them is a multiple of another. Conversely, if two of the leading terms are in the same position then one of them divides the other. □

Using Remark 3.1 we have the following consequence.

COROLLARY 3.1. *A minimal Gröbner basis of a submodule of $F[x]^q$ has at most q elements. Each basis produced during the algorithm has the same number of elements as that in the initial basis.*

It follows that in this case the procedure ord is greatly simplified. The positions of the leading terms are unchanged by the incremental step and the degrees of the leading terms are unchanged except for that of $xW[j^*]$. Thus, re-ordering takes the form of inserting this element into its correct new position.

For many applications the function α can also be quite simple. The next lemma provides a useful setting.

LEMMA 3.2. *Let $J \supseteq I$ be ideals in $F[x]$ generated by p_J, p_I and suppose that $p_I = (x - \beta)p_J$ for some $\beta \in F$. The function $\alpha : J \longrightarrow F$ defined by $\alpha(h) = \alpha(fp_J) = f(\beta)$ is F-linear and has kernel I.*

Proof. Each $h \in J$ has the form $h = fp_J$ for uniquely defined f. By the division algorithm $f = (x - \beta)q + f(\beta)$, for some q, and thus $h \in I$ if and only if $f(\beta) = 0$. It is straightforward to prove that α is linear. □

Suppose that the ideals $I^{(k)}$ are of the form

$$I^{(k)} = \langle (x - \beta_k)^{N_k} \rangle, k = 1, \ldots, p$$

and the corresponding sequences of approximating ideals are

$$I_{l_k}^{(k)} = \langle (x - \beta_k)^{l_k} \rangle, l_k = 0, \ldots, N_k.$$

If $h \in J = I_{l_k}^{(k)} \supseteq I_{l_k+1}^{(k)} = I$ is expanded as a polynomial in $x - \beta_k$ then $\alpha_{l_k}^k(h)$ can be "read off" directly as the coefficient of $(x - \beta_k)^{l_k}$.

4. Applications. The generalised algorithm can be applied to a wide range of problems. In this section we review previous work on scalar partial realisation and rational interpolation and on solving multivariable polynomial congruences, and, in more detail, present a new application to the modelling of discrete-time behaviours. In general, the input data are provided by the h_{ij}, the term order $<$ corresponds to "degree" constraints, and the $I^{(k)}, \beta_{sk}$ relate to "order" contraints and interpolation points. Of particular interest are situations where there is a unique "required solution" that can be identified as the minimal element of the solution module with respect to a certain term order defined by the problem. Such an element must lie in a Gröbner basis with respect to that term order.

4.1. Partial realisation, scalar rational interpolation. Both of these problems can be viewed as the parameterisation of the solutions of the system of congruences

$$b_1(-1) + b_2 h_k \equiv 0 \bmod (x - \beta_k)^{N_k}, k = 1, \ldots, p$$

where $h_k = \sum_{t=0}^{N_k-1} c_{kt}(x-\beta_k)^t$. Thus, $q = 2$ and $h_{1k} = -1, h_{2k} = h_k$ for all k. Here the initial module is $M_0 = A^2$. Various conditions may be imposed on the solutions, such as, $\delta b_1 < \delta b_2, \delta b_1 + \delta b_2 < N = \sum_{k=1}^p N_k, b_1, b_2$ relatively prime, $b_2(\beta_k) \neq 0$ for all k, and so on. These are used to define an appropriate term order for the solution module and in certain situations to identify a unique required solution as the minimal element in the module. The partial realisation problem corresponds to the case $p = 1$ (and $\beta_1 = 0$), while the distinct-abcissae rational interpolation problem corresponds to $N_k = 1$ for all k. See [7] for more details.

4.2. Errors-and-erasures decoding of alternant codes. The specific context of decoding was addressed in [6, 9]. The errors-only case corresponds to the solution of the single congruence

$$b_1(-1) + b_2 h \equiv 0 \bmod x^n$$

subject to the conditions $\delta b_1 \leq m_1, \delta b_2 \leq m_2, m_1 + m_2 < n$ and b_1, b_2 relatively prime. This is a partial realisation problem in which a required solution is shown to be minimal with respect to a certain term order defined relative to the parameter $r = m_1 - m_2$. The algorithm produces a Gröbner basis of the solution module containing the required solution. The generalisation to errors-and-erasures decoding involves the solution of the congruence subject to the further condition that b_2 be divisible by a fixed polynomial f. In this case initialisation is at the basis $\{(x^{\delta f+r+1}, 0), (\overline{fh}, f)\}$, where $\overline{fh}$ is the remainder of fh modulo $x^{\delta c+r+1}$. This provides an application in which the initial solution module M_0 is not A^q.

4.3. Multiple indeterminates. The main problem analysed in [8] is the determination of a Gröbner basis of the solution module of the congruence

$$b_1(-1) + \sum_{i=2}^q b_i h_i \equiv 0 \bmod I$$

where $I \subseteq F[x_1, \ldots, x_n]$ is a zero dimensional ideal. Thus $p = 1$ in this case. A recursive algorithm is given that applies in the special case where a sequence of approximating ideals can be defined with the property that for each l there is a term $\phi_l \notin I_{l+1}$ such that

$$I_l = \langle \phi_l, I_{l+1} \rangle \text{ and } x_s \phi_l \in I_{l+1}, s = 1, \ldots, n$$

For $g \in I_l$ the value $\alpha(g)$ is the coefficient of ϕ_l in the normal form of g modulo I_{l+1}. Applications to multivariable Padé approximation, Hensel codes, and algebraic geometry codes are given.

4.4. Modelling discrete-time behaviours. Antoulas ([2]) uses a behavioural approach to determine models of vector-valued discrete-time time series. In particular, *autonomous*, linear, time-invariant models are sought and it is shown that one such model θ^*, with *minimal complexity*, generates all linear time-invariant models of the time series, and also contains a minimal complexity *controllable* model. A recursive procedure for constructing this model is given. Multivariable (in the systems theory sense) partial realisation is a special case of this problem (see also Dickinson, Morph and Kailath [5] and Kuijper [10]).

The models θ under consideration are matrices with q columns whose elements are polynomials in the *forward shift* σ^{-1}. The behavior $B(\theta)$ of θ is the set $\{\mathbf{w} \in (F^q)^{\mathcal{Z}_-} : \theta(\sigma^{-1})\mathbf{w} = 0\}$. Equivalently, each $\mathbf{w}$ can be viewed as a (negative) power series, $\sum_{t \in \mathcal{Z}_-} \mathbf{w}_t x^t$, where $\mathbf{w}_t \in F^q$. The forward shift is equivalent to multiplying by x and truncating at the term of degree zero.

The time series $\mathbf{w}^{(k)}, k = 1, \ldots, p$ is zero until a finite time $-N_k$ in the past and current time is at the origin. Such a time series is in $B(\theta)$ if and only if the polynomial q-vector $\mathbf{h}^{(k)} = x^{N_k}\mathbf{w}^{(k)}$ satisfies

$$\theta(x)\mathbf{h}^{(k)} \equiv 0 \bmod x^{N_k+1}, k = 1, \ldots, p.$$

If we view each of the q components individually and consider a row $(b_1, \ldots, b_q)$ of the model θ, then θ contains $\mathbf{w}^{(k)}$ in its behaviour if and only if each row of θ satisfies

$$(4.1) \qquad \sum_{i=1}^{q} b_i h_i^{(k)} \equiv 0 \bmod x^{N_k+1}, k = 1, \ldots, p.$$

Thus a generating model corresponds to a basis of the submodule of $F[x]^q$ whose elements satisfy this system of congruences. This makes the problem amenable to solution by our techniques.

For each k, the sequence of approximating ideals can be chosen as

$$I_{l_k}^{(k)} = \langle x^{l_k} \rangle, l_k = 0, \ldots N_k.$$

The function $\alpha_{l_k}^{(k)}$ returns the coefficient of x^{l_k} in $\sum_{i=1}^{q} b_i h_i^{(k)}$. The term-over-position (TOP) term order in $F[x]^q$ is defined by $x^r \mathbf{e}_i < x^t \mathbf{e}_j$ if $r < t$ or if $r = t$ and $i < j$. With this term order our algorithm produces a Gröbner basis which corresponds to a generating model equivalent to that given in [2]. The degree of the leading term of a basis element is the *row degree* of the corresponding model row. The procedure ord need only maintain a table giving degree and position of the leading term for each basis element. These pairs are unchanged by the incremental step except for the degree of the element $\mathcal{W}[j^*]$, and so re-ordering takes the form of inserting the multiple $x\mathcal{W}[j^*]$ into its correct new position. The initial module is A

with Gröbner basis $\{e_1, \ldots, e_q\}$. The final basis produced corresponds to a model with ordered row degrees, which is row reduced by Lemma 3.1. A minimal complexity controllable model can be extracted as a subset of the basis at each stage. We illustrate the application of our algorithm in this context using an example of Antoulas ([2, Example B]).

EXAMPLE 1.   We want to determine a generating system and a controllable model of minimum complexity from the following Markov parameters $(A_i, i = 0, \ldots, 4)$ of a three-input, two-output system.

$$A_0 = 0 \qquad A_1 = \begin{pmatrix} 1 & 1 & 1 \\ 2 & 2 & 0 \end{pmatrix} \qquad A_2 = \begin{pmatrix} 2 & 2 & 1 \\ 4 & 4 & 0 \end{pmatrix}$$

$$A_3 = \begin{pmatrix} 4 & 6 & 3 \\ 8 & 10 & 2 \end{pmatrix} \qquad A_4 = \begin{pmatrix} 8 & 11 & 6 \\ 16 & 21 & 7 \end{pmatrix}.$$

The corresponding matrix of polynomials $h_i^{(k)}$ is

$$\begin{pmatrix} 1 & 0 & 0 \\ 0 & 1 & 0 \\ 0 & 0 & 1 \\ x + 2x^2 + 4x^3 + 8x^4 & x + 2x^2 + 6x^3 + 11x^4 & x + x^2 + 3x^3 + 6x^4 \\ 2x + 4x^2 + 8x^3 + 16x^4 & 2x + 4x^2 + 10x^3 + 21x^4 & 2x^3 + 7x^4 \end{pmatrix}$$

and we denote the columns of this matrix $h^{(1)}, h^{(2)}, h^{(3)}$.

We apply our algorithm using the TOP term order. After the Markov parameters A_1, A_2, A_3 have been processed we have the basis consisting of the rows of the matrix

$$\begin{pmatrix} \frac{2}{3}x & \frac{2}{3}x & -\frac{2}{3}x & \frac{2}{3} - \frac{2}{3}x & -\frac{2}{3} + x \\ \frac{1}{2}x^2 & \frac{1}{2}x^2 & x^2 & -x + x^2 & \frac{1}{4}x \\ -\frac{1}{3}x + x^3 & -\frac{1}{3}x & -\frac{2}{3}x & \frac{2}{3} - \frac{2}{3}x - x^2 & -\frac{1}{6} \\ x^3 & x^3 & 0 & 0 & -\frac{1}{2}x^2 \\ 0 & 0 & x^3 & -x^2 & \frac{1}{2}x^2 \end{pmatrix}$$

and the controllable model is derived from rows 1 and 3.

The next step is to consider the component h_1 modulo x^5. The α_j are $0, 0, -2, -2, 0$ and this gives $j^* = 3$ and the next basis

$$\begin{pmatrix} \frac{2}{3}x & \frac{2}{3}x & -\frac{2}{3}x & \frac{2}{3} - \frac{2}{3}x & -\frac{2}{3} + x \\ \frac{1}{2}x^2 & \frac{1}{2}x^2 & x^2 & -x + x^2 & \frac{1}{4}x \\ \frac{1}{3}x & \frac{1}{3}x + x^3 & \frac{2}{3}x & -\frac{2}{3} + \frac{2}{3}x + x^2 & \frac{1}{6} - \frac{1}{2}x^2 \\ 0 & 0 & x^3 & -x^2 & \frac{1}{2}x^2 \\ -\frac{1}{3}x^2 + x^4 & -\frac{1}{3}x^2 & -\frac{2}{3}x^2 & \frac{2}{3}x - \frac{2}{3}x^2 - x^3 & -\frac{1}{6}x \end{pmatrix}.$$

The new controllable model is in row 1 and row 3 (which was row 4 in the previous basis). When the two remaining series h_2, h_3 are processed, the final basis is

$$
\begin{pmatrix}
-\frac{3}{2}x + \frac{1}{2}x^2 & -\frac{3}{2}x + \frac{1}{2}x^2 & \frac{3}{2}x + x & -\frac{3}{2} + \frac{1}{2}x + x^2 & \frac{3}{2} - 2x \\
\frac{2}{3}x^2 & \frac{2}{3}x^2 & -\frac{2}{3}x^2 & \frac{2}{3}x - \frac{2}{3}x^2 & -\frac{2}{3}x + x^2 \\
\frac{1}{2}x & \frac{1}{2}x + x^3 & \frac{1}{2}x & -\frac{1}{2} + \frac{1}{2}x + x^2 & \frac{1}{4}x - \frac{1}{2}x^2 \\
-\frac{1}{3}x^2 + x^4 & -\frac{1}{3}x^2 & -\frac{2}{3}x^2 & \frac{2}{3}x - \frac{2}{3}x^2 - x^3 & \frac{1}{6}x \\
0 & 0 & x^4 & -x^3 & \frac{1}{2}x^3
\end{pmatrix}
$$

and the controllable model is in rows 1 and 3.

Finally, we give an example to illustrate the possibility of deriving a closed form solution.

EXAMPLE 2.   For modelling discrete-time behaviours a formal implementation of the algorithm allows us to "write down" a Gröbner basis with respect to Position Over Term (POT) order, defined in this 1-variable situation by $x^r \mathbf{e}_i < x^t \mathbf{e}_j$ if $i < j$ or if $i = j$ and $r < t$. In this setting the operation of the algorithm is predictable in advance and the final result can be derived directly from the input data without any computation. In the previous example this gives the basis

$$
\begin{pmatrix}
x^5 & 0 & 0 & 0 & 0 \\
0 & x^5 & 0 & 0 & 0 \\
0 & 0 & x^5 & 0 & 0 \\
-x - 2x^2 - 4x^3 - 8x^4 & -x - 2x^2 - 6x^3 - 11x^4 & -x - x^2 - 3x^3 - 6x^4 & 1 & 0 \\
-2x - 4x^2 - 8x^3 - 16x^4 & -2x - 4x^2 - 10x^3 - 21x^4 & -2x^3 - 7x^4 & 0 & 1
\end{pmatrix}.
$$

REFERENCES

[1] ADAMS, W.W., AND LOUSTAUNAU, P, *An Introduction to Gröbner bases,* Springer-Verlag, New York, Berlin, 1994.

[2] A.C. ANTOULAS, Recursive modeling of discrete-time time series, in P. Van Doren and B. Wyman (eds.) *Linear Algebra for Control Theory, IMA, Vol 62,* 1776–1802, Springer, Berlin, 1994.

[3] THOMAS BECKER, VOLKER WEISPFENNING, *Gröbner Bases A Computational Approach to Commutative Algebra* , Springer Verlag, New York, Berlin, 1993.

[4] DAVID COX, JOHN LITTLE, DONAL O'SHEA, *Ideals, varieties, and algorithms: An introduction to computational algebraic geometry and commutative algebra.,* Springer-Verlag, New York, Berlin, 1992.

[5] B.W. DICKINSON, M. MORPH, AND T. KAILATH, A minimal realization algorithm for matrix sequences, *IEEE Trans. on Automatic Control,* AC–19 (1974), 31–38.

[6] P. FITZPATRICK, On the key equation, *IEEE Trans. on Information Theory,* IT–41 (1995), 1290–1302.

[7] P. FITZPATRICK, On the scalar rational interpolation problem, *Mathematics of Control, Signals, and Systems,* 9 (1996), 352–369.

[8] P. FITZPATRICK, Solving a multivariable congruence by change of term order, *J. Symbolic Computation,* 11 (1997), 505–510.

[9] P. FITZPATRICK, Errors and erasures decoding of BCH codes, *IEE Proc.-Commun.*, 146, No. 2 (1999), 79–81.

[10] M. KUIJPER, An algorithm for constructing a minimal partial realization in the multivariable case, *Systems and Control Letters*, 31 (1997), 225–233.

ON ITERATIVE DECODING OF CYCLE CODES
OF GRAPHS

GILLES ZÉMOR*

Abstract. We analyze iterative decoding of cycle codes of graphs for the erasure channel and the binary symmetric channel. Cycle codes can achieve vanishing error-probability after decoding: furthermore, threshold probabilities can be computed exactly. We also prove that, for these codes, the asymptotical performance of iterative decoding and maximum-likelihood decoding coincide.

Key words. Graphs, Cycle codes, iterative decoding, min-sum algorithm.

AMS(MOS) subject classifications. 60C05, 68R10, 94B35, 94B70.

1. Introduction. Cycle codes of graphs are codes that have a parity-check matrix with exactly two "ones" per column. Among all codes that are amenable to iterative decoding, i.e. codes with "low-density" parity-check matrices (among which Gallager codes, Tanner codes, Turbocodes and others), cycle codes of graphs are therefore the codes with lowest possible density. Even though they turn out to be not truly practical, their study is interesting from the point of view of iterative decoding because their simpler structure makes their analysis easier than that of the other low-density codes. It is natural to hope that understanding properly iterative decoding for the simpler codes will give insight into the performance of the more sophisticated families of codes.

In [1] and [13] we studied the behaviour of maximum-likelihood decoding of cycle-codes of graphs for the binary symmetric channel. In [1], together with L. Decreusefond we showed that, for cycle codes of Δ-regular graphs, the error probability of maximum-likelihood decoding cannot vanish with the block length if the transition error probability p of the channel satisfies $p > \theta$ where

$$\theta = \frac{1}{2} \left(1 - \sqrt{1 - \frac{1}{(\Delta - 1)^2}} \right).$$

In [13], together with J-P. Tillich we showed that the cycle codes of some classes of Δ-regular Ramanujan graphs are such that the error probability of maximum-likelihood decoding vanishes with block length when $p < \theta$. The quantity θ can therefore be considered as a threshold probability for maximum-likelihood decoding.

It turns out that for $p < \theta$, iterative decoding achieves vanishing bit error probability. Therefore θ is a threshold value in more than one sense.

---

*École Nationale Supérieure des Télécommunications, Computer Science and Network Dept., 46 rue Barrault, 75634 Paris 13, France. Email: zemor@infres.enst.fr.

In this paper we wish to address these issues and take up the subject of cycle codes of graphs from the perspective of iterative decoding: we shall explain the relevant probabilistic tools from [1].

2. Cycle codes, overview. Let G be a finite, undirected, connected graph without loops or multiple edges. It is defined by its *vertex*-set V and its *edge*-set where an edge is a pair of vertices. Let r and n denote the number of vertices and edges respectively, and let the set of edges be numbered, so that it is identified with $\{1, \ldots n\}$. Identify furthermore subsets of edges with their characteristic vectors in $\{0, 1\}^n$: A *cycle* $c \in \{0, 1\}^n$ of G is a subset of edges with the property that any vertex of G is incident to an even number of edges of c. The set C of cycles of G is a subset of $\{0, 1\}^n$ stable under addition modulo 2: it is therefore a linear code of length n called the *cycle code* of G.

An incidence matrix $H = (h_{ij})$ of G is a $r \times n$ matrix whose rows are indexed by the vertices of G, whose columns are indexed by the edges, and such that $h_{ij} = 1$ whenever vertex i belongs to edge j and $h_{ij} = 0$ otherwise. An incidence matrix of G is also a parity check matrix of the cycle code C. The cycle code of a graph has therefore a parity check matrix with exactly two 1's per column: it is a low-density parity check code in the sense of Gallager [3].

It is well known that the dimension of C is $k = n - r + 1$ when the graph is connected. The study of the minimum distance d of C, i.e. the size of the smallest cycle, or *girth*, of G has been of interest to graph-theorists since the early 60's and is still very much open, but the bad news for coding-theorists is that for fixed rate $R = k/n$ and growing n, d cannot grow faster than a logarithm of n.

Here is a short proof of this in the case when G is *regular*. A graph G is said to be Δ-*regular* if every vertex is incident to exactly Δ edges, or equivalently if every vertex has exactly Δ neighbouring vertices. Note that $R = 1 - 2/\Delta + 1/n$, so that for large n, fixing the degree fixes the rate. Let v be any given vertex of the graph and consider the set of vertices at distance from v strictly less than $d/2$ in the graph. Because G does not contain a cycle of length strictly smaller than d, the subgraph induced by this set of vertices must be a tree, as represented in figure 1.

The number of vertices of G at distance 1 from v is therefore Δ, the number of vertices at distance 2 from v is $\Delta(\Delta - 1)$, and the number of vertices at distance i from v, $i < d/2$, is $\Delta(\Delta - 1)^i$. We conclude that the total number of vertices $|V|$ in the graph has to be bigger than

$$1 + \Delta + \Delta(\Delta - 1) + \cdots + \Delta(\Delta - 1)^{\lceil d/2 \rceil - 1}.$$

In other words, when n is large d cannot be significantly larger than

$$2 \log_{(\Delta - 1)} |V|.$$

Let us mention in passing that it still is a challenging open problem to know just how good this upper bound is asymptotically. Erdös and Sachs [2]

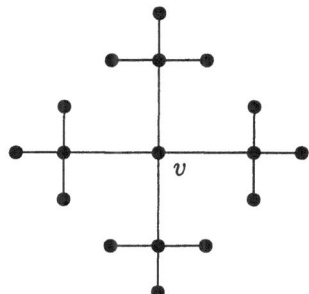

FIG. 1. *The neighbourhood of v in a Δ-regular graph (here Δ = 4) with no small cycles.*

showed with random arguments that families of Δ-regular graphs exist that satisfy $d \geq \log_{(\Delta-1)} |V|$, and Margulis [10] , and independently Lubotsky, Philips and Sarnak [8], found an algebraic construction of graphs satisfying $d \geq 4/3 \log_{(\Delta-1)} |V|$: this is at present where the gap stands between upper and lower bounds on d.

For non-regular graphs the situation is rather worse. For coding purposes the regular graphs tend to be the best and most of our efforts will be concerned with them. We shall occasionally encounter the general irregular case.

Cycle codes of graphs aroused early interest [6], but because of their poor minimum distance they were quickly discarded by the early coding theorists. Possibly this was one of the reasons Gallager focused his analysis on codes whose parity-check matrices have at least three 1's per column.

However, even though they have poor minimum distances, cycle codes of graphs can achieve vanishing error probability after decoding. This is true both for maximum-likelihood decoding and for iterative decoding. Furthermore, their performance can be analyzed rigorously and their asymptotical error-correcting capabilities computed exactly. We give a precise meaning to these statements in the next sections.

3. Decoding. Our main objective is the understanding of the asymptotical behaviour of cycle codes on the binary symmetric channel. We will also be concerned with the erasure channel which is simpler and will serve as a very useful intermediate study.

3.1. Maximum-likelihood decoding. Let us say that a vector $x = (x_1, \ldots, x_n)$ of $\mathbb{F}_2^n$ *covers* another vector $y = (y_1, \ldots, y_n)$ if $y_i = 1$ implies $x_i = 1$. Let us write $y \subset x$ to signify that x covers y. Let C be any binary linear code of length n.

The erasure channel. When submitted to the erasure channel with erasure probability p, each symbol of the transmitted codeword

$$x = (x_1, \dots, x_n)$$

is erased, independently of the others, with probability p. Let

$$\varepsilon = (\varepsilon_1, \dots, \varepsilon_n)$$

be the characteristic vector of the set of positions where an erasure occurs, and let us call it the *erasure vector*. The codeword x can be recovered with no ambiguity by the receiver if and only if the erasure vector does not cover a non-zero codeword c. The probability that this does not happen, i.e. that unrecoverable ambiguities occur, is therefore:

$$P[\varepsilon \in A] = \sum_{a \in A} p^{|a|}(1-p)^{n-|a|}$$

where $|a|$ denotes the weight of a and where

$$A = \{a \in \mathbb{F}_2^n \mid \exists c \in C, c \neq 0, c \subset a\}.$$

The binary symmetric channel. When a codeword $x = (x_1, \dots, x_n)$ is submitted to the binary symmetric channel with bit error probability p, the receiver gets $x + e$ where the *error vector* $e = (e_1, \dots, e_n) \in \mathbb{F}_2^n$ is such that $e_i = 1$ with probability p, independently of the others. Maximum-likelihood decoding consists of picking one of the codewords (there might be several) closest to $x + e$ for the Hamming distance. Decoding fails if $e \in E$ where E is defined as the set of vectors closer to a non-zero codeword than to the zero codeword, i.e.

$$E = \{e \in \mathbb{F}_2^n \mid \exists c \in C, d(e, c) < |e|\}.$$

Notice that E may be redefined as

$$E = \{e \in \mathbb{F}_2^n \mid \exists c \in C, y \subset c, |y| > |c|/2, \text{ and } y \subset e\}.$$

In other words, E is the set of vectors that "cover half the coordinate positions of some non-zero codeword". Note the similarity of this wording of E with the set A defined above for erasures. Maximum-likelihood decoding fails with probability at least

$$P[e \in E].$$

3.2. Iterative decoding. When recovering from erasures is possible, then it can be done with reasonable complexity, since this involves at worst solving a system of $n - |\varepsilon|$ linear equations in k binary variables. However, maximum-likelihood decoding for the binary symmetric channel is NP-complete [4], and even when the code C is restricted to the class of cycle codes of graphs no polynomial algorithm is known. Iterative decoding is suboptimal, but its complexity is polynomial (actually almost linear) in n. Several versions of iterative decoding schemes exist and have been extensively discussed [15]: for the sake of clarity and self-containment we give here a brief description of the one that will be most adapted to our purposes.

A. Local decisions. Here is a very general decoding idea. Let $x = (x_1, \ldots, x_n)$ be the original codeword. Let $i \in \{1, \ldots, n\}$ be a given coordinate position and let us try and decide whether $x_i = 0$ or $x_i = 1$ based on the received vector. Since complete decoding is difficult let us consider a subset $I \subset \{1, \ldots, n\}$ of coordinate positions such that $i \in I$, and let us base our decision *locally*, i.e. only on the subset I of coordinates. For any vector $y \in \mathbb{F}_2^n$, denote by y_I the shortened vector $y_I = (y_j)_{j \in I}$ and by C_I the *shortened code*

$$C_I = \{c_I, c \in C\}.$$

Now define

$$C_I(i, 0) = \{c_I \in C_I, c_i = 0\}$$
$$C_I(i, 1) = \{c_I \in C_I, c_i = 1\}.$$

Note that $C_I(i, 0)$ is a subcode of C_I and that $C_I(i, 1)$ is a coset of $C_I(i, 1)$. We shall drop the i when the coordinate i is implicit and write $C_I(0)$ and $C_I(1)$.

For the erasure channel, the situation is very similar to the global case. The value of x_i can be recovered unambiguously if and only if the shortened erasure vector ε_I covers a shortened codeword of $C_I(1)$.

For the binary symmetric channel, the situation is somewhat modified. MAP (Maximum A Posteriori Probability) decoding involves computing, conditional on the received bits in the positions of I, the probabilities $p_I(0)$ and $p_I(1)$ that $x_I \in C_I(0)$ and that $x_I \in C_I(1)$ respectively. The decoder then decides what is the value of x_i based on which of the two probabilities is largest. Note that since $p_I(0)$ and $p_I(1)$ are functions of p, the decoder must know p.

We shall rather consider the following variant for decoding. Let $v = x + e$ be the received vector. Let us decide that $x_i = 1$ if the minimum number of coordinates of $x + e$ in I that must be changed in order to obtain a codeword of $C_I(1)$ is smaller than the minimum number of coordinates that must be changed to obtain a codeword of $C_I(0)$. In other words, this means computing the Hamming distances $D_I(0) = d(v_I, C_I(0))$ and $D_I(1) = d(v_I, C_I(1))$, and deciding that $x_i = 0$ if $D_I(0) < D_I(1)$ and $x_i = 1$ if $D_I(0) > D_I(1)$.

Strictly speaking, this differs slightly from MAP decoding, but it may be argued that the events leading to different decisions for x_i are rare and concern unlikely code structures.

Note that $D_I(0)$ and $D_I(1)$ can always be computed by exhaustively examining all the codewords of C_I.

B. Iterating procedures. If I is too large, the direct computation of the quantities we just discussed may be too complex, so suppose now that $I = J \cup K$ where $i \in J$ and where $J \cap K = \{j\}$, $j \neq i$. Let us discuss briefly how, by using this decomposition of I, a decision on the value of x_i can be computed that takes into account all the coordinates in I.

For the erasure channel, it should be clear that the shortened erasure vector ε_I covers a shortened codeword of $C_I(i, 1)$ if and only if

- either x_j is unerased and ε_J covers a shortened codeword of $C_J(i, 1)$
- either x_j is erased and ε_J covers a shortened codeword of $C_J(i, 1)$ and ε_K covers a shortened codeword of $C_K(j, 1)$.

Therefore, to discover x_i it is enough to first discover (if possible) x_j by decoding C_K and then to decode C_J.

For the binary symmetric channel, we need to compute $D_I(0)$ and $D_I(1)$. We have:

$$D_I(0) = \min_{c_J \in C_J(i,0)} (d(v_J, c_J) + d(v_K, C_K(j, c_j)) - d(v_j, c_j))$$
$$D_I(1) = \min_{c_J \in C_J(i,1)} (d(v_J, c_J) + d(v_K, C_K(j, c_j)) - d(v_j, c_j)).$$

Therefore, to compute $D_I(0)$ and $D_I(1)$ we can first compute

$$d(v_K, C_K(j, 0)) \quad \text{and} \quad d(v_K, C_K(j, 1))$$

and then apply the above equalities. This strategy can be generalized if I can be decomposed into a treelike union of subsets, any two of which have an intersection of at most one element. This is now known as the "min-sum" algorithm and is essentially Tanner's algorithm B [12].

C. Cycle codes. We now turn to the case when C is the cycle code of a Δ-regular graph G. Keep in mind that a coordinate position i numbers an *edge* of the graph. To obtain a "local" set I of coordinates (edges), take the set N_1 of edges incident to *one* of the endpoints of i, together with the set N_2 of edges that have an endpoint incident to N_1 and so on, so that

$$I = N_1 \cup N_2 \cup \cdots N_h.$$

Let us call the edge i the *root* of the *neighbourhood* I and call h the *depth* of I. We shall suppose throughout that I has the tree structure depicted in figure 2, i.e. that I contains no cycle.

To justify this, note that if G is a randomly chosen Δ-regular graph with n edges, it can be proved that the expected number of cycles of a given fixed length is constant (does not grow with n). Therefore, for fixed h and growing n, almost all edges i will have tree-like neighbourhoods. Even better, there exist constructions (for example those of [8] and [10]) of

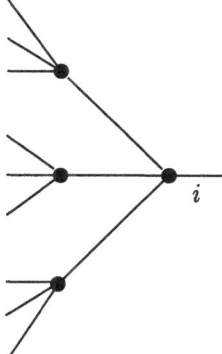

FIG. 2. *The local decoding region I associated to edge, or coordinate, i. Here the depth of I is $h = 2$.*

Δ-regular graphs with n edges and no cycle of length smaller than a linear function of $\log n$.

From now on we are not concerned anymore with algorithmic and complexity issues. By this we mean that we shall not try to analyze the iterative procedure itself, but rather, since we have just seen that it is equivalent, we shall study directly the probability that complete decoding of C_I recovers the correct value of x_i. We shall focus in particular on the behaviour, as the depth of I grows, that decoding on I fails.

4. Cycle codes and the erasure channel. In this section let us denote by $f_I(p)$ the probability that the symbol in position i is erased and that local decoding will not recover it, i.e. that the erasure vector covers a word of $C_I(1)$.

We shall prove that:

PROPOSITION 4.1.

1. *When $p < 1/(\Delta - 1)$, then $f_I(p)$ converges to zero as h tends to infinity.*
2. a. *When $p > 1/(\Delta - 1)$, then, independently of the depth h of I, $f_I(p)$ is lower bounded by a non-zero constant.*
   b. *Furthermore, "global" decoding will not fare better. The probability that the whole erasure vector covers a codeword is also lower bounded by a constant (depending only on p and not on the graph).*

The first step is to notice that the erasure vector ε_I covers a codeword of $C_I(1)$ if and only if ε_I *covers a path of I of length $h + 1$ that contains the root i.*

Proving point 1 is now easy. The probability that the erasure vector covers a path of I rooted at i of length $h + 1$ is less than the expected number of such paths, so that:

(4.1) $$f_I(p) \leq (\Delta - 1)^h p^{h+1}.$$

This clearly deals with point 1. Actually $f_I(p)$ may be estimated more precisely. If we write I_h instead of I to specify the depth of the neighbourhood I we have the relation

(4.2) $$f_{I_{h+1}}(p) = p(1 - (1 - f_{I_h}(p))^{\Delta-1}).$$

By studying (4.2) it is straightforward to see that $f_{I_h}(p)$ decreases with h (also seen directly, since a path of length $h + 1$ contains a path of length h), and that it converges to a non-zero value when $p > 1/(\Delta - 1)$. This proves point 2a. We shall later give an alternative proof of this that can be generalized in ways that (4.2) cannot.

Let us now address point 2b. The intuition behind our argument is that when $p > 1/(\Delta - 1)$ then with non-zero probability we can follow an infinite path of "erased" edges on an infinite tree rooted at i that "projects" onto a path of the finite graph G. Since G is finite, this path must wrap around itself and contain cycles.

The rest of this section is devoted to making this argument rigorous. Let μ_p be the measure on $\{0, 1\}$ defined by $\mu(\{1\}) = p$ and $\mu(\{1\}) = 1 - p = q$. We have two probability spaces: one is defined on the edge-set of G with the product measure $P_G = \mu_p^{\otimes\{1,\ldots,n\}}$, the other is defined on the edge-set $\mathcal{E}$ of an infinite Δ-regular tree Γ with the product measure $P_\Gamma = \mu_p^{\otimes\mathcal{E}}$. It is useful to view the infinite tree Γ as *the universal cover* of G. This is obtained from G by considering all possible paths in G starting at a given edge i. In this way all edges of Γ are labeled by a path of G starting at i.

Let A_h be the event consisting of all subsets of edges of the infinite tree that include a path of length $h > n$ containing i. Let us compare the probability $P_G(\varepsilon \text{ covers a cycle})$ in the finite graph G with the probability $P_\Gamma(A_h)$ in the tree Γ.

Let $\mathcal{T}$ be the collection made up of the emptyset and of all subgraphs T of G which are trees, contain the edge i, and are such that that no edge of G has both its endpoints in T. Let $T \in \mathcal{T}$ be given and let α be the number of edges of T and let β be the number of edges of G that have exactly one endpoint in T. Now consider the set of erased edges of G and let R_G be the induced connected component containing the edge i. The probability in G that $R_G = T$ is:

$$P_G(R_G = T) = p^\alpha q^\beta.$$

Because we have supposed that edges outside T have at most one endpoint in T we must have

$$\beta = (\alpha + 1)(\Delta - 2) + 2.$$

Now any tree T containing i is isomorphic to a subtree of Γ containing i. Consider any random subset ω of edges of Γ and let R_Γ be the induced connected component in Γ that contains i: the key observation is that in Γ we also have

$$P_\Gamma(R_\Gamma = T) = p^\alpha q^{(\alpha+1)(\Delta-2)+2}.$$

Therefore we have

(4.3) $$P_\Gamma(R_\Gamma \in \mathcal{T}) = P_G(R_G \in \mathcal{T}).$$

Let ω be a random set of edges of Γ and suppose $\omega \in A_h$, i.e. ω is an event which includes a path of length h containing i. Because we have chosen $h > n$ and all subtrees of G have at most n vertices we must have $R_\Gamma \notin \mathcal{T}$. Therefore $P_\Gamma(A_h) \leq P_\Gamma(R_\Gamma \notin \mathcal{T})$. Put $\eta = \lim_{h\to\infty} P_\Gamma(A_h)$: the hypothesis $p > 1/(\Delta - 1)$ implies $\eta > 0$ by point 2a, so that:

$$\eta < P_\Gamma(R_\Gamma \notin \mathcal{T}).$$

But (4.3) means that we have proved:

$$\eta < P_G(R_G \notin \mathcal{T}).$$

Now the event $(R_G \notin \mathcal{T})$ in G means that the connected component containing i is either not a tree, and therefore contains a cycle, or is a tree, but which contains two vertices joined by a non-erased edge. In other words, we have bounded away from zero (by η) the probability that the erasure vector covers, either a cycle, or a cycle with a missing edge.

Strictly speaking, this is not completely what we want, namely to bound away from zero the probability that the erasure vector covers a cycle. This technical difficulty could be dealt with directly by a more careful comparison of the relevant events on the finite graph G and on the infinite tree. However we can also deal with it with an indirect argument along the following lines. Consider the random vector $\varepsilon \cdot \lambda = (\varepsilon_1\lambda_1, \ldots, \varepsilon_n\lambda_n)$ where the λ_j are chosen independently and such that $\lambda_j = 1$ with probability π, $\lambda_j = 0$ otherwise. Choose π such that

$$p\pi = \frac{1}{\Delta - 1} + \frac{1}{2}\left(p - \frac{1}{\Delta - 1}\right).$$

We have $p\pi > 1/(\Delta - 1)$, so the probability that the vector $\varepsilon \cdot \lambda$ covers a cycle, or a cycle with a missing edge is, as we have seen, bounded away from zero. But then the probability that the remaining set of edges of ε, i.e. those j such that $\varepsilon_j = 1$ and $\lambda_j = 0$, contains the missing edge is also bounded away from zero.

5. Interlude: percolation on trees. Consider an infinite graph Γ, with edge-set $\mathcal{E}$, and pick a privileged vertex v. Choose a random subset of edges of Γ by picking each edge independently of the others with probability p. In other words we are measuring events on $\{0,1\}^{\mathcal{E}}$ with the product probability measure $P_\Gamma = \mu_p^{\otimes \mathcal{E}}$. *Percolation* theory concerns itself with the probability $f(p)$ that the random subset contains an infinite path passing through v. The theory is especially concerned with *critical* probabilities: the quantity $\theta \in (0,1)$ is a critical probability for the function f if $f(p) = 0$ for $p < \theta$ and $f(p) > 0$ for $p > \theta$. Probably the most famous result of percolation theory is that $\theta = 1/2$ is a critical probability when Γ is the infinite square lattice on $\mathbb{Z}^2$ (Kesten, [7]). A very readable account of the theory is [5].

The quantity $\theta = 1/(\Delta - 1)$ appearing in proposition 4.1 is in effect the critical probability for percolation when Γ is the infinite Δ-regular tree. Indeed, our discussion in sections 3.2 and 4 has shown that, in the case of the erasure channel, the limiting behaviour of iterative decoding for cycle codes of Δ-regular graphs is really given by $f(p)$ for the Δ-regular tree: (well, almost, because in choosing the neighbourhood I in figure 2 we have thrown away one of the branches leading away from i, but this does not affect critical probabilities).

As we have seen in section 4, proving that $f(p) = 0$ for $p < \theta$ can be achieved simply by computing expectations, namely proving that the expected size of the connected component R containing v is finite. When $p > \theta$, the expected size of R is infinite, but this is not enough in itself to ensure that the *probability* $f(p)$ that R contains an infinite path is non-zero. When Γ is the Δ-regular tree there is a straightforward way around this difficulty, which consists of studying the probabilities $f_h(p)$ that the random set of edges ω contains a path of length h starting at v. Because there is a relation between $f_{h+1}(p)$ and $f_h(p)$ (akin to (4.2)) the limit $\lim_{h\to\infty} f_h(p)$ can be computed. One finds that when $p > \theta$ this limit is positive, and so is $f(p)$.

Here is an alternative way of proving that $f(p) > 0$, which does not involve any formula of the type (4.2), and which will turn out to be especially interesting for us: it is due to R. Lyons [9].

For any vertex σ of Γ denote by $\partial\sigma$ the *depth* of σ, i.e. the length of the (unique) path from v to σ. Denote by S_h the sphere of radius h, i.e. the set of vertices of Γ of depth h. For any subset $\omega \subset \mathcal{E}$ of edges of Γ let us say that a vertex σ is *reachable* if all the edges of the path from v to σ are in ω. Define the set $R(\omega)$ of *reachable* vertices. For any vertex σ define

$$\phi(\sigma) = (\Delta - 1)^{-\partial\sigma}.$$

Finally, define the random variables

(5.1)
$$X_h = \sum_{\sigma \in S_h} \phi(\sigma) p^{-h} \mathbf{1}_{R(\omega)}(\sigma).$$

In words, the variable X_h counts the number of reachable vertices in S_h and multiplies them by the quantity $(\Delta - 1)^{-h} p^{-h}$.

The variable X_h depends only on the state (in ω or not in ω) of the edges at distance $\leq h$ from v: furthermore it is straightforward to check that the sequence (X_h) forms a martingale. Martingale theory tells us that (X_h) must converge almost surely to some variable X. It also tells us that if

$$(5.2) \qquad \sup_{h \geq 1} E[X_h^2] < \infty$$

then we must have $E[X] = E[X_h] = 1$. In that case X must be non-zero on a set of non-zero probability, so that, on a set of non-zero probability, $X_h \neq 0$ for all h, which means exactly that $R(\omega)$ is infinite and that there exists an infinite path passing through v.

Note that the function ϕ is a *unit flow*, i.e. satisfies $\phi(v) = 1$ and for any σ

$$\phi(\sigma) = \sum_{\sigma \to \tau} \phi(\tau)$$

where $\sigma \to \tau$ means that τ is a neighbour of σ of depth $\partial \tau = \partial \sigma + 1$. This makes the evaluation of $E[X_h^2]$ simple, we have:

$$E[X_h^2] = \sum_{\sigma, \tau \in S_h} \phi(\sigma) \phi(\tau) p^{-2h} P[\sigma, \tau \in R(\omega)]$$

$$= \sum_{\sigma, \tau \in S_h} \phi(\sigma) \phi(\tau) p^{-\partial(\sigma \wedge \tau)}$$

where $\sigma \wedge \tau$ means the vertex furthest away from v which is on both paths from v to σ and from v to τ. We have therefore

$$E[X_h^2] = \sum_{\psi, \partial \psi \leq h} p^{-\partial \psi} \sum_{\sigma \wedge \tau = \psi, \, \sigma, \tau \in S_h} \phi(\sigma) \phi(\tau)$$

$$(5.3) \qquad E[X_h^2] \leq \sum_{\psi, \partial \psi \leq h} p^{-\partial \psi} \phi(\psi)^2,$$

because ϕ is a flow. From the upper bound (5.3) it is now straightforward to obtain (5.2) whenever $p > 1/(\Delta - 1)$ and the desired result that there exists, with non-zero probability, an infinite path passing through v.

This argument may seem involved, but it has a number of advantages, in particular it generalizes to arbitrary, non-regular trees, for which there is no equivalent of the relation (4.2). Let X_h be defined as before by (5.1) but with an arbitrary unit flow ϕ. Inequality (5.3) still holds and can be upper bounded by introducing the quantity

$$(5.4) \qquad \mathrm{br}\Gamma = \sup_{\phi} \rho(\phi)$$

where ϕ runs over all unit flows and $\rho(\phi)$ is the radius of convergence of the series

$$\sum_{\sigma} \phi(\sigma)^2 \lambda^{\partial \sigma}$$

this last sum extending to all vertices σ of Γ. With this definition and the same argument as before we have that whenever $p > 1/\mathrm{br}\Gamma$, there exists an infinite path passing through v with non-zero probability.

R. Lyons [9] calls the quantity $\mathrm{br}\Gamma$ the *branching number* of Γ, and shows that we have

$$(5.5) \qquad \mathrm{br}\Gamma = \inf \left\{ \lambda > 0, \ \inf_{K} \sum_{\sigma \in K} \lambda^{-\partial \sigma} \right\}$$

where K runs over all *cutsets* K. A cutset K is a set of vertices of Γ such that the connected component of $\Gamma \setminus K$ is finite, and such that K is minimal with this property. Actually, Lyons shows that definitions (5.4) and (5.5) are equivalent (this involves the max-flow min-cut theorem). With definition (5.5) it is straightforward to show that whenever $p < 1/\mathrm{br}\Gamma$, the expected number of paths of length h starting at v must tend to zero with h and therefore the probability that an infinite path passes through v is zero. The quantity $1/\mathrm{br}\Gamma$ is therefore the critical probability for the function $f(p)$ for any infinite tree Γ.

The branching number is related to a simpler quantity called the *growth* $\mathrm{gr}\Gamma$, defined by:

$$\mathrm{gr}\Gamma = \liminf_{h \to \infty} M_h^{1/h}$$

where $M_h = \mathrm{card} S_h$. We have

$$\mathrm{gr}\Gamma \leq \mathrm{br}\Gamma,$$

and the inequality may be strict, see [9] for examples of such trees. However, if Γ is the universal cover of a finite graph we have $\mathrm{br}\Gamma = \mathrm{gr}\Gamma$.

6. Cycle codes and the binary symmetric channel. Consider now the binary symmetric channel and the issue of local decoding on the neighbourhood $I = I_h$ (figure 2) of an edge i. When does local decoding go wrong? We have seen that it goes wrong when the shortened error vector e_I is closer to the coset of the shortened subcode $C_I(1)$ than to $C_I(0)$. In particular, for this to happen, the error vector e_I must be closer to $C_I(1)$ than to the the 0 vector (which is in $C_I(0)$). Now the reader will easily convince himself that this happens if and only if *there exists in I a path of length $h + 1$ containing i, and with more than $(h + 1)/2$ edges in error.*

Let us consider this last event, that there exists in I a path of length $h+ 1$ containing i with more than $(h + 1)/2$ edges in error, and denote by $F_I(p)$

its probability. Our last discussion has just shown that the probability that local decoding fails is smaller than $F_I(p)$. There is a critical value for $F_I(p)$ when the depth h of I goes to infinity, which is:

$$(6.1) \qquad \theta = \frac{1}{2}\left(1 - \sqrt{1 - \frac{1}{(\Delta - 1)^2}}\right).$$

We have results similar to those of proposition 4.1, they read:

PROPOSITION 6.1.

1. *When $p < \theta$, then $F_I(p)$ converges to zero as h tends to infinity. In particular the probability that iterative decoding fails converges to zero as h tends to infinity.*

2.   a. *When $p > \theta$, then, independently of the depth h of I, $F_I(p)$ is lower bounded by a non-zero constant.*

       b. *Furthermore, "global" decoding will not fare better. The probability that the whole error vector is closer to a non-zero code-word than to the zero vector is also lower bounded by a constant (depending only on p and not on the graph).*

Proving point 1 is not difficult and is very much similar to the way point 1 of proposition 4.1 was proved. Let us say that an edge j is *extreme* in I if the unique path leading from i to j contains $h + 1$ edges (is of length $h + 1$). Let us call an edge j of I *reachable* if the path leading from i to j has more than half its edges "in error". Let us now count the expected number of reachable extreme edges. There are $(\Delta - 1)^h$ extreme edges, and the probability that a given extreme edge is reachable is

$$\sum_{\ell > (h+1)/2} \binom{h+1}{\ell} p^\ell q^{h+1-\ell}.$$

Since $p < 1/2$, the dominant term in this sum is that of the lowest ℓ, the sum is therefore equal to γ^h with γ asymptotically equivalent to $2(pq)^{1/2}$. The expected number of reachable extreme edges tends therefore to zero with h if and only if $2(\Delta - 1)(pq)^{1/2} < 1$. This yields point 1 of the proposition after some rearranging.

Assume point 2a for the moment and let us address point 2b. The same technique can be applied as in the proof of point 2b of proposition 4.1, namely comparing probabilities of events on the finite graph G and on the infinite Δ-regular tree Γ. Call R the set of vertices of G "reachable" from a given edge i: by this we mean those vertices that are at the other end of a path starting at i and with more than half its edges in error. Then consider the event that the subgraph of G induced by the set R of reachable vertices is a tree. On the one hand this is very close to the probability that i does not belong to a cycle with more than half its edges in error, and on the other hand this must be less than $1 - F_{I_h}(p)$ for h large enough.

To turn this argument into a formal proof involves some technicalities. We need to modify slightly the definition of "reachable" vertices to mean vertices that are the endpoints of path containing a fraction λ of edges in error with λ slightly larger than $1/2$, and we also need to demand that the relevant paths make up a sufficiently balanced set of edges in error and edges not in error. This is technical but not conceptually difficult: we leave out the details here and refer the interested reader to [1] for a complete proof.

Let us now finally address point 2a, the most difficult issue. The neighbourhood I can be imbedded in an infinite Δ-regular tree Γ. As in section 5, let $\mathcal{E}$ denote the set of edges of Γ and let us measure events on $\{0,1\}^{\mathcal{E}}$ with the product probability measure $P_{\Gamma} = \mu_p^{\otimes \mathcal{E}}$. As before, let v be a fixed vertex of Γ. This time, however, let us define a vertex σ to be *reachable* for $\omega \subset \mathcal{E}$ if the unique path from v to σ

$$[v, \sigma_1, \sigma_2, \dots, \sigma_h = \sigma]$$

is such that, for every ℓ, $1 \leq \ell \leq h$, at least half the edges of the path

$$[v, \sigma_1, \sigma_2, \dots, \sigma_\ell]$$

are in ω. Let $R(\omega)$ denote the set of reachable vertices and let $F(p)$ be the probability that $R(\omega)$ is infinite. It should be clear that $F_{I_h}(p) \geq F(p)$ for every h, so that we will be home when we prove:

THEOREM 6.1. *The quantity θ in (6.1) is a critical probability for $F(p)$, i.e. $F(p) = 0$ for $p < \theta$ and $F(p) > 0$ for $p > \theta$.*

That $F(p) = 0$ for $p < \theta$ is a consequence of point 1 of proposition 6.1. To prove that $F(p) > 0$ when $p > \theta$ we need an indirect method, since there is no direct way to compute $P_{\Gamma}(|R(\omega)| \geq h)$ or similar quantities. The martingale approach of section 5 turns out to be adequate. We just sketch the argument, refering the reader to [1] for details.

We need to restrict the definition of "reachable". Let $(u_\ell)_{\ell \geq 1}$ be an increasing sequence of positive integers and let $U_h = \sum_{\ell=1}^{h} u_\ell$. Let us say that a path

$$[\sigma_1, \sigma_2, \dots, \sigma_h]$$

is *heavy* if for every ℓ, $2 \leq \ell \leq h$, at least half the edges of the path

$$[\sigma_1, \sigma_2, \dots, \sigma_\ell]$$

are in ω. Let us say that a vertex σ of depth U_h is *u-reachable* if the subpaths of $[v, \sigma]$

$$[v, \dots, \sigma_{U_1}], [\sigma_{U_1}, \dots, \sigma_{U_2}] \cdots [\sigma_{U_{h-1}}, \dots, \sigma_{U_h} = \sigma]$$

are all heavy. Finally, Let $R_u^h(\omega)$ denote the set of u-reachable vertices of S_{U_h} and let $\pi(h)$ denote the probability that a vertex of depth U_h is u-reachable. Define the random variables

$$X_h = \sum_{\sigma \in S_{U_h}} \phi(\sigma) \pi(h)^{-1} \mathbf{1}_{R_u^h(\omega)}(\sigma)$$

for a unit flow ϕ. The sequence (X_h) is a martingale and by choosing $\phi(\sigma) = (\Delta - 1)^{-\partial\sigma}$ the same strategy as that of section 5 yields theorem 6.1. The proof extends naturally to arbitrary trees Γ and we have [1]:

THEOREM 6.2. *Let $F(p)$ be the probability that there exists an infinite path passing through v such that, for any vertex σ on that path, the subpath $[v, \sigma]$ has more than half its edges in ω. The quantity*

$$(6.2) \qquad \theta = \frac{1}{2}\left(1 - \sqrt{1 - \frac{1}{(\mathrm{br}\Gamma)^2}}\right)$$

is a critical probability for $F(p)$, i.e. $F(p) = 0$ for $p < \theta$ and $F(p) > 0$ for $p > \theta$.

7. Conclusion and comments.

7.1. Δ-regular graphs. On the binary symmetric channel, when $p < \theta$, for θ given by (6.1), we have shown implicitly that the min-sum iterative algorithm must converge to the right decision with error probability vanishing with block length n. This is because the probability that the min-sum algorithm fails is less than the quantity $F_{I_h}(p)$ studied in section 6. For the same reason, the sum-product algorithm, which is equivalent to MAP decoding on I must also have a vanishing probability of failure.

It is not clear however, how many bits will stay in error after iterative decoding? Only their *proportion* is proved to go to zero for $p < \theta$. Maximum-likelihood decoding may perform better in one way because it can correct *all* the bits in error, but only if the graph structure is right, i.e. it is not enough for the graph to have no small cycles. Those graphs do exist, and can be constructed, see [13].

When $p > \theta$ then we know that maximum-likelihood decoding fails with probability bounded below by a constant. A little more can be proved: if G belongs to a family (G_m) of Δ-regular graphs with girth tending to infinity with m, then the probability that maximum-likelihood decoding fails must tend to 1 with m. This is achieved by invoking Margulis's graph connectivity theorem [11] or its improvements [16, 14]. From the point of view of iterative decoding however, It is not really clear what fraction of bits will be left in error.

7.2. Irregular graphs. In that case the situation is somewhat more difficult to assess because individual bits can be protected differently. For instance one can choose a family (G_m) of graphs such that their cycle codes have fixed rate, but with a growing "handle", (a connected subgraph of degree 2). The bits corresponding to an edge in the middle of this "handle" can be corrected with a probability of error going to zero with m for any $p < 1/2$. However, if we restrict every G_m to the family of graphs with a given universal cover Γ, then the arguments of section 6 generalize and proposition 6.1 holds with the value of θ replaced by the formula (6.2). An example of a family of graphs with a given irregular universal cover is that of biregular bipartite graphs with degrees Δ_1 and Δ_2. In that case the branching number equals the growth and we have $\mathrm{br}\Gamma = \mathrm{gr}\Gamma = \sqrt{(\Delta_1 - 1)(\Delta_2 - 1)}$.

REFERENCES

[1] L. DECREUSEFOND AND G. ZÉMOR, *On the Error-Correcting Capabilities of Cycle Codes of Graphs*, Combinatorics, Probability and Computing (1997), **6**, 27–38.

[2] P. ERDÖS AND H. SACHS, *Reguläre Graphen gegebener Taillenweite mit minimaler Knotenzahl*, Wiss. Z. Univ. Halle–Wittenberg, Math.-Nat. (1963), **12**, 251–258.

[3] R.G. GALLAGER, *Low-density parity-check codes*, M.I.T. Press, 1963.

[4] M.R. GAREY, D.S. JOHNSON, AND L. STOCKMEYER, *Some simplified NP-complete graph problems*, Theor. Comput. Sci. (1976), **1**, 237–267.

[5] G. GRIMMET, *Percolation*, Springer-Verlag, 1989.

[6] S.L. HAMKIMI AND J.G. BREDESON, *Graph theoretic error-correcting codes*, IEEE Trans. on Inf. Theory (1968) IT-14, 584–591.

[7] H. KESTEN, *The critical probability of bond percolation on the square lattice equals* $\frac{1}{2}$, Communications in Mathematical Physics (1980), **74**, 41–59.

[8] A. LUBOTSKY AND R. PHILIPS AND P. SARNACK, *Ramanujan graphs*, Combinatorica (1988), **8**, 261–277.

[9] R. LYONS, *Random walks and percolation on trees*, The annals of probability (1990), **18**(3), 931–958.

[10] G.A. MARGULIS, *Explicit group-theoretical constructions of combinatorial schemes and their application to the design of expanders and concentrators*, Problemy Peredachi Informatsii (1988), **24**, 51–60.

[11] G.A. MARGULIS, *Probabilistic characteristics of graphs with large connectivity*, Problemy Peredachi Informatsii (1974), **10**, 101–108

[12] M. TANNER, *A recursive approach to Low-complexity codes*, IEEE Trans. on Inf. Theory (1981) IT-27, No. **5**, 533–547.

[13] J-P. TILLICH AND G. ZÉMOR, *Optimal cycle codes constructed from Ramanujan graphs*, Siam Journal on Discrete Math. (1997), **10**(3), 447–459.

[14] J-P TILLICH AND G. ZÉMOR, *Isoperimetric inequalities and the probability of a decoding error*, to appear in Combinatorics, Probability & Computing.

[15] N. WIBERG, *Codes and decoding on general graphs*, Ph.D. Thesis, Linköping, Sweden, 1996.

[16] G. ZÉMOR, *Threshold effects in codes*, in *First French-Israeli workshop on algebraic coding*, 1993 Lecture notes in Comput. Sci. 781, Springer-Verlag.

Part 4. Convolutional codes and codes over rings

CONVOLUTIONAL CODES OVER FINITE ABELIAN GROUPS: SOME BASIC RESULTS

FABIO FAGNANI* AND SANDRO ZAMPIERI†

Abstract. Abelian groups provide the most natural structure to represent codes over phase modulation signals. Convolutional codes over finite Abelian groups are introduced and the properties of linear encoders for this class of codes are analyzed. Through the structure theorem for finitely generated Abelian groups this analysis can be reduced to the study of of convolutional codes over the ring $\mathbb{Z}_n$. We can in this way introduce the concept of encoding group and to compare it with the more classical input group introduced in [6]. In the last part of the paper the state space realization of a convolutional code and of a convolutional encoder is investigated.

1. Introduction. Often the most convenient way to represent a linear code is by an encoder, which is an injective map whose range coincides with the code. For linear convolutional codes the encoders which are more classically considered are the so called "canonical feedback-free" encoders and the "systematic" encoders.

In recent years linear codes over rings and groups have attracted much attention, since they appear to be particularly suitable for the representation of codes over phase modulation signals [12, 13]. Encoder representation is particularly useful also for this class of codes. In the literature two different methods have been proposed to solve the encoder synthesis problem. The first method [6, 11] aims to extend the concept of canonical feedback-free encoder to this setup and to connect it with the controllability structure of the code. This strategy has the advantage that it can be applied to very general codes. Actually it can be proved that such a construction can be pursued even for timevaring linear codes over non-Abelian groups. The disadvantage is given mainly by the nonlinearity of the encoder it provides. This is the main reason why duality in this context is a difficult issue so that the canonical feedback-free syndrome former construction does not follow as a dual procedure [4]. Another disadvantage of this approach is that it is not based on polynomial matrices, which constitute a very powerful tool for the manipulation and the characterization of convolutional encoders over fields.

The second strategy in the encoder construction for convolutional codes over groups and rings, first proposed in [12, 13] and then developed in [7, 4], is based on the preliminary requirement that the encoders have to be homomorphisms. The advantage of this approach is on the one hand that the use of polynomial matrices is now allowed, and, on the other hand, that the duality for these encoders is straightforward. The main disadvantage of

---

*Dipartimento di Matematica, Politecnico di Torino, C.so Duca degli Abruzzi, 24, 10129 Torino, Italy.

†Dipartimento di Elettronica ed Informatica, Università di Padova, via Gradenigo, 6/a, 35131 Padova, Italy.

this approach is the fact that the class of codes for which the homomorphic encoders can be found is much more limited. For instance the concept of systematic encoder can be easily defined in this context [13]. However it can be shown that not every convolutional code over a ring admits such an encoder [13].

Most of the results according to this second approach, which can be found in the literature, are limited to codes characterized by an input space which is a free module [13, 12, 7, 4]. The aim of this paper is to show that it is possible to overcome this restriction and develop a more general theory.

Here we will focus our attention on some system theoretic properties of convolutional codes over finite Abelian groups and of their linear convolutional encoders. Moreover we will show the relations between some basic notions of convolutional codes and encoders, such as completeness, rationality, state space realizability. In a forthcoming paper we will investigate more specific coding theoretic properties of this class of codes, such as non-catastrophicity, minimality and systematicity. In Section 2 we will briefly recall certain classical facts for convolutional codes over finite fields. In Sections 3 and 4 we will establish a number of intrinsic characterizations for convolutional codes over finite Abelian groups, while in Section 5 we will obtain some structure results for their encoders.

2. Convolutional codes over finite fields.
We start by recalling the main definitions in the theory of convolutional codes over fields.

Let F be any field and let $F((D))$ denote the field of Laurent power series in D. Thus, $F((D))$ consists of elements of the form

$$a(D) = \sum_{k=-\infty}^{+\infty} a_k D^k, \qquad a_k \in F$$

where only finitely many coefficients with negative indices may be non-zero. The ring of formal power series $F[[D]]$, i.e. of power series having only non-negative powers, is a subring of $F((D))$. Also the polynomial rings $F[D]$, $F[D^{-1}]$ and $F[D, D^{-1}]$ are subrings of $F((D))$. With the symbol $F((D))^q$ we will indicate the $F((D))$-vector space of q-dimensional rows with entries in $F((D))$. The field of rational functions $F(D)$ is a subfield of $F((D))$. A convolutional code $\mathcal{C}$ over the field F can be defined as a subspace of $F((D))^q$ admitting a set of generators in $F(D)^q$ [14].

Notice that, if a subspace $\mathcal{C}$ of $F((D))^q$ admits a set of generators in $F(D)^q$, then it admits a basis in $F(D)^q$. Let $\{g_1(D), \ldots, g_m(D)\}$ be a basis of $\mathcal{C}$. Then

$$\mathcal{C} = \{w_1(D)g_1(D) + \cdots + w_m(D)g_m(D) \ : \ w_1(D), \ldots, w_m(D) \in F((D))\}$$

and so, defining the rational matrix

$$G(D) := \begin{bmatrix} g_1(D) \\ \vdots \\ g_m(D) \end{bmatrix} \in F(D)^{m \times q},$$

we can write

$$\mathcal{C} = \{w(D)G(D) \ : \ w(D) \in F((D))^m\}.$$

The rational matrix $G(D)$ can be seen as a map from $F((D))^m$ to $F((D))^q$ whose range coincides with the convolutional code $\mathcal{C}$. This map is called an *encoder* for $\mathcal{C}$. The advantage of rational encoders is the fact that they can be realized through a state space realization. More precisely, given a rational matrix $G(D) \in F(D)^{m \times q}$, there exist integers $n, l \in \mathbb{N}$ and matrices $A \in F^{n \times n}, B \in F^{m \times n}, C \in F^{n \times q}, D \in F^{m \times q}$ such that, given an input sequence $w(D) = \sum_k w_k D^k \in F((D))^m$ and an output sequence $v(D) = \sum_k v_k D^k \in F((D))^q$, we have that $v(D) = u(D)G(D)$ if and only if there exists a state sequence $x(D) = \sum_k x_k D^k \in F((D))^n$ for which the following relations hold

$$(1) \qquad \begin{cases} x_{k+1} &= x_k A + w_k B \\ v_{k-l} &= x_k C + w_k D \end{cases}$$

for all $k \in \mathbb{Z}$. The vector space F^n is called the state space, n is called the dimension of the state space realization and finally the integer l is called the *delay* of the state space realization.

3. Convolutional codes over finite Abelian groups. In general, a convolutional code over a finite Abelian group V is a subgroup of the Laurent power series group $V((D))$ admitting a rational encoder, a concept which can be defined as follows. The fundamental decomposition theorem for finite Abelian groups insures that V can be decomposed as the direct sum of of subgroups

$$V = V_1 \oplus \cdots \oplus V_l,$$

where V_i are subgroups of V which are isomorphic to $(\mathbb{Z}_{p_i^s})^{q_i}$ and where p_i are prime numbers in $\mathbb{Z}$. Similarly also the convolutional code $\mathcal{C}$ can be decomposed as

$$\mathcal{C} = \mathcal{C}_1 \oplus \cdots \oplus \mathcal{C}_l,$$

where each $\mathcal{C}_i$ is a convolutional code over $(\mathbb{Z}_{p_i^s})^{q_i}$. For this reason, without loss of generality, in the sequel we will present the definitions and the results assuming that $\mathcal{C}$ is a convolutional code over $(\mathbb{Z}_{p^s})^q$. In this case we can define the concept of convolutional code generalizing the definition we gave for convolutional codes over fields. To this aim we need the concept of Laurent power series with coefficients in $\mathbb{Z}_{p^s}$, which can be defined as above and that will be denoted by $\mathbb{Z}_{p^s}((D))$. This is not a field, and so $\mathbb{Z}_{p^s}((D))^q$ is not a vector space but only a module over $\mathbb{Z}_{p^s}((D))$. While the rings $\mathbb{Z}_{p^s}[[D]]$, $\mathbb{Z}_{p^s}[D]$, $\mathbb{Z}_{p^s}[D^{-1}]$ and $\mathbb{Z}_{p^s}[D, D^{-1}]$ can be defined in the obvious way, more attention must be paid to the definition of rational functions [4]. The ring of rational functions over $\mathbb{Z}_{p^s}$ can defined as follows

$$\mathbb{Z}_{p^s}(D) := \{p(D)/D^m q(D) \ : \ p(D), q(D) \in \mathbb{Z}_{p^s}[D], \ q_0 = 1, \ m \in \mathbb{N}\}.$$

Notice that according to this definition the ring of rational functions constitutes a subring of $\mathbb{Z}_{p^s}((D))$.

Definition. A convolutional code $\mathcal{C}$ over $\mathbb{Z}_{p^s}$ is a $\mathbb{Z}_{p^s}((D))$-submodule of $\mathbb{Z}_{p^s}((D))^q$ admitting a set of generators in $\mathbb{Z}_{p^s}(D)^q$.

The previous definition extends the definition given in [13, 12, 7, 4], since it does not requires that $\mathcal{C}$ is a free module, i.e. that $\mathcal{C}$ admits a rational basis. The generality of this extension will be clarified further by Theorem 3.2.

As above, if we take a set of generators $\{g_1(D), \ldots, g_m(D)\} \in \mathbb{Z}_{p^s}(D)^q$ of $\mathcal{C}$ and if we define from them the rational matrix $G(D) \in \mathbb{Z}_{p^s}(D)^{m \times q}$, we have that the rational matrix $G(D)$ can be seen as a map from $\mathbb{Z}_{p^s}((D))^m$ to $\mathbb{Z}_{p^s}((D))^q$, whose range coincides with the convolutional code $\mathcal{C}$. However, if $\mathcal{C}$ is not a free module, this map will never be an encoder for $\mathcal{C}$, since the generators $g_1(D), \ldots, g_m(D)$ will never be linearly independent and, consequently, the map associated with $G(D)$ will never be injective. In the literature devoted to linear encoder analysis, only convolutional codes admitting a rational basis have been considered. In order to overcome this difficulty we need to enlarge the class of modules which can be the input of the encoder. This is consequence of the following theorem.

THEOREM 3.1. *Given a subset $\mathcal{C}$ of $\mathbb{Z}_{p^s}((D))^q$, the following facts are equivalent.*

1. $\mathcal{C}$ *is a convolutional code over $\mathbb{Z}_{p^s}$.*
2. *There exist rational matrices $G_i(D) \in \mathbb{Z}_{p^s}(D)^{m_i \times q}$, $i = 0, 1, \ldots, s - 1$, such that*
   (a) $\mathcal{C} = \{w_0(D)G_0(D) + w_1(D)G_1(D) + \cdots + w_{s-1}(D)G_{s-1}(D) \ : \ w_i(D) \in p^i \mathbb{Z}_{p^s}((D))^{m_i}\}$;
   (b) *there exist polynomial matrices $Y_i(D) \in \mathbb{Z}_{p^s}[D, D^{-1}]^{q \times m_i}$, $i = 0, 1, \ldots, s - 1$, such that*

(2)
$$\begin{bmatrix} G_0(D) \\ G_1(D) \\ \vdots \\ G_{s-1}(D) \end{bmatrix} [Y_0(D) \quad Y_1(D) \quad \ldots \quad Y_{s-1}(D)] = I,$$

where I is the $m \times m$ identity matrix and where $m := m_0 + m_1 + \cdots + m_{s-1}$.

The integers m_i are uniquely determined by $\mathcal{C}$.

Proof. (2.$\Rightarrow$1.) is straightforward.

(1.$\Rightarrow$2.) Observe that the ring $\mathbb{Z}_{p^s}(D)$ is a principal ideal ring [4] and so matrices over this ring admits the Smith form [9].

Assume now that $\mathcal{C}$ is $\mathbb{Z}_{p^s}((D))$-submodule of $\mathbb{Z}_{p^s}((D))^q$ admitting a set of generators in $\mathbb{Z}_{p^s}(D)^q$. This implies that there exists a matrix $L(D) \in \mathbb{Z}_{p^s}(D)^{l \times q}$ such that

$$\mathcal{C} = \{u(D)L(D) \; : \; u(D) \in \mathbb{Z}_{p^s}((D))^l\}.$$

Consider the Smith form of $L(D)$

$$L(D) = A(D) \begin{bmatrix} \Lambda & 0 \\ 0 & 0 \end{bmatrix} B(D),$$

where $A(D) \in \mathbb{Z}_{p^s}(D)^{l \times l}$ and $B(D) \in \mathbb{Z}_{p^s}(D)^{q \times q}$ are unimodular matrices and Λ is the diagonal matrix

$$\Lambda = \begin{bmatrix} I_{m_0} & 0 & \cdots & 0 \\ 0 & pI_{m_1} & \cdots & 0 \\ \vdots & \vdots & \ddots & \vdots \\ 0 & 0 & \cdots & p^{s-1}I_{m_{s-1}} \end{bmatrix}$$

and where I_{m_i} is the $m_i \times m_i$ identity matrix. Let $m := m_0 + m_1 + \cdots + m_{s-1}$ and let $\bar{G}(D)$ be the submatrix of $B(D)$ formed by the first m rows of $B(D)$. Notice that

$$\begin{aligned} \mathcal{C} &= \{u(D)L(D) \; : \; u(D) \in \mathbb{Z}_{p^s}((D))^l\} \\ &= \{w(D)\Lambda\bar{G}(D) \; : \; w(D) \in \mathbb{Z}_{p^s}((D))^m\}. \end{aligned}$$

Observe now that, since $\bar{G}(D)$ is a submatrix of a unimodular matrix, there exists $\bar{Y}(D) \in \mathbb{Z}_{p^s}(D)^{q \times m}$ such that $\bar{G}(D)\bar{Y}(D) = I$. Moreover, there exists a polynomial $r(D) \in \mathbb{Z}_{p^s}[D, D^{-1}]$ such that $r(D)\bar{Y}(D)$ is a polynomial matrix. Let $Y(D) := r(D)\bar{Y}(D)$ and $G(D) := r(D)^{-1}\bar{G}(D)$. Then we have that

$$\begin{aligned} \mathcal{C} &= \{w(D)\Lambda G(D) \; : \; w(D) \in \mathbb{Z}_{p^s}((D))^m\} \\ &= \{w_0(D)\bar{G}_0(D) + w_1(D)\bar{G}_1(D) + \cdots + w_{s-1}(D)\bar{G}_{s-1}(D) \; : \\ &\qquad\qquad w_i(D) \in p^i\mathbb{Z}_{p^s}((D))^{m_i}\}, \end{aligned}$$

where the matrices $G_i(D)$ are produced by the partition of $G(D)$

(3)
$$G(D) = \begin{bmatrix} G_0(D) \\ G_1(D) \\ \vdots \\ G_{s-1}(D) \end{bmatrix}.$$

Moreover, by partitioning $Y(D)$ in a suitable way, we obtain that (2) holds true.

In order to prove the uniqueness of the indices m_i introduce the following subgroups of $\mathcal{C}$

$$\mathcal{C}_i := \{p^{i-1}v(D) \ : \ v(D) \in \mathcal{C}, \ p^i v(D) = 0\}.$$

It can be proved that $\mathcal{C}_i$ are vector spaces over $\mathbb{Z}_p((D))$ and that they are isomorphic to $\mathbb{Z}_p((D))^{m_0+\cdots+m_{i-1}}$. This implies that the indices m_i are canonically determined from $\mathcal{C}$.   ∎

Notice that the previous theorem has various important consequences. First, the theorem shows that there exists an injective rational encoder for any convolutional code $\mathcal{C}$ over $\mathbb{Z}_{p^s}$ subject to the fact that the input space is given by the Laurent series space $W((D))$, where

$$W = W(\mathcal{C}) := \mathbb{Z}_{p^s}^{m_0} \oplus p\mathbb{Z}_{p^s}^{m_1} \oplus \cdots \oplus p^{s-1}\mathbb{Z}_{p^s}^{m_{s-1}}.$$

Indeed, the rational matrix $G(D) \in \mathbb{Z}_{p^s}(D)^{m \times q}$ introduced in (3) can be seen as an injective map from $\mathbb{Z}_{p^s}((D))^m$ to $\mathbb{Z}_{p^s}((D))^q$, and, restricting this map to the submodule $W((D))$ of $\mathbb{Z}_{p^s}((D))^m$, we obtain a new map, still denoted by the symbol $G(D)$, which is injective and whose range coincides with $\mathcal{C}$. Therefore this map can be considered an encoder for $\mathcal{C}$. The Abelian group $W(\mathcal{C})$ is called the *encoding group* of $\mathcal{C}$ and depends only on the convolutional code $\mathcal{C}$.

The previous theorem ensures moreover that this encoder is non-catastrophic [5] since, by property 2b in the previous theorem, a finite support codeword in $\mathcal{C}$ must be the image of a finite support sequence in the input space $W((D))$.

Finally the previous theorem shows that a polynomial encoder can be obtained from the encoder proposed in theorem. Indeed, observe that there exists a polynomial $r(D) \in \mathbb{Z}_{p^s}[D, D^{-1}]$ such that $P(D) := r(D)G(D)$ is a polynomial matrix. It is clear that the polynomial matrix $P(D)$ provides a map form $W((D))$ to $\mathbb{Z}_{p^s}((D))^q$ which constitutes an encoder for the convolutional code $\mathcal{C}$.

We want to give now a more intrinsic characterization of convolutional codes over $\mathbb{Z}_{p^s}$. To this aim we need to introduce the concept of completeness. Given a Laurent power series $v(D) = \sum_k v_k D^k \in \mathbb{Z}_{p^s}((D))^q$, for any $a \leq b \in \mathbb{Z}$ we define the truncation

$$v(D)|_{[a,b]} := \sum_{k=a}^{b} v_k D^k.$$

Moreover, given a subset $\mathcal{C}$ of $\mathbb{Z}_{p^s}((D))^q$, we define $\mathcal{C}|_{[a,b]}$ to be the set of the truncations $v(D)|_{[a,b]}$ for all $v(D) \in \mathcal{C}$.

Definition. [16] A subset C of $\mathbb{Z}_{p^s}((D))^q$ is said to be *complete* if

$$v(D) \in \mathbb{Z}_{p^s}((D))^q \quad \text{such that} \quad v(D)|_{[k,k+L]} \in C|_{[k,k+L]}, \ \forall k \in \mathbb{Z},$$

$$\forall L \in \mathbb{N} \quad \Rightarrow \quad v(D) \in C.$$

We can now give a result which provides a nice intrinsic characterization of convolutional codes. The proof of this theorem follows easily from [4, Lemma 3].

THEOREM 3.2. *Given a subset C of $\mathbb{Z}_{p^s}((D))^q$, the following facts are equivalent.*

1. *C is a convolutional code over $\mathbb{Z}_{p^s}$.*
2. *(a) C is a $\mathbb{Z}_{p^s}$-module (linearity).*
   (b) $DC = C$ (time-invariance).
   (c) C is complete.

In fact, convolutional codes satisfy a stronger version of completeness, called *strong completeness*. This is defined as follows.

Definition. [16] A subset C of $\mathbb{Z}_{p^s}((D))^q$ is said to be *L-complete* if

$$v(D) \in \mathbb{Z}_{p^s}((D))^q \quad \text{such that} \quad v(D)|_{[k,k+L]} \in C|_{[k,k+L]},$$

$$\forall k \in \mathbb{Z} \quad \Rightarrow \quad v(D) \in C.$$

A subset C of $\mathbb{Z}_{p^s}((D))^q$ is said to be *strongly complete* if it is *L*-complete for some $L \in \mathbb{N}$.

We have the following result, whose proof can be found in [11, Prop. 2.3] or [17, Prop. 1].

THEOREM 3.3. *A convolutional code C is strongly complete.*

We noticed in Theorem 3.1 that the encoding group $W(C)$ is canonically determined starting from the convolutional code C and it constitutes the right input space for all the homomorphic encoders for C. In the canonical encoder synthesis proposed in [6] the following finite Abelian group

$$U(C) = \{v_0 \in \mathbb{Z}_{p^s}^q \ : \ v(D) = \sum_{k=0}^{+\infty} v_k D^k \in C\}$$

appeared to be the right input space for the non-homomorphic encoders proposed there. The finite Abelian group $U(C)$ is called the *input group* of the convolutional code C.

The following result establishes the relation between the input group and the encoding group of a convolutional code.

PROPOSITION 3.1. *Let $C \subseteq \mathbb{Z}_{p^s}((D))^q$ be a convolutional code over $\mathbb{Z}_{p^s}$. Then the finite Abelian groups $U(C)$ and $W(C)$ have the same cardinality.*

Proof. The proof uses techniques from symbolic dynamics. The reader is referred to [10] for the explanation of the various concepts which will be used. Consider in $\mathbb{Z}_{p^s}^q$ the discrete topology and in $(\mathbb{Z}_{p^s}^q)^{\mathbb{Z}}$ the induced product topology. Let $\bar{C}$ be the closure of C in $(\mathbb{Z}_{p^s}^q)^{\mathbb{Z}}$. Since C is complete, it can be shown that

$$C = \bar{C} \cap \mathbb{Z}_{p^s}((D))^q.$$

On the other hand, it follows from Theorem 3.1 that there exists a polynomial encoder for C

$$\psi \; : \; W((D))^q \; \to \; C,$$

where $W := W(C)$. This encoder ψ can be trivially extended to a map

$$\bar{\psi} : W^{\mathbb{Z}} \to (\mathbb{Z}_{p^s}^q)^{\mathbb{Z}}.$$

It can be shown that $\bar{\psi}(W^{\mathbb{Z}}) = \bar{C}$, while the kernel $\ker \bar{\psi}$ of $\bar{\psi}$ can be non-zero.

For closed time-invariant sets of trajectories, such as $\ker \bar{\psi}$, $W^{\mathbb{Z}}$ and $\bar{C}$, the important concept of topological entropy can be introduced. The entropy, which is denoted by the symbol $h(\cdot)$, is a sort of measure of the size of these sets. It is known [10] that the entropy of a closed time-invariant set of trajectories $\mathcal{B}$ which forms a group coincides with $\log_2 |U(\mathcal{B})|$, where $|U(\mathcal{B})|$ is the cardinality of the input group of $\mathcal{B}$. It is known [10] moreover that the following additive formula holds true

(4) $$h(W^{\mathbb{Z}}) = h(\bar{C}) + h(\ker \bar{\psi}).$$

These facts imply that $h(W^{\mathbb{Z}}) = \log_2 |W|$ and moreover, since

$$U(\ker \bar{\psi}) = U(\ker \bar{\psi} \cap W((D))) = U(\ker \psi) = \{0\},$$

we have that $h(\ker \bar{\psi}) = 0$. Finally, since $U(\bar{C}) = U(C)$, then $h(\bar{C}) = \log_2 |W|$ which yields the result. ∎

In general the Abelian groups $W(C)$ and $U(C)$ are not isomorphic as shown in the following example.

Example. Consider the convolutional code C over $\mathbb{Z}_4$ defined as follows

$$C := \{w(D)G(D) \; : \; w(D) \in \mathbb{Z}_4((D))\},$$

where $G(D) := [2 \quad D] \in \mathbb{Z}_4(D)^{1 \times 2}$. It is clear that $G(D)$ provides an encoder for C and this implies that $W(C) = \mathbb{Z}_4$. On the other hand it can be shown that $U(C) = 2\mathbb{Z}_4 \oplus 2\mathbb{Z}_4$. Therefore in this case we have that $W(C)$ and $U(C)$ are not isomorphic.

4. Generalized convolutional codes. In the following we provide also a generalized version of a convolutional code, which corresponds to a convolutional code without the requirement of rationality of the generators. We will then compare this class of codes with the class of convolutional codes.

Definition. A *generalized convolutional code* $\mathcal{C}$ over $\mathbb{Z}_{p^s}$ is a $\mathbb{Z}_{p^s}((D))$-submodule of $\mathbb{Z}_{p^s}((D))^q$.

It is interesting to notice that this generalization of convolutional codes can be introduce also in the context of convolutional codes over fields as the following example clarifies.

Example. Consider the systematic encoder $[1 \quad a(D)] \in \mathbb{Z}_2[[D]]^{1\times 2}$, where

$$a(D) = \sum_{k=0}^{+\infty} D^{2^k}.$$

Since $a(D)$ is not rational, this encoder clearly defines a generalized convolutional code. Notice that this code has infinite free distance, but, as we will clarify in the sequel, does not admit an encoder which can be realized by finite state space realization.

Also in this case, if we take a set of generators $\{g_1(D), \ldots, g_m(D)\} \in \mathbb{Z}_{p^s}((D))^q$ of $\mathcal{C}$ and if we define from them the matrix $G(D) \in \mathbb{Z}_{p^s}((D))^{m\times q}$, we have that the matrix $G(D)$ can be seen as a map from $\mathbb{Z}_{p^s}((D))^m$ to $\mathbb{Z}_{p^s}((D))^q$, whose range coincides with the convolutional code $\mathcal{C}$. However, also in this case we may have that no such maps are injective. This problem can be solved as we did for convolutional codes, i.e. by enlarging the class of modules which can be the input of the encoder. More precisely, we can find a subgroup W of $\mathbb{Z}_{p^s}^m$, and a matrix $G(D) \in \mathbb{Z}_{p^s}((D))^{m\times q}$ such that, by restricting this map to the submodule $W((D))$ of $\mathbb{Z}_{p^s}((D))^m$, we obtain a new map, still denoted by the symbol $G(D)$, which is injective and whose range coincides with $\mathcal{C}$. Therefore this map can be considered an encoder for $\mathcal{C}$. In this case $G(D)$ will be called a generalized encoder for the generalized convolutional code $\mathcal{C}$.

These facts are shown by the following Theorem where we provide also an intrinsic characterization for such class of codes. To this aim, we need to introduce first a weakened version of the concept of completeness.

Definition. A subset $\mathcal{C}$ of $\mathbb{Z}_{p^s}((D))^q$ is said to be past-complete if

$$v(D) \in \mathbb{Z}_{p^s}((D))^q \quad \text{such that} \quad v(D)|_{(-\infty,b]} \in \mathcal{C}|_{(-\infty,b]},$$

$$\forall b \in \mathbb{Z} \quad \Rightarrow \quad v(D) \in \mathcal{C}.$$

THEOREM 4.1. *Given a subset $\mathcal{C}$ of $\mathbb{Z}_{p^s}((D))^q$. Then the following facts are equivalent.*

1. C is a generalized convolutional code over $\mathbb{Z}_{p^s}$.
2. (a) C is a $\mathbb{Z}_{p^s}$-module (linearity).
   (b) $DC = C$ (time-invariance).
   (c) C is past-complete.
3. There exists a subgroup W of $\mathbb{Z}_{p^s}^m$ and a right invertible $G(D) \in \mathbb{Z}_{p^s}((D))^{m \times q}$ such that $G(D) : W((D)) \to \mathbb{Z}_{p^s}((D))^q$ provides a generalized encoder for C.

Moreover, if $G_1(D) : W_1((D)) \to \mathbb{Z}_{p^s}((D))^q$ and $G_2(D) : W_2((D)) \to \mathbb{Z}_{p^s}((D))^q$ are two generalized encoders for C, then the groups W_1 and W_2 are isomorphic.

Proof. (3.⇒2.) The fact that C is a $\mathbb{Z}_{p^s}$-module and that $DC = C$ is trivial. We have to prove that C is past-complete. Let $v(D) \in \mathbb{Z}_{p^s}((D))^q$ be such that $v(D)|_{(-\infty,t]} \in C|_{(-\infty,t]}$ for every $t \in \mathbb{N}$. Then for every $t \in \mathbb{N}$ there exists $v^t(D) \in C$ such that

$$v(D)|_{(-\infty,t]} = v^t(D)|_{(-\infty,t]}.$$

Let $G(D)$ be a right invertible matrix in $\mathbb{Z}_{p^s}((D))^{m \times q}$ providing a generalized encoder $G(D) : W((D)) \to \mathbb{Z}_{p^s}((D))^q$ for C. Notice that there exists $w^t(D) \in W((D))$ such that $v^t(D) = w^t(D)G(D)$. It is not restrictive to assume that $G(D)$ is causal. Assume moreover that the right inverse of $G(D)$ has support in $[-l, +\infty)$, with $l \geq 0$ Define $w(D) \in W((D))$ as follows

$$w_k := \begin{cases} w_k^0 & \text{if } k < -l \\ w_k^{k+l} & \text{if } k \geq -l \end{cases}$$

Observe that $v^{t+n}(D) - v^t(D)$ has support in $[t+1, +\infty)$ for all $n \in \mathbb{N}$ and this implies that $w^{t+n}(D) - w^t(D)$ has support in $[t-l+1, +\infty)$ and so $w^t(D)|_{(-\infty,t-l]} = w^{t+n}(D)|_{(-\infty,t-l]}$ for all $n \in \mathbb{N}$. ¿From this we can argue that for any $\tau \leq t$ we have that $w_\tau = w_\tau^{\tau+l} = w_\tau^{t+l}$ showing that $w(D)|_{[(-\infty,t]} = w^{t+l}(D)|_{(-\infty,t]}$. Using the causality of $G(D)$, this implies that

$$\begin{aligned} w(D)G(D)|_{(-\infty,t]} &= w^{t+l}(D)G(D)|_{(-\infty,t]} = v^{t+l}(D)|_{(-\infty,t]} \\ &= v^t(D)|_{(-\infty,t]} = v(D)|_{(-\infty,t]}. \end{aligned}$$

Since this happens for all t, we can argue that $v(D) = w(D)G(D)$ and so $v(D) \in C$.

(2.⇒1.) We have to show that C is a $\mathbb{Z}_{p^s}((D))$-submodule of $\mathbb{Z}_{p^s}((D))^q$. Let $v(D) \in C$ and assume that $v(D)$ has support in $[t_0, +\infty)$ for some $t_0 \in \mathbb{Z}$. Let $r(D) = \sum_{i=l}^{+\infty} r_i D^i \in \mathbb{Z}_{p^s}((D))$. Then, for all $t \in \mathbb{Z}$ we have that

$$(r(D)v(D))|_{(-\infty,t]} = \left(\sum_{i=l}^{t-t_0} r_i D^i v(D) \right)|_{(-\infty,t]} \in C|_{(-\infty,t]}.$$

The result now immediately follows from past-completeness.

(1.⇒3.) Observe first that $\mathbb{Z}_{p^s}((D))^q$ is a free module over the ring $\mathbb{Z}_{p^s}((D))$. It can be shown that this ring is a principal ideal ring. Indeed, observe that any Laurent power series $a(D) \in \mathbb{Z}_{p^s}((D))$ can be expressed as $a(D) = p^i(b(D) - pc(D))$, where $b(D)$ has unitary trailing coefficient. Observe moreover that

$$(b(D) - pc(D))(b(D)^{s-1} + b(D)^{s-2}pc(D) + \ldots + p^{s-1}c(D)^{s-1}) = b(D)^s$$

which is a unit in $\mathbb{Z}_{p^s}((D))$. Therefore also $(b(D) - pc(D))$ is a unit in $\mathbb{Z}_{p^s}((D))$. This shows that this ring is a principal ideal ring.

Assume now that $\mathcal{C}$ is a $\mathbb{Z}_{p^s}((D))$-submodule of $\mathbb{Z}_{p^s}((D))^q$. This implies that there exists a matrix $L(D) \in \mathbb{Z}_{p^s}((D))^{l \times q}$ such that

$$\mathcal{C} = \{w(D)L(D) \ : \ w(D) \in \mathbb{Z}_{p^s}((D))^l\}.$$

Consider the Smith form of $L(D)$, whose existence is ensured by the fact that the ring $\mathbb{Z}_{p^s}((D))$ is a principal ideal ring [9]

$$L(D) = A(D) \begin{bmatrix} \Lambda & 0 \\ 0 & 0 \end{bmatrix} B(D),$$

where $A(D) \in \mathbb{Z}_{p^s}((D))^{l \times l}$ and $B(D) \in \mathbb{Z}_{p^s}((D))^{q \times q}$ are unimodular matrices and Λ is the diagonal matrix

$$\Lambda = \begin{bmatrix} I_{m_0} & 0 & \cdots & 0 \\ 0 & pI_{m_1} & \cdots & 0 \\ \vdots & \vdots & \ddots & \vdots \\ 0 & 0 & \cdots & p^{s-1}I_{m_{s-1}} \end{bmatrix},$$

and where I_{m_i} is the $m_i \times m_i$ identity matrix. Let $m := m_0 + m_1 + \cdots + m_{s-1}$ and let $G(D)$ be the submatrix of $B(D)$ formed by the first m rows of $B(D)$. Consider moreover the Abelian group

$$W := \bigoplus_{i=1}^{s-1} p^i \mathbb{Z}_{p^s}^{l_i}.$$

It is easy to verify that $G(D)$ is right invertible, since it is a portion of a unimodular matrix, and that it induces a generalized encoder for $\mathcal{C}$ with input space $W((D))$.

In order to prove the uniqueness of the Abelian group W we have to show that the indices l_i, which determine W, can be canonically determined from $\mathcal{C}$. This fact can be proved using the same argument used in the proof of Theorem 3.1.   ∎

Notice that, as observed in the previous theorem, also for generalized convolutional codes the concept of encoding group is well defined.

We complete the comparison between generalized convolutional codes and convolutional codes by giving the following result. We need to introduce the concept of canonical state space of a generalized convolutional code. Given a generalized convolutional code $\mathcal{C} \subseteq \mathbb{Z}_{p^s}((D))^q$, we define

$$X(\mathcal{C}) := \frac{\mathcal{C}}{\mathcal{C}_+ \oplus \mathcal{C}_-},$$

where $\mathcal{C}_+ := \mathcal{C} \cap \mathbb{Z}_{p^s}[[D]]^q$ and where $\mathcal{C}_- := \mathcal{C} \cap D^{-1}\mathbb{Z}_{p^s}[D^{-1}]^q$. This is called the *canonical state space* of $\mathcal{C}$. This is a group which plays a fundamental role in the analysis of the complexity of the trellis representations of group codes [6].

THEOREM 4.2. *Let $\mathcal{C} \subseteq \mathbb{Z}_{p^s}((D))^q$ be a generalized convolutional code over $\mathbb{Z}_{p^s}$. Then the following facts are equivalent.*

1. *$\mathcal{C}$ is a convolutional code over $\mathbb{Z}_{p^s}$.*
2. *The canonical state space $X(\mathcal{C})$ is a finite Abelian group.*

Proof. (1.$\Rightarrow$2.) In the proof of this implication we use the fact that a convolutional code is strongly complete and so it is L-complete for some $L \in \mathbb{N}$. Consider the the homomorphism

$$\pi : \mathcal{C}_{|[0,L]} \to X(\mathcal{C})$$

defined as follows: given $m(D) \in \mathcal{C}_{|[0,L]}$, let $v(D) \in \mathcal{C}$ be such that $v(D)|_{[0,L]} = m(D)$ and define $\pi(m(D)) := v(D) + \mathcal{C}_- + \mathcal{C}_+$. This is a good definition. Actually, if $v'(D) \in \mathcal{C}$ is such that $v'(D)|_{[0,L]} = m(D)$, then $\delta(D) := v(D) - v'(D) \in \mathcal{C}$ and $\delta(D)_{[0,L]} = 0$ and so, by L-completeness, the signal $v_1(D)$ such that $v_1(D)|_{(-\infty,-1]} = \delta(D)|_{(-\infty,-1]}$ and $v_1(D)|_{[0,+\infty)} = 0$ is in $\mathcal{C}$. It is clear that $v_1(D) \in \mathcal{C}_-$ and $v_2(D) := \delta(D) - v_1(D) \in \mathcal{C}_+$. Therefore $\delta(D) \in \mathcal{C}_- + \mathcal{C}_+$. The map π is clearly surjective and so $X(\mathcal{C})$ is isomorphic to $\mathcal{C}_{|[0,L]}/\ker \pi$ that is a finite Abelian group.

(2.$\Rightarrow$1.) To prove this implication we need a preliminary fact. Consider the decreasing sequence of Abelian groups

$$(5) \qquad \mathcal{C}_i := \{v(D) \in \mathcal{C} \: : \: v(D)|_{[0,i]} = 0\}, \qquad i \in \mathbb{N}.$$

Notice that

$$\mathcal{C}_- \subseteq \mathcal{C}_i|_{(-\infty,-1]} \subseteq \mathcal{C}|_{(-\infty,-1]}.$$

This implies that

$$\frac{\mathcal{C}_i|_{(-\infty,-1]}}{\mathcal{C}_-} \subseteq \frac{\mathcal{C}|_{(-\infty,-1]}}{\mathcal{C}_-}$$

and the right hand side is known [6, State space theorem] to be isomorphic to $X(\mathcal{C})$. This fact implies that the sequence of Abelian groups (5) must become stationary, i.e., there exists $L \in \mathbb{N}$ such that $\mathcal{C}_i = \mathcal{C}_L$ for all $i \geq L$.

To prove the implication we need only to prove that C is complete. Assume that $v(D) \in \mathbb{Z}_{p^s}((D))^q$ is such that $v(D)|_{[k,k+N]} \in C|_{[k,k+N]}$ for all $k \in \mathbb{Z}$ and $N \in \mathbb{N}$. We want to show that $v(D) \in C$. It is not restrictive to assume that $v(D) \in \mathbb{Z}_{p^s}[[D]]^q$. Then we have that $v(D)|_{[-L,n]} \in C|_{[-L,n]}$ for all $n \in \mathbb{N}$. Therefore for every $n \in \mathbb{N}$ there exists $v_n(D) \in C$ such that $v_n(D)|_{[-L,n]} = v(D)|_{[-L,n]}$. Notice that

$$(6) \qquad D^L v_n(D) \in C_L = C_i \, , \qquad \forall i \geq L.$$

Define $v_{1n}(D) := v_n(D)|_{(-\infty,-1]}$ and $v_{2n}(D) := v_n(D) - v_{1n}(D)$. From (6) we can argue that $v_{1n}(D)|_{(-\infty,m]} \in C|_{(-\infty,m]}$ for all $m \in \mathbb{N}$, and this, by past completeness, implies that $v_{1n}(D) \in C$ and that $v_{2n}(D) \in C$. Observe finally that

$$v(D)|_{(-\infty,n]} = v_{2n}(D)|_{(-\infty,n]} \in C|_{(-\infty,n]}$$

which again, by past completeness, implies that $v(D) \in C$.   ∎

5. State space realization of encoders. In this last section we will consider the problem of state space realization of encoders. More precisely we will show how to characterize non-rational, rational and polynomial encoders in terms of their state space realizations.

Let W and V be finite Abelian groups. An element

$$(7) \qquad N(D) = \sum_{k=l}^{+\infty} N_k D^k \in \mathrm{Hom}(W,V)((D))$$

naturally induces a time-invariant homomorphism

$$(8) \qquad \begin{aligned} N(D) \quad : \quad & W((D)) \quad \rightarrow \quad V((D)) \\ & w(D) \quad \mapsto \quad v(D) = N(D)(w(D)) \end{aligned}$$

by letting

$$(9) \qquad v_k := \sum_{i=l}^{+\infty} N_i(w_{k-i}).$$

Operators defined in this way are called *convolution operators*. If $N(D)$ in (7) is non-zero, then we can assume that $N_l \neq 0$. The non-negative integer $\max\{0, l\}$ is called the *anticipation* of the convolution operator. Convolution operators with zero anticipation are called *causal*.

It is clear that, when $W = \mathbb{Z}_{p^s}^m$ and $V = \mathbb{Z}_{p^s}^q$, then $\mathrm{Hom}(W,V)((D)) = \mathbb{Z}_{p^s}((D))^{m \times q}$ and $N(D) \in \mathrm{Hom}(W,V)((D))$ operates on elements in $\mathbb{Z}_{p^s}((D))^m$ by right multiplication.

When $N(D) \in \mathrm{Hom}(W,V)[D, D^{-1}]$, it is called a *polynomial convolution operator*. A convolution operator $N(D) \in \mathrm{Hom}(W,V)((D))$ is said to be *rational* if there exist polynomial convolution operator $P(D) \in \text{Hom}(W, V)[D, D^{-1}]$ and polynomial $r(D) \in \mathbb{Z}[D, D^{-1}]$ having unitary trailing coefficient such that

$$r(D)N(D) = P(D).$$

The class of rational convolution operators is denoted by the symbol $\text{Hom}(W, V)(D)$. Also in this case it is not difficult to verify that, when $W = \mathbb{Z}_{p^s}^m$ and $V = \mathbb{Z}_{p^s}^q$, then $\text{Hom}(W, V)(D) = \mathbb{Z}_{p^s}(D)^{m \times q}$.

Let W and V be finite Abelian groups and let $N(D) : W((D)) \to V((D))$ be a convolution operator. A *state-space realization* of $N(D)$ is a quadruple (X, f, g, l), where X is a set called the *state space*, f and g are maps

$$f : X \times W \longrightarrow X$$
$$g : X \times W \longrightarrow V$$

and $l \in \mathbb{N}$ is called the *anticipation* of the realization, such that, given $w(D) \in W((D))$ and $v(D) \in V((D))$, we have that $v(D) = N(D)(w(D))$ if and only if there exists a sequence $(x_k) \in X^{\mathbb{Z}}$ for which the following relations hold

$$(10) \qquad \begin{cases} x_{k+1} & = & f(x_k, w_k) \\ v_{k-l} & = & g(x_k, w_k) \end{cases}$$

for all $k \in \mathbb{Z}$. When the state space X is also an Abelian group and f and g are homomorphisms, (X, f, g, l) is called a *linear state-space realization* of $N(D)$.

In the following proposition we show that all convolution operators admit a state space realization.

PROPOSITION 5.1. *Let W and V be finite Abelian groups and let $\psi : W((D)) \to V((D))$ be a time invariant homomorphism. The following conditions are equivalent*

1. ψ is a convolution operator with anticipation l.

2. ψ admits a linear state-space realization with anticipation l.

Proof. (1.⇒2.) This implication is based on the existence of a canonical realization of input/output maps [8]. Assume that ψ is a convolution operator with anticipation l. Then the map $\bar{\psi}$ mapping $w(D)$ into $D^l \psi(w(D))$ is causal. It is clear that, in order to prove this implication, it is sufficient to show that $\bar{\psi}$ admits a linear state-space realization with zero anticipation.

Consider the following quotient group

$$(11) \qquad X(\bar{\psi}) := \frac{D^{-1}W[D^{-1}]}{D^{-1}W[D^{-1}] \cap \bar{\psi}^{-1}(D^{-1}V[D^{-1}])}$$

and define $f_{\bar{\psi}} : X(\bar{\psi}) \times W \to X(\bar{\psi})$ and $g_{\bar{\psi}} : X(\bar{\psi}) \times W \to V$ in the following way: let $a \in W$ and $x \in X(\bar{\psi})$ and let $w(D) \in D^{-1}W[D^{-1}]$ be any of the representatives of x. Let $\tilde{w}(D) \in D^{-1}W[D^{-1}]$ be such that $\tilde{w}_k = w_{k+1}$ for all $k \leq -2$ and $\tilde{w}_{-1} := a$. Let $\tilde{x}$ be the state associated with $\tilde{w}(D)$ and $v(D) := \psi(\tilde{w}(D))$. We define

$$f_{\tilde{\psi}}(x, a) := \tilde{x}$$
$$g_{\tilde{\psi}}(x, a) := v_0.$$

Standard considerations show that $(X(\tilde{\psi}), f_{\tilde{\psi}}, g_{\tilde{\psi}}, 0)$ is a linear state-space realization for $\tilde{\psi}$ and so, letting $X(\psi) := X(\tilde{\psi})$, $f_\psi := f_{\tilde{\psi}}$ and $g_\psi := g_{\tilde{\psi}}$, we obtain that $(X(\psi), f_\psi, g_\psi, l)$ is a linear state-space realization for ψ.

(2.$\Rightarrow$1.) Assume that ψ admits a linear realization with delay l. Define $\tilde{\psi}$ from ψ as above. Then $\tilde{\psi}$ admits a linear realization with zero anticipation. We now prove that $\tilde{\psi}$ is a causal convolution operator which implies that ψ is a convolution operator with anticipation l. First observe that, if $w(D) \in W[[D]]$, then $\tilde{\psi}(w(D)) \in V[[D]]$. Define recursively, $x \in X[[D]]$ by

$$\text{(12)} \qquad \begin{cases} x_{k+1} &= f(x_k, w_k) \ \ k \geq 0 \\ x_0 &= 0. \end{cases}$$

Then define $v(D) \in V[[D]]$ by

$$v_k = g(x_k, w_k).$$

It is easy to verify that $w(D), x(D), v(D)$ satisfy the state equations (10) and this implies that $\tilde{\psi}(w(D)) = v(D) \in V[[D]]$.

For each $k \in \mathbb{Z}$ consider the homomorphism $N_k \in \text{Hom}(W, V)$ defined as follows. Take $a \in W$ and consider it as an element in $W((D))$. Let $v(D) := \tilde{\psi}(a)$ and let $N_k(a) := v_k$. From $N_k \in \text{Hom}(W, V)$ we can define the causal convolution operator $N(D) := \sum_k N_k D^k$. We want to show that the maps $\tilde{\psi}$ and $N(D)$ coincide. Observe that, by linearity and time invariance, these maps coincides over the subset $W[D, D^{-1}]$ of $W((D))$. Observe moreover that, if $w(D) \in W((D))$, then

$$\tilde{\psi}(w(D)) = \tilde{\psi}(w(D)|_{(-\infty, k]}) + \tilde{\psi}(w(D)|_{[k+1, +\infty)}).$$

By causality of $\tilde{\psi}$ we have that the support of $\tilde{\psi}(w(D)|_{[k+1, +\infty)})$ is contained in $[k+1, +\infty)$ and so

$$\tilde{\psi}(w(D))_{(-\infty, k]} = \tilde{\psi}(w(D)|_{(-\infty, k]})|_{(-infty, k]}.$$

In a similar way it can be shown that

$$N(D)(w(D))_{(-\infty, k]} = N(D)(w(D)|_{(-\infty, k]})|_{(-\infty, k]}.$$

Observe finally that, since $w(D)|_{(-\infty, k]} \in W[D, D^{-1}]$, then

$$N(D)(w(D)|_{(-\infty, k]}) = \tilde{\psi}(w(D)|_{(-\infty, k]}).$$

Therefore $N(D)(w(D)$ and $\bar{\psi}(w(D))$ coincide on $(-\infty, k]$ for every k, and so they must coincide. ∎

For a convolutional operator $\psi : W((D)) \to V((D))$ the linear state-space realization $(X(\psi), f_\psi, g_\psi, l)$ introduced in the proof of (1.⇒2.) of Proposition 5.1 is called the *canonical realization* and possesses an important minimality property. To illustrate this consider another state space realization (X, f, g, l) of ψ

$$(13) \qquad \begin{cases} x_{k+1} &= f(x_k, w_k) \\ v_k &= g(x_k, w_k) \end{cases} \quad \forall k \in \mathbb{Z},$$

where we don't necessarily assume that f and g are homomorphisms. Consider the set

$$X_0 := \Big\{ z \in X \ : \ \exists \ w(D) \in W((D)), v(D) \in V((D)), (x_k) \in X^{\mathbb{Z}}$$
$$\text{satisfying (13) and } x_0 = z \Big\}$$

and define

$$\pi \ : \ X_0 \ \to \ X(\psi)$$

as follows: given a $z \in X_0$, suppose that $w(D) \in W((D)), v(D) \in V((D))$, $(x_k) \in X^{\mathbb{Z}}$ satisfy (13) and that $x_0 = z$. Then define

$$\pi(z) := w(D)|_{(-\infty, -1]} + D^{-1} W[D^{-1}] \cap \bar{\psi}^{-1}(D^{-1} V[D^{-1}]).$$

It is easy to see that π is a well-defined surjective map. This shows that the cardinality of $X(\psi)$ must be less than or equal to the cardinality of the state space X.

It may happen that the convolution operator admits only state realization with infinite cardinality state space. The following proposition shows that only rational convolution operators admits state realization with finite cardinality state space. This result is quite classical [2, 3, 15]. Nonetheless, for aims of completeness, we prefer to give a sketch of the proof of this result, following [3].

PROPOSITION 5.2. *Let W and V be finite Abelian groups and let $\psi : W((D)) \to V((D))$ be a time invariant homomorphism. The following conditions are equivalent*

    1. ψ is a rational convolution operator.

    2. ψ admits a linear state-space realization with a finite Abelian group as state space.

Proof. Notice first that, as in Proposition 5.1, we can assume without loss of generality that ψ is causal convolution operator.

(2.⇒1.) By hypothesis we have that the canonical state space $X(\psi)$ is a finite Abelian group and so it is a finitely generated $\mathbb{Z}$-module. The multiplication by D^{-1} naturally induces a homomorphism π from $X(\psi)$ into itself. This implies that $X(\psi)$ is a $\mathbb{Z}[D^{-1}]$-module. Applying Nakayama's lemma [1, Proposition 2.4], there exists a polynomial

$$r(D) = D^{-n} + \sum_{i=0}^{n-1} r_i D^{-i} \in \mathbb{Z}[D^{-1}]$$

such that

(14) $r(D)X(\psi) = 0.$

We want to show that $r(D)\psi \in \mathrm{Hom}(W, V)[D, D^{-1}]$. To this aim, it is enough to prove that for any finitely supported $w(D) \in W[D, D^{-1}]$ we have that $r(D)\psi(w(D))$ is finitely supported. Notice that we can assume without loss of generality that $w(D) \in D^{-1}W[D^{-1}]$. Equation (14) implies that $r(D)w(D) \in \psi^{-1}(D^{-1}V[D^{-1}])$ and so

$$r(D)\psi(w(D)) = \psi(r(D)w(D)) \in D^{-1}V[D^{-1}]$$

which must be finitely supported.

   (1.⇒2.)   Since ψ is rational, then there exists a $r = \sum_{i=0}^{n} r_i D^{-i} \in \mathbb{Z}[D^{-1}]$ having unitary trailing coefficient such that $r(D)\psi \in \mathrm{Hom}(W, V)$ $[D, D^{-1}]$. It is possible to choose $r(D) \in \mathbb{Z}[D^{-1}]$ such that $r(D)\psi \in \mathrm{Hom}(W, V)[D^{-1}]$. This implies that, for any $w(D) \in D^{-1}W[D^{-1}]$, we have that $r(D)\psi(w(D)) \in D^{-1}V[D^{-1}]$. Consider the homomorphism

$$\mu \quad : \qquad\qquad X(\psi) \qquad\qquad\qquad \to \qquad\qquad V^n$$
$$\quad w(D) + D^{-1}W[D^{-1}] \cap \psi^{-1}(D^{-1}V[D^{-1})) \quad \mapsto \quad \psi(w(D))|_{[0, n-1]}$$

It is easy to see that this homomorphism is well defined. It is injective. Indeed, let $w(D) \in D^{-1}W[D^{-1}]$ and $v(D) := \psi(w(D))$ and assume that $v(D)|_{[0, n-1]} = 0$. Since $r(D)v(D) = \psi(r(D)w(D)) \in D^{-1}V[D^{-1}]$, we have that

$$\sum_{i=0}^{n} r_i v_{k+i} = 0 \quad \forall k \geq 0.$$

This together with the fact that $v(D)_{|[0, n-1]} = 0$ implies that $v(D)|_{[0, +\infty)} = 0$. The fact that μ is injective and that the codomain of μ is finite implies that $X(\psi)$ must be finite.   ∎

   Also polynomial shift operators admit a nice characterization as shown in the following theorem. For proving it we need the following lemma.

   LEMMA 5.1.   *Let X be a finite Abelian group which is a $\mathbb{Z}[[D^{-1}]]$-module. Then there exists $n \in \mathbb{N}$ such that $D^{-n}X = 0$*
Proof.   Consider the ideal of annihilators

$$\mathcal{A} := \{s(D) \in \mathbb{Z}[[D^{-1}]] \ : \ s(D)X = 0\}.$$

Since X is finite and since it is $\mathbb{Z}[D^{-1}]$-module, applying Nakayama's lemma [1, Proposition 2.4], there exists in $\mathcal{A}$ a polynomial

$$r(D) = r_0 + r_1 D^{-1} + \ldots + r_{n-1} D^{-n+1} + D^{-n} \in \mathbb{Z}[D^{-1}].$$

Fix any prime number $p \in \mathbb{Z}$. The polynomial $r(D)$ can be decomposed as $r(D) = h(D) + pg(D)$, where $h(D) = h_m D^{-m} + h_{m+1} D^{-m-1} + \ldots$ has coefficient h_m, $m \leq n$, which is not divisible by p. Then there exist $\alpha, \beta \in \mathbb{Z}$ such that $\alpha h_m + \beta p = 1$. This implies that

$$\alpha r(D) = \alpha(h_m D^{-m} + h_{m+1} D^{-m-1} + \ldots) + p\alpha g(D) =$$
$$= (D^{-m} + \alpha h_{m+1} D^{-m-1} + \ldots) + p(\alpha g(D) - \beta D^{-m})$$

The first term of the previous sum is invertible in $\mathbb{Z}[[D^{-1}]]$. Taking into account that $m \leq n$, this implies that $D^{-n} + pq(D) \in \mathcal{A}$ for some $q(D) \in \mathbb{Z}[[D^{-1}]]$. Notice now that for any $x \in X$ we have that $(D^{-n} + pq(D))x = 0$ and so we have that $D^{-n}x = py$ for some $y \in X$ depending on p. Since this is true for any prime number p, this implies that $D^{-n}x = 0$ and so $D^{-n}X = 0$. ∎

PROPOSITION 5.3. *Let W and V be finite Abelian groups and let $\psi : W((D)) \to V((D))$ be a time invariant homomorphism. The following conditions are equivalent.*
  1. *ψ is a polynomial convolution operator.*
  2. *ψ can be extended to a time invariant homomorphism $\tilde{\psi} : W^{\mathbb{Z}} \to V^{\mathbb{Z}}$ admitting a linear state space realization with a finite Abelian group as state space.*

Proof. (1.$\Rightarrow$2.) It is clear that a polynomial convolution operator ψ can be extended to a time invariant homomorphism $\tilde{\psi} : W^{\mathbb{Z}} \to V^{\mathbb{Z}}$. This operator admits a canonical state space realization with state space

$$(15) \qquad X(\tilde{\psi}) := \frac{D^{-1}W[[D^{-1}]]}{D^{-1}W[[D^{-1}]] \cap \tilde{\psi}^{-1}(D^{-1}V[[D^{-1}]])}$$

The fact that this Abelian group is finite can be shown using the same arguments used in the proof of the implication (1.$\Rightarrow$2.) of the previous proposition.

(2.$\Rightarrow$1.) Since the state realization of $\tilde{\psi}$ is a state realization also of ψ, by Proposition 5.2 we have that ψ is a rational convolution operator. The fact that it is a convolution operator allows us to assume without loss of generality that ψ is causal. Consider the canonical state space $X(\tilde{\psi})$ of $\tilde{\psi}$. By the same arguments used above it can be shown that the state space $X(\tilde{\psi})$ of the canonical state space realization has smaller cardinality than every other state space realization of $\tilde{\psi}$. This shows that $X(\tilde{\psi})$ has finite cardinality. To prove the assertion we need to prove the following intermediate fact.

Fact: $X(\bar{\psi})$ is a $\mathbb{Z}[[D^{-1}]]$-module.

Since $D^{-1}W[[D^{-1}]]$ is clearly a $\mathbb{Z}[[D^{-1}]]$-module, it is enough to prove that $\mathcal{W} := D^{-1}W[[D^{-1}]] \cap \bar{\psi}^{-1}(D^{-1}V[[D^{-1}]])$ is a $\mathbb{Z}[[D^{-1}]]$-module. Let $s(D) \in \mathbb{Z}[[D^{-1}]]$ and $w(D) \in \mathcal{W}$. We have to show that $s(D)w(D) \in \mathcal{W}$. Let $v(D) := \bar{\psi}(w(D))$. Notice that the canonical state space realization $(X(\bar{\psi}), f_{\bar{\psi}}, g_{\bar{\psi}}, 0)$ of $\bar{\psi}$ can be introduced as done in the proof of the implication $(1.\Rightarrow 2.)$ of Proposition 5.1. The following sequence of states in the canonical state space $X(\bar{\psi})$

$$x_k := D^k w(D)|_{(-\infty, k-1]} + \mathcal{W},$$

verify the state space equations

(16)
$$\begin{cases} x_{k+1} &= f_{\bar{\psi}}(x_k, w_k) \\ v_k &= g_{\bar{\psi}}(x_k, w_k) \end{cases} \quad \forall k \in \mathbb{Z}.$$

Notice that $x(D) := \sum x_k D^k \in D^{-1}X(\bar{\psi})[[D^{-1}]]$. Define

$$\bar{x}(D) := s(D)x(D) \in D^{-1}X(\bar{\psi})[[D^{-1}]]$$
$$\bar{w}(D) := s(D)w(D) \in D^{-1}W(\bar{\psi})[[D^{-1}]]$$
$$\bar{v}(D) := s(D)v(D) \in D^{-1}V[[D^{-1}]] .$$

It is easy to verify that the sequences $(\bar{x}_k) \in X(\bar{\psi})^{\mathbb{Z}}$, $(\bar{w}_k) \in W^{\mathbb{Z}}$, $(\bar{v}_k) \in V^{\mathbb{Z}}$ still satisfy the state space equations (16) and so we have that $\bar{v}(D) = \bar{\psi}(\bar{w}(D))$. Since $\bar{v}(D) \in V[[D^{-1}]]$, then $w(D) \in \mathcal{W}$.

We want to show finally that ψ is a polynomial convolutional operator. For this it is enough to prove that, if $w(D) \in W((D))$ is such that $w(D)|_{[-n,0]} = 0$, then $\psi(w(D)) = \bar{\psi}(w(D))$ is zero at zero. Decompose $w(D) = w_1(D) + w_2(D)$ such that $w_1(D) \in D^{-n-1}W[[D^{-1}]]$, $w_2(D) \in DW[[D]]$. Observe that from the fact proved above and by Lemma 5.1 we have that $D^n w_1(D) \in D^{-1}W[[D^{-1}]]$. Since $D^{-n}X(\bar{\psi}) = 0$, we can argue that $\bar{\psi}(w_2(D)) = D^{-n}\bar{\psi}(D^n w_2(D)) \in D^{-1}V[[D^{-1}]]$. On the other hand, by causality of $\bar{\psi}$, we have that $\bar{\psi}(w_1(D)) \in DV[[D]]$. These two facts imply that $\bar{\psi}(w(D)) = \bar{\psi}(w_1(D)) + \bar{\psi}(w_2(D))$ is zero at zero. ∎

6. Conclusion. In this paper the class of convolutional codes over finite Abelian groups are analyzed. Using the structure theorem we could reduce to investigating convolutional codes over the ring $\mathbb{Z}_{p^s}$. Since rational functions over $\mathbb{Z}_{p^s}$ form a principal ideal ring, using the Smith canonical form, for any convolutional code we could define a rational encoder which is defined over an Abelian group, called the encoding group of the convolutional code. We compared this group with the more classical input group, introduced in [6]. We provided also a system theoretic characterization of the class of convolutional codes in terms of completeness. We introduced moreover the concept of generalized convolutional code which is a code generated by a non-rational encoder and we provided also a system theoretic characterization of this class of codes in terms of past-completeness. We showed that convolutional codes are exactly the generalized convolutional codes having finite canonical state space [6]. Finally we considered the problem of state realization of polynomial, rational and non-rational encoders.

The analysis of more specific classes of convolutional codes and encoders, such as systematic, minimal and basic convolutional codes or encoders, is the subject of our present investigation.

REFERENCES

[1] M.F ATIYAH AND I.G. MACDONALD. *Commutative Algebra*. Addison-Wesley, 1969.
[2] R. BROCKETT AND A.S. WILLSKY. Finite group homomorphic sequential systems. *IEEE Trans. Automatic Control*, AC-**17**:483–490, 1972.
[3] R. DEB. JOHNSTON. *Linear systems over various rings*. PhD thesis, Massachusetts Institute of Technology, 1973.
[4] F. FAGNANI AND S. ZAMPIERI. System theoretic properties of convolutional codes over rings. 1999. Submitted for publication.
[5] G.D. FORNEY. Convolutional codes I: Algebraic structure. *IEEE Trans. Information Theory*, IT-**16**:720–738, 1970.
[6] G.D. FORNEY AND M.D. TROTT. The dynamics of group codes: State spaces, trellis diagrams and canonical encoders. *IEEE Trans. Information Theory*, IT-**39**:1491–1513, 1993.
[7] R. JOHANNESSON, Z. WAN, AND E. WITTENMARK. Some structural properties of convolutional codes over rings. *IEEE Trans. Information Theory*, IT-**44**:839–845, 1998.
[8] R.E. KALMAN, P.L. FALB, AND M.A. ARBIB. *Topics in Mathematical System Theory*. McGraw Hill, 1969.
[9] I. KAPLANSKI. Elementary divisors and modules. *Trans. of Amer. Math. Society*, **66**:464–491, 1979.
[10] D. LIND AND B. MARCUS. *Symbolic Dynamics and Coding*. Cambridge Univ., 1995.
[11] H.A. LOELIGER AND T. MITTELHOLZER. Convolutional codes over groups. *IEEE Trans. Inf. Theory*, IT-**42**:1660–1686, 1996.
[12] J.L. MASSEY AND T. MITTELHOLZER. Convolutional codes over rings. In *Proc. Joint Swedish-Soviet Int. Workshop on Inform. Theory*, pp. 14–18, Gotland, Sweeden, 1989.
[13] J.L. MASSEY AND T. MITTELHOLZER. Systematicity and rotational invariance of convolutional codes over rings. In *Proc. Int. Workshop on Alg. and Combinatorial Coding Theory*, pp. 154–158, Leningrad, 1990.
[14] R.J. MCELIECE. The algebraic theory of convolutional codes. In V.S. Pless and W.C. Hoffman, editors, *Handbook of Coding Theory*, Volume 1, Elsevier, 1998.
[15] E.D. SONTAG. Linear systems over commutative rings: A survey. *Ricerche di Automatica*, **7**:1–34, 1976.
[16] J.C. WILLEMS. Models for dynamics. *Dynamics Reported*, **2**:171–269, 1988.
[17] S. ZAMPIERI AND S.K. MITTER. Linear systems over Noetherian rings in the behavioural approach. *Journal of Math. Systems, Est., and Cont.*, **6**:235–238, 1996.

SYMBOLIC DYNAMICS AND CONVOLUTIONAL CODES

BRUCE KITCHENS*

Abstract. Convolutional codes and their encoders are examined in a symbolic dynamics setting. Two previously unrelated areas of dynamics are used. The first is the Adler-Coppersmith-Hassner formulation of the finite memory channel coding problem. The second is the Kitchens-Schmidt use of algebraic duality to define and examine algebraic dynamical systems. Convolutional codes and their encoders are seen as lying in the intersection of these two areas. Codes, encoders, memory, distances and other invariants are examined in this framework.

1. Symbolic dynamics. In this section we will very briefly discuss subshifts of finite type, continuous maps between them and two equivalence relations. The results we will use are stated but not proved. For a thorough discussion of these ideas see [Kit98] or [LM95].

Begin with a finite symbol set $\{1, \ldots, n\}$ and form the two-sided sequence space $\{1, \ldots, n\}^{\mathbb{Z}}$. Define a metric on the space by $d(x, y) = 0$ if $x = y$ and $d(x, y) = 1/2^N$ with N the integer where $x_i = y_i$ for $|i| < N$ and $x_N \neq y_N$ or $x_{-N} \neq y_N$. When $n > 1$ the space with this metric topology is compact, totally disconnected and homeomorphic to the usual middle thirds Cantor set. Define the shift homeomorphism, σ, from the space to itself by $\sigma(x)_i = x_{i+1}$. The space $\{1, \ldots, n\}^{\mathbb{Z}}$ together with the map σ is a dynamical system and is called the *full shift on n symbols* or just the *full n-shift*.

A *subshift of finite type* or *topological Markov shift* is a closed, shift-invariant subset of a full shift which is defined by a finite set of admissible blocks. Let $L \subseteq \{1, \ldots, n\}^k$, for some k, be a list of *admissible blocks*. The list L defines a space which is

$$\Sigma = \{x \in \{1, \ldots, n\}^{\mathbb{Z}} : [x_i, x_{i+1}, \ldots, x_{i+k-1}] \in L \text{ for all } i \in \mathbb{Z}\}.$$

Let Σ have the topology defined by the metric on $\{1, \ldots, n\}^{\mathbb{Z}}$, then the space Σ together with the shift map is a dynamical system and is called the subshift of finite type defined by L. A subshift of finite type that contains a point whose orbit under σ is dense in the entire space is said to be *transitive* or *irreducible*.

Let $\Sigma \subseteq \{1, \ldots, n\}^{\mathbb{Z}}$ be a subshift of finite type and let

$$\mathcal{W}(\Sigma, \ell) = \{[x_0, \ldots, x_{\ell-1}] : x \in \Sigma\}.$$

These are *words of length ℓ* of Σ. A word $[x_0, \ldots, x_{\ell-1}] \in \mathcal{W}(\Sigma, \ell)$ has a follower and a predecessor set. The *follower set* is

$$f([x_0, \ldots, x_{\ell-1}]) = \{j \in \{1, \ldots, n\} : [x_0, \ldots, x_{\ell-1}, j] \in \mathcal{W}(\Sigma, \ell+1)\}.$$

*IBM Watson Research Center, P.O. Box 218, Yorktown Heights, NY 10598; brucek@us.ibm.com.

Correspondingly, the *predecessor set* is

$$p([x_0, \dots, x_{\ell-1}]) = \{j \in \{1, \dots, n\} : [j, x_0, \dots, x_{\ell-1}] \in \mathcal{W}(\Sigma, \ell+1)\}.$$

Suppose Σ is a subshift of finite type. There will exist an ℓ so that for any $k \geq 0$ and any word $[x_{-k}, \dots, x_{-1}, x_0, \dots, x_{\ell-1}]$ in $\mathcal{W}(\Sigma, \ell+k)$

$$f([x_0, \dots, x_{\ell-1}]) = f([x_{-k}, \dots, x_{-1}, x_0, \dots, x_{\ell-1}]).$$

Note that the corresponding statement will necessarily hold for the same ℓ and the predecessor sets. If ℓ is the minimal integer for which these conditions hold we say that Σ is an $\ell - step$ subshift of finite type. This means that the subshift of finite type has a memory of only ℓ steps and the symbols preceding those ℓ symbols do not affect which symbols can occur next. A one-step subshift of finite type is defined by a *transition matrix*. For a one-step subshift of finite type $\Sigma \subseteq \{1, \dots, n\}^{\mathbb{Z}}$ define an $n \times n$, $\{0, 1\}$ matrix A by

$$A_{ij} = \begin{cases} 1 & \text{if } j \in f(i) \\ 0 & \text{if } j \notin f(i) \end{cases}.$$

Then

$$\Sigma = \{x : A_{x_i x_{i+1}} = 1 \text{ for all } i \in \mathbb{Z}\}.$$

We denote the one-step subshift of finite type defined by the transition matrix A by Σ_A. A one-step subshift of finite type is transitive or irreducible (as defined above) if and only if its transition matrix is irreducible (in the matrix theoretic sense).

Let Σ be a subshift of finite type. The *topological entropy* of Σ is defined to be

$$h(\Sigma) = \lim_{\ell \to \infty} \frac{1}{\ell} \log |\mathcal{W}(\Sigma, \ell)|,$$

where $| \cdot |$ denotes the cardinality of the set. This limit always exists. Topological entropy was originally called *capacity* by C. Shannon. If A is an irreducible transition matrix then it has a unique eigenvalue of largest modulus. It is a nonnegative real number and is called the *Perron value of A*. For an irreducible transition matrix A it can be seen that $h(\Sigma_A) = \log \lambda$, where λ is the Perron value of the matrix A.

Let Σ and Σ' be two subshifts of finite type. A map φ from Σ to Σ' is a *block map* if it is defined by a rule φ' from some $\mathcal{W}(\Sigma, 2\ell+1)$ to $\mathcal{W}(\Sigma', 1)$ where $\varphi(x)_i = \varphi'([x_{i-\ell}, \dots, x_{i+\ell}])$ for all $x \in \Sigma$ and all $i \in \mathbb{Z}$. Such a map is said to be a $(2\ell+1)$-block map. A block map is a continuous map and it commutes with the shifts on the domain and range. The converse of this statement is also true and the two statements together are the *Curtis-Hedlund-Lyndon Theorem*: A map between two subshifts of finite type is continuous and commutes with the shifts if and only if it is a block map.

A continuous, onto, shift-commuting map from one subshift of finite type to another is called a *factor map* and a shift-commuting homeomorphism between two subshifts of finite type is called a *topological conjugacy*. In the case of a factor map the image subshift of finite type is said to be a *factor* of the domain subshift of finite type and in the case of a conjugacy the two subshifts of finite type are said to be *topologically conjugate*. It is easy to see that an ℓ-step subshift of finite type is topologically conjugate to a one-step subshift of finite type with symbols $\mathcal{W}(\Sigma, \ell)$. There are a number of important properties of factor maps between irreducible subshifts of finite type. Here we briefly mention some of them. A discussion of these properties and proofs of the results can be found in Chapter 4 of [Kit98]. Factor maps between irreducible subshifts of finite type fall into two classes. A factor map is said to be *finite-to-one* if it is uniformly bounded to one. When this happens there will be a d so that each point in the image subshift of finite type with a dense forward and backward σ orbit has exactly d preimages under the factor map. The integer d is the *degree* of the map. A factor map which is not finite-to-one is said to be *infinite-to-one* and in this case each point in the image subshift of finite type with a dense forward and backward σ orbit has uncountably many preimages under the factor map. If there is a factor map between two irreducible subshifts of finite type then the factor map is finite-to-one if the two subshifts of finite type have the same topological entropy and infinite-to-one if the topological entropy of the domain is greater than the topological entropy of the range.

There is a particular type of factor map which is especially useful for coding purposes. These are the *resolving maps*. Suppose φ is a one-block factor map between the one-step subshifts of finite type Σ_A and Σ_B. Then φ is said to be *right resolving* if for each symbol $i \in \mathcal{W}(\Sigma_A, 1)$ the map φ defines a bijection between $f(i) \subseteq \mathcal{W}(\Sigma_A, 1)$ and $f(\varphi(i)) \subseteq \mathcal{W}(\Sigma_B, 1)$, where we consider φ both as a map on symbols and on points. We say a factor map is *left resolving* if the analogous statement holds between predecessor sets. A resolving map is always a finite-to-one factor map.

We have said that two subshifts of finite type are topologically conjugate if there is a shift commuting homeomorphism between them and we have seen that any subshift of finite type is topologically conjugate to a one-step subshift of finite type. But, it is not known how to determine whether or not two subshifts of finite type are topologically conjugate from their transition matrices. In fact, it is not known whether or not this is algorithmically decidable. However, there is another important equivalence relation which is easily decidable. We say two subshifts of finite type, Σ_A and Σ_B, are *finitely equivalent* if there is a third subshift of finite type, Σ_C, and finite-to-one factor maps from Σ_C to Σ_A and to Σ_B. Then, there is the following result by W. Parry [Par77]: Two irreducible subshifts of finite type are finitely equivalent if and only if they have the same topological entropy. The proof actually produces a stronger result, it shows Σ_C can be made irreducible, that one of the factor maps can be made left resolving and the other right resolving.

2. Channel coding. In this section we very briefly discuss an application of symbolic dynamics to channel coding. The application is due to R. Adler, D. Coppersmith and M. Hassner [ACH83]. The proof of the crucial theorem (Theorem 2.1) is not included here but can be found in [ACH83] and in either [Kit98] or [LM95]. The problem is to code a stream of unconstrained data into a sequence of channel constrained symbols in such a way that the original data can be recovered in an efficient manner. The unconstrained data is modeled by a full n-shift and the channel constrained data by a subshift of finite type Σ. The topological entropy of the full n-shift is $\log n$ and let the topological entropy of Σ be $\log \lambda$. The value λ is the Perron value of a transition matrix for Σ. Let p/q be any rational number so that $p \log n \leq q \log \lambda$. We will see that it is possible to construct an encoder, embodied by a finite state machine, which takes blocks of length p of the unconstrained data to blocks of length q of the channel constrained symbols. The encoder is constructed in such a way that the decoder is a block map. The block map condition means that an error in the encoded channel sequence will only propagate to a uniformly bounded number of decoded data symbols. The construction is based on the proof of the following theorem.

THEOREM 2.1. *Suppose Σ is an irreducible subshift of finite type with topological entropy $\log \lambda \geq \log n$. Then there is an irreducible subshift of finite type $\Sigma' \subseteq \Sigma$ and a right resolving factor map φ from a one-step representation of Σ' to $\{1, \dots, n\}^{\mathbb{Z}}$.*

The proof of this theorem is constructive and it solves the channel coding problem. A right resolving map can be "locally inverted" by a finite state machine. This follows because for each symbol $i \in \mathcal{W}(\Sigma', 1)$ the one-block map φ defines a bijection between $f(i)$ and $f(\varphi(i))$. The finite state machine has the symbols of Σ' as internal states and outputs. The inputs to the finite state machine are the symbols $\{1, \dots, n\}$. Fix any symbol i_0 as the initial state, then a sequence $x_0 x_1 \dots$ of data is encoded by starting at the initial state i_0 and using the bijection between $f(i_0)$ and $f(\varphi(i_0))$ to encode $x_0 \in f(\varphi(i_0))$ to the corresponding symbol $z_0 \in f(i_0)$. Then continue to $x_1 \in f(x_0)$ and $z_1 \in f(z_0)$ and so forth. This encoding is then decoded by the block map φ.

Suppose we have our original channel coding problem with $\{1, \dots, n\}^{\mathbb{Z}}$, Σ and p/q. Then $\{1, \dots, n\}^{\mathbb{Z}}$ acted on by σ^p is the full shift on the blocks $\{1, \dots, n\}^p$ and Σ acted on by σ^q is a subshift of finite type on the symbols $\mathcal{W}(\Sigma, q)$ and has topological entropy $q \log \lambda$. Now apply the previous observation to produce the desired encoder and decoder.

3. Algebraic shifts. In this section we present results about shift spaces which have a group structure. These results can be found in [Kit87]. Let G be a finite group and form the sequence space $G^{\mathbb{Z}}$. Then it is a compact zero-dimensional group where the group operation is defined coordinate by coordinate and the shift map is a continuous group automorphism. The space $G^{\mathbb{Z}}$ together with the shift automorphism is called a *full group shift*. Suppose $\Sigma \subseteq G^{\mathbb{Z}}$ is a closed shift invariant subgroup. Then Σ must be a subshift of finite type. Such a subgroup together with the shift automorphism is a *group shift* or a *Markov subgroup*. The fundamental result is the following theorem [Kit87].

THEOREM 3.1. *If Σ is a group shift then it is topologically conjugate to a full shift cross a finite group with an automorphism and if Σ is irreducible then it is topologically conjugate to a full shift.*

In the proof of the theorem the topological conjugacy is constructed in a concrete manner. If a factor map between two group shifts is also a group homomorphism then it is said to be an *algebraic factor map* and if a topological conjugacy is also a group isomorphism then it is said to be an *algebraic conjugacy* and the two groups shifts are said to be *algebraically conjugate*. Note that Theorem 3.1 says that any group shift is topologically conjugate to a full shift cross a finite group with an automorphism. It does **not** say that is algebraically conjugate to the direct sum of a full group shift and a finite group with an automorphism. In [Kit87] there is an example of a group shift that is topologically conjugate to a full shift but is not algebraically conjugate to a full group shift.

We also have the following useful proposition.

PROPOSITION 3.1. *A finite-to-one algebraic factor map between irreducible group shifts must be exactly d-to-one on every point for some d. The group shifts can be recoded so the factor map is both left and right resolving.*

Suppose G is a finite group and we form $G^{\mathbb{Z}^d}$, the space of d-dimensional arrays with entries in G. It is a compact zero-dimensional group when the group operation is defined coordinate by coordinate. There are d commuting automorphisms $\sigma_1, \ldots, \sigma_d$, where σ_i is the shift along the ith coordinate axis. This space with the shifts is the *full d-dimensional group shift*. If $\Lambda \subseteq G^{\mathbb{Z}^d}$ is a closed, shift invariant subgroup then it is a d-dimensional subshift of finite type. Such subshifts are *d-dimensional group shifts* or *d-dimensional Markov subgroups*. When $d > 1$ a group shift need **not** be topologically conjugate to a full shift cross a finite group with automorphisms. Suppose X is a compact abelian group. A *group character* for X is a group homomorphism χ from X into the unit circle in the complex plane. The set of a group characters for X is itself a group using pointwise multiplication as an operation. It is the *character group of X*. A good reference for the theory of character groups is [HR63]. To each group shift is associated its character group which is a quotient of a Laurent polynomial ring and there is a developing theory of relating the dynamics of the group shift to the algebra of the character group. A description of these ideas, results, and problems can be found in [Sch95].

4. Algebra. The main results that follow about convolutional codes rely on the Structure Theorem for Finitely Generated Modules over a Principal Ideal Domain. An excellent treatment and the relationship to polynomial matrices can be found in [Jac85]. Let $\mathcal{R}$ be a principal ideal domain and $\mathcal{M}$ a module over $\mathcal{R}$. A *cyclic module* is one with a single generator. If $x \in \mathcal{M}$ we denote by $\mathcal{R}x$ the cyclic module generated by x. Let $x \in \mathcal{M}$ and define the *ring annihilator* of x to be *ann* $x = \{a \in \mathcal{R} : ax = 0\}$. Note that any two generators for a cyclic module have the same ring annihilator. The *torsion submodule of* $\mathcal{M}$ is *tor* $\mathcal{M} = \{x \in \mathcal{M} : ax = 0 \text{ for some } a \neq 0\}$. A module with no torsion elements is said to be *torsion free*. A module $\mathcal{M}$ is a *free module* if it has a linearly independent generating set. If $\mathcal{M}$ is a free module over a commutative ring then the cardinality of any two linearly independent generating sets is the same. This cardinality is the *rank of* $\mathcal{M}$.

Now we state the Structure Theorem for Finitely Generated Modules over a Principal Ideal Domain.

THEOREM 4.1. *If $\mathcal{R}$ is a principal ideal domain and $\mathcal{M}$ a finitely generated module over $\mathcal{R}$ then $\mathcal{M}$ is a direct sum of cyclic modules*

$$M = \mathcal{R}x^1 \oplus \mathcal{R}x^2 \oplus \cdots \oplus \mathcal{R}x^s$$

satisfying

$$ann\ x^1 \subseteq ann\ x^2 \subseteq \cdots \subseteq ann\ x^s.$$

Moreover, the set of ring annihilators is unique for any such decomposition.

This is a generalization of the familiar Structure Theorem for finitely generated abelian groups.

A consequence of this decomposition is that there is an r such that *ann* $x^i = \{0\}$ for $i \leq r$ and *ann* $x^i \neq \{0\}$ for $i > r$. Then $\bigoplus \mathcal{R}x^i$, $i \leq r$ is a free module and *tor* $\mathcal{M} = \bigoplus \mathcal{R}x^i$, $i > r$. We will call such a set of generators $\{x^1, \ldots x^s\}$ a *base* for $\mathcal{M}$. Note also that if $\mathcal{M}$ is a finitely generated module over a principal ideal domain then it is a free module if and only if it is torsion free. The relevant point is that a polynomial or Laurent polynomial ring in one variable with coefficients in a field is a principal ideal domain.

Now we will see how the above applies. Let $\mathbb{F}$ be a finite field, then the full group shift $(\mathbb{F}^n)^{\mathbb{Z}}$ is an infinite dimensional vector space over $\mathbb{F}$ and σ is a vector space automorphism. Rather than focusing on the character group of $(\mathbb{F}^n)^{\mathbb{Z}}$ we will focus on its *dual vector space*. A *linear functional* on a vector space X over a field $\mathbb{F}$ is a vector space homomorphism from X into $\mathbb{F}$. The set of all linear functionals forms the dual space. We denote the dual space of a vector space X by $\widehat{X}$. The dual space has more structure than the character group and we will take advantage of it. The dual space of $(\mathbb{F}^n)^{\mathbb{Z}}$ is isomorphic to $\mathbb{F}[u^{\pm 1}]^n$ and multiplication by u is the dual automorphism to the shift. If X is a vector space and $Y \subseteq X$ is a closed vector subspace then the *annihilator* of Y in $\widehat{X}$ is

$$Y^{\perp} = \{ f \in \widehat{X} : f(y) = 0 \text{ for all } y \in Y \}.$$

In what follows it will be more useful to consider the sequence space $(\mathbb{F}^n)^{\mathbb{Z}}$ as module over $\mathbb{F}[\sigma^{\pm 1}]$ where the shift σ acts as expected. This module has torsion elements and is not finitely generated. However, the set

$$W^s(\bar{0}) \cap W^u(\bar{0}) = \{ x : d(\sigma^i(x), \bar{0}) \to 0 \text{ as } i \to \pm\infty \}$$
$$= \{ x : x_i = 0, |i| \geq r \text{ for some } r \},$$

where $\bar{0}$ denotes the sequence of all 0's, is a free, rank n submodule which is topologically dense in $(\mathbb{F}^n)^{\mathbb{Z}}$. This is the submodule of sequences with *finite support* which we denote by $\mathcal{F}((\mathbb{F}^n)^{\mathbb{Z}})$. In this setting the dual vector space of $(\mathbb{F}^n)^{\mathbb{Z}}$ is $\mathbb{F}[u^{\pm 1}]^n$ and it is considered as a free, rank n, module over the Laurent polynomial ring $\mathbb{F}[u^{\pm 1}]$.

The next propositions are stated but proofs are not included since they are formulations of standard theorems about character groups in the setting of dual spaces.

PROPOSITION 4.1. *Let $\Sigma \subseteq (\mathbb{F}^n)^{\mathbb{Z}}$ be a closed $\mathbb{F}[\sigma^{\pm 1}]$ submodule. Then:*

1. *Σ's annihilator submodule $\Sigma^{\perp} \subseteq \mathbb{F}[u^{\pm 1}]^n$ is free with rank $(n-k)$, for some $k \leq n$ and so is isomorphic to $\mathbb{F}[u^{\pm 1}]^{n-k}$;*
2. *Σ's dual module $\widehat{\Sigma} \simeq \mathbb{F}[u^{\pm 1}]^n/\Sigma^{\perp}$ has rank k over $\mathbb{F}[u^{\pm 1}]$.*

PROPOSITION 4.2. *Let $\Sigma \subseteq (\mathbb{F}^n)^{\mathbb{Z}}$ be a closed submodule whose annihilator submodule has rank $(n-k)$ and suppose its dual module $\widehat{\Sigma} \simeq \mathbb{F}[u^{\pm 1}]^n/\Sigma^{\perp}$ is free. Then:*

1. *the dual module $\widehat{\Sigma} \simeq \mathbb{F}[u^{\pm 1}]^n/\Sigma^{\perp} \simeq \mathbb{F}[u^{\pm 1}]^k$;*
2. *the module $\mathbb{F}[u^{\pm 1}]^n \simeq \Sigma^{\perp} \oplus \widehat{\Sigma} \simeq \Sigma^{\perp} \oplus \mathbb{F}[u^{\pm 1}]^n/\Sigma^{\perp}$;*
3. *the dual module $\widehat{(\Sigma^{\perp})}$ is isomorphic to a closed submodule of $(\mathbb{F}^n)^{\mathbb{Z}}$ whose dual module is isomorphic to the free module $\Sigma^{\perp}$;*
4. *the full group shift $(\mathbb{F}^n)^{\mathbb{Z}} \simeq \Sigma \oplus \widehat{(\Sigma^{\perp})}$;*

PROPOSITION 4.3. *Let $\Sigma \subseteq (\mathbb{F}^n)^{\mathbb{Z}}$ be a closed submodule whose annihilator submodule has rank $(n-k)$ and suppose its dual module $\widehat{\Sigma} \simeq \mathbb{F}[u^{\pm 1}]^n/\Sigma^{\perp}$ is free. If $\{f_1, \ldots, f_k\}$ is a base for $\widehat{\Sigma} \simeq \mathbb{F}[u^{\pm 1}]^n/\Sigma^{\perp}$ then the map*

$$f : \Sigma \to (\mathbb{F}^k)^{\mathbb{Z}}$$

defined by

$$f(x)_i = (f_1(\sigma^i(x)), \ldots, f_k(\sigma^i(x)))$$

is a module isomorphism.

An interesting point to consider is the formulation of Proposition 4.3 when the dual module of Σ is not free.

5. Convolutional codes.

Definition. An irreducible subshift $C \subseteq (\mathbb{F}^n)^Z$ which is a vector space over $\mathbb{F}$ is a *convolutional code*.

Definition. The *free distance* d_{free} of a convolutional code is the minimal number of nonzero symbols appearing in a nonzero element of the code.

EXAMPLE 1. *Let $\mathbb{F}_2$ be the finite field with two elements and $C \subseteq (\mathbb{F}_2^2)^Z$ be defined by the labeled graph in Figure 1. Then C is a convolutional code with free distance 5.*

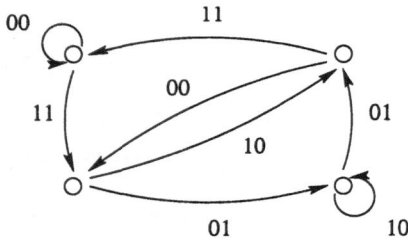

FIG. 1. *Transitions for convolutional code.*

A convolutional code $C \subseteq (\mathbb{F}^n)^Z$ is a vector space over $\mathbb{F}$ and so it is also a closed submodule of $(\mathbb{F}^n)^Z$ over the principal ideal domain $\mathbb{F}[\sigma^{\pm 1}]$. We will think of a convolutional code as a module and apply the results from Section 4. By Proposition 4.1 we know that there is a one-to-one correspondence between closed submodules of $(\mathbb{F}^n)^Z$ and their annihilator submodules in $\mathbb{F}[u^{\pm 1}]^n$. Next we see how the irreducibility of a submodule in $(\mathbb{F}^n)^Z$ relates to the algebra of its corresponding submodule in $\mathbb{F}[u^{\pm 1}]^n$.

PROPOSITION 5.1. *A convolutional code is a closed submodule of $(\mathbb{F}^n)^Z$ whose dual module is free.*

Proof. This follows immediately from 3.1 or by applying the reasoning used to prove Proposition 4.3.   □

This sets up a one-to-one correspondence between convolutional codes contained in $(\mathbb{F}^n)^Z$ and submodules $\mathcal{M} \subseteq \mathbb{F}[u^{\pm 1}]^n$ whose quotient modules $\mathbb{F}[u^{\pm 1}]^n / \mathcal{M}^\perp$ are free.

From Propositions 4.1 we also know that the dual module of every convolutional code $C \subseteq (\mathbb{F}^n)^Z$ has rank $k \leq n$.

Definition. A convolutional code $C \subseteq (\mathbb{F}^n)^Z$ whose dual module has rank k is an (n, k) *convolutional code*.

Let C be an (n, k) convolutional code then (C, σ) is a dynamical system and we define the finite support submodule of C to be

$$\mathcal{F}(C) = \{x \in C : d(\sigma^i(x), \bar{0}) \to 0 \text{ as } i \to \pm\infty\}$$
$$= \{x \in C : x_i = 0, |i| \geq r \text{ for some } r\}.$$

The next two propositions follow immediately.

PROPOSITION 5.2. *The module $\mathcal{F}(\mathcal{C})$ is a free, rank k submodule of $\mathcal{C}$ and is topologically dense in $\mathcal{C}$.*

PROPOSITION 5.3. *The topological entropy of an (n,k) convolutional code $\mathcal{C}$ is $h(\mathcal{C}) = k \log |\mathbb{F}|$.*

EXAMPLE 2. *The code described in Example 1 is a (2,1) convolutional code with free distance 5.*

From Proposition 5.1 and Proposition 4.3 we obtain the next result.

PROPOSITION 5.4. *If $\mathcal{C}$ is an (n,k) convolutional code then it is isomorphic to $(\mathbb{F}^k)^{\mathbb{Z}}$ and consequently $\mathcal{C}$ is an ℓ-step subshift of finite type for some ℓ.*

Definition. If a convolutional code is an ℓ-step subshift of finite type then it is said to have *memory ℓ*.

In Proposition 5.11 we see that this definition of the memory of a convolutional code agrees with the definition in [McE98] formulated with polynomial generator matrices.

EXAMPLE 3. *The code described in Examples 1 and 2 has memory two.*

If $\mathcal{C} \subseteq (\mathbb{F}^n)^{\mathbb{Z}}$ is an (n,k) convolutional code then its annihilator submodule $\mathcal{C}^{\perp} \subseteq \mathbb{F}[u^{\pm 1}]^n$ is a free module of rank $(n-k)$. Let $\{p_1, \ldots, p_{n-k}\} \subseteq \mathbb{F}[u^{\pm 1}]^n$ be a base for $\mathcal{C}^{\perp}$. Then $(p_1, \ldots, p_{n-k})$ defines a map from $(\mathbb{F}^n)^{\mathbb{Z}}$ to $\mathbb{F}^{n-k}$ and

$$\mathcal{C} = \{x \in (\mathbb{F}^n)^{\mathbb{Z}} : (p_1(\sigma^i(x)), \ldots, p_{n-k}(\sigma^i(x))) = (0, \ldots, 0), \ \forall i \in \mathbb{Z}\}.$$

The map $(p_1, \ldots, p_{n-k})$ is a *parity check map* and the matrix with polynomial entries that represents the map is a *parity check matrix*.

Definition. Let $\mathcal{C}$ be an (n,k) convolutional code and $\{x^1, \ldots, x^r\} \subseteq \mathcal{C}$. Define $\langle x^1, \ldots, x^r \rangle$ to be the topological closure in $\mathcal{C}$ of the module $(\mathbb{F}[u^{\pm 1}])x^1 + \cdots + (\mathbb{F}[u^{\pm 1}])x^r$.

PROPOSITION 5.5. *Let $\mathcal{C}$ be an (n,k) convolutional code, $x \in \mathcal{F}(\mathcal{C})$ and $\alpha \in \mathbb{F}[\sigma^{\pm 1}]$. Then $(\mathbb{F}[u^{\pm 1}])(\alpha x) \subseteq (\mathbb{F}[u^{\pm 1}])x$ with equality if and only if α is invertible in $(\mathbb{F}[u^{\pm 1}])$. However, when $\alpha \neq 0$ we have $\langle \alpha x \rangle = \langle x \rangle$.*

Proof. The first statement is clear. The final statement follows because $\mathbb{F}$ has characteristic p for some p. This means a shift of $\alpha^{p^t}(x)$ approaches $x \in \mathcal{C}$ and so $x \in \langle \alpha x \rangle$. □

PROPOSITION 5.6. *Let $\mathcal{C}$ be an (n,k) convolutional code and $x \in \mathcal{F}(\mathcal{C})$ then $\langle x \rangle$ is an $(n,1)$ convolutional code.*

Proof. By definition $\langle x \rangle$ is a topologically closed $\mathbb{F}[\sigma^{\pm 1}]$ submodule of $\mathcal{C}$. The only question is whether it has a dense σ orbit. When $x \in \mathcal{F}(\mathcal{C})$ a point in $\langle x \rangle$ with a dense orbit is easily constructed. □

Note If $x \in \mathcal{C}$ but $x \notin \mathcal{F}(\mathcal{C})$ then $\langle x \rangle$ may not be a convolutional code. For example let x be the point of σ period three defined by the block 110 in $(\mathbb{F}_2)^{\mathbb{Z}}$. Then $\langle x \rangle$ is a four point space and is not irreducible.

Suppose $\mathcal{C}$ and $\mathcal{C}'$ are convolutional codes contained in $(\mathbb{F}^n)^{\mathbb{Z}}$. Then $\mathcal{C} + \mathcal{C}'$ is a topologically closed submodule of $(\mathbb{F}^n)^{\mathbb{Z}}$. It can be seen to be closed by observing that $C + C'$ contains the limit point of a convergent sequence which lies in the set. If $C \cap C' = \{\bar{0}\}$ we denote the sum by $C \oplus C'$.

PROPOSITION 5.7. *Let $C \subseteq (\mathbb{F}^n)^{\mathbb{Z}}$ be an (n, k) convolutional code and C' be an (n, k') convolutional code also contained in $(\mathbb{F}^n)^{\mathbb{Z}}$. Suppose $C \cap C' = \{\bar{0}\}$. Then $C \oplus C' \subseteq (\mathbb{F}^n)^{\mathbb{Z}}$ is an $(n, k + k')$ convolutional code with memory equal to the larger of the memories of C and C'.*

Proof. Clearly, $C \oplus C'$ is a topologically closed submodule of C. It is also easily seen to be irreducible. It is an $(n, k + k')$ code because it is a direct sum. The memory of $C \oplus C'$ cannot be any greater than the larger of the memories and it is seen to be no less by thinking of $\mathcal{W}(C, \ell)$ and $\mathcal{W}(C', \ell)$ as subsets of $\mathcal{W}(C \oplus C', \ell)$ and considering their follower sets. □

Definition. Let C be an (n, k) convolutional code and let $\{x^1, \dots, x^r\} \subseteq \mathcal{F}(C)$. The set $\{x^1, \dots, x^r\}$ is a *generating set for C* if $\langle x^1, \dots, x^r \rangle = C$.

If C is an (n, k) convolutional code and $\{x^1, \dots, x^k\}$ is a base for $\mathcal{F}(C)$ then it is a generating set for C. Conversely, any generating set for C must have at least k elements.

PROPOSITION 5.8. *Let C be an (n, k) convolutional code and let $\{x^1, \dots, x^k\}$ be a generating set for C. Then*

$$C = \langle x^1 \rangle \oplus \cdots \oplus \langle x^k \rangle.$$

Proof. This is a consequence of Propositions 5.6 and 5.7. □

Definition. Let C be an (n, k) convolutional code and $\{x^1, \dots, x^k\}$ a generating set for C. Each $\langle x^j \rangle$ is an $(n, 1)$ convolutional code and so has a memory which we denote by $m(x^j)$. The set has *combined memory* $m(\{x^1, \dots, x^k\}) = m(x^1) + \cdots + m(x^k)$. For convenience we order any generating set for C with k elements so that $m(x^1) \leq m(x^2) \leq \cdots \leq m(x^k)$ and we say $(m(x^1), \dots, m(x^k))$ are the *indices* of the generating set.

PROPOSITION 5.9. *Let C be an (n, k) convolutional code and let $\{x^1, \dots, x^k\}$ be a generating set for C. Then the memory of the code C is $m(k)$.*

Proof. This is also a consequence of Proposition 5.7. □

PROPOSITION 5.10. *Let C be an (n, k) convolutional code. Also, let $\{x^1, \dots, x^k\}$ and $\{y^1, \dots, y^k\}$ be two generating sets for C whose combined memories are minimal over all generating sets for C. Then the indices for the sets are the same.*

Proof. Let $\{x^1, \dots, x^k\}$ and $\{y^1, \dots, y^k\}$ be generating sets for C with minimal combined memories. Suppose $m(x^r) < m(y^r)$ and $m(x^i) = m(y^i)$ for $i < r$. Consider the convolutional codes $\langle x^1, \dots, x^r \rangle$ and $\langle y^{r+1}, \dots y^k \rangle$. Since the memory of $\langle x^1, \dots, x^r \rangle$ is less than the memory of $\langle y^j \rangle$ for each $j = r + 1, \dots k$ we use Proposition 5.7 to see that

$$\langle x^1, \dots, x^r \rangle \cap \langle y^{r+1}, \dots y^k \rangle = \{0\}.$$

Applying Proposition 5.7 again shows

$$\langle x^1, \dots, x^r \rangle \oplus \langle y^{r+1}, \dots y^k \rangle = \langle x^1, \dots, x^r, y^{r+1}, \dots, y^k \rangle$$

is an (n, k) convolutional code. By Proposition 5.3 we see that its topological entropy is $k \log |\mathbb{F}|$ and so it is equal to C. Then

$$\{x^1, \ldots, x^r, y^{r+1}, \ldots y^k\}$$

is also a generating set for C with

$$m(\{x^1, \ldots, x^r, y^{r+1}, \ldots y^k\}) < m(\{y^1, \ldots, y^k\}).$$

This contradicts the supposition $\{y^1, \ldots, y^k\}$ has the minimal combined memory.   □

Definition. The Forney indices are the indices of any base with minimal combined memory and are denoted by $(f_1, \ldots, f_k)$. The *degree* of a convolutional code C is *deg* $C = f_1 + \cdots f_k$ or the minimal combined memory of the code.

Definition. An (n, k) convolutional code with degree m and free distance d is said to be an (n, k, m, d) code.

EXAMPLE 4. *The convolutional code in Examples 1, 2 and 3 has a single Forney index which is two. This makes it a $(2, 1, 2, 5)$ convolutional code.*

PROPOSITION 5.11. *The definition of the Forney indices agrees with the definition given in [McE98] using polynomial generator matrices. This means the memory as previously defined is equal to the kth Forney index and so also agrees with the definition of memory given there.*

6. Encoders and decoders. In coding theory an (n, k) *convolutional encoder* is defined by four matrices A, B, C, and D with entries in $\mathbb{F}$ and dimensions

$$A \text{ is } \ell \times \ell,$$
$$B \text{ is } k \times \ell,$$
$$C \text{ is } \ell \times n,$$
$$B \text{ is } k \times n.$$

The convolutional encoder may have up to $|\mathbb{F}|^\ell$ internal states. It converts an input sequence $u \in (\mathbb{F}^k)^{\mathbb{N}}$ into an encoded sequence $x \in (\mathbb{F}^n)^{\mathbb{N}}$. It works by specifying that the internal state at time zero is $s(0) = 0$ and for $i \geq 0$,

$$s(i + 1) = s(i)A + u(i)B$$
$$x(i) = s(i)C + u(i)D.$$

This means both the output and the next internal state are linear functions of both the input and the current internal state. The set of all possible outputs defines the convolutional code C.

EXAMPLE 5. *The following matrices describe an encoder for the* $(2, 1, 2, 5)$ *code of Section 5*

$$A = \begin{bmatrix} 0 & 1 \\ 0 & 0 \end{bmatrix} \quad B = \begin{bmatrix} 1 & 0 \end{bmatrix} \quad C = \begin{bmatrix} 1 & 0 \\ 1 & 1 \end{bmatrix} \quad D = \begin{bmatrix} 1 & 1 \end{bmatrix}.$$

If C and C' are two convolutional codes with a finite equivalence between them given by $\varphi : \Sigma \to C$ and $\psi : \Sigma \to C'$ where Σ is a convolutional code and φ and ψ are module homomorphisms then we say that there is an *algebraic finite equivalence* between them. This leads to the next observation.

PROPOSITION 6.1. *A convolutional encoder is an algebraic finite equivalence between* $(\mathbb{F}^k)^{\mathbb{Z}}$ *and an* (n, k) *convolutional code* $C \subseteq (\mathbb{F}^n)^{\mathbb{Z}}$.

Proof. First we see that an encoder defines an algebraic finite equivalence. Suppose we have a convolutional encoder defined by the matrices A, B, C and D. Define a new convolutional code Σ whose alphabet consists of the internal states of the encoder. There is one transition from state s to s' for each input u with $s' = sA + uB$. The homomorphisms $\varphi : \Sigma \to C$ and $\psi : \Sigma \to (\mathbb{F}^k)^{\mathbb{Z}}$ are two block maps with ψ defined by mapping the transition from a state s to state s' to the input u which gave rise to the transition in the equation $s' = sA + uB$. The map φ is then defined by mapping the transition to the output x defined by the equation $x = sC + uD$.

Next suppose $\Sigma \subseteq (\mathbb{F}^{\ell})^{\mathbb{Z}}$ is an (ℓ, k) convolutional code and $\varphi : \Sigma \to C$ and $\psi : \Sigma \to (\mathbb{F}^k)^{\mathbb{Z}}$ are algebraic factor maps. By Proposition 3.1 we can recode so that ψ is right resolving. Then we apply the discussion about channel encoders in section 2. The encoder takes input sequences from $(\mathbb{F}^k)^{\mathbb{Z}}$, uses ψ to lift them to sequences in Σ just as in Section 2, and then uses φ to convert the sequences in Σ to output sequences in C. The sequence in Σ is the sequence of internal states. □

EXAMPLE 6. *The encoder of Example 5 describes an algebraic finite equivalence where* $\Sigma \subseteq (\mathbb{F}_2^2)^{\mathbb{Z}}$ *and the two-block algebraic factor maps* $\varphi :$ $\Sigma \to C$ *and* $\psi : \Sigma \to (\mathbb{F}_2)^{\mathbb{Z}}$ *are defined by the diagram in Figure 2. The code* Σ *is defined by the graph and the maps are defined by the labelings on the edges with the pair representing* ψ/φ.

A crucial point is that if the map φ is an isomorphism then the decoder is the block map defined by $\psi \circ \varphi^{-1}$ just as for the codes described in Section 2. If the map φ is not an isomorphism then the encoding may or may not be inverted by a block map. If it is not inverted by a block map then a single error may be propagated forever. When $\psi \circ \varphi^{-1}$ is not a block map the encoder is said to be *catastrophic*. By Proposition 5.4 we see the following.

PROPOSITION 6.2. *If* C *is any convolutional code then there exists a noncatastrophic encoder for* C.

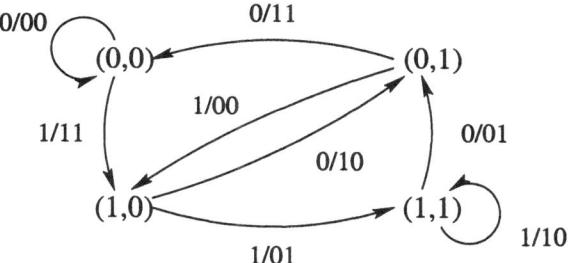

FIG. 2. *Transitions for Σ and the two-block maps.*

7. Dual codes. Let $\mathcal{C} \subseteq (\mathbb{F}^n)^{\mathbb{Z}}$ be an (n, k) convolutional code with memory ℓ. Then $\mathcal{W}(\mathcal{C}, \ell + 1)$ is a vector subspace of $(\mathbb{F}^n)^{\ell+1}$ and

$$\mathcal{C} = \{x \in (\mathbb{F}^n)^{\mathbb{Z}} : [x_i, \ldots, x_{i+\ell}] \in \mathcal{W}(\mathcal{C}, \ell + 1) \text{ for all } i \in \mathbb{Z}\}.$$

Let $\mathcal{W}(\mathcal{C}, \ell + 1)^{\perp}$ be the annihilator vector space of $\mathcal{W}(\mathcal{C}, \ell + 1)$ in $(\mathbb{F}^n)^{\ell+1}$ and

$$\mathcal{A} = \{x \in (\mathbb{F}^n)^{\mathbb{Z}} : [x_i, \ldots, x_{i+\ell}] \in \mathcal{W}(\mathcal{C}, \ell + 1)^{\perp} \text{ for all } i \in \mathbb{Z}\}$$

be the subshift of finite type defined by $\mathcal{W}(\mathcal{C}, \ell + 1)^{\perp}$.

PROPOSITION 7.1. *The subshift of finite type $\mathcal{A} \subseteq (\mathbb{F}^n)^{\mathbb{Z}}$ defined above is an $(n, n - k)$ convolutional code.*
Proof. The character submodule of $\mathcal{A}$ is isomorphic to $\mathcal{C}^{\perp}$ which is a free rank $(n - k)$ module over $\mathbb{F}[u^{\pm 1}]$. □

PROPOSITION 7.2. *Let $\mathcal{C} \subseteq (\mathbb{F}^n)^{\mathbb{Z}}$ be an (n, k) convolutional code and $\mathcal{A} \subseteq (\mathbb{F}^n)^{\mathbb{Z}}$ be the $(n, n - k)$ convolutional code defined as above from $\mathcal{C}$. Then:*

    1. $(\mathbb{F}^n)^{\mathbb{Z}} \simeq \mathcal{C} \oplus \mathcal{A}$;
    2. $\mathbb{F}[u^{\pm 1}]^n \simeq \mathcal{C}^{\perp} \oplus \mathcal{A}^{\perp}$;
    3. $\widehat{\mathcal{C}} \simeq \mathcal{A}^{\perp}$;
    4. $\widehat{\mathcal{A}} \simeq \mathcal{C}^{\perp}$.

Definition. The $(n, n - k)$ convolutional code $\mathcal{A}$ defined from $\mathcal{C}$ is the Wyner-Ash *dual code of* $\mathcal{C}$.

PROPOSITION 7.3. *If $\mathcal{C} \subseteq (\mathbb{F}^n)^{\mathbb{Z}}$ is an (n, k) convolutional code with memory ℓ then the Wyner-Ash dual code also has memory ℓ.*
Proof. This follows by the symmetry of the duality. □

8. Descriptions. There are three standard ways to describe a convolutional code and they have their analogues in the description of algebraic symbolic systems. This is summarized below.

PROPOSITION 8.1. *Let $\mathcal{C} \subseteq (\mathbb{F}^n)^{\mathbb{Z}}$ be an (n, k) convolutional code. It may be described in any of the following ways.*

1. *A polynomial generator matrix which corresponds to a set of generators $\{x^1, \ldots, x^k\} \subseteq \mathcal{F}(\mathcal{C})$ with $\langle x^1, \ldots, x^k \rangle = \mathcal{C}$.*

2. *A parity check matrix which corresponds to a set of generators $\{p_1, \ldots, p_{n-k}\} \subseteq \mathbb{F}[u^{\pm 1}]^n$ for the annihilator submodule $\mathcal{C}^{\perp}$.*

3. *A convolutional encoder which corresponds to an algebraic finite equivalence between $\mathcal{C}$ and $(\mathbb{F}^k)^{\mathbb{Z}}$.*

REFERENCES

[ACH83]   ROY ADLER, DON COPPERSMITH, AND MARTIN HASSNER. Algorithms for sliding block codes. *IEEE Transactions on Information Theory*, pp. 5–22, 1983.

[For99]   G. DAVID FORNEY, JR. The dynamics of group codes: Dual abelian group codes and systems. *Preprint*, 1999.

[HR63]   EDWIN HEWITT AND KENNETH ROSS. *Abstract Harmonic Analysis*. Academic Press and Springer, 1963.

[Jac85]   NATHAN JACOBSON. *Basic Algebra I*. Freeman, 1985.

[Kit87]   BRUCE KITCHENS. Expansive dynamics on zero-dimensional groups. *Ergodic Theory and Dynamical Systems*, 7:249–261, 1987.

[Kit98]   BRUCE KITCHENS. *Symbolic Dynamics, One-sided, Two-sided and Countable State Markov Shifts*. Springer, 1998.

[LM95]   DOUGLAS LIND AND BRIAN MARCUS. *Symbolic Dynamics and Coding*. Cambridge, 1995.

[McE98]   ROBERT McELIECE. The algebraic theory of convolutional codes. In V.S. Pless and W.C. Hoffman, editors, *Handbook of Coding Theory*, Volume 1. Elsevier, 1998.

[Par77]   WILLIAM PARRY. A finitary classification of topological markov chains and sofic systems. *Bulletin of the London Mathematical Society*, pp. 86–92, 1977.

[Ros99]   JOACHIM ROSENTHAL. Connections between linear systems and convolutional codes. *Preprint*, 1999.

[Sch95]   KLAUS SCHMIDT. *Dynamical Systems of Algebraic Origin*. Burkhäuser, 1995.

LINEAR CODES AND THEIR DUALS OVER ARTINIAN RINGS

THOMAS MITTELHOLZER*

Abstract. Linear codes over commutative artinian rings R are considered. For a linear functional-based definition of duality, it is shown that the class of length-n linear block codes over R should consist of projective submodules of the free module R^n. For this class, the familiar duality properties from the field case can be generalized to the ring case. In particular, the MacWilliams identity is derived for linear codes over any finite commutative ring. Duals of convolutional codes are also considered, and it is shown that for convolutional codes over commutative artinian rings, the duality property holds for a code and its dual as well as for the local description of the code by its canonical trellis section and its dual trellis section.

1. Introduction. Extensive work has been done on duality properties of linear codes in the framework of character-based (or Pontryagin) duality [1–4]. Recently duality of modules over a finite ring R was studied in [5], where it was shown that if R is a quasi-Frobenius ring (i.e., if R is injective as an R-module) then the duality concept based on Pontryagin duality via characters is equivalent to the duality concept based on linear functionals. In that study, the underlying category of codes consisted of all finitely generated R-modules.

In this paper, we restrict the category of codes to projective modules that are submodules of R^n. As a consequence of this reduction, one can extend the class of rings for which the usual duality properties hold. In particular, this approach provides duality relations for linear codes over commutative artinian rings.

The linear functional-based *dual (or orthogonal)* of an R-submodule $M \subset R^n$ is defined via orthogonality in R^n, i.e., $M^\perp = \{f \in Hom(R^n, R) : f(\mathbf{m}) = 0 \text{ all } \mathbf{m} \in M\}$,[1] which is equivalent to

$$(1) \qquad M^\perp = \{\mathbf{x} \in R^n : \mathbf{x} \cdot \mathbf{m}^T = 0, \text{ all } \mathbf{m} \in M\}.[2]$$

The motivation to study linear functional-based duality is twofold. First, it allows one to obtain the desired duality properties for codes over some infinite rings, for which character-based duality is not suitable. Second, in the case of finite rings, both duality concepts apply, however, they do not agree, in general, i.e., different duality properties may hold, which is illustrated by the following example.

EXAMPLE 1. *The commutative ring $R = GF(2)[x, y]/(x^2, y^2, xy)$ is artinian but not quasi-Frobenius (i.e., is not injective as an R-module).*

---

*IBM Research, Zurich Research Laboratory, Säumerstrasse 4, CH-8803 Rueschlikon, Switzerland; Email: tmi@zurich.ibm.com.

[1] $Hom(R^n, R)$ denotes the set of all R-homomorphims from R^n to R.

[2] $\mathbf{m}$ is considered a row vector and $\mathbf{m}^T$ denotes the transpose.

Let $C \subset R^2$ be the free rank-1 module generated by the systematic encoding matrix

$$G = [1 \quad 1 + x + y],$$

i.e, the code C equals

$$\{[0\ 0], [1\ 1+x+y], [1+x+y\ 1], [x\ x], [y\ y], [1+x\ 1+y], [1+y\ 1+x], [x+y\ x+y]\}.$$

The parity check matrix for this code is equal to the generator matrix, $H = G$; that is, the code is self-dual, $C^\perp = C$, when using the linear functional duality concept.

    Pontryagin duality is based on the notion of a continuous character, that is of a continuous group-homomorphism from the additive group of R to the multiplicative group of the complex number field $\mathbf{C}$. Let $\hat{R}$ denote the set of all such homomorphisms.

    Regarded as an additive group, R is isomorphic to $\mathbf{Z}_2^3$ and, therefore, $\hat{R} \cong \mathbf{Z}_2^3$ as groups. An explicit description of the character group of R is obtained by fixing some isomorphism, for instance,

$$\phi : \mathbf{Z}_2 \cdot 1 \oplus \mathbf{Z}_2 \cdot x \oplus \mathbf{Z}_2 \cdot y \to R$$

given by $\phi(r_1, r_x, r_y) = r_1 + r_x x + r_y y$. Every $a = a_1 + a_x x + a_y y$ then defines a character χ_a, which operates on an element $r = r_1 + r_x x + r_y y$, by

$$\chi_a(r) = (-1)^{a_1 \cdot r_1 + a_x \cdot r_x + a_y \cdot r_y}.$$

Thus, the group-isomorphism ϕ induces a group-isomorphism $\alpha : R \to \hat{R}$. Note that α is not a natural isomorphism; actually, there is no natural isomorphism between R and $\hat{R}$, which is in contrast to the fact that R and $\hat{\hat{R}}$ are naturally isomorphic (see Chap. 1.7 in [6]).

    The character-based (or Pontryagin) dual of the code C is defined by orthogonality

$$(2) \quad C^\vdash = \{[\chi^{(1)}, \chi^{(2)}] \in \hat{R}^2 : \chi^{(1)}(c_1) \cdot \chi^{(2)}(c_2) = 1, \ \text{all} \ [c_1\ c_2] \in C\}.$$

    When regarded as a subgroup of R^2, the Pontryagin dual $\alpha^{-1}(C^\vdash)$ equals

$$\{[0\ 0], [1\ 1], [x\ 1+x], [y\ 1+y], [1+x\ x], [1+y\ y], [1+x+y\ 1+x+y], [x+y\ x+y]\}.$$

Note that $\alpha^{-1}(C^\vdash)$ is not an R-module and $C^\perp \neq \alpha^{-1}(C^\vdash)$. Moreover, the complete weight enumerators for the code C and the dual $\alpha^{-1}(C^\vdash)$ are different; they are given by the following two polynomials in the 8 indeterminates $Z_0, Z_1, Z_x, Z_y, Z_{1+x}, Z_{1+y}, Z_{x+y}, Z_{1+x+y}$

$$A(\mathbf{Z}) = Z_0^2 + 2Z_1 Z_{1+x+y} + Z_x^2 + Z_y^2 + 2Z_{1+x} Z_{1+y} + Z_{x+y}^2$$

$$B(\mathbf{Z}) = Z_0^2 + Z_1^2 + 2Z_x Z_{1+x} + 2Z_y Z_{1+y} + Z_{1+x+y}^2 + Z_{x+y}^2$$

respectively. It is interesting that although $A(\mathbf{Z}) \neq B(\mathbf{Z})$, the corresponding weight enumerator polynomials with respect to Hamming weight (which are obtained by setting $Z_0 = 1$ and the 7 other indeterminates to Z) are equal, i.e., $A_H(Z) = 1 + 7Z^2 = B_H(Z)$.

The MacWilliams identities always hold for the character-based definition of duals for 'linear' codes defined over finite abelian groups [3]; in particular, they hold for the polynomials $A(\mathbf{Z})$ and $B(\mathbf{Z})$ as well as for $A_H(Z)$ and $B_H(Z)$, when considering the dual pair C and $C^\vdash$. For the dual pair C and $C^\perp$ of codes based on linear functionals, the generalized MacWilliams identities do not hold because $C = C^\perp \neq \alpha^{-1}(C^\vdash)$ but $A(\mathbf{Z}) \neq B(\mathbf{Z})$. However, in the special case of the weight enumerator polynomials $A_H(Z)$, $B_H(Z)$ with respect to Hamming weight, the identity (6) as given below holds.

Whether $C^\perp$ and $\alpha^{-1}(C^\vdash)$ are isomorphic as R-modules depends on the choice of the isomorphism ϕ. By choosing instead of ϕ another (suitable) isomorphism ϕ', one can obtain an induced isomorphism α' for which $C^\perp = \alpha'^{-1}(C^\vdash)$. To obtain this equality, one needs to choose the isomorphism ϕ' as a function of the code C and there is no (fixed) canonical map ϕ, which induces a natural R-isomorphism between a code and its Pontryagin dual. In particular, one can show that if one chooses as a second code $\tilde{C}$ with generator matrix $\tilde{G} = [1 \quad x]$ then there is no isomorphism ϕ for which one obtains simultanous R-isomorphisms $C^\perp \cong \alpha^{-1}(C^\vdash)$ and $\tilde{C}^\perp \cong \alpha^{-1}(\tilde{C}^\vdash)$. Thus, for finite commutative rings, duality based on linear functionals and character-based duality are not equivalent, in general.

After setting a suitable framework for duals over commutative artinian rings in the next section, a generalization of the MacWilliams identities from the field to the ring case will be formulated and proved in Section 3 for weight enumerators with respect to Hamming distance. In Section 4, the duality results are extended from linear block codes to convolutional codes.

2. Duality properties of codes over artinian rings. Let R be an abelian[3] artinian ring. According to the structure theorem for such rings [7], R can be written as a finite direct sum of local rings R_i, i.e.,

$$(3) \qquad\qquad R = R_1 \oplus R_2 \oplus \ldots \oplus R_e.$$

PROPOSITION 2.1. *Let R be a commutative artinian ring with a decomposition (3). Then,*

(i) R has e maximal ideals, which are of the form

$$m_i = R_1 \oplus \ldots \oplus R_{i-1} \oplus m_i' \oplus R_{i+1} \oplus \ldots \oplus R_e$$

where m_i' denotes the maximal ideal of R_i;

---

[3]The terms abelian and commutative are used interchangeably.

(ii) the localization of R at the maximal ideal m_i is isomorphic to R_i, i.e., $R_i \cong R_{m_i}$.

Proof. (i) is clear because m'_i is maximal in R_i.

To prove (ii), we consider the homomorphism $i : R_i \to R_{m_i}$ given by sending an element $r^{(i)} \in R_i$ to $r^{(i)}/1$ in $R_{m_i} \triangleq \{r/s : r \in R, s \in R \setminus m_i\}$ and show that it is an isomorphism. The homomorphism i is injective because the annihilator of an element $r^{(i)} \neq 0$ is an ideal that is contained in the maximal ideal m_i and, hence, there is no $s \in R \setminus m_i$ such that $s \cdot r^{(i)} = 0$. To show surjectivity, consider an arbitrary element $r/s \in R_{m_i}$. Using (3), one has $r = (r_1, \ldots, r_i, \ldots, r_e)$ and $s = (s_1, \ldots, s_i, \ldots, s_e)$, where $s_i \in R_i \setminus m'_i$. As m'_i is a maximal ideal, s_i is invertible in R_i with inverse s'_i. The element $s'_i r_i$ is mapped onto r/s, i.e., there is an element $t \in R \setminus m_i$ such that $t(r - ss'_i r_i) = 0$, viz., $t = (0, \ldots, 1, \ldots, 0)$ with a single component 1 at position i. $\square$

For arbitrary R-modules there is no duality relation, in general. A suitable subcategory, for which duality holds, is the class of all finitely generated projective R-modules. A projective R-module is a direct summand in a free R-module (see Chap. 3.10 in [7]) and, thus, it can be regarded as a generalization of a free module. In the following, we recall some results on projective modules and show that they satisfy the desired duality relations. Using (3), every R-module M can be decomposed as

$$M = R \otimes_R M = M_1 \oplus M_2 \oplus \ldots \oplus M_e$$

where $M_i = R_i \otimes_R M$ is the tensor product of the R-modules R_i and M. The following proposition results by applying a well-known local-global result in commutative algebra (see [8] Chap. I.3, Corollary 3.4) to the special case of artinian rings and by using $R_i \cong R_{m_i}$ (cf. Proposition 2.1 (ii)).

PROPOSITION 2.2. *Let M be a finitely generated module over R. Then, M is R-projective if and only if M_i is R_i-projective for all $i = 1, 2, \ldots, e$.*

Remark. The R_i-modules M_i are actually free because the rings R_i are local, i.e., R_i contains exactly one maximal ideal [7].

The following lemma for projective modules over commutative artinian rings is crucial for the orthogonality (or duality) property of linear functional-based duals.

LEMMA 2.1. *Suppose $M \subset R^n$ is projective. Then there exists a projective submodule Q of R^n such that*

$$(4) \qquad\qquad R^n = M \oplus Q.$$

Proof. Using the results above, one can assume without loss of essential generality that R is local artinian and, hence, consider M to be free over R. For this case, (4) was proved in Appendix II of [9]. $\square$

It is well known in commutative algebra that when localizing a finitely generated projective module U at a prime ideal $\wp$, one obtains an $R_\wp$-module $U_\wp$ that is free (see e.g. Chap. 7.7 in [7]). This allows one to define the $\wp$-rank $rk_\wp(U)$ of a finitely generated projective module U, which is given by the cardinality of a basis of the localization $U_\wp$.

PROPOSITION 2.3. *Let R be commutative artinian and suppose that U and V are submodules of R^n, which are projective. Then*

(i) $U^\perp$ *is projective and* $(U^\perp)^\perp = U$; *moreover,* $rk_\wp(U^\perp) = n - rk_\wp(U)$ *for any prime ideal $\wp$.*

(ii) $(U + V)^\perp = U^\perp \cap V^\perp$.

(iii) $U + V$ *and* $U \cap V$ *are projective and for any prime ideal $\wp$, the following rank formula holds*

$$rk_\wp(U \cap V) + rk_\wp(U + V) = rk_\wp(U) + rk_\wp(V).$$

Remark. If R is artinian but not local then even if U and V are free the intersection $U \cap V$ need not be free. A simple example of this fact is obtained by letting R be the ring $\mathbf{Z}_6$ of integers modulo 6 and by considering the rank-1 $\mathbf{Z}_6$-submodules U and V of R^2 generated by $[2\ 3]$ and $[1\ 0]$, respectively. Then, $U \cap V = \{[0\ 0], [2\ 0], [4\ 0]\}$ is projective but not free. Thus, when R is commutative artinian, the class of free modules need not be closed under the intersection operation, but the class of projective modules is.

Proof. Using Proposition 2.2, one can assume that R is local and, hence, that the projective modules U and V are free with $rk(U) = k$. Proof of (i): Choose a basis $\mathbf{g}_1, \ldots, \mathbf{g}_k$ for U. Let G be the corresponding generator matrix. By the above Lemma, there is a complementary free module U' such that

(5) $$R^n = U \oplus U'.$$

In other words, the basis of U can be extended by $n - k$ n-tuples $\mathbf{q}_{k+1}, \ldots, \mathbf{q}_n$ to form a basis of R^n. Let Q be the $(n-k) \times n$ matrix with $\mathbf{q}_{k+1}, \ldots, \mathbf{q}_n$ as rows. The $n \times n$ matrix consisting of the submatrices G and Q is invertible, i.e.,

$$\begin{bmatrix} G \\ Q \end{bmatrix} \cdot [\, K^T \quad H^T \,] = \begin{bmatrix} I_k & 0 \\ 0 & I_{n-k} \end{bmatrix} = \begin{bmatrix} K \\ H \end{bmatrix} \cdot [\, G^T \quad Q^T \,],$$

where I_k denotes the $k \times k$ identity matrix and H and K are $(n-k) \times n$ and $k \times n$ matrices, respectively. It follows that the rows of H form a basis for $U^\perp$, hence $U^\perp$ is free of rank $n - k$ and, therefore, $rk_\wp(U^\perp) = n - rk_\wp(U)$. Similarly, the rows of G form a basis for $(U^\perp)^\perp$ and, therefore, $(U^\perp)^\perp = U$. Proof of (ii):

$$(U + V)^\perp = \{\mathbf{x} \in R^n : \mathbf{x} \cdot \mathbf{y}^T = 0, \text{ for all } \mathbf{y} \in U \text{ or } \mathbf{y} \in V\}$$
$$= \{\mathbf{x} : \mathbf{x} \cdot \mathbf{u}^T = 0, \text{ for all } \mathbf{u} \in U\} \cap \{\mathbf{x} : \mathbf{x} \cdot \mathbf{v}^T = 0, \text{ for all } \mathbf{v} \in V\}$$
$$= U^\perp \cap V^\perp.$$

Proof of (iii): (5) implies $V \cap U \oplus V \cap U' = V$. Thus, $V \cap U$ is a direct summand in the free module V and, hence, projective. Using (i), it follows similarly that $V^\perp \cap U^\perp$ is projective. Now (ii) and (i) imply that $U + V$ is projective.

To show the rank formula, we make use again of the above Lemma to conclude that there is a free R-module V' such that $R^n = V \oplus V'$. One easily verifies that $U + V = U \cap V + U \cap V' + V \cap U'$ and that the right-hand side is actually a direct sum. Thus,

$$(U \cap V) \oplus (U + V) \cong U \cap V \oplus U \cap V' \oplus U \cap V \oplus V \cap U' \cong U \oplus V$$

and this implies the rank formula

$$rk(U \cap V) + rk(U + V) = rk(U \oplus V) = rk(U) + rk(V),$$

which also holds after further localization at a prime ideal $\wp$.   ☐

In terms of category theory [7], the result of Proposition 2.3 can be expressed by saying that $\perp$ is a (contravariant) duality functor from the category of projective submodules $U \subset R^n$ onto itself. Here, the morphisms considered must be defined for the entire space R^n. Duality means that the functor that results from applying $\perp$ twice is naturally equivalent to the identity functor.

EXAMPLE 2. *The commutative ring $R = GF(2)[x, y]/(x^2, y^2, xy)$ is artinian but not quasi-Frobenius. The non-projective R-module $U = \{0, x\}$ has as dual $U^\perp = \{0, x, y, x + y\}$. But $U \neq (U^\perp)^\perp = U^\perp$ and, therefore, $\perp$ is not a duality functor on the category of finitely generated R-modules. This shows that the restriction to projective modules is essential.*

The two notions of duality (1) and (2), which are based on linear functionals and on characters, coincide for finite commutative quasi-Frobenius rings [5]. In the case of other rings, the two duality notions can be considered complementary. For example, the ring of integers $\mathbf{Z}$ is not quasi-Frobenius (not injective) and not artinian. The linear functional based functor does not give the desired duality properties but Pontryagin duality does.

A complementary example is given by the field of rational numbers $\mathbf{Q}$. The rationals are not locally compact. Another example is the ring of rational functions $\mathbf{Z}_4(t)$ over the integers modulo 4 (cf. Section 4.1). This ring is a subgroup in the bi-infinite direct product group $\mathbf{Z}_4^{\mathbf{Z}}$. Putting the discrete topology on the component groups $\mathbf{Z}_4$, the infinite product $\mathbf{Z}_4^{\mathbf{Z}}$ is a compact group with respect to the product topology; the subgroup $\mathbf{Z}_4(t)$ is neither open nor closed and, hence, not locally compact. Similarly, one can show that if $R(D)$ is regarded as a subgroup of the Laurent sequence space $R((D))$, it is not locally compact. For these two examples, Pontryagin duality does not provide the desired duality property but duality based on linear functionals does.

3. MacWilliams identities. In the following discussion it is assumed that the ring is finite and commutative; in particular, R is artinian and the results of the preceding section apply. The class of codes considered are projective modules $U \subset R^n$. The spectrum of a code U and its dual $U^\perp$ is given by

$$A_i = |\{\mathbf{u} \in U : w_H(\mathbf{u}) = i\}|$$
$$B_i = |\{\mathbf{v} \in U^\perp : w_H(\mathbf{v}) = i\}|,$$

where $w_H(\mathbf{u})$ denotes the Hamming weight of a codeword $\mathbf{u}$ and $|V|$ denotes the cardinality of a set V.

The MacWilliams identities give a relation between the weight coefficients A_i and B_i.

THEOREM 3.1. *(MacWilliams identity) Let R be a finite commutative ring and let A_i and B_i be the weight coefficients of an R-linear code U and its dual $U^\perp$, respectively. Then, the following polynomial identity holds:*

$$(6) \qquad \sum_{i=0}^{n} B_i X^i = \frac{1}{|U|} \sum_{j=0}^{n} A_j (1 - X)^j \{1 + (|R| - 1)X\}^{n-j}.$$

Remark. From Example 1 it is clear that the generalized MacWilliams identity does not hold for generalized distance measures.

Proof. The proof goes along the lines of the original proof No. 1 in [10]. Setting $X = 1/(1 + Y)$ in (6), one obtains

$$\sum_{i=0}^{n} B_i (1 + Y)^{n-i} = \frac{1}{|U|} \sum_{j=0}^{n} A_j Y^j (Y + |R|)^{n-j}.$$

Expanding and comparing coefficients of Y^ℓ yields

$$(7) \qquad \sum_{i=0}^{n-\ell} B_i \binom{n-i}{\ell} = \frac{|R|^{n-\ell}}{|U|} \sum_{j=0}^{\ell} A_j \binom{n-j}{n-\ell}.$$

It is sufficient to show (7) for all $\ell = 0, 1, \ldots, n$.

For each subset $s = \{s_1, \ldots, s_\ell\} \subset \{1, \ldots, n\}$ of cardinality ℓ, we define a free rank-ℓ submodule $F_s \subset R^n$ with support in s:

$$F_s = \{\mathbf{x} \in R^n : supp(\mathbf{x}) \subseteq s\} = Re_{s_1} \oplus \ldots \oplus Re_{s_\ell},$$

where $e_1, \ldots, e_n$ is the standard basis for R^n and $supp(\mathbf{x})$ denotes the support, i.e., the indices with non-zero components, of $\mathbf{x}$. Let $t = \{1, \ldots, n\} \backslash s$ be the complementary set of s. Then, clearly $F_s^\perp = F_t$. Using Proposition 2.3, one obtains

$$(U + F_s)^\perp = U^\perp \cap F_t$$
$$rk_m(U + F_s) = rk_m(U^\perp \cap F_t)^\perp = n - rk_m(U^\perp \cap F_t)$$
$$rk_m(U) + rk_m(F_s) = rk_m(U \cap F_s) + rk_m(U + F_s),$$

where m is one of the e maximal ideals of R. As $rk_m(F_s) = \ell$, the latter two equations imply

$$(8) \qquad n - \ell + rk_m(U \cap F_s) - rk_m(U) = rk_m(U^\perp \cap F_t)$$

or, equivalently,

$$(9) \qquad |R_m|^{n-\ell+rk_m(U \cap F_s)-rk_m(U)} = |R_m|^{rk_m(U^\perp \cap F_t)},$$

where R_m denotes the localization at the maximal ideal m.

For fixed cardinality ℓ, consider pairs $(s, \mathbf{u})$, where $\mathbf{u} \in U \cap F_s$. For each choice of s, there are $|U \cap F_s| = \prod_m |R_m|^{rk_m(U \cap F_s)}$ such pairs, where the product is over all the e maximal ideals m of R. Considering all possible choices for s, the total number of such pairs is

$$\sum_{\substack{s \subset \{1,\ldots,n\} \\ |s| = \ell}} \prod_m |R_m|^{rk_m(U \cap F_s)}.$$

A second way of counting these pairs is as follows. For each $\mathbf{u} \in U$ of weight j, there are $n - j$ zero components. Thus, any subset $t = \{t_1, \ldots, t_{n-\ell}\} \subset \{1, \ldots, n\} \setminus supp(\mathbf{u})$ of cardinality $n - \ell$ defines a complementary set s, which can be paired with $\mathbf{u}$. There are $\binom{n-j}{n-\ell}$ choices for t or s, respectively. There are A_j codewords of weight j in U, hence

$$\sum_{\substack{s \subset \{1,\ldots,n\} \\ |s| = \ell}} \prod_m |R_m|^{rk_m(U \cap F_s)} = \sum_{j=0}^{\ell} A_j \binom{n-j}{n-\ell}.$$

Applying the same argument to $U^\perp$, one obtains

$$\sum_{\substack{t \subset \{1,\ldots,n\} \\ |t| = n-\ell}} \prod_m |R_m|^{rk_m(U^\perp \cap F_t)} = \sum_{i=0}^{n-\ell} B_i \binom{n-i}{\ell}.$$

Using (9) and the fact that the complementary sets s and t are in one-to-one correspondence yields

$$\sum_{i=0}^{n-\ell} B_i \binom{n-i}{\ell} = \sum_{\substack{t \subset \{1,\ldots,n\} \\ |t| = n-\ell}} \prod_m |R_m|^{rk_m(U^\perp \cap F_t)}$$

$$= \sum_{\substack{t \subset \{1, \ldots, n\} \\ |t| = n - \ell}} \prod_m |R_m|^{n - \ell + rk_m (U \cap F_s) - rk_m (U)}$$

$$= \prod_m |R_m|^{n - \ell - rk_m (U)} \sum_{j=0}^{\ell} A_j \begin{pmatrix} n - j \\ n - \ell \end{pmatrix}$$

$$= \frac{|R|^{n - \ell}}{|U|} \sum_{j=0}^{\ell} A_j \begin{pmatrix} n - j \\ n - \ell \end{pmatrix}.$$

$\square$

EXAMPLE 3. [2 2] *generates a* $\mathbf{Z}_6$*-module* $U = \{[0\ 0], [2\ 2], [4\ 4]\}$, *which is projective but not free. The dual code is*

$$U^{\perp} = \{[0\ 0], [3\ 0], [0\ 3], [3\ 3], [2\ 4], [4\ 2], [1\ 5], [5\ 1], [1\ 2], [2\ 1], [4\ 5], [5\ 4]\}.$$

The weight enumerator polynomials of U *and* $U^{\perp}$ *are* $A(X) = 1 + 2X^2$ *and* $B(X) = 1 + 2X + 9X^2$, *respectively. The MacWilliams identities are readily verified:*

$$B(X) = \frac{1}{3} \sum_{j=0}^{2} A_j (1 - X)^j (1 + 5X)^{2-j}.$$

4. Duality of convolutional codes over rings.

4.1. Dual and convolutional dual. An (n, k) convolutional code over a field F can be regarded as a block code over the field of rational functions $F(D)$ [11]. For a commutative ring R, we can define the ring of rational functions similarly as in the field case by

$$R(D) = \left\{ \frac{f(D)}{D^m s(D)} : f(D), s(D) \in R[D], s(0) = 1, m \in \mathbf{Z} \right\}.$$

A *convolutional code over* R is a projective $R(D)$-submodule of $R(D)^n$. For coding purposes, one considers the standard embedding of $R(D)$ in the ring of formal Laurent series $R((D))$, which allows one to view a codeword of an (n, k) code over $R(D)$ as a code sequence with components in R^n. The following proposition is crucial to extend the duality results from linear block codes to convolutional codes.

PROPOSITION 4.1. *If* R *is commutative artinian, then so is* $R(D)$.

Proof. Owing to the structure theorem for commutative artinian rings (3), the ring of rational functions decomposes as

$$R(D) = R_1(D) \oplus R_2(D) \oplus \ldots \oplus R_e(D).$$

Thus, one can assume that R is local, artinian with a nilpotent maximal ideal $\wp$, say $\wp^t = 0$ but $\wp^{(t-1)} \neq 0$. We will show that $R(D)$ is local and has the nilpotent maximal ideal $\wp R(D)$. By the structure theorem mentioned above, this implies that $R(D)$ is artinian.

As $\wp$ is nilpotent, it is clear that $\wp R(D)$ is also nilpotent. Thus, it remains to show that $R(D)$ is local, which will be proved by showing that $R(D)\backslash\wp R(D)$ consists of invertible elements.

Let $r(D) = \frac{f(D)}{s(D)} \in R(D)\backslash\wp R(D)$. As $s(D)$ is invertible it is sufficient to show that $f(D) \in R[D]\backslash\wp R[D]$ is also invertible. Write

$$f(D) = a(D) + b(D),$$

where $a(D) \in (R\backslash\wp)[D]$ and $b(D) \in \wp R[D]$. By successive multiplication of $a(D) + b(D)$ by complementary binomial-like terms, one obtains

$$f(D) \cdot (a(D) - b(D)) = a^2(D) - b^2(D)$$
$$f(D) \cdot (a(D) - b(D))(a^2(D) + b^2(D)) = a^4(D) - b^4(D)$$
$$\text{etc.}$$

Continuing in this way, one finds a polynomial $h(D)$ such that $f(D)\cdot h(D) = a^{2^\ell}(D) - b^{2^\ell}(D)$ for some ℓ, such that $t \leq 2^\ell$. As $\wp^t = 0$, it follows that $b^{2^\ell}(D)$ vanishes and, therefore, $f(D)\cdot h(D) = a^{2^\ell}(D)$. By construction, the trailing coefficient a_T of $a(D)$, i.e., the first non-zero coefficient of lowest order, is a unit of R because it lies in $R\backslash\wp$. Therefore, the trailing coefficient of $a^{2^\ell}(D)$, which equals a_T^ℓ, is also a unit. Thus, $f(D) \cdot h(D) = a^{2^\ell}(D)$ and, a fortiori, $f(D)$ is invertible in $R(D)$. □

Remark. For a commutative artinian ring, the structure theorem implies that R contains only a finite number of maximal ideals $m_1, m_2, \ldots, m_e$. Let $Y = m_1 \cup m_2 \cup \ldots \cup m_e$ and note that $R\backslash Y$ is a multiplicatively closed set. The above proof shows that for any polynomial $f(D) \in R[D]\backslash Y[D]$, there exists a polynomial $h(D)$ such that the trailing coefficient of $f(D)h(D)$ is 1. This implies that one can use the multiplicatively closed set $R[D]\backslash Y[D]$ as denominators and the ring of rational functions can be characterized as

$$(10) \qquad R(D) = \left\{ \frac{f(D)}{s(D)} : f(D) \in R[D], s(D) \in R[D]\backslash Y[D] \right\}.$$

Following [12], we define the *convolutional dual* of a convolutional code $C \subset R(D)^n$ as in the block code case:

$$C^{\perp_c} = \{\mathbf{x}(D) \in R^n(D) : \mathbf{c}(D)\cdot\mathbf{x}(D)^T = 0, \text{ all } \mathbf{c}(D) \in C\}.$$

COROLLARY 4.1. *Let R be a commutative, artinian ring. Then, every convolutional code C over R has a convolutional dual $C^{\perp_c}$, which satisfies $(C^{\perp_c})^{\perp_c} = C$. Moreover, if $C \subset R(D)^n$ is free of rank k, then $C^{\perp_c}$ has rank $n - k$.*

EXAMPLE 4. *Consider the ring of integers* $\mathbf{Z}$, *which is not an artinian ring, and let* C *be the convolutional code over* $\mathbf{Z}$ *that is generated by the* 1×1 *generator matrix*

$$G(D) = [2 + D].$$

One can show that 1 *is not a codeword and, thus,* $C \neq \mathbf{Z}(D)$. *As* $\mathbf{Z}$ *has no zero divisors,* $x(D)G(D) = 0$ *implies* $x(D) = 0$ *and, therefore, the convolutional dual equals* $C^{\perp_c} = 0$. *Now, the twofold dual is* $(C^{\perp_c})^{\perp_c} = \mathbf{Z}(D)$ *and, hence, the duality property does not hold.*

In the field case, duality for convolutional codes is based on the non-degenerate pairing (cf. [4] and Appendix in [13]) of formal Laurent series $R((D))$ and formal anti-Laurent series $R((D^{-1}))$

(11) $$\pi : R((D)) \times R((D^{-1})) \to R$$

defined by $\pi(D^{-a}(s_0 + s_1 D + s_2 D^2 + \ldots), D^b(t_0 + t_{-1} D^{-1} + t_{-2} D^{-2} + \ldots)) = \sum_i s_{i+a} t_{-i-b}$. The rational functions $R(D)$ or $R(D^{-1})$ can be expanded around zero or around infinity to yield a formal Laurent series or anti-Laurent series, respectively. By restricting the pairing π to the rational functions, one obtains a pairing for $R(D) \times R(D^{-1})$.

We can now define the dual of a convolutional code using the above pairing, which can be extended from the scalar case to the case of n codeword components in the usual way. Let $C \subset R^n(D) \subset R^n((D))$ be a convolutional code, then the dual code lies in $R^n(D^{-1}) \subset R^n((D^{-1}))$ and is defined by

(12) $$C^{\perp} = \{ \mathbf{x}(D^{-1}) \in R^n(D^{-1}) : \pi(\mathbf{c}(D), \mathbf{x}(D^{-1})) = 0, \text{ all } \mathbf{c}(D) \in C \}.$$

Note that in contrast to the field case, $R(D) = R(D^{-1})$ does not hold, in general (cf. the above example). However, for commutative artinian rings, this equality holds.

PROPOSITION 4.2. *Let* C *be a convolutional code over a commutative artinian ring* R. *Then,*
   (i) $R(D) = R(D^{-1})$;
   (ii) $C_{rev}^{\perp} = C^{\perp_c}$,
      where $C_{rev}^{\perp}$ *denotes the time reversal of the dual* $C^{\perp}$.

Proof. (i) follows from (10) and the fact the a polynomial $s(D) = s_0 + s_1 D + \ldots + s_\ell D^\ell$ is contained in $R[D] \backslash Y[D]$ if and only if $D^{-\ell} s(D) = s_0 D^{-\ell} + s_1 D^{-\ell+1} + \ldots + s_\ell$ is contained in $R[D^{-1}] \backslash Y[D^{-1}]$.

To show (ii), one can assume without loss of essential generality that R is local. The proof of proposition 4.1 implies that $R(D)$ is also local. It is well known [7] that a projective module over a local ring is free and, therefore, C is a free $R(D)$-module. Thus, C has some generator matrix $G(D)$.

Consider a codeword $\mathbf{x}(D)$ of the convolutional dual, which by definition satisfies $G(D)\mathbf{x}(D)^T = 0$ or, equivalently, $\mathbf{g}_i(D)\mathbf{x}(D)^T = 0$ for all the rows $\mathbf{g}_i(D)$ of the generator matrix. The dual codeword can be expanded into a formal Laurent series $\mathbf{x}(D) = D^{-a}(\mathbf{x}_0 + \mathbf{x}_1 D + \mathbf{x}_2 D^2 + \ldots)$. Let $x(D^{-1})_{\mathrm{rev}} = D^a(\mathbf{x}_0 + \mathbf{x}_1 D^{-1} + \mathbf{x}_2 D^{-2} + \ldots)$ be the time reversal of $\mathbf{x}(D)$.

One then checks that the condition $\mathbf{g}_i(D)\mathbf{x}(D)^T = 0$ is equivalent to the condition $\pi(D^\ell \cdot \mathbf{g}_i(D), \mathbf{x}(D^{-1})_{\mathrm{rev}}) = 0$, for all integer shifts ℓ. Thus, any codeword $\mathbf{x}(D)$ of the convolutional dual $C^{\perp_c}$ gives rise to a code-word $\mathbf{x}(D^{-1})_{\mathrm{rev}}$ in $C^\perp$ and, conversely, every codeword $\mathbf{x}(D^{-1})_{\mathrm{rev}} \in C^\perp$ determines a time-reversed codeword $\mathbf{x}(D) \in C^{\perp_c}$. This proves claim (ii).
□

Remark. By part (i) of Proposition 4.2, the rational functions $R(D)$ have both a formal Laurent series expansions around zero as well as around infinity. Thus, the codewords of $C^\perp \subset R^n(D^{-1}) \subset R^n((D^{-1}))$ can also be considered elements of $R^n(D) \subset R^n((D))$; but note that a codeword will have different expansions around zero or infinity, in general.

4.2. Locally defined duals. In the framework of character-based duality, one can define a convolutional code and its dual locally by the canonical trellis section and the dual trellis section [14]. This local description of the dual is of particular interest because it can be generalized to codes defined over graphs [4]. In this section, we will consider this local duality description in the context of linear functional-based duals. To motivate these duality results, we start with an example. We assume that the reader is familiar with the notion of a trellis as defined, for instance, in [9] or [15].

EXAMPLE 5. *Consider the ring $R = GF(2)[x,y]/(x^2,y^2,xy)$, which is not a quasi-Frobenius ring. Let C be the convolutional (2,1) code generated by the generator matrix*

$$G(D) = [1 \quad 1 + xD].$$

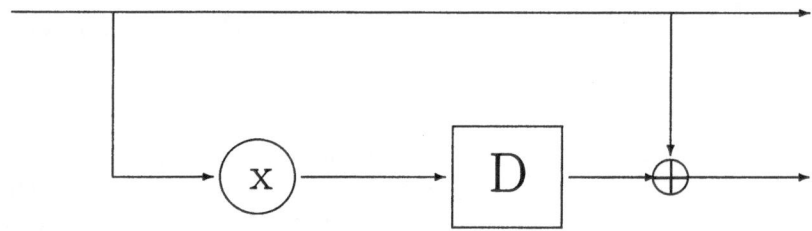

FIG. 1. *2-state realization of the generator matrix $G(D)$.*

This generator matrix has a minimal 2-state realization as shown in Fig. 1. The scalar multiplication by the ring element x as well as the addition is with respect to the arithmetic of the ring R. Note that the delay cell will contain only the values 0 and x, thus, this is indeed a 2-state realization with state space $S = \{0, x\}$. A branch of the trellis corresponding to this realization is a triple $b = (s, v_1 v_2, s')$, where $s \in S$ is the current *state, $s' \in S$ is the next state, and $v_1 v_2$ are the two outputs produced by the given realization, where $v_i \in R$. The set of all branches B at a fixed time instant of this time-invariant trellis consists of all the triples*

$$
\begin{array}{llll}
(0, 00, 0) & (0, xx, 0) & (0, yy, 0) & (0, x + yx + y, 0) \\
(0, 11, x) & (0, 1 + x1 + x, x) & (0, 1 + y1 + y, x) & (0, 1 + x + y1 + x + y, x) \\
(x, 0x, 0) & (x, x0, 0) & (x, yx + y, 0) & (x, x + yy, 0) \\
(x, 11 + x, x) & (x, 1 + x1, x) & (x, 1 + y1 + x + y, x) & (x, 1 + x + y1 + y, x).
\end{array}
$$

Note that B forms a group under componentwise composition of the triples $b = (s, v_1 v_2, s')$.

As $G(D)G(D)^T = 0$, the code C is self-dual, i.e., $C = C^{\perp c}$. The dual code $C^{\perp}$, which is the time reversal of $C^{\perp c}$, has a 2-state minimal trellis that is the time reversal of the trellis for C. The state space $\tilde{S}$ of the dual code can be endowed with an algebraic structure as follows. Let $f_x : \{0, x\} \to R$ be the inclusion map, i.e., $f_x(0) = 0$ and $f_x(x) = x$, and set $\tilde{S} = \{0, f_x\}$. If we define $(f_x + f_x)(s) = f_x(s) + f_x(s)$, then $\tilde{S}$ is isomorphic to the cyclic group of order 2, hence, isomorphic to S. The branch group $B^{\perp}$ for the time-reverse trellis with state space $\tilde{S}$ consists of triples $b^{\perp} = (f, w_1 w_2, f')$, where $f, f' \in \tilde{S}$ and $w_i \in R$. The set of all branches in $B^{\perp}$ is given by

$$
\begin{array}{llll}
(0, 00, 0) & (0, xx, 0) & (0, yy, 0) & (0, x + yx + y, 0) \\
(f_x, 11, 0) & (f_x, 1 + x1 + x, 0) & (f_x, 1 + y1 + y, 0) & (f_x, 1 + x + y1 + x + y, 0) \\
(0, 0x, f_x) & (0, x0, f_x) & (0, yx + y, f_x) & (0, x + yy, f_x) \\
(f_x, 11 + x, f_x) & (f_x, 1 + x1, f_x) & (f_x, 1 + y1 + x + y, f_x) & (f_x, 1 + x + y1 + y, f_x).
\end{array}
$$

We now define a pairing $\langle ., . \rangle$ on $(S \times R^2 \times S) \times (\tilde{S} \times R^2 \times \tilde{S})$ with values in R by

$$
\langle (s, v_1 v_2, s'), (f, w_1 w_2, f') \rangle = f(s) + v_1 \cdot w_1 + v_2 \cdot w_2 - f'(s').
$$

This pairing gives the following orthogonality characterization of the dual branch group

$$
B^{\perp} = \{ b^{\perp} \in \tilde{S} \times R^2 \times \tilde{S} : \langle b, b^{\perp} \rangle = 0 \text{ for all } b \in B \}.
$$

To formulate duality locally, we need the notion of the canonical trellis of a convolutional code. We briefly review the relevant concepts, as presented e.g. in [9]. The *canonical state space* of a convolutional code $C \subset R^n(D) \subset R^n((D))$ (or $C \subset R^n(D^{-1}) \subset R^n((D^{-1})))$[4] is given by the R-quotient module

$$
S(C) = C/(C_- + C_+),
$$

---

[4]The definition actually applies to any time-invariant code that is a subgroup in the sequence space of all bi-infinite sequences [9].

where $C_+ = C \cap R[[D]]$ consists of all causal codewords and $C_- = C \cap D^{-1}R[[D^{-1}]]$ consists of all anti-causal codewords.[5] Every codeword $\mathbf{x}(D)$ determines an equivalence class, i.e., an element in the quotient module $S(C)$; this equivalence class will be denoted by $[\mathbf{x}(D)]$. A trellis section of the canonical (time-invariant) trellis is determined by the *canonical branch set* of C, which is a subgroup of $S(C) \times R^n \times S(C)$ characterized by

$$B(C) = \{([\mathbf{x}(D)], \mathbf{x}_0, ([D^{-1}\mathbf{x}(D)])) : \mathbf{x}(D) \in C\},$$

where $\mathbf{x}_0$ is the time-0 component of the codeword $\mathbf{x}(D)$. The dual code $C^\perp \subset R^n(D^{-1})$ has a canonical branch group lying in the ambient space $S(C^\perp) \times R^n \times S(C^\perp)$, which is determined by

$$B(C^\perp) = \{([\mathbf{y}(D^{-1})], \mathbf{y}_0, ([D^{-1}\mathbf{y}(D^{-1})])) : \mathbf{y}(D^{-1}) \in C^\perp\}.$$

The key to obtain a duality relation for the above branch groups is based on a pairing of the canonical state spaces. This pairing is defined as follows:

$$
\begin{aligned}
\rho : \quad & S(C) \times S(C^\perp) && \longrightarrow && R \\
& ([\mathbf{x}(D)], [\mathbf{y}(D^{-1})]) && \mapsto && \sum_{i=-\infty}^{-1} \mathbf{x}_i \cdot \mathbf{y}_i^T,
\end{aligned}
$$

(13)

where $\mathbf{x}_i$ and $\mathbf{y}_i$ denote the time-i components of the codewords.

LEMMA 4.1. *The pairing ρ is well defined.*

Proof. Note that the sum in (13) contains only finitely many non-zero terms because the codewords of C are regarded as formal Laurent series and the codewords of the dual $C^\perp$ are formal anti-Laurent series.

It remains to be shown that the sum does not depend on the choice of representatives for the states. We will show that the sum does not depend on the choice of $\mathbf{x}(D)$ to represent the state $[\mathbf{x}(D)]$. The independence of the sum from the representative for the dual state can be proved in a similar manner and is left to the reader.

As ρ is an R-homomorphism with respect to the first argument, it is sufficient to check that $\rho([\mathbf{x}(D)], [\mathbf{y}(D^{-1})]) = 0$ holds for any codeword $[\mathbf{x}(D)]$ that goes through the zero-state at time zero, which is equivalent to the condition $\mathbf{x}(D) \in C_- + C_+$. Let $\mathbf{x}_-(D)$ and $\mathbf{x}_+(D)$ be the anti-causal and causal part, respectively, of the codeword $\mathbf{x}(D)$. It is clear that $\rho([\mathbf{x}_-(D) + \mathbf{x}_+(D)], [\mathbf{y}(D^{-1})]) = \rho([\mathbf{x}_-(D)], [\mathbf{y}(D^{-1})])$ because the sum in (13) runs only through the negative time components. As $\mathbf{x}_-(D)$ and $\mathbf{y}(D^{-1})$ are dual codewords, they are orthogonal and the desired result follows from

$$0 = \sum_{i=-\infty}^{\infty} \mathbf{x}_i^{(-)} \cdot \mathbf{y}_i^T = \sum_{i=-\infty}^{-1} \mathbf{x}_i^{(-)} \cdot \mathbf{y}_i^T = \rho([\mathbf{x}_-(D)], [\mathbf{y}(D^{-1})]),$$

---

[5] $R[[D]]$ denotes the ring of formal power series.

where $x_i^{(-)}$ denote the components of the anti-causal codeword $\mathbf{x}_-(D)$. □

We can now define a pairing $\langle ., . \rangle_\rho$ for the ambient spaces $S(C) \times R^n \times S(C)$ and $S(C^\perp) \times R^n \times S(C^\perp)$ of the canonical branch groups $B(C)$ and $B(C^\perp)$ of the code and its dual, respectively. This pairing takes values in R and is given by

$$\langle (s, \mathbf{x}, s'), (f, \mathbf{y}, f') \rangle_\rho = \rho(s, f) + \mathbf{x} \cdot \mathbf{y}^T - \rho(s', f').$$

THEOREM 4.1.   *Let C be a convolutional code over a commutative artinian ring R. The pairing $\langle ., . \rangle_\rho$ provides a duality relation between the canonical branch group $B(C)$ of the code C and the canonical branch group $B(C^\perp)$ of the dual code $C^\perp$. In particular,*

(i) $B(C^\perp) = \{(f, \mathbf{y}, f') : < (s, \mathbf{x}, s'), (f, \mathbf{y}, f') >_\rho = 0$, *for all* $(s, \mathbf{x}, s') \in B(C)\}$;

(ii) $B(C) = \{(s, \mathbf{x}, s') : < (s, \mathbf{x}, s'), (f, \mathbf{y}, f') >_\rho = 0$, *for all* $(f, \mathbf{y}, f') \in B(C^\perp)\}$.

Proof. (i) We write $B^\perp$ for the right-hand side of statement (i). Every branch in the branch group $b^\perp \in B(C^\perp)$ is determined by a code sequence $\mathbf{y}(D^{-1})$ in the dual code $C^\perp$, viz., $b^\perp = ([\mathbf{y}(D^{-1})], \mathbf{y}_0, [D^{-1}\mathbf{y}(D^{-1})])$. Let $b = ([\mathbf{x}(D)], \mathbf{x}_0, [D^{-1}\mathbf{x}(D)])$ be an arbitrary branch of $B(C)$. The definition of ρ implies

$$\rho([\mathbf{x}(D)], [\mathbf{y}(D^{-1})]) + \mathbf{x}_0 \cdot \mathbf{y}_0^T = \rho([D^{-1}\mathbf{x}(D)], [D^{-1}\mathbf{y}(D^{-1})]).$$

This shows that $B(C^\perp)$ is contained in $B^\perp$.

To prove the equality $B(C^\perp) = B^\perp$, consider a sequence $\mathbf{y}(D^{-1}) \in R^n((D^{-1}))$ that is generated by the trellis with branch group $B^\perp$. By definition of $B^\perp$, this sequence is orthogonal to C. Hence, all such sequences $\mathbf{y}(D^{-1})$ that come from rational functions lie in $C^\perp$. Therefore, the code $C_{B^\perp}$ in $R^n(D^{-1})$ that is generated by $B^\perp$ is contained in $C^\perp$ and the first part of the proof implies $C_{B^\perp} = C^\perp$. Now, as $B(C^\perp) \subset B^\perp$, and $B(C^\perp)$ and $B^\perp$ generate the same code and have the same state space $S(C^\perp)$, it follows that $B(C^\perp) = B^\perp$.

(ii) We write B for the right-hand side of statement (ii). A similar argument as in part (i) of the proof shows that $B(C) \subset B$ and that the code C_B, which is generated by B, is contained in $(C^\perp)^\perp$. Now, the fact that R is artinian implies $(C^\perp)^\perp = C$ by Corollary 1 and Proposition 5. The same arguments as in part (i) show that $B(C) = B$.   □

If R is artinian, then $S(C)$ and $S(C^\perp)$ are also artinian. Hence, $B(C) = B$ and $B(C^\perp) = B^\perp$ generate minimal trellises (cf. [9]). The minimality criterion (Theorem 5.2 in [9]) implies that B and $B^\perp$ have no nontrivial branches of the form $(s, 0, 0)$. We now show that ρ is non-degenerate. Suppose that there is a state $s \in S(C)$ such that $\rho(s, f) = 0$ for all $f \in S(C^\perp)$. This is equivalent to having a branch $(s, 0, 0) \in B$. But B is minimal and, hence, one must have $s = 0$. Similarly, $\rho(s, f) = 0$ for all $s \in S(C)$ implies $f = 0$. This proves the following proposition.

PROPOSITION 4.3. *If R is commutative and artinian, then the pairing ρ is non-degenerate. In particular, ρ induces the following two R-homomorphisms,[6] which are one-to-one:*

$$(14) \qquad \begin{array}{cccc} \nu: & S(C^\perp) & \longrightarrow & S(C)^* \\ & f & \mapsto & \rho(.,f) \end{array}$$

$$(15) \qquad \begin{array}{cccc} \mu: & S(C) & \longrightarrow & S(C^\perp)^*. \\ & s & \mapsto & \rho(s,.) \end{array}$$

COROLLARY 4.2. *For a convolutional code C over a commutative, artinian ring R one has*

(i) *the dual $C^\perp$ is generated by the branch group*

$$B^* = \{(f,\mathbf{y},f') \in S(C)^* \times R^n \times S(C)^* : f(s) + \mathbf{x} \cdot \mathbf{y}^T - f'(s') = 0$$
$$\textit{for all } (s,\mathbf{x},s') \in B(C)\}.$$

Moreover, $B^* \cong B(C^\perp)$.

(ii) $$B(C) = \{(s,\mathbf{x},s') \in S(C) \times R^n \times S(C) : f(s) + \mathbf{x} \cdot \mathbf{y}^T - f'(s') = 0$$
$$\textit{for all } (f,\mathbf{y},f') \in B^*\}.$$

Proof. (i) As the map ν in (14) is one-to-one, an isomorphic image of $S(C^\perp)$ is contained in $S(C)^*$ and, therefore, B^* generates $C^\perp$ or a larger code. This code must be $C^\perp$ because there is no larger code in $R(D^{-1})^n$ that is orthogonal to C.

To prove $B^* \cong B(C^\perp)$, we will show that B^* is a minimal branch group, i.e., it generates a minimal time-invariant trellis for the dual $C^\perp$. To apply the minimality criterion in [9], we make the trellis state-trim by omitting unused states, i.e., states that are not connected by any branch.

If B^* were not minimal, then by Theorem 5.1 (i) in [9], either there exists a negative-time, nontrivial, semi-infinite, zero-label path

$$(16) \qquad \ldots,(f_{-3},\mathbf{0},f_{-2}),(f_{-2},\mathbf{0},f_{-1}),(f_{-1},\mathbf{0},f_0)$$

or a positive-time, nontrivial, semi-infinite, zero-label path

$$(17) \qquad (f_0,\mathbf{0},f_1,),(f_1,\mathbf{0},f_2,),(f_2,\mathbf{0},f_3,),\ldots$$

through the trellis generated by B^* with $f_0 \neq 0$. We will show that $f_0 \neq 0$ is not possible by assuming the first case (16); the proof for the other case (17) is similar.

As R is commutative artinian, the code C and the trellis generated by $B(C)$ are strongly controllable and, therefore, any state $s_0 = s \in S(C)$ at

---

[6] $M^* = Hom(M,R)$ denotes the dual of an R-module M.

time 0 can be reached from the zero state, say $s_{-\ell} = 0$, by some length-ℓ path

$$(s_{-\ell}, \mathbf{x}_{-\ell}, s_{-\ell+1}), (s_{-\ell+1}, \mathbf{x}_{-\ell+1}, s_{-\ell+2}), \ldots, (s_{-1}, \mathbf{x}_{-1}, s_0 = s)$$

through this trellis. From the definition of B^* it follows that

$$f_0(s) = f_0(s_0) = f_{-1}(s_{-1}) = \ldots = f_{-\ell}(s_{-\ell}) = f_{-\ell}(0) = 0.$$

Thus, f_0 maps every element of $S(C)$ to zero and, therefore, $f_0 = 0$, which concludes the proof of (i).

(ii) Let $(B^*)^\perp$ denote the right-hand side of the equation in statement (ii). Clearly, every branch $(s, \mathbf{x}, s') \in B(C)$ satisfies the orthogonality condition $f(s) + \mathbf{x}\mathbf{y}^T - f'(s') = 0$ for all $(f, \mathbf{y}, f') \in B^*$, hence, $B(C) \subset (B^*)^\perp$.

Conversely, as the map ν in (14) is one-to-one, an isomorphic image of $B(C^\perp)$ is contained in B^* and, thus, $(B^*)^\perp \subset B(C)$, by Theorem 4.1(ii).
□

Corollary 4.2 is the linear functional-based duality counterpart of Theorem 2 in [14]. Note that $S(C^\perp) \not\cong S(C)^*$, in general, which is illustrated in the example below. However, in the case of finite quasi-Frobenius rings, linear functional-based and character-based duality are equivalent and $S(C^\perp) \cong S(C) \cong S(C)^*$ (cf. [5]).

EXAMPLE 5. *(continued) The state space* $S = \{0, x\}$ *is an R-module and has as dual R-module*

$$S^* = \{f_0, f_x, f_y, f_{x+y}\},$$

where $f_a(x) = a$, *for* $a = 0, x, y, x + y$. *The branch group* B^* *as given in Corollary 4.2 coincides with the previously defined dual branch group* $B^\perp$. *Hence, B^* is not state-trim, i.e., there are no branches that start or end in the states f_y or f_{x+y} of S^*.*

One can show that $S^* \cong \mathbf{Z}_2 \times \mathbf{Z}_2$ *as additive groups and, moreover, that* $(S^*)^*$ *contains 16 elements. Thus, the duality property* $(S^*)^* \cong S$ *does not hold but B and B^* are nevertheless dual to each other by Corollary 4.2.*

5. Conclusions. A framework for studying duality properties of linear codes over commutative artinian rings was developed. As in the field case, duality can be based on linear functionals rather than on characters if the class of 'linear' codes is chosen appropriately. In the block code case, this class of 'linear' length-n block codes over R consists of all projective submodules of R^n, and for convolutional codes the rational functions trick [11] was used, i.e., convolutional codes are regarded as block codes over the ring of rational functions $R(D)$. This class of 'linear' block and convolutional codes is closed under the operations of taking the dual (or orthogonal) of a code, the intersection, and the sum of two codes, provided R is commutative artinian.

Linear functional-based duality is equivalent to character-based duality for linear codes over finite quasi-Frobenius rings [5] but, in general, the two duality concepts are shown to be complementary in the sense that for certain infinite rings there are instances where only one of the two duality concepts provides the desired duality properties. An example was given to illustrate that even for some finite ring, the two duality concepts are not equivalent. Despite this fact, it was shown that for a finite commutative ring the MacWilliams identity holds for linear block codes and their linear functional-based duals (the corresponding result for character-based duality is well known [3]).

Two notions of linear functional-based duals of convolutional codes were considered, the dual and the convolutional dual [12], which were shown to be time reversals of each other in the case that the underlying ring is commutative artinian. In the framework of character-based duality, the local (trellis-based) characterization of a dual convolutional code [14] has proved to be suitable to define duality of linear codes over graphs [4]. It is shown that a trellis-section-based characterization of the dual also applies to linear functional-based duality, which can be generalized to codes over graphs. The main result concerning local duality is the orthogonality (duality) relation between the canonical branch group of a code C and its dual branch group as given in Corollary 4.2. This orthogonality relation is remarkable in view of the fact that the state spaces $S(C)$ and $S(C)^*$ of these branch groups do not satisfy the duality property $S(C)^{**} \cong S(C)$, in general.

Acknowledgments. This paper has benefited from comments by G.D. Forney, Jr., and an anonymous reviewer.

REFERENCES

[1] G.D. FORNEY, JR., "The Dynamics of Group Codes: Dual Group Codes and Systems," Manuscript, Sep. 1994.

[2] G.D. FORNEY, JR., "Controllability, Observability, and Duality in Behavioral Group Systems," *Proc. 34th Conf. Dec. Ctrl. (New Orleans)*, Vol. 3, Dec. 1995, pp. 3259–3264.

[3] T. ERICSON AND V. ZINOVIEV, "On Fourier-invariant partitions of finite abelian groups and the MacWilliams identity for group codes," Prob. Peredachi Informatsii, **32**, 137–143, 1996.

[4] G.D. FORNEY, JR., "Group Codes and Behaviors," *Proc. MTNS '98* (Padova, Italy); Zurich: Birkhäuser, 1998, pp. 1–20.

[5] J.A. WOOD, "Duality for Modules over Finite Rings and Applications to Coding Theory," *American J. of Math.*, Vol. **121.3**, June 1999, pp. 555–575.

[6] W. RUDIN, *Fourier Analysis on Groups*, Wiley, 1962.

[7] N. JACOBSON, *Basic Algebra II*, Freeman, San Francisco, 1980 .

[8] T.Y. LAM, *Serre's Conjecture*, LNM 635, Springer, 1978.

[9] H.-A. LOELIGER AND T. MITTELHOLZER, "Convolutional Codes over Groups, *IEEE Trans. Information Th.*, Vol. **42**(6), Nov. 1996, pp. 1660–1686.

[10] J. MACWILLIAMS, "A Theorem on the Distribution of Weights in a Systematic Code," *Bell Syst. Tech. J.*, Vol. **42**, pp. 79–94, 1963.

[11] JAMES L. MASSEY, *Coding Theory*, in *Handbook of Applicable Mathematics* (Ed. W. Ledermann), Vol. V, Part B, *Combinatorics and Geometry* (Ed. W. Ledermann, S. Vajda). Chichester & New York: Wiley, 1985.

[12] ROLF JOHANNESSON AND KAMIL ZIGANGIROV, *Fundamentals of Convolutional Coding*, IEEE Press, New York, 1999.

[13] G.D. FORNEY, JR., "Structural Analysis of Convolutional Codes via Dual Codes," *IEEE Trans. Inform. Th.*, Vol. IT-19, July 1973, pp. 512–518.

[14] T. MITTELHOLZER, "Convolutional Codes over Groups: A Pragmatic Approach," *Proc. 33rd Ann. Allerton Conf. on Communication, Control, and Computing* (Monticello, IL, Oct. 4–6, 1995), pp. 380–381.

[15] A. VARDY, *Trellis Structure of Codes*, in *Handbook of Coding Theory, Part II*, Eds. V.S. Pless and W.C. Huffman. North-Holland, Elsevier Science, 1998.

UNIT MEMORY CONVOLUTIONAL CODES WITH MAXIMUM DISTANCE

ROXANA SMARANDACHE*

Abstract. Unit memory codes and in particular, partial unit memory codes are reviewed. Conditions for the optimality of partial unit memory codes with degree $k - 1$ are given, where optimal codes are the codes having the maximum free distance among all codes of the same parameters k, n and degree μ. A binary construction of unit memory codes with $\mu = k - 1$ is discussed for the cases that satisfy the optimality conditions. This construction is generalized for codes over fields of characteristic $p > 2$.

Key words. Unit memory convolutional codes, MDS-convolutional codes.

1. Introduction. Maximum distance separable (MDS) convolutional codes, introduced first in [10], are convolutional codes characterized through the property that their free distance is maximum among all codes of the same parameters n, k and degree μ. The free distance of an MDS code represents a bound for the free distance of all the other codes with the same parameters. We call this bound the generalized Singleton bound, since in the case of block codes it obviously reduces to the Singleton bound. The existence of MDS convolutional codes was established in [10] by using elements of algebraic geometry and an input-state-output representation of convolutional codes. The proof was existential and could not provide a method of construction. A constructive proof was given later in [12]. The construction starts from a large Reed Solomon block code and therefore needs a fairly large finite field.

The question of finding a lower bound for the field size and MDS codes attaining this bound was now raised. Since unit memory codes have the simplest representation among the codes of nonzero memory we started our study by analyzing the conditions these codes need to satisfy in order to be maximum. We also took a new approach. We started with the binary field and came to study the binary partial unit memory codes with degree $\mu = k - 1$, this being the only nontrivial case where binary MDS codes exist.

Binary partial unit memory codes were studied in the literature by Lauer [4] and Justesen [3] who showed that in some situations a unit memory code performs better than the codes having the same rate and degree but memory larger than 1. Some constructions given in [4] are the inspi-

*Department of Mathematics, University of Notre Dame, Notre Dame, Indiana 46556-5683, USA, *e-mail:* Smarandache.1@nd.edu. The author was supported by NSF grant DMS-96-10389 and by a fellowship from the Center of Applied Mathematics at the University of Notre Dame.

ration for the idea of this paper. In [3] quasi-cyclic unit memory codes are studied and some constructions and computer search results are presented. Furthermore some of the basic structural properties are discussed, such as noncatastrophicity, minimality conditions, distance measures, properties that we will use in this paper. Therefore we chose to use the same language as in these papers, only mentioning what this means in terms of the language of the MDS papers [10, 12]. For the development of this paper this is quite enough.

The paper consists of 6 sections, the first two being introductory and the last one an appendix section containing material that we will heavily use. In Section 3 we state some equivalent conditions for binary PUM codes to be optimal and in Section 4 we give a method of construction for this type of codes. Section 5 generalizes the binary construction to the case where the field has characteristic larger than 2. We add examples of both methods in Section 6.

2. Unit memory codes. Let $\mathbb{F}$ denote a finite field. A unit memory encoder is defined through the following encoding scheme:

$$(2.1) \qquad v_t = u_t G_0 + u_{t-1} G_1$$

where $u_t \in \mathbb{F}^k$ is the k information tuple at time $t, t = 0, 1, \dots$ and $v_t \in \mathbb{F}^n$ is the n-tuple denoting the encoded vector at time t. By convention $u_t = 0$ for $t < 0$. The matrices G_0 and G_1 are defined over the field $\mathbb{F}$ and have size $k \times n$. We assume that G_0 has rank k.

Then a rate k/n *unit memory code* (UMC) is the set of all sequences generated by an encoder (G_0, G_1), with G_0 of rank k, satisfying the above encoding rule.

The code can also be defined through the compressed $k \times n$ matrix

$$G_0 + DG_1,$$

where D defines the time delay operator.

Following [4] we will say that two unit memory encoders: (G_0, G_1) and (G_0', G_1') are *equivalent* if there exists a nonsingular matrix T such that $G_0' = TG_0$ and $G_1' = TG_1$. Two equivalent encoders generate the same code. An encoder is called *catastrophic* if an information sequence with infinitely many nonzero information vectors produces an encoded sequence with finitely many nonzero encoded vectors. Thus two equivalent encoders are either both catastrophic or both noncatastrophic. We have the following criteria from [3]:

THEOREM 2.1. *[3] A UM encoder (G_0, G_1) is catastrophic if and only if there exists an $s \times k$ matrix P of rank $s, s > 0$ and a nonsingular $s \times s$ matrix Q such that*

$$QPG_0 = PG_1$$

Following the lines of [4] and [3] we define the *degree* μ of the encoder to be the rank of G_1. We consider the degree to be the third important parameter of a convolutional code $\mathcal{C}$ and in [10, 12] we define it in the general case of the convolutional codes of memory greater or equal to 1, as the maximal degree of the $k \times k$ full size minors of $G(D)$. (See [9] for details). If G_∞ denotes the high order coefficient matrix of a polynomial matrix $G(D)$, then we have that every code $\mathcal{C}$ of rate k/n has a $k \times n$ generator matrix $G(D)$ whose matrix G_∞ has rank k and whose row degrees are non-increasing or non-decreasing. The degree μ is in this case equal to the sum of the row degrees of the encoder $G(D)$ and we say that $G(D)$ is in *column proper form*. In the literature the degree μ is sometimes called the *total memory* of the code (see [5]) or *state-complexity*.

We therefore have that for any PUM code generated by (G_0, G_1) there exists an encoder (TG_0, TG_1) with T nonsingular such that the first $k - \mu$ rows of TG_1 are zero. We say that this encoder is in *standard form*. We say that a standard form encoder (G_0, G_1) is *minimal* if among all encoders, G_1 has the smallest number of nonzero rows. We have from [3]:

THEOREM 2.2.   *[3] A noncatastrophic UM encoder (G_0, G_1) of the form:*

$$\left[\begin{array}{cc} G_0 & G_1 \end{array}\right] = \left[\begin{array}{cc} G_0' & 0 \\ G_0'' & G_1'' \end{array}\right],$$

where G_0'', G_1'' have μ rows, is minimal if and only if:

$$\mathrm{rank}\left[\begin{array}{c} G_1'' \\ G_0' \end{array}\right] = k.$$

Unit memory codes having $\mu < k$ are called *partial unit memory codes* (PUM) since the encoder requires only μ memory cells for storage.

There are several distance functions that are important when deciding on the decoding properties of a convolutional code. The free distance d_{free} of the code, defined as the minimum Hamming weight of the nonzero encoded sequences having minimum weight, seems to be the most important. We will define also the jth column distances d_j^c and the jth row distances d_j^r. We follow the approach of [1, 2].

The jth *order column distance* d_j^c is defined as the minimum of the weights of the truncated codewords $v_{[0,j]} := (v_0, v_1, \ldots, v_j)$ resulting from an information sequence $u_{[0,j]} := (u_0, u_1, \ldots, u_j)$ with $u_0 \neq 0$. The tuple $d\mathbf{P} = [d_0^c, d_1^c]$ is called the *distance profile*. The limit $d_\infty^c = \lim_{j \to \infty} d_j^c$ exists and we have $d_0^c \leq d_1^c \leq \ldots \leq d_\infty^c$.

The jth *row distance* d_j^r is defined as the minimum of the weights of all the finite codewords $v_{[0,j+1]} := (v_0, v_1, \ldots, v_{j+1})$ resulting from an

information sequence
$u_{[0,j]} := (u_0, u_1, \ldots, u_j) \neq 0$. The limit $d^r_\infty = \lim_{j \to \infty} d^r_j$ exists and, if the encoder is noncatastrophic, we have (see [11, 2] for details):

$$(2.2) \qquad d^c_0 \leq d^c_1 \leq \ldots \leq d^c_\infty = d_{free} = d^r_\infty \leq \ldots \leq d^r_1 \leq d^r_0.$$

In terms of state space descriptions d^r_∞ is equal to the minimal weight of a nonzero trajectory which starts from and returns to the all zero state. d^c_∞ is equal to the minimal weight of a nonzero trajectory which starts from and not necessarily returns to the all zero state. Also it follows that for a non-catastrophic encoder the minimal weight codewords are generated by finite information sequences, so the free distance can be computed from the weights of finite encoded sequences.

We will discuss now the PUM codes with degree $\mu = k - 1$ and we will search for conditions they need to satisfy in order that they are optimal among the codes with the same parameters, in the sense that they attain the maximum distance possible. We will work first over the binary field and then over larger finite fields.

We have the following obvious bound on the free distance of an PUM code with degree $\mu = k - 1$:

THEOREM 2.3. *Let C be a rate k/n PUM convolutional code of degree $\mu < k$ generated by a minimal encoder (G_0, G_1) over $\mathbb{F}$:*

$$(2.3) \qquad \begin{bmatrix} G_0 & G_1 \end{bmatrix} = \begin{bmatrix} g_1 & \cdots & g_n & 0 & \cdots & 0 \\ & G'_0 & & & G'_1 & \end{bmatrix}$$

Then the free distance

$$d_{free} \leq n - k + \mu + 1.$$

This bound is a particular case of the more general bound studied in [10] and [12]:

LEMMA 2.1. *Let C be a convolutional code of rate k/n and degree μ and let $G(D)$ be a polymnomial encoder in row proper form.*

Let ν denote the smallest value of the row degrees of $G(D)$. Let l be the number of row degrees having the value equal to ν. Then the free distance must satisfy:

$$(2.4) \qquad d_{free} \leq n(\nu + 1) - l + 1.$$

We called this bound the generalized Singleton Bound and we showed in [10] that there are codes attaining this bound, over sufficiently large finite fields. We called them MDS convolutional codes. In [12] we were able to give a concrete construction of MDS codes, starting from some Reed Solomon block codes.

Therefore we know that there are PUM codes of degree $k-1$ attaining the maximum bound, there is n, over some finite field with enough elements. We will call them PUM-MDS codes with $\mu = k-1$. We will take a different approach from [12] in the construction of such codes. We will start with the field $\mathbb{F}_2$ and discuss the cases when maximum distance codes exist, and also we will give a construction in these specific cases. Then we will generalize the construction for fields $\mathbb{F}_p, p > 2$ obtaining constructions in some other cases not covered yet.

We conclude the section with a simple theorem that tells us how to obtain k'/n rate PUM codes of degree $\mu = k' - 1$ and maximum distance $d_{free} = n$ from k/n rate PUM-codes of degree $\mu = k - 1$ and $d_{free} = n$ maximum, where $k' < k$.

THEOREM 2.4. *Let C be a PUM code of rate k/n generated by the minimal encoder (G_0, G_1) with $\mu = k - 1$. Let $(\bar{G}_0, \bar{G}_1) \in \mathbb{F}^{(k-1) \times 2n}$ be the matrix obtained from (G_0, G_1) by omitting any of the last $k - 1$ rows of (G_0, G_1). If C has free distance n, then the same is true for the code $\bar{C}$ generated by the encoder $(\bar{G}_0, \bar{G}_1)$.*

Proof. The theorem follows from the inclusion $\bar{C} \subseteq C$. □

3. Partial unit memory codes over $\mathbb{F}_2$. If G_0, G_1 generate a k/n PUM code of degree $\mu = k-1$ with maximum distance n over $\mathbb{F}_2$, then the matrices need to have the following form:

$$(3.1) \quad [\, G_0 \quad G_1 \,] = \begin{bmatrix} 1 & \cdots & 1 & 0 & \cdots & 0 \\ & G_0' & & & G_1' & \end{bmatrix}, \text{ with } \mathrm{rank}(G_1') = k - 1,$$

where G_0', G_1' need to satisfy some conditions that make the encoder G_0, G_1 noncatastrophic and minimal and the code generated by it through (2.1) optimal, i.e. MDS.

REMARK 3.1. It can be easily shown that if $2k - 1 \leq n$, the code is noncatastrophic provided that the matrix:

$$(3.2) \quad \begin{bmatrix} & G_1' & \\ 1 & \cdots & 1 \\ & G_0' & \end{bmatrix} \text{ has full rank } 2k - 1.$$

That assures the minimality as well.

For the next theorem we will need the following definition:

DEFINITION 3.1. A block code (k, n) is called *equidistant* if all nonzero codewords have the same weight d_{min}.

If a code is equidistant and G an arbitrary $k \times n$ encoder, then the entries of G have the property that all $\mathbb{F}$-linear combinations of its rows have the same weight d_{min}. Such a matrix will be called an equidistant matrix.

Then we have the following theorem:

THEOREM 3.1. *Let (G_0, G_1) of the form (3.1) generate an UM-MDS code over $\mathbb{F}_2$. Then:*

  1. n is even

  2. G_0', G_1' generate equidistant $(k-1, n)$ block codes.

  Proof. Let $u \in \mathbb{F}_2^{k-1}$, $u \neq 0$ arbitrarily chosen. Let $x = \text{wt}[uG_0']$, $y = \text{wt}[uG_1']$. We need to prove that $x = y = n/2$. Let $u_1, u_{k+1} \in \mathbb{F}_2$.

Since $d_1^r = n$ we have that the weight of

$$(u_1, u, u_{k+1}) \begin{bmatrix} 1 & \cdots & 1 & 0 & \cdots & 0 \\ & G_0' & & & G_1' & \\ & & & 1 & \cdots & 1 \end{bmatrix}$$

is greater or equal to n. By giving different values to u_1, u_{k+1} we have:

$x + y \geq n, \quad n - x + y \geq n, \quad x + n - y \geq n, \quad n - x + n - y \geq n$

$\Rightarrow \quad x = y, \quad x + y = n$. Hence, we obtain that n is even and that

$$x = y = n/2,$$

which means that G_0', G_1' generate equidistant $(k-1, n)$ block codes. $\square$

In the same way we proved that n is even we can prove that $2^{k-1} \mid n$. Hence $n = 2^{k-1} j$.

Actually we have the following straight forward lemma:

LEMMA 3.1. *The matrices G_0' and G_1' generate equidistant $(k, 2^{k-1})$ block codes if and only if the matrix*

$$\begin{bmatrix} G_0 & G_1 \end{bmatrix}$$

given through (3.1) is a generator matrix for a $(k+1, 2^k)$ equidistant block code.

  Proof. Suppose G_0', G' are equidistant. If $u = (u_1, \ldots, u_k) \in \mathbb{F}_2^k$ then uG_0 and uG_1 have the weight either n and respectively 0, if $(u_2, \ldots, u_k) = 0$, or $n/2$, if not. Hence $\begin{bmatrix} G_0 & G_1 \end{bmatrix}$ is equidistant as well.

The other implication was just proved by the previous theorem 3.1. $\square$

We therefore have a stronger statement:

THEOREM 3.2. *Suppose (G_0, G_1) of the form (3.1) generate a PUM code over $\mathbb{F}_2$. Suppose $2k - 1 \leq n$ and that condition (3.2) is satisfied (therefore the code is noncatastrophic). Then $\mathcal{C}$ is a noncatastrophic PUM-MDS convolutional code over $\mathbb{F}_2$ if and only if*

  1. $n = 2^{k-1} j$.

  2. G_0' and G_1' generate equidistant $(k-1, n)$ block codes.

In other words this statement gives us all the k/n MDS-PUM codes for $k \geq 4$ (so that $2k - 1 \leq 2^{k-1}$).

Proof. Theorem 3.1 gives us the necessity implication. We still need to prove the sufficiency of the two conditions. From *2.* we have that $d_0^r = n$. Let $u_1, u_{k+1} \in \mathbb{F}_2$ and $u, v \in \mathbb{F}_2^{k-1}$, so that $(u_1, u) \neq 0$. The weight

$$
\text{wt } (u_1, u, u_{k+1}, v)
\begin{bmatrix}
1 & \cdots & 1 & 0 & \cdots & 0 & & & & \\
& G_0' & & & G_1' & & & & & \\
& & & 1 & \cdots & 1 & 0 & \cdots & 0 & \\
& & & & G_0' & & & G_1' & &
\end{bmatrix} \geq
$$

$$
\geq \begin{cases}
\text{wt } (u_1, u)G_0 + \text{wt } (v \cdot G_1') \geq n, & \text{if } v \neq 0 \\
\\
\text{wt } (u_1, u)G_0 + \text{wt } (u, u_{k+1}) \begin{bmatrix} & G_1' & \\ 1 & \cdots & 1 \end{bmatrix} \geq n, & \text{if } v = 0
\end{cases},
$$

because of condition (3.2). Hence $d_r^1 = n$. In the same way we show $d_i^r = n$. Also because of (3.2) we have:

$$
d_0^c \geq n/2, \ d_1^c \geq n/2 + 1, \ d_2^c \geq n/2 + 2, \ldots, d_{n/2}^c \geq n/2 + n/2 = n \Rightarrow
$$

$$
\Rightarrow d_{free} = n.
$$

Hence the code is MDS.

The noncatastrophicity is implied by the full rank condition on the $(2k - 1, n)$ matrix. Due to this condition an infinite weight input can not produce a finite output. $\square$

Therefore in order to construct rate $\frac{k}{2^{k-1}j}$ UPM codes with degree $\mu = k - 1$ and maximum distance over $\mathbb{F}_2$, it is enough to construct rate $\frac{k}{2^{k-1}}$, $\mu = k - 1$, $d_{free} = n$, MDS codes and concatenate them j times. From this, using the Theorem 2.4 we get PUM-MDS codes of rate $\frac{i}{2^{k-1}j}$, $1 \leq i \leq k$.

4. A binary construction of partial unit memory codes with maximum free distance. For the construction of PUM codes having maximal distance n over $\mathbb{F}_2$ we use an idea found in [4] but we will have a slightly different approach.

For that we need to introduce the following natural association:

REMARK 4.1. Through the following isomorphism of vector spaces:

$$
(4.1) \qquad
\begin{array}{ccc}
\mathbb{F}_2[X]/(X^{2^k} - 1) & \longrightarrow & \mathbb{F}_2^{2^k} \\
a_0 + a_1 X + \ldots + a_{2^k-1} X^{2^k-1} & \longmapsto & (a_0, a_1, \ldots, a_{2^k-1}),
\end{array}
$$

any scalar encoded sequence in a PUM code $(v_0, v_1, v_2, \ldots)$, given through (2.1), where $v_i \in \mathbb{F}_2^{2^k}$ can be viewed as a polynomial encoded sequence:

$(v_0(X), v_1(X), v_2(X), \ldots)$, where all $v_i(X)$ are polynomials of degree at most $2^k - 1$.

Using the above isomorphism (4.1) we can also define an association between polynomial matrices $k \times 1$ and their coefficient matrices $k \times 2^k$:

(4.2)

$$A = \begin{bmatrix} a_{1,0} & a_{1,1} & \cdots & a_{1,2^k-1} \\ a_{2,0} & a_{2,1} & \cdots & a_{2,2^k-1} \\ \cdots & \cdots & \cdots & \cdots \\ a_{k,0} & a_{k,1} & \cdots & a_{k,2^k-1} \end{bmatrix} \mapsto$$

$$\mapsto A(X) := \begin{bmatrix} a_{1,0} + a_{1,1}X + \ldots + a_{1,2^k-1}X^{2^k-1} \\ a_{2,0} + a_{2,1}X + \ldots + a_{2,2^k-1}X^{2^k-1} \\ \cdots \\ a_{k,0} + a_{k,1}X + \ldots + a_{k,2^k-1}X^{2^k-1} \end{bmatrix}.$$

With this association we have that

$$\text{wt}\,[(u_1, \ldots, u_k)A] = \text{wt}\,[(u_1, \ldots, u_k)A(X)], \forall (u_1, \ldots, u_k) \in \mathbb{F}^k.$$

It follows from Definition 3.1 and the above associations that an equidistant scalar matrix has the property that the associated polynomial matrix through (4.2) has all the polynomial entries of weight 2^{k-1} and any $\mathbb{F}_2$-linear combination of those polynomials gives another polynomial of the same weight. The weight of a polynomial is defined as the sum of the Hamming weights of all the coefficients.

Therefore instead of looking for $(k-1) \times (2^{k-1} - 1), k \geq 2$ equidistant scalar matrices G'_0, G'_1 we could instead look for $(k-1) \times 1$ polynomial matrices with the equivalent property. For this we will heavily use Lemmas 7.1 and 7.2 in the appendix. These lemmas will provide us such polynomial matrices. We have the following theorem:

THEOREM 4.1. *Let G'_0, G'_1 be $(k-1) \times (2^{k-1} - 1), k \geq 4$, scalar matrices associated to*

(4.3) $$G'_0(X) := \begin{bmatrix} P_1(X) \\ P_2(X) \\ \cdots \\ P_{k-1}(X) \end{bmatrix}, \qquad G'_1(X) := \begin{bmatrix} Q_1(X) \\ Q_2(X) \\ \cdots \\ Q_{k-1}(X) \end{bmatrix},$$

where all polynomials $P_i(X)$, $Q_j(X), i, j = \overline{1, k-1}$, have degree less or equal to $2^{k-1} - 2$. Then the rate $\frac{k}{2^{k-1}}$ PUM convolutional code generated by G_0, G_1 of the form in (3.1) is a noncatastrophic MDS code over $\mathbb{F}_2$ (i.e. it has maximal distance n) if and only if:

   1. Any $\mathbb{F}_2$-linear combination of polynomials $P_1(X), \ldots, P_{k-1}(X)$ and any $\mathbb{F}_2$-linear combination of polynomials $Q_1(X), \ldots, Q_{k-1}(X)$ have weight 2^{k-1}.

2. *The polynomials $P_1(X), \ldots, P_{k-1}(X), Q_1(X), \ldots, Q_{k-1}(X)$ are linearly independent.*

Proof. The linear independence of the polynomials is equivalent to the noncatastrophicity of the code, condition given by 3.2 and the fact that all polynomials have degree strictly less than $2^{k-1} - 1$. $\square$

The following lemma will give an inductive construction of PUM codes with maximal distance n over $\mathbb{F}_2$:

THEOREM 4.2. *Let $P_1(X), \ldots, P_{k-1}(X)$ be polynomials of degree less or equal to $2^{k-1} - 2$ and weight 2^{k-2}. Moreover, suppose that any linear combination of the $k - 1$ polynomials has also weight 2^{k-2}. Then the following polynomials:*

(4.4) $\quad P_1(X)(X^{2^{k-1}} + 1), \quad \ldots, \quad P_{k-1}(X)(X^{2^{k-1}} + 1), \quad (X+1)^{2^{k-1}-1}$

form a set of k polynomials with the property that any linear combination of the polynomials has degree less than 2^k and weight 2^{k-1}.

The same weight property holds for the set of k polynomials :

(4.5) $\quad P_1(X)(X^{2^{k-1}} + 1), \quad \ldots, \quad P_{k-1}(X)(X^{2^{k-1}} + 1), \quad [X(X+1)]^{2^{k-1}-1}$

Moreover if $Q_1(X), \ldots, Q_{k-1}(X)$ form also a set of $k - 1$ polynomials of degree less or equal to $2^{k-1} - 2$ with the same property that any linear combination of the polynomials has weight 2^{k-2} and if the polynomials

(4.6) $\quad P_1(X), \quad \ldots, \quad P_{k-1}(X), \quad Q_1(X), \quad Q_2(X), \quad \ldots, \quad Q_{k-1}(X)$

and $1 + X + X^2 + \ldots + X^{2^{k-1}-2}$

are $\mathbb{F}_2$-linearly independent, then the polynomials:

(4.7) $\quad P_1(X)(X^{2^{k-1}} + 1), \ldots, P_{k-1}(X)(X^{2^{k-1}} + 1), (X+1)^{2^{k-1}-1},$

$Q_1(X)(X^{2^{k-1}} + 1), \ldots, Q_{k-1}(X)(X^{2^{k-1}} + 1), [X(X+1)]^{2^{k-1}-1}$
are $\mathbb{F}_2$-linearly independent.

Proof. Let $P(X) = u_1 P_1(X) + u_2 P_2(X) + \ldots + u_{k-1} P_{k-1}(X), u_i \in \mathbb{F}_2, \forall i = \overline{1, k-1}$ be a linear combination of $P_1(X), P_2(X), \ldots, P_{k-1}(X)$. A linear combination of the new k polynomials has the form:

$$u(X+1)^{2^{k-1}-1} + P(X)(X^{2^{k-1}} + 1) = u(X+1)^{2^{k-1}-1} + P(X)(X+1)^{2^{k-1}} =$$

$$= (X+1)^{2^{k-1}-1}(u+P(X)(X+1)), \text{ with } u \in \mathbb{F}_2, \text{ or, in the second situation:}$$

$$u[X(X+1)]^{2^{k-1}-1} + P(X)(X^{2^{k-1}} + 1) =$$

$$= (X+1)^{2^{k-1}-1}(uX^{2^{k-1}-1} + P(X)(X+1)).$$

If $u = 0$ we obtain $P(X)(X+1)^{2^{k-1}}$ that has weight twice the weight of $P(X)$ as we stated before in the lemma 7.4. If $u = 1$ we use the weight retaining property (7.3):

$$\text{wt}\left[(X+1)^{2^{k-1}-1}(u + P(X)(X+1))\right] \geq$$

$$\geq \text{wt}\left[(X+1)^{2^{k-1}-1}\right] \cdot \text{wt}\left[(u + P(X)(X+1)) \bmod (X+1)\right] = 2^{k-1}.$$

The second case goes the same way.

For the second part let $Q(X) = v_1 Q_1(X) + \ldots + v_{k-1} Q_{k-1}(X), v_i \in \mathbb{F}_2$, $\forall i = \overline{1, k-1}$ be a linear combination of $Q_1(X), Q_2(X), \ldots, Q_{k-1}(X)$. Let

$$(X+1)^{2^{k-1}-1}(u+P(X)(X+1))+(X+1)^{2^{k-1}-1}(vX^{2^{k-1}-1}+Q(X)(X+1)) =$$

$$= (X+1)^{2^{k-1}-1}(u + vX^{2^{k-1}-1} + (Q(X)+P(X))(X+1)) = 0, \ u, v \in \mathbb{F}_2,$$

be a linear combination of the new polynomials that is equal to zero. It implies $u = v$ and we obtain:

$$u(1 + X^{2^{k-1}-1}) + (Q(X) + P(X))(X+1) = 0 \Leftrightarrow$$

$$u(1 + X + X^2 + \ldots + X^{2^{k-1}-2}) + Q(X) + P(X) = 0,$$

which leads to $u = u_1 = \ldots = u_{k-1} = v_1 = \ldots = v_{k-1} = 0$ because of (4.7). That gives the linear independence of the new polynomials. $\square$

Basically, Theorem 4.2 says that if we have two equidistant matrices G_0' and G_1' of sizes $(k-1) \times (2^{k-1} - 1)$, $k \geq 4$ associated to the polynomial matrices $G_0'(X), G_1'(X)$ through (4.3), where the sets of polynomials $P_1(X), \ldots, P_{k-1}(X)$ and $Q_1(X), \ldots, Q_{k-1}(X)$ satisfy the conditions in Theorem 4.2, we can inductively construct equidistant matrices of size $j \times (2^j - 1)$, $j \geq k$.

For example, if we take 1 ($k = 2$), multiply it with $(X^2 + 1)$ and add the extra polynomial $1 + X$, respectively $X(1 + X)$ we obtain the 2×4 matrices:

$$G_0'(X) = \left[\begin{array}{c} 1+X \\ 1+X^2 \end{array}\right], \ G_1'(X) = \left[\begin{array}{c} (1+X)X \\ (1+X^2) \end{array}\right],$$

and the 3×8 matrices, after the next step:

$$G_0'(X) = \left[\begin{array}{c} (1+X)^3 \\ (1+X)^5 \\ (1+X)^6 \end{array}\right], \ G_1'(X) = \left[\begin{array}{c} (1+X)^3 X^3 \\ (1+X)^5 X \\ (1+X)^6 \end{array}\right].$$

Of course this is not a good choice, since the polynomials obtained are not linearly independent, the code generated in this way being catastrophic. Therefore we have to change somehow these matrices in order to have the properties of Theorem 4.2. We will keep the matrix $G_0'(X)$ and change the matrix $G_1'(X)$ by multiplying the entries with different powers of X modulo $X^7 + 1$. The following choice for $G_1'(X)$:

$$G_1'(X) = \begin{bmatrix} (1+X)^3 \cdot X^4 mod\ (X^7 - 1) \\ (1+X)^5 \cdot X^3 mod\ (X^7 - 1) \\ (1+X)^6 \cdot X^3 mod\ (X^7 - 1) \end{bmatrix}.$$

together with the $G_0'(X)$ constructed above will satisfy the condition of the theorem. Hence we could use the polynomial entries of $G_0'(X), G_1'(X)$ for the inductive construction of Theorem 4.2. We have the following construction theorem:

THEOREM 4.3. *Let* $P_1 = (1+X)^3$, $P_2 = (1+X)^5$, $P_3 = (1+X)^6$ *and* $Q_1 = (1+X)^3 \cdot X^4 mod\ (X^7 - 1)$, $Q_2 = (1+X)^5 \cdot X^3 mod\ (X^7 - 1)$, $Q_3 = (1+X)^6 \cdot X^3 mod\ (X^7 - 1)$.

Applying Theorem 4.2 inductively we obtain rate $\frac{k}{2^{k-1}}$ noncatastrophic convolutional codes that have maximal free distance 2^{k-1} over $\mathbb{F}_2$, for all $k \geq 4$.

REMARK 4.2. The rate $\frac{k}{2^{k-1}}$ code constructed above has the matrix G_0' associated to the following polynomial matrix:

$$G_0'(X) = \begin{bmatrix} (X+1)^{i_1} \\ (X+1)^{i_2} \\ \dots \\ (X+1)^{i_{k-1}} \end{bmatrix}$$

with $i_1, i_2, \dots, i_{k-1}$ nonnegative integer strictly less than 2^{k-1} of weight $k - 2$, where we defined the weight of an integer in (7.1). We could apply (7.2) to show directly that the matrix G_0' generates an equidistant $(k - 1, 2^{k-1})$ block code. We will use this direct approach rather than the inductive one, in the following section, for constructing MDS convolutional codes of rate k/n where n is odd. Of course we will have to use a larger field.

5. Constructions of partial unit memory codes with maximum free distance over $\mathbb{F}_p$.

Let $\mathbb{F}_p$ be the field with p elements. Let $k \geq 1$, $n = p^{k-1}$.

THEOREM 5.1. *Let* G_0, G_1 *be the* $k \times n$ *scalar matrices associated to the following polynomial matrices:*

$$G_0(X) = \begin{bmatrix} (X+1)^{i_0} \\ (X+1)^{i_1} \\ \dots \\ (X+1)^{i_{k-1}} \end{bmatrix}, \quad G_1(X) = \begin{bmatrix} 0 \\ (X+1)^{j_1} \\ \dots \\ (X+1)^{j_{k-1}} \end{bmatrix}$$

with $i_0 = (p-1) + (p-1)p + \ldots + (p-1)p^{k-2} = p^{k-1} - 1 = n - 1$,
$I := \{i_1, \ldots i_{k-1}\}$ *the set of all nonnegative integers with radix-p form (see Lemma 7.1) having one component equal to $p-2$ and the other $k-2$ components equal to $p-1$, and $J := \{j_1, \ldots, j_{k-1}\}$, the set of all nonnegative integers having one component equal to 0 and the other $k-2$ components equal to $p-1$. Both sets have $\binom{k-1}{k-2} = k-1$ elements. Then the convolutional code generated by G_0, G_1 over $\mathbb{F}_p$ is noncatastrophic and MDS.*

Proof. We compute d_0^c and d_1^c.

By (7.1) we have: $\mathrm{wt}[(X+1)^{i_l}] = \begin{cases} (p-1)p^{k-2}, & l \neq 0 \\ p^{k-1}, & l = 0 \end{cases}$ and $\mathrm{wt}[(X+1)^{j_l}] = p^{k-2}$.

Let $u = (u_0, \ldots, u_{k-1}) \in \mathbb{F}_p^k, u \neq 0$. Then:

$$\mathrm{wt}[uG_0] = \mathrm{wt}[uG_0(X)] = \mathrm{wt}\left[\sum_{l=0}^{k-1} u_l(X+1)^{i_l}\right] \geq$$

$$\geq \mathrm{wt}\left[(X+1)^{i_{min}}\right] \geq (p-1)p^{k-2},$$

by (7.2). We denoted by i_{min} the smallest of all integers $i_l, l \in \{0, \ldots, k-1\}$ with, $u_l \neq 0$. Therefore $d_0^c \geq (p-1)p^{k-2}$ and since there is a row of this weight we have:

$$d_0^c = (p-1)p^{k-2}.$$

For d_1^c we do the same. Let $u = (u_0, \ldots, u_{2k-1}), \in \mathbb{F}_p^{2k-1}, u \neq 0$. If $(u_1, \ldots, u_{2k-1}) = 0$, we obtain the codeword associated to $u_0(X+1)^{i_0}$ which has weight p^{k-1} by the choice of i_0. If $(u_1, \ldots, u_{2k-1}) \neq 0$ then the weight

$$\mathrm{wt}\, u \begin{bmatrix} G_1 \\ G_0 \end{bmatrix} = \mathrm{wt}\, u \begin{bmatrix} G_1(X) \\ G_0(X) \end{bmatrix} =$$

$$\mathrm{wt}\left[\sum_{s=1}^{k-1} u_l(X+1)^{j_s} + \sum_{l=1}^{k-1} u_{l+k-1}(X+1)^{i_l}\right] \geq p^{k-2},$$

by (7.2), since all the powers i_l, $l = \overline{0, k-1}$ differ from j_s, $s == \overline{1, k-1}$. Then

$$d_1^c \geq (p-1)p^{k-2} + p^{k-2} = p^{k-1} = n.$$

Therefore $d_1^c = d_{free} = n$ and the code is noncatastrophic and MDS. $\square$

REMARK 5.1. Since the main fact we used was that the sum

$$\mathrm{wt}[(X+1)^{i_l}] + \mathrm{wt}[(X+1)^{j_s}] \geq p^{k-1},$$

for any $l = \overline{0, k-1}$, $s = \overline{1, k-1}$, we could use instead in the construction the sets I and J, with I and J formed by all nonnegative integers having the radix-p form with one component equal to $p-i$, respectively $i-2$, and the rest $k-2$ components equal to $p-1$, for all i such that $p-i > i-2$, i.e. for all $i = \overline{2, \lfloor \frac{p+2}{2} \rfloor}$. The weights $wt[(X+1)^{i_l}] = \begin{cases} (p-i+1)p^{k-2}, & l \neq 0 \\ p^{k-1}, & l = 0 \end{cases}$ and $wt[(X+1)^{j_l}] = (i-1)p^{k-2}$ have also the sum greater than p^{k-1}. Also the sets I and J formed like this have both $k-1$ elements as it is needed.

6. Examples. We will give here two concrete examples to show how Theorems 4.3 and 5.1 are applied, the rate 4/8 and 3/9. After that we will discuss also the cases $k = 2, k = 3$ that have not been covered by the binary theorem. We will use here the polynomial matrix representation $G(D) = G_0 + DG_1$.

EXAMPLE 1.   We already showed in the previous section how to choose the matrices G_0, G_1 of sizes 4×8 over $\mathbb{F}_2$. In conformity with Theorem reffinal we have that the polynomial matrix $G(D) = G_0 + DG_1$, given by:

$$\begin{bmatrix} 1 & 1 & 1 & 1 & 1 & 1 & 1 & 1 \\ 1+D & 1 & 1 & 1 & D & D & D & 0 \\ 1+D & 1+D & 0 & D & 1+D & 1 & 0 & 0 \\ 1+D & 0 & 1+D & D & 1 & D & 1 & 0 \end{bmatrix}$$

generates a rate 4/8 PUM convolutional code of degree $\mu = 3$ and maximum distance 8.

EXAMPLE 2.   Let G_0, G_1 be the 3×9 matrices over $\mathbb{F}_3$ associated to the following polynomial matrices:

$$\begin{bmatrix} (X+1)^8 \\ (X+1)^5 \\ (X+1)^7 \end{bmatrix}, \begin{bmatrix} 0 \\ (X+1)^2 \\ (X+1)^6 \end{bmatrix}.$$

The convolutional code generated by

$$G(D) = \begin{bmatrix} 1 & -1 & 1 & -1 & 1 & -1 & 1 & -1 & 1 \\ 1+D & -1 & 1 & 1-D & -1 & 1 & D & 0 & 0 \\ 1+D & 1-D & D & -1 & -1 & 0 & 1 & 1 & 0 \end{bmatrix}$$

is noncatastrophic, has degree $\mu = 2$ and maximum distance 9.

EXAMPLE 3.   We will construct rate $2/n$ PUM convolutional codes that are MDS and noncatastrophic, for all $n \geq 3$.

1. In the case n even we can do the construction over the binary field.

Let:

$$G(D)=\begin{bmatrix} \underbrace{\begin{matrix} 1 & 1 & 1 & 1 \\ 1+D & 1 & 0 & D \end{matrix}}_{j \text{ times}} \end{bmatrix}, G(D)=\begin{bmatrix} \underbrace{\begin{matrix} 1 & 1 & 1 & 1 \\ 1+D & 1 & 0 & D \end{matrix}}_{j \text{ times}} & \begin{matrix} 1 & 1 \\ 1+D & 0 \end{matrix} \end{bmatrix},$$

for $n = 4j$, respectively $n = 4j + 2$. Then the $2/n$ code generated by $G(D)$ is noncatastrophic and has distance n over $\mathbb{F}_2$. The code has the column distances: $d_0^c = n/2$, $d_1^c = n = d_{free}$ in both cases.

2. The cases where n is odd requires more field elements. It turns out that a field with 3 elements is enough for a construction. Therefore, over $\mathbb{F}_3$ we obtain:

$$G(D)=\begin{bmatrix} \underbrace{\begin{matrix} 1 & 1 & 1 & 1 \\ 1+D & 1 & 0 & D \end{matrix}}_{j \text{ times}} & \begin{matrix} 1 \\ D+2 \end{matrix} \end{bmatrix}, G(D)=\begin{bmatrix} \underbrace{\begin{matrix} 1 & 1 & 1 & 1 \\ 1+D & 1 & 0 & D \end{matrix}}_{j \text{ times}} & \begin{matrix} 1 & 1 & 1 \\ 2+D & 1+D & 0 \end{matrix} \end{bmatrix},$$

for $n = 4j + 1$, respectively $n = 4j + 3$. The column distances are in both cases:

$d_0^c = \lfloor n/2 \rfloor + 1$, $d_1^c = n = d_{free}$.

EXAMPLE 4. The construction Theorem 3.2 can not be applied in the case of $k = 3$. It turns out that any choice of binary matrices G_0, G_1 we take gives a catastrophic encoder. Therefore there is no noncatastrophic PUM convolutional code of rate $3/4$, degree 2, having distance 4. The smallest field we can construct such a $3/4$, code with degree 2, distance 4 is $\mathbb{F}_3$. Taking

$$G_0 = \begin{bmatrix} 1 & 1 & 1 & 1 \\ 0 & 1 & 1 & 0 \\ 0 & 1 & 0 & 1 \end{bmatrix}, \; G_1 = \begin{bmatrix} 0 & 0 & 0 & 0 \\ 0 & 1 & 0 & 1 \\ 1 & 1 & 0 & 0 \end{bmatrix}$$

we obtain an MDS code but over a field of characteristic $p \neq 2$. The column distances are $d_0^c = 2$, $d_1^c = 2$, $d_2^c = 3$, $d_3^c = 3$, $d_4^c = 4 = d_{free}$.

7. Appendix. We state here some results that we need along the paper. For more details see [6].

LEMMA 7.1.   [6] Let $c \in \mathbb{F}, c \neq 0$ and let $i \geq 1$ with radix-p form $[i_0, i_1, \ldots, i_{m-1}]$, i.e. $i = i_0 + i_1 p + \ldots + i_{m-1} p^{m-1}$. Then:

(7.1) $$\text{wt}[(X + c)^i] = \prod_{j=0}^{m-1} (i_j + 1).$$

In particular, for $p = 2$,

(7.2) $$\text{wt}[(X + 1)^i] = 2^{\text{wt}(i)},$$

where $\mathrm{wt}(i)$ *is the number of* 1's *in* $\{i_0, i_1, \ldots, i_{m-1}\}$.

LEMMA 7.2.   *[6] Let* I *be any nonempty finite set of nonnegative integers with least integer* i_{min} *and let*

$$P(X) = \sum_{i \in I} b_i (X - c)^i,$$

where $c, b_i \in \mathbb{F}$, *all nonzero. Then:*

(7.3) $$\mathrm{wt}[P(X)] \geq \mathrm{wt}[(X + c)^{i_{min}}].$$

LEMMA 7.3.   *[6] For any polynomial* $P(X)$ *over* $\mathbb{F}$, *any* $c \in \mathbb{F}, c \neq 0$, *and any nonnegative integers* n *and* N,

(7.4) $\mathrm{wt}\left[P(X)(X^n + c)^N\right] \geq \mathrm{wt}\left[(X + c)^N\right] \mathrm{wt}\left[P(X) mod\ (X^n - c)\right].$

The following lemma gives a very obvious result that we need for the constructions. It could be also seen as a corollary to Lemma 7.3:

LEMMA 7.4.   *If* $P(X)$ *is a polynomial over* $\mathbb{F}_2$ *of degree less or equal to* $2^k - 1$, *then the weight* $\mathrm{wt}\left[P(X)(X^{2^k} + 1)\right] = 2wt[P(X)]$.

REFERENCES

[1] R. JOHANNESSON AND K. ZIGANGIROV. Distances and distance bounds for convolutional codes – an overview. In *Topics in Coding Theory. In honour of L. H. Zetterberg.*, Lecture Notes in Control and Information Sciences # 128, pages 109–136. Springer Verlag, 1989.

[2] R. JOHANNESSON AND K.SH. ZIGANGIROV. *Fundamentals of Convolutional Coding.* IEEE Press, New York, 1999.

[3] J. JUSTESEN, E. PAASKE, AND M. BALLAN. Quasi-cyclic unit memory convolutional codes. *IEEE Trans. Inform. Theory*, IT-36(3):540–547, 1990.

[4] G.S. LAUER. Some optimal partial-unit-memory codes. *IEEE Trans. Inform. Theory*, 25:240–243, 1979.

[5] S. LIN AND D.J. COSTELLO. *Error Control Coding: Fundamentals and Applications.* Prentice-Hall, Englewood Cliffs, NJ, 1983.

[6] J.L. MASSEY, D.J. COSTELLO, AND J. JUSTESEN. Polynomial weights and code constructions. *IEEE Trans. Inform. Theory*, IT-19(1):101–110, 1973.

[7] R.J. McELIECE. The algebraic theory of convolutional codes. In V. Pless and W.C. Huffman, editors, *Handbook of Coding Theory*, Volume 1, pages 1065–1138. Elsevier Science Publishers, Amsterdam, The Netherlands, 1998.

[8] PH. PIRET. *Convolutional Codes, an Algebraic Approach.* MIT Press, Cambridge, MA, 1988.

[9] J. ROSENTHAL, J.M. SCHUMACHER, AND E.V. YORK. On behaviors and convolutional codes. *IEEE Trans. Inform. Theory*, 42(6, Part 1):1881–1891, 1996.

[10] J. ROSENTHAL AND R. SMARANDACHE. Maximum distance separable convolutional codes. *Appl. Algebra Engrg. Comm. Comput.*, 10(1):15–32, 1999.

[11] J. ROSENTHAL AND E.V. YORK. BCH convolutional codes. *IEEE Trans. Inform. Theory*, 45(6):1833–1844, 1999.

[12] R. SMARANDACHE, H. GLUESING-LUERSSEN, AND J. ROSENTHAL. Constructions of MDS-convolutional codes. Submitted to IEEE Trans. Inform. Theory, August 1999.

BASIC PROPERTIES OF MULTIDIMENSIONAL
CONVOLUTIONAL CODES*

PAUL WEINER†

Abstract. Let $\mathbb{F}$ be a finite field, and let $\mathcal{D} = \mathbb{F}[z_1, ..., z_m]$ be a polynomial ring in m indeterminates over $\mathbb{F}$. In this paper, we define an m-dimensional convolutional code of length n to be a $\mathcal{D}$-submodule, $\mathcal{C}$, of the free module $\mathcal{D}^n$. Using this point of view, a multidimensional convolutional code may be regarded as the row space of a polynomial matrix (the number of columns gives the length of the code).

We will consider some of the algebraic properties of multidimensional convolutional codes, seeing that their structure is different from the structure of one-dimensional convolutional codes.

We will also look at distance properties of multidimensional convolutional codes. In this regard, we will apply some ideas regarding monomial orders to give a code construction with a lower distance bound; additionally, we will show that for dimension 2 or higher, arbitrarily large distance may be achieved by convolutional codes that are the row spaces of 1×1 polynomial matrices.

1. Introduction. According to one point of view, a (one-dimensional) convolutional code is a submodule of $\mathbb{F}[z]^n$ where $\mathbb{F}[z]$ is a polynomial ring in one indeterminate over a finite field. We generalize this to using a polynomial ring in several indeterminates. Multidimensional convolutional codes, especially in the two-dimensional case, using a Laurent polynomial ring have been considered by Fornasini and Valcher [3, 11, 4]. The algebraic theory of multidimensional convolutional codes is similar over a polynomial ring or over a laurent polynomial ring, but there are some differences in the notion of catastrophicity of a code–see Remark 4.17 below or [12]. Multidimensional convolutional codes have been considered by the current author in his dissertation [12], and some of the material below follows along the lines of that dissertation. In [9] in this volume, Rosenthal includes a survey of different possible frameworks for convolutional codes (although [9] is about what here we would call one-dimensional convolutional codes, the ideas do generalize to the higher-dimensional case).

Using a module theoretic point of view gives (multidimensional) convolutional codes and (multidimensional) behavioral systems as dual objects. This module-system duality is developed in detail by Oberst [7]. We will not pursue this duality here, though in addition to the paper by Oberst, the interested reader may also see [12, Section 2.6], or [5].

We begin by setting some notation that will be used throughout this paper. Let $\mathbb{F} = \mathbb{F}_q$ be the finite field with q elements. Let $\mathbb{N}$ be the set of nonnegative integers. Let $\mathcal{D} = \mathbb{F}[z_1, ..., z_m]$ be the polynomial ring in m indeterminates over $\mathbb{F}$. The free $\mathcal{D}$-module, $\mathcal{D}^k$, of k-component row

---

*This work was supported in part by a fellowship from the Center for Applied Mathematics, University of Notre Dame.

†Department of Mathematics; Saint Mary's University of Minnesota; Winona, MN 55987, USA; *e-mail:* pweiner@smumn.edu.

vectors with entries in $\mathcal{D}$ will be called the m-dimensional (m-D for short) message space of length k. Elements of $\mathcal{D}^k$ carry m-D information as the following example illustrates.

Example 1.1 *Consider a very small black and white picture on a 4×4 pixel field. Each pixel can be white (0) or black (1). Here is the picture.*

$$
\begin{array}{cccc}
\bullet_1 & \bullet_0 & \bullet_0 & \bullet_1 \\
\bullet_0 & \bullet_0 & \bullet_0 & \bullet_1 \\
\bullet_1 & \bullet_1 & \bullet_1 & \bullet_0 \\
\bullet_1 & \bullet_0 & \bullet_1 & \bullet_1
\end{array}
$$

This picture can be represented as a polynomial in $\mathcal{D} = \mathbb{F}_2[x, y]$ (often in the 2-D case we use x and y as indeterminates instead of z_1 and z_2). In this context the picture is $1 + y + x^2 + xy + x^3 + x^2 y + y^3 + x^3 y^2 + x^3 y^3$. This is obtained by associating the monomial $x^i y^j$ with the pixel whose cartesian coordinates are (i, j) where the lower left pixel has cartesian coordinates $(0, 0)$. Then each monomial gets as a coefficient its color in the form of 0 or 1.

We may also have a color picture with more than two colors. In this case we attach to each pixel (monomial) a string of k bits (allowing for 2^k different colors). We then form k polynomials as above, one from the first bit on each pixel, one from the second bit, and so on. Then the picture is represented by a vector of k polynomials. For instance the vector $\begin{bmatrix} 1 + y & x + y & 1 + x + xy \end{bmatrix}$ corresponds to the picture

$$
\begin{array}{cc}
\bullet_{110} & \bullet_{001} \\
\bullet_{101} & \bullet_{011}
\end{array}
$$

Note that this polynomial vector may also be expressed as

$$
\begin{bmatrix} 1 & 0 & 1 \end{bmatrix} + \begin{bmatrix} 0 & 1 & 1 \end{bmatrix} x + \begin{bmatrix} 1 & 1 & 0 \end{bmatrix} y + \begin{bmatrix} 0 & 0 & 1 \end{bmatrix} xy.
$$

This idea extends to higher dimensions where m-D data is finitely supported on $\mathbb{N}^m$. Then we need monomials in m indeterminates to represent pixels. I.e., the monomial $z_1^{i_1} \cdot \ldots \cdot z_m^{i_m}$ corresponds to the pixel $(i_1, \ldots, i_m) \in \mathbb{N}^m$. Thus m-D information is carried by polynomial vectors in m indeterminates.

Our next goal is to introduce error protection. To do this we will inject the message space $\mathcal{D}^k$ into a larger space.

Definition 1.2 *An m-dimensional code of length n over $\mathbb{F}$ is a subset $C \subseteq \mathcal{D}^n$. An element $w \in C$ is a codeword.*

A code C is linear if it is an $\mathbb{F}$-linear subset of $\mathcal{D}^n$ (i.e., C is closed under addition and under scalar multiplication by elements of $\mathbb{F}$). C is *right shift invariant if it is closed under multiplication by* z_i, $i = 1, \dots, m$ *(i.e., $z_i C \subseteq C$ for $i = 1, \dots, m$).*

It follows at once that a code C is linear and right shift invariant if and only if C is a $\mathcal{D}$-submodule of $\mathcal{D}^n$.

Definition 1.3 *An m-dimensional convolutional code of length n is a $\mathcal{D}$-submodule of the free module $\mathcal{D}^n$.*

Remark 1.4 *A 0-dimensional convolutional code of length n is a linear block code of length n. A 1-dimensional convolutional code is a convolutional code in the sense used by York [13].*

2. Generator matrices, free codes, and encoders. Any m-D convolutional code, C, of length n is finitely generated as a $\mathcal{D}$-module because $\mathcal{D}$ is a noetherian ring. So there is a positive integer l and a matrix $G \in \mathcal{D}^{l \times n}$ such that $C = \mathrm{rowspace}_{\mathcal{D}}(G) = \mathcal{D}^l \cdot G$. Such a matrix G is a *generator matrix* of C. Note that C is the image of the map $\mathcal{D}^l \xrightarrow{\cdot G} \mathcal{D}^n$. For this reason we write $C = \mathrm{im}(G)$ for the convolutional code C that has G as a generator matrix. We may also call G an *image representation* for C. Every convolutional code C has an image representation.

Definition 2.1 *Given a polynomial matrix $G \in \mathcal{D}^{l \times n}$, the rank of the convolutional code $C = \mathrm{im}(G)$ is the largest integer k for which there is a nonzero $k \times k$ minor of G. We write $\mathrm{rank}(C) = k$.*

Equivalently, letting $Q(\mathcal{D})$ be the field of fractions of $\mathcal{D}$, $\mathrm{rank}(C)$ is the dimension of the vector subspace spanned by the rows of G in $Q(\mathcal{D})^n$ over the field $Q(\mathcal{D})$.

If the rank of C is k, then the rate of C is $\frac{k}{n}$.

We note that the rank and rate of a convolutional code C are independent of the generator matrix chosen for C.

Definition 2.2 *The m-dimensional convolutional code C is free if it is free as a $\mathcal{D}$-module. If C is free of rate k/n, then $C = \mathrm{im}(G)$ for some $G \in \mathcal{D}^{k \times n}$ (i.e., C has a full row rank generator matrix). In this case we say G is an encoder for C.*

Remark 2.3 *If $m = 1$, then $\mathcal{D}$ is a principal ideal domain (PID), and so any submodule of $\mathcal{D}^n$ is free. Thus any 1-D convolutional code is free. This fails in dimension ≥ 2 (see Example 2.8 below).*

If C is free of rate k/n with encoder $G \in \mathcal{D}^{k \times n}$, then the map

$$\begin{array}{ccc} \mathcal{D}^k & \xrightarrow{\cdot G} & \mathcal{D}^n \\ v & \longmapsto & v \cdot G \end{array}$$

is an injective map with image C. So this maps the message space $\mathcal{D}^k$ bijectively to the code C. This justifies calling G an encoder for C.

Recall that a matrix $U \in \mathcal{D}^{k \times k}$ is *unimodular* if $\det(U)$ is a nonzero element of the ground field $\mathbb{F}$. Equivalently, U is unimodular if and only if U is invertible as an element of $\mathcal{D}^{k \times k}$.

Proposition 2.4 *Let* $G \in \mathcal{D}^{l \times n}, G_1 \in \mathcal{D}^{l_1 \times n}$ *(not necessarily of full row rank). Let* $C = \mathrm{im}(G)$ *and* $C_1 = \mathrm{im}(G_1)$. $C_1 \subseteq C$ *if and only if there exists a matrix* $T \in \mathcal{D}^{l_1 \times l}$ *such that* $G_1 = T \cdot G$.

Proof. Observe that $C_1 \subseteq C$ if and only if $\mathrm{rowspace}_{\mathcal{D}}(G_1) \subseteq$ $\mathrm{rowspace}_{\mathcal{D}}(G)$. This in turn occurs if and only if every row of G_1 can be written as a $\mathcal{D}$-linear combination of the rows of G. This is equivalent to the statement of the proposition. $\square$

Corollary 2.5 *Let* $G, G_1 \in \mathcal{D}^{k \times n}$ *be of full row rank. Let* $C = \mathrm{im}(G)$ *and* $C_1 = \mathrm{im}(G_1)$. $C = C_1$ *if and only if there exists a unimodular matrix* $U \in \mathcal{D}^{k \times k}$ *such that* $G_1 = U \cdot G$.

Proof. Suppose $C = C_1$. By Proposition 2.4 there are matrices $U, V \in \mathcal{D}^{k \times k}$ such that $G_1 = U \cdot G$ and $G = V \cdot G_1$. Hence $G = V \cdot U \cdot G$. Since G is of full row rank, $V \cdot U = \mathrm{Id}_k$. So U is unimodular.

The converse follows immediately from Proposition 2.4. $\square$

We conclude this section with some examples to clarify the above ideas. All of the following examples will use the binary field, $\mathbb{F} = \mathbb{F}_2$ as the ground field and will be 2-D with $\mathcal{D} = \mathbb{F}[x, y]$.

Example 2.6 $G = [\ 1 + xy + x^3 \quad x + y^2\ ]$. *The code* $C = \mathrm{im}(G)$ *is a free rate 1/2 code with encoder* G.

Example 2.7 *Let* $G = \begin{bmatrix} xy & x + x^2 y & x^2 \\ y + xy & 1 + x + xy + x^2 y & x + x^2 \end{bmatrix}$. *The code* $C = \mathrm{im}(G)$ *is of rate 1/3 (the reader can easily verify that all 2×2 minors of G are 0). C is free. An encoder for C is* $G_1 = [\ y \quad 1 + xy \quad x\ ]$. *Note that* $G = \begin{bmatrix} x \\ 1 + x \end{bmatrix} G_1$ *and* $G_1 = [\ 1 \quad 1\ ]G$. *So by Proposition 2.4,* $C = \mathrm{im}(G_1)$.

Example 2.8 *Let* $G = \begin{bmatrix} x & x^2 & xy \\ y & xy & y^2 \end{bmatrix}$. *The code* $C = \mathrm{im}(G)$ *is rate 1/3 but not free. That is, C has no 1×3 generator matrix. To see this, suppose* $G_1 = [\ f \quad g \quad h\ ]$ *is a 1×3 generator matrix of C. Then the rows of G,* $[\ x \quad x^2 \quad xy\ ]$ *and* $[\ y \quad xy \quad y^2\ ]$, *must both be in* $\mathrm{im}(G_1)$. *So we must have* $f \mid x$ *and* $f \mid y$. *Hence* $f = 1$. *From this it follows that* $G_1 = [\ 1 \quad x \quad y\ ]$.

In fact, we then have $C \subseteq \text{im}(G_1)$. *However,* $[\ 1\ \ x\ \ y\ ] \notin C$ *since no* $\mathcal{D}$-*linear combination of the rows of* G *produces* $[\ 1\ \ x\ \ y\ ]$. *So* C *is properly contained in but not equal to the free code* $\text{im}(G_1)$.

3. Parity check matrices and orthogonal codes.

Definition 3.1 *Let* C *be an* m-D *convolutional code of length* n. *A matrix* $H \in \mathcal{D}^{j \times n}$ *is a* parity check matrix *for* C *if* $C = \{w \in \mathcal{D}^n : wH^t = 0\}$. *Note that this gives* C *as the left kernel of the matrix* H^t, *and for that reason we may call* H^t *a* kernel representation *of* C *and write* $C = \text{ker}(H^t)$.

It turns out that not every m-D convolutional code has a kernel representation. We will have more to say about this later in this section.

Definition 3.2 *Let* C *be an* m-D *convolutional code of length* n. *An element* $p \in \mathcal{D}^n$ *is a* parity check vector *for* C *if* $wp^t = 0$ *for all codewords* $w \in C$. *The zero-vector in* $\mathcal{D}^n$ *is the* trivial *parity check vector.*

Remark 3.3 *Suppose* $C = \text{im}(G)$, $G \in \mathcal{D}^{l \times n}$. *The vector* $p \in \mathcal{D}^n$ *is a parity check vector for* C *if and only if* $Gp^t = 0$.

Example 3.4 *Let* $\mathcal{D} = \mathbb{F}_2[x, y, z]$. *Let* $G = \begin{bmatrix} 1 & x & yz & xy \\ 0 & 1 & x & z \end{bmatrix}$. *Let* $C = \text{im}(G)$. *The vector* $p = [\ x^2 + yz\ \ x\ \ 1\ \ 0\ ]$ *is a parity check vector for* C *since* $Gp^t = 0$.

Definition 3.5 *The set of all parity check vectors of the convolutional code* C *form the* orthogonal code, $C^{\perp}$, *of* C. *That is*

$$C^{\perp} = \{p \in \mathcal{D}^n\ :\ wp^t = 0 \text{ for all } w \in C\}.$$

Remark 3.6 *For a convolutional code* C, *the orthogonal code,* $C^{\perp}$, *is also a submodule of* $\mathcal{D}^n$. *So* $C^{\perp}$ *is itself an* m-D *convolutional code of length* n.

Moreover, if $C = \text{im}(G)$, *then* $C^{\perp} = \text{ker}(G^t)$. *So automatically the orthogonal code of any* m-D *convolutional code has a kernel representation.*

The following lemma gives some standard properties of the orthogonal code.

Lemma 3.7 *Let* C, C_1, *and,* C_2 *be* m-D *convolutional codes of length* n.
(i) *If* $C_1 \subseteq C_2$, *then* $C_2^{\perp} \subseteq C_1^{\perp}$.
(ii) $C \subseteq C^{\perp\perp}$.
(iii) $C^{\perp} = C^{\perp\perp\perp}$.

We use the preceding lemma to derive conditions under which a convolutional code has a parity check matrix.

Proposition 3.8 *Let C be an m-D code of length n. The following conditions are equivalent.*

(i) *C has a parity check matrix.*

(ii) *C is the orthogonal code of some code (i.e., $C = C_1^\perp$ for some convolutional code C_1).*

(iii) *$C = C^{\perp\perp}$.*

Proof.

(i) $\Rightarrow$ (ii): Suppose C has a parity check matrix H. So $C = \ker(H^t)$. Let $C_1 = \operatorname{im}(H)$. So by Remark 3.6 we have $C_1^\perp = \ker(H^t) = C$. So C is an orthogonal code.

(ii) $\Rightarrow$ (iii): Suppose $C = C_1^\perp$ for some convolutional code C_1. Then

$$C^{\perp\perp} = C_1^{\perp\perp\perp} = C_1^\perp = C.$$

(iii) $\Rightarrow$ (i): Suppose $C = C^{\perp\perp}$. Let $C^\perp$ have generator matrix H. So $C^\perp = \operatorname{im}(H)$. Then $C^{\perp\perp} = \ker(H^t)$, and so $C = \ker H^t$. Therefore H is a parity check matrix for C. $\square$

Definition 3.9 *If the convolutional code C satisfies the equivalent conditions of the preceding proposition, then $C^\perp$ is called the dual code of C.*

The following result giving a necessary and sufficient condition for a free m-D convolutional code to have a parity check matrix is proved in [12, Theorem 3.3.8].

Proposition 3.10 *Suppose C is a free m-D convolutional code of rate k/n. C has a parity check matrix if and only if C has a minor prime encoder $G \in \mathcal{D}^{k \times n}$. (To say that $G \in \mathcal{D}^{k \times n}$ is minor prime means that the $k \times k$ minors of G have no nonunit common factors in $\mathcal{D}$.)*

It is possible for an m-D code to be free with a dual code, but to have the dual code not be free.

Example 3.11 *Let $\mathcal{D} = \mathbb{F}_2[x, y, z]$. Let $G = [y \ \ z \ \ x]$. Let $C = \operatorname{im}(G)$. So C is a free 3-D convolutional code of rate 1/3. C has a dual code because its encoder, G, is minor prime. However the dual code has generator matrix*

$$H = \begin{bmatrix} x & 0 & y \\ 0 & x & z \\ z & y & 0 \end{bmatrix}.$$

This code has rate 2/3 but is not free (see [15] or [12, Example 3.4.8]) for a proof of this).

There are however classes of free codes for which the dual code exists and is also free.

Definition 3.12 *Let $G \in \mathcal{D}^{k \times n}$ be of full row rank. G is zero prime if the ideal of $\mathcal{D}$ generated by the $k \times k$ minors of G is all of $\mathcal{D}$. Or equivalently, if $1 \in \mathcal{D}$ is a $\mathcal{D}$-linear combination of the $k \times k$ minors of G.*

Proposition 3.13 *Suppose C is a free m-D convolutional code of rate k/n. If C has a zero prime encoder $G \in \mathcal{D}^{k \times n}$, then C has a free dual code.*

Proof. By the Quillen-Suslin Theorem (see for instance [14]), by adding rows, G can be completed to a unimodular matrix in $\mathcal{D}^{n \times n}$. That is, there exists $P \in \mathcal{D}^{(n-k) \times n}$ such that $\left[\dfrac{G}{P} \right] \in \mathcal{D}^{n \times n}$ is unimodular. This matrix then has an inverse which we partition as $[\ Q^t \ | \ H^t \ ]$ where $Q \in \mathcal{D}^{k \times n}$, and $H \in \mathcal{D}^{(n-k) \times n}$. We then have

$$\left[\frac{G}{P} \right] [\ Q^t \ | \ H^t \ ] = \mathrm{Id}_n = [\ Q^t \ | \ H^t \ ] \left[\frac{G}{P} \right].$$

This in turn gives us five equations:
(i) $GQ^t = \mathrm{Id}_k$,
(ii) $GH^t = 0_{k \times (n-k)}$,
(iii) $PQ^t = 0_{(n-k) \times k}$,
(iv) $PH^t = \mathrm{Id}_{n-k}$, and
(v) $Q^t G + H^t P = \mathrm{Id}_n$.

We use these equations to show that $C = \ker(H^t)$, thus showing that H is a parity check matrix for C and that C has a dual code. Since $GH^t = 0$ we have that $C = \mathrm{im}(G) \subseteq \ker(H^t)$. For the reverse inclusion, let $w \in \ker(H^t)$. So $wH^t = 0$. But also from (v) above,

$$w(Q^t G + H^t P) = w\mathrm{Id}_n = w.$$

We then have $w = wQ^t G \in \mathrm{im}(G) = C$. So $\ker(H^t) \subseteq C$. So indeed $C = \ker(H^t)$, and so C has a dual code by Proposition 3.8 and Definition 3.9.

Next we show that the dual code $C^\perp$ is equal to $\mathrm{im}(H)$. We know that $C^\perp = \ker(G^t)$ (Remark 3.6). Hence we must show that $\ker(G^t) = \mathrm{im}(H)$. This follows along the exact same lines as the previous argument, using the transposed forms of the equations (i)-(v) above.

Finally the dual code $C^\perp = \mathrm{im}(H)$ is free because H consists of the last $n - k$ rows of the unimodular matrix $\left[\dfrac{Q}{H} \right]$. $\square$

Remark 3.14 *Another situation in which the dual code (if there is one) is free is in the case of dimension 2 (or lower) when any orthogonal code is free (see [12, Proposition 3.4.5] for a proof of this statement).*

4. Weight and distance. In this section we consider the notions of weight and distance for multidimensional convolutional codes. These will generalize the definitions of Hamming weight and distance for a block code or for a one-dimensional convolutional code.

We begin with some notation. Let $\alpha = (\alpha_1, \ldots, \alpha_m) \in \mathbb{N}^m$ be a multi-index of length m. By $\mathbf{z}^\alpha$ we mean the monomial $z_1^{\alpha_1} \cdot \ldots \cdot z_m^{\alpha_m}$. Then a polynomial $f \in \mathcal{D}$ may be written as $f = \sum_{\alpha \in \mathbb{N}^m} f_\alpha \mathbf{z}^\alpha$ where $f_\alpha \in \mathbb{F}$. Similarly a polynomial vector $w = [w_1 \ \ldots \ w_n] \in \mathcal{D}^n$ may be written as $w = \sum_{\alpha \in \mathbb{N}^m} w_\alpha \mathbf{z}^\alpha$ where $w_\alpha \in \mathbb{F}^n$.

Definition 4.1 *Let $a \in \mathbb{F}^n$. The weight of a, $\mathrm{wt}(a)$, is the number of nonzero entries of a.*

Let $f \in \mathcal{D}$. The weight of f, $\mathrm{wt}(f)$, is the number of nonzero terms of f.

Let $w = [w_1 \ \ldots \ w_n] \in \mathcal{D}^n$. The weight of w is given by $\mathrm{wt}(w) = \sum_{j=1}^n \mathrm{wt}(w_j)$. Equivalently if $w = \sum_{\alpha \in \mathbb{N}^m} b_\alpha \mathbf{z}^\alpha$ where $b_\alpha \in \mathbb{F}^n$, then $\mathrm{wt}(w) = \sum_{\alpha \in \mathbb{N}^m} \mathrm{wt}(b_\alpha)$.

Example 4.2 *If $w = [1 + x^2 \quad 1 + x^2 + xy \quad y + xy]$, then also w can be written as $w = [1 \ 1 \ 0] + [0 \ 0 \ 1]y + [1 \ 1 \ 0]x^2 + [0 \ 1 \ 1]xy$.*

We have $\mathrm{wt}(w) = 7$, and this may be obtained by counting nonzero terms in the first expression for w or by counting the total number of nonzero vector entries in the second.

Definition 4.3 *Given two elements, w, $\tilde{w} \in D^n$, the (Hamming) distance between them is given by $\mathrm{dist}(w, \tilde{w}) = \mathrm{wt}(w - \tilde{w})$.*

Given any m-D code C (not necessarily convolutional) of length n, the distance of C is defined as

$$\mathrm{dist}(C) = \min\{\mathrm{dist}(w, \tilde{w}) \ : \ w, \tilde{w} \in C, \ w \neq \tilde{w}\}.$$

For a convolutional code $\mathcal{C}$, we have that

$$\mathrm{dist}(\mathcal{C}) = \min\{\mathrm{wt}(w) \ : \ w \in \mathcal{C}, \ w \neq 0\}.$$

This is because $\mathrm{dist}(w, \tilde{w}) = \mathrm{wt}(w - \tilde{w})$ and $w - \tilde{w} \in \mathcal{C}$ whenever w, $\tilde{w} \in \mathcal{C}$.

The Hamming distance is a metric on $\mathcal{D}^n$ called the *Hamming metric*. The following result is standard in coding theory. It follows immediately from the definition of the Hamming distance and from properties of a metric space.

Proposition 4.4 *Let $C \subseteq \mathcal{D}^n$ be a convolutional code with $d = \text{dist}(C)$. Let $t = \lfloor \frac{d-1}{2} \rfloor$ where $\lfloor x \rfloor$ denotes the greatest integer that is less than or equal to x. Let $y \in \mathcal{D}^n$. If $w \in C$ is a codeword with $\text{dist}(w, y) \le t$, then w is the unique codeword nearest to y (with respect to the Hamming metric). We say C can correct up to t errors.*

4.1. The block code bound. Next we derive an upper bound on the distance of an m-D convolutional code. This generalizes to the m-D case a 1-D result that may be found in the paper [6] by Lee.

Lemma 4.5 *Let $G \in \mathcal{D}^{k \times n}$. Let $C = \text{im}(G)$ be the m-D convolutional code generated by G. Write*

$$G = \sum_{i=1}^{r} G_i \mathbf{z}^{\alpha_i},$$

where $G_i \in \mathbb{F}^{k \times n}$, and $\alpha_i \in \mathbb{N}^m$ is a multi-index with $\alpha_i \ne \alpha_j$ for $i \ne j$. (That is, G may be written with only r different monomials among its entries.)

Let $\hat{G} = [G_1 \mid G_2 \mid \ldots \mid G_r] \in \mathbb{F}^{k \times (rn)}$. Let C be the linear block code generated by $\hat{G}$ (i.e., C is the subspace of $\mathbb{F}^{rn}$ spanned by the rows of $\hat{G}$). We allow the possibility that $\dim(C) < k$, but we observe that in any event, $C = \mathbb{F}^k \cdot G$.

Then $\text{dist}(C) \le \text{dist}(C)$.

Proof. Suppose C has a word of weight p. That is, there exists $v \in \mathbb{F}^k$ such that $\text{wt}(v\hat{G}) = p$. We note that

$$v\hat{G} = [vG_1 \mid vG_2 \mid \ldots \mid vG_r] \in \mathbb{F}^{(rn)}.$$

Hence $p = \text{wt}(v\hat{G}) = \text{wt}(vG_1) + \text{wt}(vG_2) + \cdots + \text{wt}(vG_r)$.

We may also consider $v \in \mathcal{D}^k$. Then since v and G_i have all entries in $\mathbb{F}$, we have

$$
\begin{aligned}
\text{wt}(vG) &= \text{wt}(vG_1 \mathbf{z}^{\alpha_1} + \cdots + vG_r \mathbf{z}^{\alpha_r}) \\
&= \text{wt}(vG_1 \mathbf{z}^{\alpha_1}) + \cdots + \text{wt}(vG_r \mathbf{z}^{\alpha_r}) \\
&= \text{wt}(vG_1) + \cdots + \text{wt}(vG_r) \\
&= p.
\end{aligned}
$$

Now C has a nonzero codeword of weight $\text{dist}(C)$, and so C also has a codeword of weight $\text{dist}(C)$. The lemma follows immediately from this. $\square$

The next proposition follows immediately from Lemma 4.5.

Proposition 4.6 (The block code bound) *Suppose G is as in Lemma 4.5 above. Suppose every block code of rate $\frac{k}{rn}$ has distance less than or equal to d_0. Then $C = \text{im}(G)$ has distance less than or equal to d_0.*

The next example is an application of the block code bound. It requires the fact that any binary linear block code, C, of rate $\frac{2}{3r}$ has distance at most $2r$. To see this, fix one of the $3r$ coordinates of C. By linearity, either all four codewords of C have a 0 in this coordinate, or else two have a 0 and two have a 1 in this coordinate. Hence the number of 1's in all four codewords together is at most $2 \cdot 3r = 6r$. Of course the zero codeword has no 1's in it. So by a variant of the pigeonhole principle, not all of the three nonzero codewords can have weight more than $2r$. Hence $\text{dist}(C) \leq 2r$.

Example 4.7 *Suppose $G \in \mathcal{D}^{2 \times 3}$ can be written using just r different monomials. Let $C = \text{im}(G)$. Then $\text{dist}(C) \leq 2r$.*

We will come back to this example later, showing that the distance upper bound of $2r$ can be reached by a code with a generator matrix of polynomials of degree at most 1, provided that $m \geq r - 1$. (See Example 4.22 below.)

4.2. Monomial orders. Our later results rely on properties of monomial orders which we review here. More details on monomial orders may be found in [1, 2].

Definition 4.8 *A monomial order on the indeterminates $z_1, \ldots, z_m$ is a total order, $\prec$, on the the set of monomials formed with these indeterminates such that*
  (i) $1 \prec \mathbf{z}^\alpha$ for all $\alpha \in \mathbb{N}^m$ with $\alpha \neq (0, \ldots, 0)$,
  (ii) If $\alpha, \beta, \gamma \in \mathbb{N}^m$ with $\mathbf{z}^\alpha \prec \mathbf{z}^\beta$, then $\mathbf{z}^\alpha \mathbf{z}^\gamma \prec \mathbf{z}^\beta \mathbf{z}^\gamma$.
Regarding such a total order, $\prec$, we use $\preceq, \succ, \succeq$ in the usual way.

Next we give some examples of monomial orders.

Example 4.9 (Lexicographic order) *Choose any ordering of linear monomials, say $z_1 \succ z_2 \succ \ldots \succ z_m$ Then for $\alpha = (\alpha_1, \ldots, \alpha_m), \beta = (\beta_1, \ldots, \beta_m) \in \mathbb{N}^m, \mathbf{z}^\alpha \prec \mathbf{z}^\beta$ if $\alpha_i < \beta_i$ where i is the index of the greatest linear monomial (with respect to the chosen linear monomial ordering) at which α and β disagree.*
  For instance, $z_1^3 z_2 z_5^3 \prec z_1^3 z_2^2$.

Example 4.10 (Degree lexicographic order) *As in the previous example, an ordering for linear monomials is chosen. Then monomials are first ordered by total degree (the total degree of a monomial is the sum of the exponents on its variables). In the case of a tie in total degree, lexicographic ordering is used as a tiebreaker. So for instance if $z_1 \succ z_2 \succ \ldots \succ z_m$, then with degree lexicographic order we have $z_1^2 z_2 \prec z_2^4 \prec z_1 z_2^3$. (By contrast, with lexicographic order these monomials are ordered by $z_2^4 \prec z_1 z_2^3 \prec z_1^2 z_2$.)*

Example 4.11 (Weight order) *Let $r_1, r_2, \ldots, r_m$ be positive numbers (weights). Define the weight of the monomial $\mathbf{z}^\alpha = z_1^{\alpha_1} \cdot \ldots \cdot z_m^{\alpha_m}$ to be* $weight(\mathbf{z}^\alpha) = \alpha_1 r_1 + \cdots + \alpha_m r_m$.

We define the weight order, $\prec$, relative to the weight vector $(r_1, \ldots, r_m)$ by $\mathbf{z}^\alpha \prec \mathbf{z}^\beta$ if $weight(\mathbf{z}^\alpha) < weight(\mathbf{z}^\beta)$. In the case of two different monomials having equal weights, ties may be broken by applying some fixed lexicographic order. For our purposes it will suffice to always use the lexicographic order with $z_1 \succ z_2 \succ \ldots \succ z_m$ to break ties.

So for instance we may have weights $r_1 = 1, r_2 = 5/3$ for monomials in $\mathbb{F}[x, y]$. Then $weight(x^2 y^4) = 2 \cdot 1 + 4 \cdot \frac{5}{3} = \frac{26}{3}$ and $weight(x^5 y^2) = \frac{25}{3}$. So $x^5 y^2 \prec x^2 y^4$. In this example we have $weight(x^5) = weight(y^3)$. However, using our tiebreaking convention, $y^3 \prec x^5$.

Note that degree lexicographic order is a special case of a weight order— namely with the weight vector $(1, 1, \ldots, 1)$.

A monomial order allows us to order the terms of a polynomial, and in particular it allows us to identify a *leading term* and a *trailing term* (which are respectively the terms with the greatest and least monomial parts relative to the given order). We will denote the leading and trailing terms of a polynomial, f, by $LT(f)$ and $TT(f)$. We observe that different monomial orders may give rise to different leading terms and different trailing terms. We will call a term of the polynomial f an *extreme term* if it is the leading or trailing term of f with respect to some monomial order.

We now state some simple properties of monomial orders.

Lemma 4.12 *Let $\prec$ be a monomial order for monomials in $z_1, \ldots, z_m$, and let $\alpha, \beta, \gamma, \delta \in \mathbb{N}^m$. If $\mathbf{z}^\alpha \preceq \mathbf{z}^\beta$, and $\mathbf{z}^\gamma \preceq \mathbf{z}^\delta$, then $\mathbf{z}^\alpha \mathbf{z}^\gamma \preceq \mathbf{z}^\beta \mathbf{z}^\delta$. Equality holds only if $\alpha = \beta$ and $\gamma = \delta$.*

Corollary 4.13 *Suppose f and g are polynomials. For a fixed monomial order, we have $LT(f \cdot g) = LT(f) \cdot LT(g)$, and $TT(f \cdot g) = TT(f) \cdot TT(g)$.*

In multiplying polynomials f and g, a *term product* of $f \cdot g$ is a term of f times a term of g.

The following proposition, which follows easily from Lemma 4.12 and Corollary 4.13, gives an essential property of extreme terms of a product of polynomials.

Proposition 4.14 *Any extreme term of $f \cdot g$ is formed by a single term product. That is only one term product of $f \cdot g$ has the monomial part of the given extreme term. In particular, while a general nonzero term product of $f \cdot g$ may be cancelled by some other term product(s), a term product of the form $LT(f)LT(g)$ or $TT(f)TT(g)$ (where these terms are with respect to a fixed monomial order) cannot be cancelled by any other term products.*

Example 4.15 *Over $\mathbb{F}_2[x, y]$, let $f = 1 + x + y + y^2$ and $g = x + x^2 + xy + y^2$. There are only four possible orders for the monomials involved in f and g. We give them here along with the corresponding leading and trailing terms of f and g.*

order	$LT(f)$	$LT(g)$	$TT(f)$	$TT(g)$
$1 \prec x \prec y \prec x^2 \prec xy \prec y^2$	y^2	y^2	1	x
$1 \prec y \prec x \prec y^2 \prec xy \prec x^2$	y^2	x^2	1	x
$1 \prec x \prec x^2 \prec y \prec xy \prec y^2$	y^2	y^2	1	x
$1 \prec y \prec y^2 \prec x \prec xy \prec x^2$	x	x^2	1	y^2

So the extreme terms of $f \cdot g$ are $y^4, x^2y^2, x^3, x,$ and y^2. We may apply Proposition 4.14 to this example: There must be at least five terms in the product $f \cdot g$ since there are five extreme terms.

4.3. Weight of a rate $1/1$ code. In this subsection we show that for dimension two (or higher), there are rate $1/1$ codes of arbitrarily high distance. This is in sharp contrast to the 1-D case where the distance of a rate $1/1$ code is at most 2 (this is because a polynomial $p(x)$ in one indeterminate over a finite field divides $x^N - 1$ for some positive integer N that depends on $p(x)$).

We will show in dimension two (or higher) that for any positive integer d, we can find a polynomial $p(\mathbf{z})$ with d distinct leading terms. Then by Proposition 4.14, the rate $1/1$ code $\mathcal{C} = \text{im}([p])$ will have distance at least d.

To do this, consider the two-dimensional situation where $\mathcal{D} = \mathbb{F}[x, y]$. Fix a target distance d. Choose $d - 1$ distinct rational numbers

$$\frac{a_1}{b_1} < \frac{a_2}{b_2} < \cdots < \frac{a_{d-1}}{b_{d-1}}.$$

(Here a_i, b_i are positive integers such that the fraction $\frac{a_i}{b_i}$ is reduced to lowest terms.)

Next define positive integers

$$
\begin{aligned}
A_1 &= a_1 + a_2 + a_3 + \cdots + a_{d-1} & B_1 &= 0 \\
A_2 &= a_2 + a_3 + \cdots + a_{d-1} & B_2 &= b_1 \\
A_3 &= a_3 + \cdots + a_{d-1} & B_3 &= b_1 + b_2 \\
&\;\;\vdots & &\;\;\vdots \\
A_{d-1} &= a_{d-1} & B_{d-1} &= b_1 + b_2 + \cdots + b_{d-2} \\
A_d &= 0 & B_d &= b_1 + b_2 + \cdots + b_{d-1}
\end{aligned}
$$

Define a polynomial $p = x^{A_1}y^{B_1} + \cdots + x^{A_d}y^{B_d}$. Now consider monomial weight orders with weight vectors of the form $(1, r)$.

Case 1: $0 < r < \frac{a_1}{b_1}$.

With respect to this weight order, $x^{A_1}y^{B_1}$ is the leading term of p. To see this, note that we have

$$0 < r < \frac{a_1}{b_1} < \frac{a_2}{b_2} < \ldots < \frac{a_{d-1}}{b_{d-1}}.$$

So $rb_i < a_i$ for $i = 1, 2, \ldots, d-1$. Hence for $i > 1$ we have

$$
\begin{aligned}
\text{weight}(x^{A_i}y^{B_i}) &= A_i + rB_i \\
&= a_i + a_{i+1} + \cdots + a_{d-1} + rb_1 + rb_2 + \cdots + rb_{i-1} \\
&< a_i + a_{i+1} + \cdots + a_{d-1} + a_1 + a_2 + \cdots + a_{i-1} \\
&= \text{weight}(x^{A_1}y^{B_1}).
\end{aligned}
$$

So $x^{A_1}y^{B_1}$ is a leading term of p.

Case 2: $\frac{a_{j-1}}{b_{j-1}} < r < \frac{a_j}{b_j}$ $(j = 2, 3, \ldots, d-1)$.

With respect to this weight order, $x^{A_j}y^{B_j}$ is the leading term of p: we have

$$\frac{a_1}{b_1} < \ldots < \frac{a_{j-1}}{b_{j-1}} < r < \frac{a_j}{b_j} < \ldots < \frac{a_{d-1}}{b_{d-1}}.$$

So $a_i < rb_i$ for $1 \le i \le j-1$ and $rb_i < a_i$ for $j \le i \le d-1$.

Then for $1 \le i \le j-1$ we have

$$
\begin{aligned}
\text{weight}(x^{A_i}y^{B_i}) &= a_i + \cdots + a_{j-1} + a_j + \cdots + a_{d-1} + rb_1 + rb_2 + \cdots + rb_{i-1} \\
&< rb_i + \cdots + rb_{j-1} + a_j + \cdots + a_{d-1} + rb_1 + rb_2 + \cdots + rb_{i-1} \\
&= \text{weight}(x^{A_j}y^{B_j}).
\end{aligned}
$$

A similar argument shows that for $j+1 \le i \le d-1$, we also have that $\text{weight}(x^{A_i}y^{B_i}) < \text{weight}(x^{A_j}y^{B_j})$.

This establishes that in this case, $x^{A_j}y^{B_j}$ is the leading term of p.

Case 3: $r > \frac{a_{d-1}}{b_{d-1}}$.

With respect to this weight order we have that $x^{A_d}y^{B_d}$ is the leading term of p. The details are similar to the previous cases and are left to the reader.

Example 4.16 *We use the rational numbers*

$$\frac{1}{3} < \frac{1}{2} < \frac{2}{3} < \frac{1}{1} < \frac{3}{2}.$$

The corresponding polynomial based on the above construction is

$$p = x^8 + x^7y^3 + x^6y^5 + x^4y^8 + x^3y^9 + y^{11}.$$

Using weight vectors $(1, \frac{1}{4}), (1, \frac{2}{5}), (1, \frac{3}{5}), (1, \frac{3}{4}), (1, \frac{4}{3}), (1, 2)$ *gives each of the six terms of p as a leading term. Hence letting $\mathcal{C} = \text{im}([p])$, we have $\text{dist}(\mathcal{C}) = 6$.*

It is interesting in this example to look at the exponents on the monomials as points in $\mathbb{N}^2$. That is, consider the points $(8,0), (7,3), (6,5), (4,8),$ $(3,9),$ and $(0,11)$. In the diagram below, the lower left '-' marks the point $(0,0) \in \mathbb{N}^2$. The '+' marks are at the points that are used as exponents in the polynomial p.

```
+  -  -  -  -  -  -  -  -
-  -  -  -  -  -  -  -  -
-  -  -  +  -  -  -  -  -
-  -  -  -  +  -  -  -  -
-  -  -  -  -  -  -  -  -
-  -  -  -  -  -  -  -  -
-  -  -  -  -  -  +  -  -
-  -  -  -  -  -  -  -  -
-  -  -  -  -  -  -  +  -
-  -  -  -  -  -  -  -  -
-  -  -  -  -  -  -  -  -
-  -  -  -  -  -  -  -  +
```

If we connect the marked points in order of increasing x-coordinate with line segments, and then join the last point to the first, we get a convex polygon. This will always occur when the above construction is carried out.

Remark 4.17 *In the one-dimensional case, a rate 1/1 convolutional code is always catastrophic unless the generator polynomial is a monomial. (Here we refer to a catastrophic code as one for which an infinite weight message word may produce a finite weight code word. So for instance the code generated by $G = [1 + z]$ is catastrophic since the infinite weight message word $v = [1 + z + z^2 + \cdots]$ produces the finite weight codeword $w = vG = [1]$.) Such one-dimensional codes are still useful, and in fact cyclic redundancy check codes (CRC codes) are rate 1/1 one-dimensional convolutional codes that are used for error protection in internet transmissions [9, Example 7.3].*

In the case of higher dimensional convolutional codes, a rate 1/1 code may be noncatastrophic even if the generator matrix consists of a polynomial that is not a monomial. For instance the two-dimensional code generated by $G = [x + y]$ is noncatastrophic (to see this, consider any infinite weight message word $p = p(x, y)$, a formal power series in x and y with infinitely many nonzero terms. We may write p as $p = p_0 + p_1 + p_2 + \cdots$ where p_i is the homogeneous part of p of degree i–that is all terms of p of total degree i. Then $pG = p_0 G + p_1 G + \cdots$, which gives a decomposition of pG by total degree–$p_i G$ is the degree $i + 1$ part of pG. Now since p has infinite weight, infinitely many of the p_i have positive weight. Then the the corresponding homogeneous parts, $p_i G$, of pG have positive weight, and *since there are infinitely many such homogeneous parts, pG has infinite weight).*

This notion of catastrophicity is different if one uses Laurent polynomials rather than polynomials. The interested reader may consult [3] and [12, Chapter 5] for further details.

4.4. Unit Memory Codes. In this subsection we will define multidimensional unit memory codes and give a code construction with a lower distance bound.

The following definition may be regarded as a generalization to the m-D case of the concept of unit memory encoders, described in papers by Thommesen and Justesen [10] and by Lee [6].

Definition 4.18 *A unit memory code is an m-D convolutional code that has a generator matrix with all polynomials of first degree. A unit memory encoder is a matrix $G \in \mathcal{D}^{k \times n}$ of rank k, all the entries of which are of first degree.*

A unit memory code has a generator matrix of the form $G = G_0 + z_1 G_1 + \cdots + z_m G_m$ where $G_i \in \mathbb{F}^{l \times n}$ $(0 \leq i \leq m)$ is a matrix with entries in the ground field $\mathbb{F}$. We have had good results considering a special type of unit memory codes. We give the motivation for and description of this below.

Piret gives a 1-D code construction in [8] in which he forms a parity check matrix of the form $H_0 + z H_1$ where H_0, H_1 are $(n - k) \times n$ parity check matrices of Reed-Solomon codes. The code construction we suggest here was inspired by Piret's construction, but it is different in several ways. First, we define the generator matrix directly; second, we do not restrict ourselves to using Reed-Solomon codes; and third, we generalize our construction to the m-D case.

Let $C_0, C_1, \ldots, C_m$ be rate k/n linear block codes with generator matrices

$$G_0, G_1, \ldots, G_m \in \mathbb{F}^{k \times n}.$$

Let $\mathrm{dist}(C_i) = d_i$, $i = 0, 1, \ldots, m$. Let

$$G = G_0 + z_1 G_1 + \cdots + z_m G_m \in \mathcal{D}^{k \times n}.$$

Let $\mathcal{C}$ be the m-D convolutional code generated by G.

First we prove that with this construction G is a unit memory encoder.

Lemma 4.19 *Let G and G_i $(0 \leq i \leq m)$ be as in the above construction. G is of rank k, and consequently G is an encoder for a free rate $\frac{k}{n}$ m-D convolutional code.*

Proof. We must show that G has a nonzero $k \times k$ minor. Now G_0 is the generator matrix of a rate $\frac{k}{n}$ block code and hence has a nonzero $k \times k$ minor. Then writing $G = G(z_1, \ldots, z_m)$ we have $G(0, \ldots, 0) = G_0$, and so the corresponding $k \times k$ minor of G must be nonzero. $\square$

Next we give an important lower bound on the distance of the code $\mathcal{C}$ in terms of the distances of its constituent block codes. The proof of the next proposition relies on extending some of the ideas on monomial orders from Subsection 4.2. In particular Proposition 4.14 holds in the case where the coefficients of the polynomial f are vectors in $\mathbb{F}^k$ and coefficients of the polynomial g are full rank scalar matrices in $\mathbb{F}^{k \times n}$. That is, a term product of the form $\mathrm{LT}(f)\mathrm{LT}(g)$ or $\mathrm{TT}(f)\mathrm{TT}(g)$ is nonzero and cannot be cancelled by any other term product(s).

Proposition 4.20 *Using the notation immediately preceding Lemma 4.19, we have* $\mathrm{dist}(\mathcal{C}) \geq d_0 + d_1 + \cdots + d_m$.

Proof. Let v be any nonzero element of $\mathcal{D}^k$. We must show that

$$\mathrm{wt}(v \cdot G) \geq d_0 + d_1 + \cdots + d_m.$$

Choose a lexicographic monomial ordering that gives z_1 as the greatest linear monomial. Let $v_1 \mathbf{z}^{\alpha_1}$ be the leading term of v with respect to this order (here $v_1 \in \mathbb{F}^k$, $\alpha_1 \in \mathbb{N}^m$). Then since $z_1 G_1$ is the leading term of G with respect to this ordering, we have, by Proposition 4.14, that $v_1 G_1 \mathbf{z}^{\alpha_1} z_1$ is an extreme term of the product $v \cdot G$. The weight of this term is at least d_1 because $v_1 G_1$ is a nonzero codeword of C_1, and we know that $\mathrm{dist}(C_1) = d_1$.

We can repeat this argument m times, the jth time choosing a lexicographic monomial ordering that gives z_j as the greatest linear monomial. In this manner we find m distinct terms of $v \cdot G$ with weights at least $d_1, d_2, \ldots, d_m$ respectively.

Next using any lexicographic monomial ordering, we apply the same argument to a trailing term. Because the trailing term of G is G_0, we find that there is a term of $v \cdot G$ (distinct from the m terms constructed above) whose coefficient is a nonzero codeword of C_0 and hence whose weight is at least d_0.

We conclude that $\mathrm{wt}(v \cdot G) \geq d_0 + d_1 + \cdots + d_m$. It follows that $\mathrm{dist}(\mathcal{C}) \geq d_0 + d_1 + \cdots + d_m$. $\square$

We conclude this section with some examples illustrating these ideas. The next example shows that the inequality in Proposition 4.20 may be strict.

Example 4.21 *In this example we construct a 1-D unit memory code using* $\mathcal{D} = \mathbb{F}_2[x]$. *Let*

$$G_0 = \begin{bmatrix} 1 & 1 & 1 \\ 1 & 1 & 0 \end{bmatrix} \quad and \quad G_1 = \begin{bmatrix} 1 & 0 & 1 \\ 0 & 1 & 0 \end{bmatrix}.$$

Then $d_0 = d_1 = 1$ are the distances of the linear block codes generated by G_0 and G_1. Also $G = G_0 + xG_1 =$ $\begin{bmatrix} 1+x & 1 & 1+x \\ 1 & 1+x & 0 \end{bmatrix}$. *Let $\mathcal{C}$ be the 1-D convolutional code generated by G. By Proposition 4.20, dist$(\mathcal{C}) \geq 1 + 1 = 2$.*

In fact we claim that dist$(\mathcal{C}) = 3$. First note that for any $v = [f \quad g] \in \mathcal{D}^2$ we have $vG = [(1+x)f + g \quad f + (1+x)g \quad (1+x)f]$. Now if contrary to our claim we have dist$(\mathcal{C}) < 3$, then for some nonzero $v = [f \quad g] \in \mathcal{D}^2$, wt$(vG) \leq 2$. Hence at least one of the three components of vG must be 0.

Case 1: $(1+x)f + g = 0$. Then $g = (1+x)f$, and $vG = [0 \quad x^2 f \quad (1+x)f]$. Since $f \neq 0$ (else $v = 0$), vG has at least three nonzero terms. That is wt$(vG) \geq 3$.

Case 2: $f + (1+x)g = 0$. Then $f = (1+x)g$, and so $vG = [x^2 g \quad 0 \quad (1+x^2)g]$. Then since $g \neq 0$ we have wt$(vG) \geq 3$.

Case 3: $(1+x)f = 0$. Then $f = 0$. So $vG = [g \quad (1+x)g \quad 0]$. Again wt$(vG) \geq 3$.

So in all cases, under the assumption that wt$(vG) < 3$, we obtain the contradiction that wt$(vG) \geq 3$. So we must conclude that dist$(\mathcal{C}) \geq 3$. Finally since wt$([0 \quad 1]G) = 3$, we have that dist$(\mathcal{C}) = 3$, establishing the claim.

Recall the block code bound (Proposition 4.6). After Example 4.7 we promised to give for any r an example of a rate 2/3 code using r different monomials that has distance $2r$, thus meeting the block code bound.

Example 4.22 *Fix r and let $m = r - 1$. Let*

$$G = \begin{bmatrix} 1 + z_1 + \cdots + z_{r-1} & 1 + z_1 + \cdots + z_{r-1} & 0 \\ 0 & 1 + z_1 + \cdots + z_{r-1} & 1 + z_1 + \cdots + z_{r-1} \end{bmatrix}.$$

Observe that G is a unit memory encoder and may be written as $G = G_0 + z_1 G_1 + \cdots + z_{r-1} G_{r-1}$ where $G_i = \begin{bmatrix} 1 & 1 & 0 \\ 0 & 1 & 1 \end{bmatrix}$ for $0 \leq i \leq r - 1$. For each i, dist$(\text{im}(G_i)) = 2$. So by Proposition 4.20 we have that dist$(\text{im}(G)) \geq 2r$. Thus dist$(\text{im}(G))$, which actually has distance equal to $2r$, meets the block code bound by Example 4.7.

Further work on unit memory multidimensional convolutional codes, including a decoding algorithm, may be found in [12, Chapter 4].

Conclusion. In this paper some of the basics of multidimensional convolutional codes were considered. These are a nontrivial generalizaion of one-dimensional convolutional codes, with a more complex algebraic structure. They also have many interesting distance properties.

The subject of multidimensional convolutional codes is largely open, and much work remains to be done, including finding efficient implementation for encoding and decoding multidimensional convolutional codes. Also it may prove fruitful to consider connections between multidimensional convolutional codes and algebraic geometry.

As more investigations of multidimensional convolutional codes are carried out, more interesting questions will come to light.

Acknowledgement. I would like to thank an anonymous referee for suggesting improvements to this paper and for noticing some typographical errors.

REFERENCES

[1] W.W. ADAMS AND P. LOUSTAUNAU. *An Introduction to Gröbner Bases.* Volume 3 of Graduate Studies in Mathematics. American Mathematical Society, Rhode Island, 1994.

[2] D. COX, J. LITTLE, AND D.O. O'SHEA. *Ideals, Varieties and Algorithms.* Undergraduate Texts in Mathematics. Springer Verlag, New York, 1992.

[3] E. FORNASINI AND M.E. VALCHER. Algebraic aspects of 2D convolutional codes. *IEEE Trans. Inform. Theory,* IT-40(4):1068–1082, 1994.

[4] E. FORNASINI AND M.E. VALCHER. Multidimensional systems with finite support behaviors: Signal structure, generation, and detection. *SIAM J. Control Optim.,* 36(2):760–779, 1998.

[5] H. GLUESING-LUERSSEN, J. ROSENTHAL, AND P.A. WEINER. Duality between multidimensional convolutional codes and systems. *Advances in Mathematical Systems Theory,* F. Colonius, U. Helmke, D. Praetzel-Wolters, F. Wirth (eds.), Birkhäuser, Boston, 135–150, 2000.

[6] L.N. LEE. Short unit-memory byte-oriented binary convolutional codes having maximal free distance. *IEEE Trans. Inform. Theory* IT-22(3):349–352, 1976.

[7] U. OBERST. Multidimensional constant linear systems. *Acta Appl. Math,* 20:1–175, 1990.

[8] PH. PIRET. A convolutional equivalent to Reed-Solomon codes. *Philips J. Res.,* 43(3-4):441–458, 1988.

[9] J. ROSENTHAL. Connections between linear systems and convolutional codes. *Codes, Systems and Graphical Models,* J. Rosenthal and B. Marcus, eds. IMA Volumes in Mathematics and its Applications, Springer-Verlag, 2000.

[10] C. THOMMESEN AND J. JUSTESEN. Bounds on distances and error exponents of unit memory codes. *IEEE Trans. Inform. Theory,* IT-29(5):637-649, 1983.

[11] M.E. VALCHER AND E. FORNASINI. On 2D finite support convolutional codes: an algebraic approach. *Multidim. Sys. and Sign. Proc.,* 5:231–243, 1994.

[12] P. WEINER. *Multidimensional Convolutional Codes.* PhD thesis, University of Notre Dame, 1998. Available at http://www.nd.edu/~rosen/preprints.html.

[13] E.V. YORK. *Algebraic Description and Construction of Convolutional Codes, a Systems Theory Point of View.* PhD thesis, University of Notre Dame, 1997. Available at http://www.nd.edu/~rosen/preprints.html.

[14] D.C. YOULA AND P.F. PICKEL. The Quillen-Suslin theorem and the structure of n-dimensional elementary polynomial matrices. *IEEE Trans. Circuits and Systems,* 31(6):513–517, 1984.

[15] E. ZERZ Primeness of multivariate polynomial matrices. *Systems & Control Letters,* 29:139–145, 1996.

Part 5. Symbolic Dynamics and Automata Theory

LENGTH DISTRIBUTIONS AND REGULAR SEQUENCES

FRÉDÉRIQUE BASSINO[*], MARIE-PIERRE BÉAL[*], AND
DOMINIQUE PERRIN[*]

Abstract. This paper presents a survey on length distributions of regular languages. The accent is on problems in coding theory and the relation with symbolic dynamics.

Key words. Regular sequences, finite automata, prefix codes, bifix codes, symbolic dynamics, zeta functions.

1. Introduction. The notion of a length distribution for a formal language is a simple one: it is the generating series $u(z) = \sum_{n \geq 0} u_n z^n$ of the number of words of each length. This series carries important information concerning a formal language since it measures in a sense the size of the language. It is moreover appropriate in the case of coding. In fact, a length-preserving encoding defines a one-to-one correspondence between words. The two sets of words in such a correspondence will have the same length distribution.

It is a classical result that the length distribution of a formal language carries also some information concerning the structure of the language, in the sense that algebraic operations on series correspond to operations on formal languages. Thus, as we shall see below in more detail, length distributions which are rational series correspond to regular languages.

This correspondence between operations on series and on sets is the basis of the method of generating series in enumerative combinatorics. Numerous examples of applications can be found in the book of Graham, Knuth and Pataschnik [23].

We present here a survey on length distributions of formal languages with emphasis on the problems related to coding and finite automata. We insist on the following general problem: given a family $\mathcal{F}$ of sets of words, characterize the length distributions of the elements of $\mathcal{F}$. For example, the length distributions of prefix codes on k-symbols are the sequences satisfying Kraft's inequality

$$\sum_{n \geq 0} u_n k^{-n} \leq 1,$$

i.e. $u(1/k) \leq 1$.

Our emphasis is on the property of regularity which is the definability by a finite automaton. This places our work at the intersection between

---
[*]Institut d'Électronique et d'Informatique Gaspard-Monge, Université de Marne la Vallée, 5, Boulevard Descartes, Champs-sur-Marne, 77454 Marne la Vallée Cedex 2, France. http://www-igm.univ-mlv.fr/

coding theory and automata theory. For example, one of the main results presented here is a finite-state version of Kraft-McMillan's theorem characterizing the length distributions of regular prefix codes.

We also make connexions with the field of symbolic dynamics. This is natural since the basic notion of symbolic dynamics, namely the conjugacy of subshifts is based on a one-to-one correspondence between paths in finite graphs, giving rise to an invariance of the length distributions.

Our paper is organized as follows. The first sections (Sections 2,3) present the basic notions on automata and formal series used in the paper. In Section 4, we present the finite-state version of Kraft-McMillan theorem mentioned above. The particular case of bifix codes is studied in Section 5. The last section (Section 6) presents several interconnected notions concerning subshifts of finite type and circular codes.

2. Length distributions. We consider the set A^* of all words on a given alphabet A. A subset of A^* is often called a *formal language*. For sets $X, Y \subset A^*$, we denote

$$\begin{aligned} X + Y &= X \cup Y, \\ XY &= \{xy \mid x \in X, y \in Y\}, \\ X^* &= \{x_1 x_2 \cdots x_n \mid x_i \in X, n \geq 0\} \end{aligned}$$

We say that the pair (X, Y) is unambiguous if for each $z \in XY$ there is at most one pair $(x, y) \in X \times Y$ such that $z = xy$.

We say that a set of nonempty words X is a *code* if for each $x \in X^*$ there is at most one sequence $(x_1, x_2, \ldots, x_n)$ with $x_i \in X$ such that $x = x_1 x_2 \cdots x_n$ (one also says that X is uniquely decipherable). A particular case of a code is a *prefix code*. It is a set of words X such that no element of X is a prefix of another one. It is easy to see that such a set is either reduced to the empty word or does not contain the empty word and is then a code.

The *length distribution* of a set of words X is the sequence $u_X = (u_n)_{n \geq 0}$ with

$$u_n = \operatorname{Card}(X \cap A^n).$$

We denote by u_X the formal series

$$u_X(z) = \sum_{n \geq 0} u_n z^n.$$

which is the ordinary generating series of the sequence u_X.

For example, the length distribution of $X = A^*$ is $u(z) = \frac{1}{1 - kz}$ where $k = \operatorname{Card}(A)$.

The *entropy* of a formal language X is

$$h(X) = \log(1/\rho),$$

where ρ is the radius of convergence of the series $u_X(z)$. It is well defined provided X is infinite and thus ρ is finite. If the alphabet A has k elements, we have $h(X) \le \log k$.

The following result relates the basic operations on sets with operations on series.

PROPOSITION 2.1. *The following properties hold for any subsets* X, Y *of* A^*.

(i) *If* $X \cap Y = \emptyset$, *then* $u_{X+Y} = u_X + u_Y$.

(ii) *If the pair* (X, Y) *is unambiguous, then* $u_{XY} = u_X u_Y$.

(iii) *If* X *is a code, then* $u_{X^*} = 1/(1 - u_X)$.

Proof. The first two formulae are clear. If X is a code, every word in X^* has a unique decomposition as a product of words in X. This implies that

$$u_{X^n} = (u_X)^n$$

and thus,

$$u_{X^*} = 1 + u_X + \cdots + u_{X^n} + \cdots = 1/(1 - u_X).$$

□

EXAMPLE 1. *The set* $X = \{b, ab\}$ *is a prefix code. The series* u_{X^*} *is*

$$u_{X^*}(z) = \frac{1}{1 - z - z^2}.$$

Let $(F_n)_{n \ge 0}$ *be the sequence of Fibonacci numbers defined by* $F_0 = 0$, $F_1 = 1$, *and* $F_{n+2} = F_{n+1} + F_n$. *It follows from the recurrence relation that*

$$\frac{z}{1 - z - z^2} = \sum_{n \ge 0} F_n z^n.$$

Consequently, $u_{X^*}(z) = \sum_{n \ge 0} F_{n+1} z^n$. *It can also be proved by a combinatorial argument that the number of words of length* n *in* X^* *is* F_{n+1}.

There are several variants of the generating series considered above. One may first define

$$p_X(z) = \sum_{n \ge 0} \frac{u_n}{k^n} z^n,$$

where $k = \mathrm{Card}(A)$. The coefficients of z^n in $p_X(z)$ is the probability for a word of length n to be in the set X. The relation between u_X and p_X is simple since $p_X(z) = u_X(z/k)$. Another variant of the generating series is the *exponential generating series* of the sequence $(u_n)_{n \ge 0}$ defined as

$$e(z) = \sum_{n \ge 0} \frac{u_n}{n!} z^n.$$

We will also use the zeta function of a sequence $(u_n)_{n \ge 1}$ defined as

$$\zeta(z) = \exp \sum_{n \ge 1} \frac{u_n}{n} z^n.$$

3. Regular distributions. In this section, we describe the connection between the notions of a regular language and a rational series. We prove the classical result (Theorem 3.4) characterizing the regular sequences as the length distributions of regular languages. We mention finally the possible extension to more general classes of formal languages, such as the context-free languages. These results are well-known in the theory of automata and we include them here for the sake of the reader's convenience.

A word on the terminology used here. We use constantly the term *regular* where a richer terminology is often used. In particular, what we call here a regular sequence is, in Eilenberg's terminology, an N-rational sequence (see [20], [33] or [16]). A regular set is also called a *rational* or *recognizable* set.

3.1. Regular sequences. A sequence $u = (u_n)_{n \geq 0}$ of integers is *regular* if there exists a finite graph G and two sets of vertices I, T of G such that for all $n \geq 0$,

$$u_n = \text{Card}(P(n, I, T)),$$

where $P(n, I, T)$ is the set of paths of length n from a vertex of I to a vertex of T. The graph G is one in which multiples edges are allowed (sometimes called a multigraph). We say that the graph G *recognizes* the sequence u.

An equivalent definition of regular sequences is obtained by considering nonnegative matrices.

PROPOSITION 3.1. *A sequence $u = (u_n)_{n \geq 0}$ of integers is regular iff there exists a nonnegative matrix $M \in \mathbb{N}^{k \times k}$ and two vectors $l, c \in \mathbb{N}^k$ such that*

$$u_n = l M^n c,$$

where l is considered as a row vector and c as a column vector.

Proof. Let u be a regular sequence defined by a graph G on the set $\{1, \ldots, k\}$ of vertices. We choose M to be the adjacency matrix of G, i.e. for each pair v, w of vertices, $M_{v,w}$ is the number of edges from v to w. Let l be the row vector defined by $l_v = 1$ if $v \in I$ and 0 otherwise. Let c be the column vector defined by $c_v = 1$ if $v \in T$ and 0 otherwise. The number of paths of length n from a vertex of I to a vertex of T is for each $n \geq 1$ equal to $l M^n c$.

Conversely, let G be the graph with adjacency matrix M. Since the family of regular sequences is closed under addition, we may suppose that the vectors l, c have $0, 1$ coefficients. We can then consider l, c as the characteristic vectors of sets I, T of vertices. It is then obvious that the graph thus constructed recognizes u. □

EXAMPLE 2. *Let G be the graph of Figure 1. The number of paths of length n from vertex $i = 1$ to vertex $t = 2$ is the Fibonacci number F_n.*

FIG. 1. *The Fibonacci graph.*

Accordingly, let M be the matrix

$$M = \begin{pmatrix} 1 & 1 \\ 1 & 0 \end{pmatrix}.$$

The same sequence is defined by the equation

$$F_n = \begin{bmatrix} 1 & 0 \end{bmatrix} M^n \begin{bmatrix} 0 \\ 1 \end{bmatrix}.$$

We say that a sequence u of integers is *rational* if $u(z) = p(z)/q(z)$ for some polynomials $p(z), q(z)$ with integer coefficients. The following result is classical.

THEOREM 3.1. *Any regular sequence u of nonnegative integers is rational.*

Proof. Let (l, M, c) be such that $u_n = lM^n c$. We have

$$u(z) = \sum_{n \geq 0} lM^n cz^n = l\left(\sum_{n \geq 0} (Mz)^n\right)c = l(I - Mz)^{-1}c.$$

The result follows since the coefficients of $(I - Mz)^{-1}$ are rational fractions.
□

EXAMPLE 3. *The generating function of the Fibonacci sequence is*

$$F(z) = \frac{z}{1 - z - z^2}.$$

The converse of Theorem 3.1 is not true. We have actually the following result, due to Jean Berstel (see [20] or [16]).

THEOREM 3.2. *For any regular sequence u, there is an integer p such that the set of poles of minimal modulus is the set of complex numbers $\rho\varepsilon$ where ρ is the radius of convergence of u and $\varepsilon^p = 1$ for some $p \geq 1$.*

In particular, the radius of convergence is a pole.

The following example (from [20] Example 6.1, Chapter VIII) shows the existence of rational series with non-negative integer coefficients which are not regular.

EXAMPLE 4. *Let $0 < \theta < \pi/2$ be such that $\cos\theta = a/c$ with $0 < a < c$ and $c \neq 2a$. The sequence*

$$u_n = c^{2n} \cos^2 n\theta$$

is rational but not regular (poles: $1, e^{2i\theta}, e^{-2i\theta}$).

A sequence u is a *merge* of sequences

$$u^{(0)}, \ldots, u^{(p-1)}$$

if for $n \geq 0, 0 \leq i < p$,

$$u_{pn+i} = u_n^{(i)}.$$

We say that a pole of a rational series is *dominating* if it is strictly less than the modulus of all other ones. The following result is due to Soittola (see [33]).

THEOREM 3.3. *A sequence of non-negative integers is regular iff it is an merge of rational sequences with a dominating pole.*

EXAMPLE 5. *The sequence*

$$1, 1, 2, 1, 4, 2, 8, 3, 16, 5, \ldots$$

is the merge of the sequence of powers of 2 and the Fibonacci sequence.

A third equivalent definition of regular sequences is possible. One can indeed show that a series $u(z)$ is regular iff it can be obtained by a finite number of operations of sum, product and star with

$$u^*(z) = \frac{1}{1 - u(z)},$$

starting from polynomials with nonnegative integer coefficients. An expression of this form is usually called a *regular expression.*

EXAMPLE 6. *The sequence* $(0, 1, 3, 8, 21, \ldots)$ *formed of the Fibonacci numbers of even index is regular. Indeed we have*

$$F_{2n} = lM^{2n}c$$

with the triple (l, M, c) *of Example 2. We have*

$$M^2 = \begin{pmatrix} 2 & 1 \\ 1 & 1 \end{pmatrix},$$

and thus F_{2n} *is the number of paths of length n from 1 to 2 in the graph of Figure 2. The series* $s(z) = \sum_{n \geq 0} F_{2n} z^n$ *can accordingly be written*

$$s(z) = z(2z + z^2 z^*)^* = \frac{z(1-z)}{1 - 3z + z^2}.$$

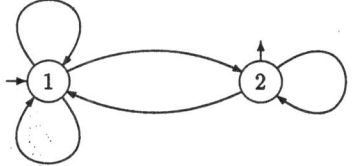

FIG. 2. *One every other Fibonacci number.*

3.2. Finite automata. We present here a brief introduction to the concepts used in automata theory. For a general reference, see [31] or [20].

An *automaton* over the alphabet A is composed of a set Q of *states*, a set $E \subset Q \times A \times Q$ of *edges* or *transitions* and two sets $I, T \subset Q$ of *initial* and *terminal* states.

A *path* in the automaton $\mathcal{A}$ is a sequence

$$(p_1, a_1, p_2), (p_2, a_2, p_3), \ldots, (p_n, a_n, p_{n+1})$$

of consecutive edges. Its label is the word $x = a_1 a_2 \cdots a_n$. A path is *successful* if it starts in an initial state and ends in a terminal state. The set *recognized* by the automaton is the set of labels of its successful paths.

An automaton is *deterministic* if, for each state p and each letter a, there is at most one edge which starts at p and is labeled by a. The term *right resolving* is also used.

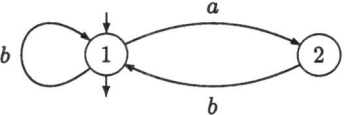

FIG. 3. *Golden mean automaton.*

EXAMPLE 7. *Let $\mathcal{A}$ be the automaton given in Figure 3 with 1 as unique initial and terminal state. It recognizes the set X^* where X is the prefix code $X = \{b, ab\}$.*

A set of words X over A is *regular* if it can be recognized by a finite automaton.

It is a classical result that a set of words is regular iff it can be obtained by a finite number of operations union, product and star, starting form the finite sets.

The following result is also classical.

PROPOSITION 3.2. *Every regular set can be recognized by a finite deterministic automaton having a unique initial state.*

Proof. Let $\mathcal{A} = (Q, E, I, T)$ be a finite automaton over A recognizing a set X. Let $\mathcal{B} = (\mathcal{R}, F, \{I\}, \mathcal{T})$ be the automaton defined as follows. Its states are the subsets

$$Q(u) = \{q \in Q \mid i \xrightarrow{u} q \text{ for some } i \in I\}$$

for all u in A^*. Since Q is finite, there is a finite number of subsets $Q(u)$. The edges of $\mathcal{B}$ are all triples

$$(Q(u), a, Q(ua)).$$

The set of terminal states is

$$\mathcal{T} = \{U \in \mathcal{R} \mid U \cap T \neq \emptyset\}.$$

It is easy to verify that $\mathcal{B}$ is deterministic and recognizes X. $\square$

THEOREM 3.4. *The length distributions of regular sets are the regular sequences.*

Proof. Let X be a regular set. By Proposition 3.2, it can be recognized by a deterministic automaton $\mathcal{A}$. Since $\mathcal{A}$ is deterministic, there is at most one path with given label, origin and end. Thus the number of paths of length n from the initial state to a terminal state is equal to the number u_n of words of X of length n.

Conversely, let u be a regular sequence enumerating the paths in a graph G from I to T. We consider the graph G as an automaton with all edges with distinct labels. Let X be the set of labels of paths from I to T. The sequence u is the length distribution of the set X. $\square$

EXAMPLE 8. *If $X = a^*b$, then*

$$u_X(z) = \frac{z}{1 - z}.$$

3.3. Beyond regular sequences. There are several natural classes of series beyond the rational ones. The algebraic series are those satisfying an algebraic equation. More generally, the hypergeometric series are those such that the quotient of two successive terms is given by a rational fraction (see [23]).

The class of algebraic series is linked with the class of context-free sets (see [21]). A typical example of a context-free set is the set of words on the binary alphabet $\{a, b\}$ having as many a's as b's. We compute below its length distribution which is an algebraic series.

EXAMPLE 9. *The set of words on $A = \{a, b\}$ having an equal number of occurrences of a and b is a submonoid of A^* generated by a prefix code D. Since any word of D^* of length $2n$ is obtained by choosing n positions among $2n$, we have*

$$u_{D^*}(z) = \sum_{n \geq 0} \binom{2n}{n} z^{2n}.$$

By a simple application of the binomial formula, we obtain

$$u_{D^*}(z) = (1 - 4z^2)^{-\frac{1}{2}}.$$

This follows indeed, using the simple identity

$$\binom{-\frac{1}{2}}{n} = \frac{1}{(-4)^n}\binom{2n}{n}.$$

We have $u_D(z) = 1 - 1/u_{D^*}(z)$ and thus

$$u_D(z) = 1 - \sqrt{1 - 4z^2}.$$

Thus $u_D(z)$ is an algebraic series, solution of the equation

$$f^2 - 2f + 4z^2 = 0.$$

4. A finite-state version of the Kraft-McMillan theorem.

Let X be a prefix code on an alphabet with k symbols. It is classical that its length distribution $u = (u_n)_{n\geq 1}$ satisfies Kraft's inequality

$$\sum_{n\geq 1} u_n k^{-n} \leq 1,$$

or equivalently $u(1/k) \leq 1$. The number $u(1/k)$ can actually be interpreted as the probability that a long enough word has a prefix in X.

There is also a connexion with the notion of entropy. Actually, if X is a prefix code, the entropy of X^* is equal to $\log(1/\rho)$ where ρ is the solution of the equation $u_X(\rho) = 1$. Thus Kraft's inequality expresses the fact that $h(X^*) \leq \log k$.

Conversely, Kraft-McMillan's theorem states that for any such sequence $u = (u_n)_{n\geq 1}$, there exists a prefix code X on a k-symbol alphabet such that $u = u_X$.

Let us briefly describe the proof. We suppose by induction to have already built a prefix code X formed of words of length at most $n - 1$ with length distribution $(u_1, u_2, \ldots, u_{n-1})$ on the alphabet $A_k = \{0, 1, \ldots, k - 1\}$. We have

$$\sum_{i=1}^{n} u_i k^{-i} \leq 1,$$

and thus

$$\sum_{i=1}^{n} u_i k^{n-i} \leq k^n.$$

This allows us to choose u_n words on the alphabet A_k of length n without a prefix in X. For the sake of a complete description of the construction, we have to specify the choice made at each step among the words of length

n which do not have already a prefix in X. A possible policy is to choose the earlier ones in the alphabetic order.

The equality case in Kraft's inequality corresponds to a particular class of prefix codes often called *complete*. A prefix code X on the alphabet A is complete if any word on A has either a prefix in X or is a prefix of a word of X.

The notion of a prefix code is related to the notion of a tree. A prefix code on k symbols corresponds to a k-ary tree. The length distribution of the prefix code is the enumerative sequence of the leaves of the tree. We call it the *length distribution* of the tree. Usually, the interest is focused on finite trees, as in Huffman algorithm for example.

We are interested here in the case of infinite trees and, more especially of regular trees arising from prefix codes which are regular, in the sense defined above. The notion of a regular tree can also be defined directly as an infinite tree with only a finite number of non-isomorphic subtrees.

By Theorem 3.4, if X is regular, then the sequence u_X is also regular. The following result shows that conversely the conjunction of the two conditions (of being regular and to satisfy Kraft's inequality) is sufficient to ensure the existence of a regular prefix code on a k-symbol alphabet.

THEOREM 4.1. *A sequence u of integers is the length distribution of a regular prefix code on k symbols iff*

(i) *it is regular.*

(ii) *it satisfies Kraft's inequality $u(1/k) \leq 1$.*

The essence of this result is a constructive method allowing one to build the regular prefix code X given the sequence u.

Two simple methods come to mind at first glance. The first one is to apply directly the proof of the Kraft's theorem. The following example shows that the result need not be a regular set, although the sequence u is itself regular.

EXAMPLE 10. *Let $u(z) = z^2/(1 - 2z^2)$. Since $u(1/2) = 1/2$, we may apply the Kraft construction to build a binary tree with length distribution u. The result is the set*

$$X = \bigcup_{n \geq 0} 01^n 0\{0, 1\}^n$$

which is not regular.

The second method takes into account the hypothesis that the sequence is regular. It will fail in its naive version but the solution is a refinement of this idea. Let G be a graph such that u_n is the number of paths of length n from I to T. We can normalize the graph G to obtain a graph such that $I = \{i\}$, $T = \{t\}$ and that no edge goes out of t. We label each edge in such a way that edges with a common start have different labels. The set recognized by the automaton thus constructed is a prefix code with length distribution equal to u.

The trouble is that the number of symbols used may well be larger than k as shown by the following example.

EXAMPLE 11. *Let u be the regular sequence given by the graph of Figure 4 on the left with $i = 1$ and $t = 4$. We have also $u(z) = 3z^2/(1-z^2)$. Furthermore $u(1/2) = 1$ and thus u satisfies Kraft's equality. However there are four edges going out of vertex 2 and the method described above fails to build a binary prefix code. A solution on $A = \{a, b\}$ is the regular prefix code*

$$X = (aa)^*(ab + ba + bb).$$

The corresponding automaton is given on Figure 4 on the right.

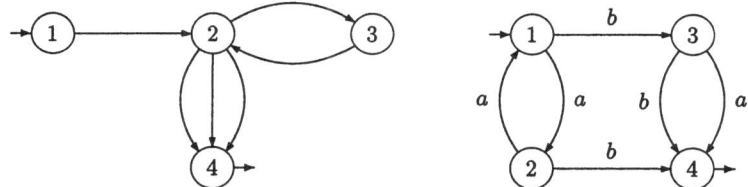

FIG. 4. *Graphs recognizing $u(z) = 3z^2/(1 - z^2)$.*

The proof of Theorem 4.1 consists in building a new graph with all vertices of outdegree at most k. It relies on a transformation called the *multiset construction* described in [8]. The proof uses the following combinatorial lemma also used in symbolic dynamics by Adler and Marcus [28],[2], and quoted in [4] as a nice variant of the pigeon-hole principle.

LEMMA 4.1. *Let $k_1, k_2, \ldots, k_n$ be positive integers. Then there is a subset $S \subset \{1, 2, \ldots, n\}$ such that $\sum_{s \in S} k_s$ is divisible by n.*

The graph obtained is shown in an example below.

EXAMPLE 12. *Let*

(4.1)
$$u(z) = \frac{z^2}{1 - z^2} + \frac{z^2}{1 - 5z^3}.$$

We have $u(1/2) = 1$. A regular binary tree with length distribution u is given in Figure 5 (note that, by convention, a vertex labeled v has its sons represented only once on the figure. Thus, for example the vertex labeled 1 on the right has the same sons as the root. The leaves of the tree are indicated by a black box).

To check that the length distribution is equal to u, one may compute from the graph the following regular expression of u and check by an elementary computation (possibly with the help of a symbolic computation system) that it is equal to u.

$$u(z) = (z^6)^*(2z^2 + z^4 + 2z^5 + z^6 + (z^2 + 3z^5)(5z^3)^*3z^3).$$

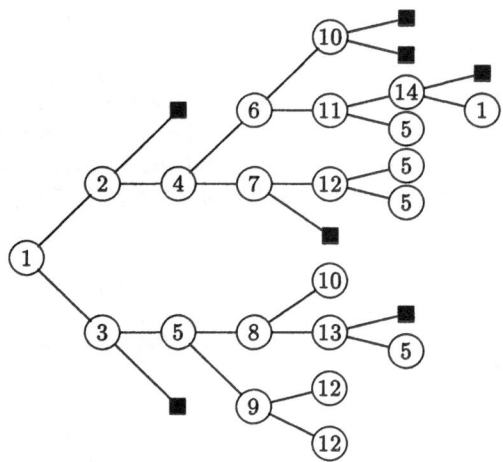

FIG. 5. *Regular binary tree with length distribution u.*

(note for a reader unfamiliar with regular expressions: the first factor $(z^6)^*$ corresponds to the vertex labeled 1 at level 6 of the tree. The term $2z^2 + z^4 + 2z^5 + z^6$ corresponds to the leaves reached by a path which does not use a vertex labeled 5. The factor $(z^2 + 3z^5)(5z^3)^*$ corresponds to the paths from the root to a vertex labeled 5. Finally, the factor $3z^3$ corresponds to the direct paths from 5 to a leaf.)

This example (suggested to us by Christophe Reutenauer) shows an interesting feature of this problem. In fact, from the point of view of regular expressions, the difficult operation in this problem is the sum. It would be a simple matter to build a rational tree for each term of the sum in the expression (12) (see Example 11). The difficulty would then be to merge these two trees to obtain one corresponding to the sum.

A curious consequence of Theorem 4.1 is the following property of regular sequences.

COROLLARY 4.1. *Let $k \geq 2$ be an integer and let u be regular sequence such that $u(1/k) \leq 1$ and $u(0) = 0$. Then there exist k regular sequences $u_1, \ldots, u_k$ such that $u_i(1/k) \leq 1$ and*

$$u(z) = \sum_{i=1}^{k} z u_i(z).$$

Proof. It is a simple consequence of Theorem 4.1. Indeed, if X is a regular prefix code on the k element alphabet A, then $X = \sum_{a \in A} a X_a$ where each X_a is a regular prefix code on the alphabet A. □

We don't know of a direct proof of this result.

5. Bifix codes. We investigate here the length distributions of a particular class of prefix codes, called bifix. Several other classes of prefix codes could give rise to a similar study (for a description to these classes, see [19]).

The definition of a suffix code is symmetric to the definition of a prefix code. It is a set of words X such that no element of X is a suffix of another one. The notion of a complete suffix code is also symmetric. A *bifix code* is a set X of words which is both a prefix and a suffix code.

Any set of words of fixed length is obviously a bifix code but there are more complicated examples.

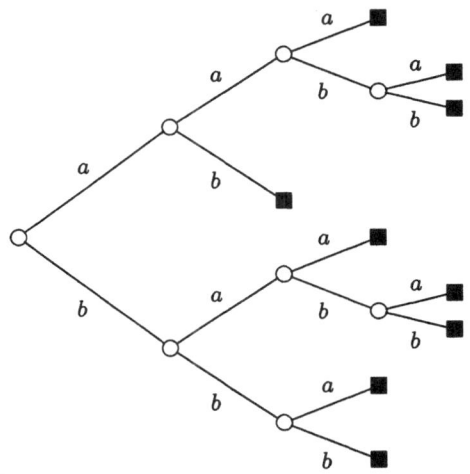

FIG. 6. *The bifix code X.*

EXAMPLE 13. *The set*

$$X = \{aaa, aaba, aabb, ab, baa, baba, babb, bba, bbb\}$$

is a complete prefix code pictured in Figure 6. It is also a complete suffix code as one may check by reading its words backwards.

Surprisingly, it is an open problem to characterize the length distributions of bifix codes. The following simple example shows that they are more constrained than those of prefix codes.

EXAMPLE 14. *The sequence $u(z) = z + 2z^2$ is not realizable as the length distribution of a bifix code on a binary alphabet although $u(1/2) = 1$. Indeed, one of the symbols has to be in X, say a. Then bb is the only word of length 2 that can be added.*

The following nice partial result is due to Ahlswede, Balkenhol and Khachatrian [3]. We state the result for a binary alphabet. It can be readily generalized to k symbols but it presents less interest.

THEOREM 5.1. *For any integer sequence u such that*

$$u(1/2) \leq 1/2,$$

there is a bifix code X such that $u = u_X$.

Proof. The proof is by induction. We suppose that we have already built a bifix code X formed of words of length at most $n - 1$ with length distribution $(u_1, u_2, \ldots, u_{n-1})$. We have

$$\sum_{i=1}^{n} u_i 2^{-i} \leq 1/2,$$

and thus

$$2 \sum_{i=1}^{n} u_i 2^{n-i} \leq 2^n.$$

Finally, we obtain

$$u_n \leq 2^n - 2 \sum_{i=1}^{n-1} u_i 2^{n-i}.$$

The expression of the right handside is at most equal to the number of elements of the set $A^n - XA^* - A^*X$. Thus, we can choose u_n words of length n which do not have a prefix or a suffix in X. This proves the result by induction. □

The authors of [3] formulate the interesting conjecture that Theorem 5.1 is still true if the hypothesis $u(1/2) \leq 1/2$ is replaced by $u(1/2) \leq 3/4$.

There are known additional conditions imposed on length distributions of bifix codes. For example, one has the following result, originally due to Schützenberger (see [14]).

THEOREM 5.2. *If X is a finite complete bifix code on k symbols, then $u_X(1/k) = 1$ and $\frac{1}{k}u'_X(1/k)$ is an integer.*

The number $\frac{1}{k}u'_X(1/k)$ can be interpreted as the average length of the words of X. Indeed

$$zu'_X(z) = \sum_{x \in X} |x| z^{|x|}.$$

EXAMPLE 15. *For the bifix code of Example 13, we have*

$$u_X(z) = z^2 + 4z^3 + 4z^4$$

and thus

$$u'_X(z) = 2z + 12z^2 + 16z^3.$$

Hence $\frac{1}{2}u'_X(1/2) = 3$. The conditions of Theorem 5.2 show directly that the sequence of Example 14 is not realizable. Indeed, it satisfies the first condition but not the second one. The conditions of Theorem 5.2 are not sufficient. Indeed, if $u(z) = z + 4z^3$ we have $u(1/2) = 1$ and $u'(1/2) = 4$ although it is clearly impossible that $u = u_X$ for a bifix code X.

6. Zeta functions, subshifts of finite type and circular codes.
In this section, we present a number of results on interrelated objects which
are connected with cyclic permutation of words. We begin with notions
classical in symbolic dynamics (see [25] or [24] for a general reference; see
[13] or [22] for the link with finite automata).

6.1. Subshifts of finite type. A *subshift* is a set of biinfinite words
on a finite alphabet A which avoids a given set F of forbidden words. It is
a topological space as a closed subset of the space $A^{\mathbb{Z}}$ of functions from $\mathbb{Z}$
into the set A. The *full shift* on A is the set of all biinfinite words on A. It
corresponds to the case $F = \emptyset$.

A *sofic* subshift is the set of biinfinite labels of paths in a finite automaton. A sofic subshift is called *irreducible* if the automaton can be
chosen strongly connected. A *subshift of finite type* is the set of biinfinite
words avoiding a finite set of finite words. Any subshift of finite type is
sofic but the converse is not true. The *edge shift* of a finite graph G is the
set S_G of biinfinite paths in G (viewed as biinfinite sequences of edges). It
is a subshift of finite type.

The *shift* σ is the function on a subshift S which maps a point x to
the point $y = \sigma(x)$ whose ith coordinate is $y_i = x_{i+1}$.

A *morphism* from a subshift S into a subshift T is a function $f : S \to T$
which is continuous and invariant under the shift. A bijective morphism is
called a *conjugacy*. Any subshift of finite type is conjugate to some edge
shift.

The *entropy* $h(S)$ of a subshift S is the entropy of the formal language
formed by the finite blocks occurring in words of S. It can be shown
that the entropy is a topological invariant, in the sense that two conjugate
subshifts have the same entropy.

While the entropy is a measure of number of forbidden words, it is
possible to study the number of minimal forbidden words. It gives rise to
another invariant of subshifts [11], [12].

An integer p is a *period* of a point $x = (a_n)_{n \in \mathbb{Z}}$ if $a_{n+p} = a_n$ for all
$n \in \mathbb{Z}$. Equivalently, p is a period of x if $\sigma^p(x) = x$. The *zeta function* of
a subshift S, is defined as the series

$$\zeta(S) = \exp \sum_{n \geq 1} \frac{p_n}{n} z^n$$

where p_n is the number of words with period n in S. It is also a topological
invariant, since a point of period n is mapped by a conjugacy on a point
of the same period.

The following result due to Bowen and Lanford [18] is classical (see
[25]).

PROPOSITION 6.1. *Let G be a finite graph and let M be the adjacency
matrix of G. Then*

$$\zeta(S_G) = \det(I - Mz)^{-1}.$$

Proof. We first have for each $n \geq 1$

$$\mathrm{Tr}(M^n) = p_n$$

since the coefficient (i, j) of M^n is the number of paths from i to j. Thus

$$\zeta(S_G) = \exp \sum_{n \geq 1} \frac{p_n}{n} z^n$$

$$= \exp \sum_{n \geq 1} \frac{\mathrm{Tr}(M^n)}{n} z^n$$

$$= \exp \mathrm{Tr}(\log(I - Mz)^{-1})$$

$$= \det(I - Mz)^{-1}$$

since, by the formula of Jacobi, $\exp \mathrm{Tr} = \det \exp$. $\square$

EXAMPLE 16. *Let S be the edge shift of the graph G of Figure 7. We have*

$$M = \begin{bmatrix} 1 & 1 & 0 \\ 0 & 0 & 1 \\ 1 & 0 & 0 \end{bmatrix}.$$

Consequently

$$\zeta(S) = \frac{1}{1 - z - z^3}.$$

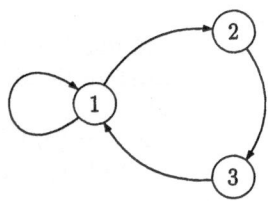

FIG. 7. *A subshift of finite type.*

Let S be a subshift of finite type and let p_n be the number of points with period n. Let q_n be the number of points with least period n. Since q_n is a multiple of n, we also denote $q_n = n l_n$. We have then the formula expressing the zeta function as an infinite product using the integers l_n as exponents.

$$\zeta(S) = \Pi_{n \geq 1} (1 - z^n)^{-l_n},$$

as one may verify using $p_n = \sum_{d|n} d l_d$ and the definition of $\zeta(S)$.

A classical result, related with what follows, is the following statement, known as Krieger's embedding theorem.

THEOREM 6.1. *Let S, T be two subshifts of finite type. There exists an injective morphism $f : S \to T$ with $f(S) \neq T$ iff*

  1. $h(S) < h(T)$
  2. *for each $n \geq 1$, $q_n(S) \leq q_n(T)$ where $q_n(S)$ (resp. $q_n(T)$) is the number of points of S (resp. T) of least period n.*

The following result is the basis of many applications of symbolic dynamics to coding. It is due to Adler, Coppersmith and Hassner [2].

THEOREM 6.2. *If S is an irreducible subshift of finite type such that $h(S) \geq \log k$, it is conjugate to a subshift of finite type S_G where the graph G has outdegree at least k.*

The proof is based on a state-splitting algorithm using approximate eigenvectors and Lemma 4.1. This result is part of a number of constructions leading to sliding block codes used in magnetic recording (see [29], [9] or [25]). It gives at the same time the following result.

THEOREM 6.3. *It S is a subshift of finite type such that $h(S) \leq \log k$, then there is a graph G of outdegree at most k such that S is conjugate to S_G.*

There is a connexion between this theorem and Theorem 4.1. Let indeed u be a regular sequence of integers such that $u(1/k) \leq 1$. Let G be a normalized graph recognizing u (in the sense of Section 4). Let $\bar{G}$ be the graph obtained by merging the initial and terminal vertex. Then $h(S_{\bar{G}}) \leq \log k$. We can apply Theorem 6.3 to obtain a graph H with outdegree at most k such that S_G and S_H are conjugate. This gives the conclusion of Theorem 4.1 provided the initial-terminal vertex did not split in the construction. The following examples show both cases (for details, see [6] and [7]).

EXAMPLE 17. *Let G be the graph of Figure 4. The splitting of vertex 2 gives a graph of outdegree 2. A normalization gives the automaton on the right.*

EXAMPLE 18. *The sequence of Example 12 is recognized by a graph G such that $\bar{G}$ has three cycles of length 2. The solution as a binary tree has only two cycles of length 2 and thus could not be obtained by state-splitting.*

6.2. Circular codes. A *circular word*, or necklace, is the equivalence class of a word under cyclic permutation. For a word w, we denote by $\bar{w}$ the circular word represented by w.

Let X be a set of words and $w = x_1 x_2 \cdots x_n$ with $x_i \in X$. The set of cyclic permutations of the sequence $(x_1, x_2, \ldots, x_n)$ is called a factorization of the circular word $\bar{w}$.

A *circular code* is a set X of words such that the factorization of circular words is unique.

EXAMPLE 19. *The set $X = \{a, aba\}$ is a circular code. Indeed, the position of the symbols b determines uniquely the occurrences of aba.*

EXAMPLE 20. *The set $X = \{ab, ba\}$ is not a circular code. Indeed, the circular word $\bar{w}$ for $w = abab$ has two factorizations namely (ab, ab) and (ba, ba).*

The following characterization is useful (see [14]).

PROPOSITION 6.2. *A set X is a circular code if and only if it is a code and for all $u, v \in A^*$,*

$$uv, vu \in X^* \Rightarrow u, v \in X^*$$

EXAMPLE 21. *We obtain another way to prove that the set $X = \{ab, ba\}$ is not a circular code. Indeed, otherwise we would have $a, b \in X^*$ which is contradictory.*

Let X be a finite code. The *flower automaton* of X, denoted $\mathcal{A}_X$, is the following automaton. The set of its states is

$$Q = \{(u, v) \in A^+ \times A^+ \mid uv \in X\} \cup (1, 1)$$

The transitions are of the form $(u, av) \overset{a}{\to} (ua, v)$ or $(1, 1) \overset{a}{\to} (a, v)$ or $(u, a) \overset{a}{\to} (1, 1)$. The unique initial and final state is $(1, 1)$.

EXAMPLE 22. *The flower automaton of the circular code $\{a, aba\}$ is pictured in Figure 8.*

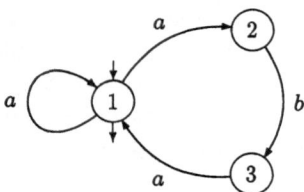

FIG. 8. *The flower automaton of $\{a, aba\}$.*

The following result is easy to prove.

PROPOSITION 6.3. *The flower automaton $\mathcal{A}_X$ recognizes X^*. The code X is circular iff for each word w, there is at most one cycle with label w.*

We now study the length distributions of circular codes. Let X be a circular code and let $u(z) = (u_n)_{n \geq 1}$ be its length distribution. For each $n \geq 1$, let p_n be the number of words w of length n such that $\bar{w}$ has a factorization in words of X.

PROPOSITION 6.4. *The sequences $(p_n)_{n \geq 1}$ and $(u_n)_{n \geq 1}$ are related by*

(6.1)
$$\exp \sum_{n \geq 1} \frac{p_n}{n} z^n = \frac{1}{1 - u(z)}.$$

Proof. Each (p_n) depends only on the first n terms of the sequence (u_n). It is therefore possible to suppose that the sequence (u_n) is finite, i.e. that the code X is finite. Let $\mathcal{A}$ be the flower automaton of X. Let S be the subshift of finite type associated with the graph of $\mathcal{A}$. Then p_n is the number of elements of period n in S. Indeed, each word w such that $\bar{w}$ has a factorization is counted exactly once as the label of a cycle in $\mathcal{A}$. We have also

$$\det(I - Mz) = 1 - u(z).$$

Thus, the result follows from Proposition 6.1. □

The explicit relation between the numbers u_n and p_n is the following. For each $i \geq 1$, let $u^{(i)} = (u_n^{(i)})_{n\geq 1}$ be the length distribution of X^i. Equivalently, $u_n^{(i)}$ is the coefficient of degree n of $u(z)^i$. Then for each $n \geq 1$

$$p_n = \sum_{i=1}^{n} \frac{n}{i} u_n^{(i)}.$$

We also have for each $n \geq 1$

$$(6.2) \qquad p_n = nu_n + \sum_{i=1}^{n-1} p_i u_{n-i}.$$

This formula can be easily deduced from Formula (6.1) by taking the logarithmic derivative of each side of the formula. It shows directly that for any sequence $(u_n)_{n\geq 1}$ of nonnegative integers, the sequence p_n defined by Formula (6.1) is formed of nonnegative integers.

Formula (6.2) is known as Newton's formula in the field of symmetric functions. Actually, the numbers u_n can be considered, up to the sign, as elementary symmetric functions and the p_n as the sums of powers (see [26]). The link between Witt vectors and symmetric functions was established in [34].

Let $p_n = \sum_{d|n} dl_d$. Then l_n is the number of non-periodic circular words of length n with a factorization. In terms of generating series, we have

$$(6.3) \qquad \exp \sum_{n\geq 1} \frac{p_n}{n} z^n = \prod_{n\geq 1} (1 - z^n)^{-l_n}.$$

Putting together Formulae (6.1) and (6.3), we obtain

$$(6.4) \qquad \frac{1}{1 - u(z)} = \prod_{n\geq 1} (1 - z^n)^{-l_n}.$$

For any sequence $(u_n)_{n\geq 1}$ of nonnegative integers, the sequence $l = (l_n)_{n\geq 1}$ thus defined is formed of nonnegative integers. This can be proved either by a direct computation or by a combinatorial argument since any sequence u of nonnegative integers is the length distribution of a circular code on a large enough alphabet. We denote $l = \phi(u)$ and we say that l is the ϕ-transform of the sequence u.

We denote by $\varphi_n(k)$ the number of non-periodic circular words of length n on k symbols. The numbers $\varphi_n(k)$ are called the *Witt numbers*. It is clear that the sequence $(\varphi_n(k))_{n \geq 1}$ is the ϕ-transform of the sequence $(k^n)_{n \geq 1}$.

The corresponding particular case of Identity (6.4)

$$1 - kz = \prod_{n \geq 1} (1 - z^n)^{\varphi_n(k)}$$

is known as the *cyclotomic identity*.

The following arrays display a tabulation of the Witt numbers for small values of n and k.

n	$\varphi_n(2)$	$\varphi_n(3)$	$\varphi_n(4)$
1	2	3	4
2	1	3	6
3	2	8	20
4	3	18	60
5	6	48	204
6	9	116	670
7	18	312	2340
8	30	810	8160
9	56	2184	29120
10	99	5880	104754

The value $\varphi_3(4) = 20$ is famous because of the genetic code: there are precisely 20 amino-acids coded by words of length 3 over a 4-symbol alphabet A,C,G,U.

For any sequence $a = (a_n)_{n \geq 1}$, let

$$p_n = \sum_{d | n} d a_d^{n/d}.$$

The pair (a, p) is called a *Witt vector* (see [30]). The numbers p_n are the *ghost components*. In terms of generating series, one has

$$\exp \sum_{n \geq 1} \frac{p_n}{n} z^n = \prod_{n \geq 1} (1 - a_n z^n)^{-1}.$$

The following result is due to Schützenberger (see [14]).

THEOREM 6.4. *Let $u = (u_n)_{n \geq 1}$ be a sequence of nonnegative integers and let $l = (l_n)_{n \geq 1}$ be the ϕ-transform of u. The sequence $(u_n)_{n \geq 1}$ is the length distribution of a circular code on k symbols iff for all $(n \geq 1)$*

$$l_n \leq \varphi_n(k).$$

Several complements to Theorem 6.4 appear in [5]. In particular, the relation with Kraft's inequality is studied. The equality case in Kraft's inequality is characterized in terms of the sequence of inequalities above.

There is a connexion between Theorem 6.4 and Krieger's embedding theorem (Theorem 6.1), in the sense that Theorem 6.4 gives a simple proof of Theorem 6.1 in a particular case. Actually, let us consider the particular case of subshift of finite type, called a *renewal system*.

A renewal system S is the edge shift of a graph G made up of cycles sharing exactly one vertex. Such a graph is determined by the sequence $u = (u_i)_{1 \leq i \leq n}$ where u_i is the number of loops with length i. Let T_k be the full shift on k symbols. Suppose that the pair formed by S and T_k satisfies the hypotheses of Krieger's theorem. The number $q_n(S)$ of points of least period n is nl_n where $l = (l_n)_{n \geq 1}$ is the ϕ-transform of the sequence u and $q_n(T_k) = n\varphi_n(k)$. Thus, the sequence u satisfies the hypotheses of Theorem 6.4. Consequently, there is circular code X such that $u_X = u$. The flower automaton of X defines an embedding of S_G into the full shift on k symbols. This gives an alternative proof of Krieger's theorem in this case.

It would be interesting to have a proof of Krieger's theorem along the same lines in the general case.

To close this section, we mention the following open problem: If the sequence u is regular and satisfies the inequalities

$$l_n \leq \varphi_n(k) \qquad (n \geq 1),$$

where $l = \phi(u)$, does there exist a rational circular code on k symbols such that $u = u_X$?

6.3. Zeta functions. Theorem 6.1 admits the following generalization due to Reutenauer [32].

THEOREM 6.5. *The zeta function of a sofic subshift is regular.*

We have seen already (Theorem 6.1) that the zeta function of a subshift of finite type is a rational fraction, and indeed the inverse of a polynomial. The stronger statement that it is regular follows from the following formula allowing to compute $\det(I - Mz)$ when M is the adjacency matrix of a $n \times n$ graph G. One has

$$\det(I - Mz) = (1 - v_1(z)) \cdots (1 - v_n(z)),$$

where $v_i(z)$ is the length distribution of the set of first returns to state i using only states $\{i, i+1, \ldots, n\}$ (see [10]).

The proof that the zeta function of a sofic subshift is rational is a result of Manning and Bowen [27], [17]. For an exposition, see [25] or [10]. A generalization appears in [15].

Acknowledgments. The authors wish to thank for the help received during the preparation of this paper. We are indebted to Julia Abrahams for the reference of the work of Ahlswede et al. and several other recent references concerning bifix codes (see [1]). The link between length distributions of circular codes and symmetric functions was disclosed to us by Jacques Désarménien and Jean-Yves Thibon. We also thank Véronique Bruyère for improving our work.

REFERENCES

[1] J. ABRAHAMS, *Code and parse trees for lossless source encoding*, in Compression and Complexity of Sequences 1997, B.C. et al., ed., IEEE Computer Society, 1998, pp. 145–171.

[2] R.L. ADLER, D. COPPERSMITH, AND M. HASSNER, *Algorithms for sliding block codes*, IEEE Trans. Inform. Theory, IT-29 (1983), pp. 5–22.

[3] R. AHLSWEDE, B. BALKENHOL, AND L. KHACHATRIAN, *Some properties of fix-free codes*, Tech. Rep. 039, University Bielefeld, 1997.

[4] M. AIGNER AND G.M. ZIEGLER, *Proofs from The Book*, Springer-Verlag, 1998.

[5] F. BASSINO, *Generating functions of circular codes*, Adv. in Appl. Math, 22 (1999), pp. 1–24.

[6] F. BASSINO, M.-P. BÉAL, AND D. PERRIN, *Enumerative sequences of leaves in rational trees*, in ICALP'97, no. 1256 in Lecture Notes in Computer Science, Springer-Verlag, 1997, pp. 76–86.

[7] ———, *Enumerative sequences of leaves and nodes in rational trees*, Theoret. Comput. Sci. (1999), pp. 41–60.

[8] ———, *A finite state version of version of Kraft-McMillan theorem*, SIAM J. Comput. (2000). To appear.

[9] M.-P. BÉAL, *Codage Symbolique*, Masson, 1993.

[10] ———, *Puissance extérieure d'un automate déterministe, application au calcul de la fonction fonction zêta d'un système sofique*, RAIRO Inform. Théor. Appl., 29 (1995), pp. 85–103.

[11] M.-P. BÉAL, F. MIGNOSI, AND A. RESTIVO, *Minimal forbidden words and symbolic dynamics*, in STACS'96, C. Puech and R. Reischuk, eds., Vol. 1046 of Lecture Notes in Computer Science, Springer-Verlag, 1996, pp. 555–566.

[12] M.-P. BÉAL, F. MIGNOSI, A. RESTIVO, AND M. SCIORTINO, *Forbidden words in symbolic dynamics*, Tech. Rep. 99-15, I.G.M., Université de Marne-la-Vallée, 1999. To appear in Adv. in Appl. Math.

[13] M.-P. BÉAL AND D. PERRIN, *Symbolic dynamics and finite automata*, in Handbook of Formal Languages, G. Rosenberg and A. Salomaa, eds., Vol. 2, Springer-Verlag, 1997, ch. 10.

[14] J. BERSTEL AND D. PERRIN, *Theory of Codes*, Academic Press, 1985.

[15] J. BERSTEL AND C. REUTENAUER, *Zeta functions of formal languages*, Trans. Amer. Math. Soc., 321 (1990), pp. 533–546.

[16] ———, *Rational Series and their Languages*, Springer-Verlag, 1998.

[17] R. BOWEN, *On Axiom A diffeomorphisms*, in AMS-CBMS Reg. Conf., Vol. 35, Providence, 1978.

[18] R. BOWEN AND O.E. LANFORD, *Zeta functions of restrictions of the shift transformation*, in Proc. Symp. Pure Math. AMS, Vol. 14, 1970, pp. 43–50.

[19] V. BRUYÈRE AND M. LATTEUX, *Variable-length maximal codes*, in Proc. 23rd International Colloquium on Automata, Languages and Programming (ICALP'96), F. Meyer and B. Monien, eds., Vol. 1099, Springer-Verlag, 1996, pp. 24–47.

[20] S. EILENBERG, *Automata,Languages and Machines*, Vol. A, Academic Press, 1974.

[21] P. FLAJOLET, *Analytic models and ambiguity of context-free languages*, Theoret. Comput. Sci., 49 (1987), pp. 283–309.

[22] G.D. FORNEY, B.H. MARCUS, N.T. SINDHUSHAYANA, AND M. TROTT, *A multilingual dictionary: System theory, coding theory, symbolic dynamics and automata theory*, in Proceedings of Symposia in Applied Mathematics, no. 50, 1995, pp. 109–138.

[23] R.L. GRAHAM, D. KNUTH, AND O. PATASCHNIK, *Concrete Mathematics*, Addison Wesley, 1988.

[24] B.P. KITCHENS, *Symbolic Dynamics*, Springer-Verlag, 1997.

[25] D.A. LIND AND B.H. MARCUS, *An Introduction to Symbolic Dynamics and Coding*, Cambridge, 1995.

[26] I.G. MACDONALD, *Symmetric Functions and Hall Polynomials*, Oxford University Press, 1995.

[27] A. MANNING, *Axiom A difeomorphisms hava rational zeta functions*, Bull. London Math. Soc., 3 (1971), pp. 215–220.

[28] B.H. MARCUS, *Factors and extensions of full shifts*, Monats. Math, 88 (1979), pp. 239–247.

[29] B.H. MARCUS, R.M. ROTH, AND P.H. SIEGEL, *Constrained systems and coding for recording channels*, in Handbook of Coding Theory, V. S. Pless and W. C. Huffman, eds., Vol. II, North Holland, 1998, ch. 20, pp. 1635–1764.

[30] N. METROPOLIS AND G.-C. ROTA, *Witt vectors and the algebra of necklaces*, Advances in Math., 50 (1983), pp. 95–125.

[31] D. PERRIN, *Finite automata*, in Handbook of Theoretical Computer Science, J. van Leeuwen, ed., Vol. B, Elsevier, 1990, ch. 1.

[32] C. REUTENAUER, $\mathbb{N}$-*rationality of zeta functions*, Adv. in Appl. Math., 29 (1997), pp. 1–17.

[33] A. SALOMAA AND M. SOITTOLA, *Automata Theoretic Properties of Formal Power Series*, Springer-Verlag, 1978.

[34] T. SCHARF AND J.-Y. THIBON, *On Witt vectors and symmetric functions*, Algebra Colloq., 3 (1996), pp. 231–238.

HANDELMAN'S THEOREM ON POLYNOMIALS WITH POSITIVE MULTIPLES

VALERIO DE ANGELIS* AND SELIM TUNCEL†

Abstract. For a polynomial p in several variables and a face F of its Newton polytope, let p_F denote the polynomial consisting of the terms of p that lie in F, with the coefficients given by p. Handelman's theorem states that p has a polynomial multiple with positive coefficients if and only if no p_F has a zero with strictly positive coordinates. We give a short and self-contained account of its proof.

Key words. polynomials, positivity, positive coefficients, ordered groups, states.

AMS(MOS) subject classifications. Primary 06F25, 37A99, 37B10.

1. Introduction. Let $R = \mathbb{R}[x_1^{\pm}, \dots, x_k^{\pm}]$ be the ring of Laurent polynomials in the variables $x_1, \dots, x_k$, and let $R^+ = \mathbb{R}^+[x_1^{\pm}, \dots, x_k^{\pm}]$ be the sub-semiring consisting of polynomials with nonnegative coefficients. For $a = (a_1, \dots, a_k) \in \mathbb{Z}^k$, write $x^a = x_1^{a_1} \cdots x_k^{a_k}$ and denote the coefficient of x^a in $p \in R$ by p_a. Then $p = \sum_{a \in \mathbb{Z}^k} p_a x^a$ and p_a are nonzero for only finitely many $a \in \mathbb{Z}^k$. Let $\text{Log}(p) = \{a \in \mathbb{Z}^k : p_a \neq 0\}$. The *Newton polytope* $N(p)$ is the rational convex hull of $\text{Log}(p)$. Denote the collection of nonempty faces of $N(p)$ by $\mathcal{F}(p)$. For $F \in \mathcal{F}(p)$ let p_F be the sum of $p_a x^a$ over $a \in F \cap \text{Log}(p)$, and call p_F the *F-face* of p. Note that in the case $F = N(p)$ we have $p_F = p$.

Handelman's Theorem [3]. *For $p \in R$ the following are equivalent.*

(i) *There exists $q \in R$ such that $qp \in R^+ \setminus \{0\}$.*

(ii) *There exists $q \in R^+$ such that $\frac{p(1, \dots, 1)}{|p(1, \dots, 1)|} qp \in R^+ \setminus \{0\}$.*

(iii) *We have $p_F(\alpha) \neq 0$ for all $F \in \mathcal{F}(p)$ and $\alpha \in (0, \infty)^k$.*

Call $p \in R$ *numerically positive* if it is the case that $p(\alpha) > 0$ for all $\alpha \in (0, \infty)^k$. Clearly, elements of R^+ are numerically positive. It is also easy to see that each of (i), (ii), (iii) implies that either p is numerically positive or $-p$ is numerically positive. The number $p(1, \dots, 1)/|p(1, \dots, 1)|$ in (ii) equals 1 or -1, depending on whether p or $-p$ is numerically positive.

Handelman's theorem has been applied in situations related to symbolic dynamics. (See [1, 4] for examples.) Its proof relies on a result from the theory of partially ordered Abelian groups [2] and takes up a significant portion of a Memoir [3]. We will give a short and self-contained account. We emphasize that this is not a new proof but, rather, an economical account of the original proof. It will be clear that similar results are valid for polynomial rings, as well as Laurent polynomial rings, whether the coefficients are restricted to $\mathbb{R}$, $\mathbb{Q}$ or $\mathbb{Z}$.

*Department of Mathematics, Xavier University, New Orleans, LA 70125.
†Department of Mathematics, University of Washington, Seattle, WA 98195.

2. Pure states. Let $f \in R^+$. Define S to be the subring of $R\left[\frac{1}{f}\right]$ generated by $\{\frac{x^a}{f} : a \in \mathrm{Log}(f)\}$, and let S^+ be the subsemiring generated by the same set. Then $S = S^+ - S^+$. For $g, h \in S$ we write $g \leq h$ if $h - g \in S^+$. Every element $g \in S$ is of the form $g = r/f^n$, with $n \in \mathbb{N}$, $r \in R$ and $\mathrm{Log}(r) \subset \mathrm{Log}(f^n)$. This expression is not unique since $r/f^n = rf^m/f^{n+m}$. An element r/f^n of S lies in S^+ if and only if $rf^m \in R^+$ for some $m \in \mathbb{N}$. The constant polynomial $1 = f/f$ belongs to S^+. It has the property that for every $\frac{r}{f^n} \in S$ there exists $K \in \mathbb{N}$ such that $\frac{r}{f^n} \leq K1$.

An additive group homomorphism $\phi : S \to \mathbb{R}$ will be called a *state* if $\phi(S^+) \subset \mathbb{R}^+$ and $\phi(1) = 1$. It is easy to see that states form a convex subset $\mathcal{S}$ of $\mathbb{R}^S$. An extreme point of $\mathcal{S}$ is said to be *pure*.

We have here examples of objects studied in the theory [2] of partially ordered Abelian groups: Let G be an (additive) Abelian group endowed with a translation invariant partial order $\leq$, and put $G^+ = \{g \in G : g \geq 0\}$. An element $u \in G^+$ is called an *order unit* if for every $g \in G$ there exists $K \in \mathbb{N}$ such that $g \leq Ku$. Given an order unit u, let $\mathcal{S}(G, u)$ denote the set of group homomorphisms $\phi : G \to \mathbb{R}$ with $\phi(G^+) \subset \mathbb{R}^+$ and $\phi(u) = 1$. An element $\phi \in \mathcal{S}(G, u)$ is a (normalized) *state*. Note, in particular, that states are order-preserving. The set $\mathcal{S}(G, u) \subset \mathbb{R}^G$ is convex, and its extreme points are said to be *pure*.

In the case we are considering, 1 is an order unit of S and $\mathcal{S} = \mathcal{S}(S, 1)$. Our first lemma is a special case of a general result [2]; we postpone its proof until section 4. We write $g > 0$ when $g \in S^+$ and $g \neq 0$.

LEMMA 2.1. *Let $g \in S$. If $\phi(g) > 0$ for every pure state ϕ, then $g > 0$.*

LEMMA 2.2. *Every pure state $\phi \in \mathcal{S}$ is multiplicative.*

Proof. Let ϕ be an extreme point of $\mathcal{S}$. Since $S = S^+ - S^+$ and ϕ is additive, it suffices to check that we have $\phi(gh) = \phi(g)\phi(h)$ for $g, h \in S^+$. Fix $g \in S^+$, and $K \in \mathbb{N}$ such that $g \leq K1$. Then $0 \leq \phi(g) \leq K$. For any $h \in S^+$ there exists $L \in \mathbb{N}$ with $h \leq L1$. If $\phi(g) = 0$ we have $0 \leq \phi(gh) \leq \phi(Lg) = L\phi(g) = 0$, and therefore $\phi(gh) = 0 = \phi(g)\phi(h)$. If $\phi(g) = K$ then, since $h, K1 - g \geq 0$, we have

$$0 \leq \phi((K1 - g)h) \leq \phi(L(K1 - g)) = L\phi(K1 - g) = 0,$$

so that

$$0 = \phi((K1 - g)h) = K\phi(h) - \phi(gh) = \phi(g)\phi(h) - \phi(gh).$$

Now assume $0 < \phi(g) < K$. Define $\psi \in \mathcal{S}$ by letting $\psi(h) = \phi(gh)/\phi(g)$ for $h \in S$. For $h \in S^+$ the inequality $\phi(gh) \leq \phi(Kh)$ implies that, with $0 < t \equiv \phi(g)/K < 1$, we have $t\psi(h) \leq \phi(h)$. This means that $\phi - t\psi \geq 0$ on S^+, so that $\theta = \frac{1}{1-t}(\phi - t\psi)$ is an element of $\mathcal{S}$. Since $\phi = t\psi + (1-t)\theta$ and ϕ is extreme, we conclude that $\psi = \phi$; that is, $\phi(gh) = \phi(g)\phi(h)$ for all $h \in S$. $\square$

For pure $\phi \in S$ define $L(\phi) = \{a \in \mathrm{Log}(f) : \phi(\frac{x^a}{f}) > 0\}$, and let $F(\phi)$ be the rational convex hull of $L(\phi)$.

LEMMA 2.3. *For pure $\phi \in S$ the set $F(\phi)$ is a face of the Newton polytope $N(f)$, and $L(\phi) = F(\phi) \cap \mathrm{Log}(f)$.*

Proof. Clearly $F(\phi) \subset N(f)$. Suppose $w \in F(\phi)$, $u, v \in N(f)$, $t \in (0,1) \cap \mathbb{Q}$ and $w = tu + (1-t)v$. We need to show $u, v \in F(\phi)$. Write

$$u = \sum_{a \in \mathrm{Log}(f)} \alpha_a a, \quad v = \sum_{a \in \mathrm{Log}(f)} \beta_a a, \quad w = \sum_{a \in L(\phi)} \gamma_a a,$$

with $\alpha_a, \beta_a, \gamma_a \in \mathbb{Q}^+$ and $\sum_{a \in \mathrm{Log}(f)} \alpha_a = \sum_{a \in \mathrm{Log}(f)} \beta_a = \sum_{a \in L(\phi)} \gamma_a = 1$. Find $N \in \mathbb{N}$ such that all $\widetilde{\alpha}_a = Nt\alpha_a$, $\widetilde{\beta}_a = N(1-t)\beta_a$, $\widetilde{\gamma}_a = N\gamma_a$ are integers. Since ϕ is multiplicative (Lemma 2.2), we then have

$$\prod_{a \in L(\phi)} \left[\phi\left(\frac{x^a}{f}\right)\right]^{\widetilde{\gamma}_a} = \prod_{a \in \mathrm{Log}(f)} \left[\phi\left(\frac{x^a}{f}\right)\right]^{\widetilde{\alpha}_a} \cdot \prod_{a \in \mathrm{Log}(f)} \left[\phi\left(\frac{x^a}{f}\right)\right]^{\widetilde{\beta}_a}.$$

The left-hand side is positive by the definition of $L(\phi)$. So, the right-hand is also positive, and we have $\alpha_a = \beta_a = 0$ unless $a \in L(\phi)$. Hence $u, v \in F(\phi)$ and $F(\phi)$ is a face of $N(f)$.

It is clear that $L(\phi) \subset F(\phi) \cap \mathrm{Log}(f)$. For the reverse inclusion, suppose the element $w = \sum_{a \in L(\phi)} \gamma_a a$ above actually belongs to $\mathrm{Log}(f)$. Then we have

$$\left[\phi\left(\frac{x^w}{f}\right)\right]^N = \prod_{a \in L(\phi)} \left[\phi\left(\frac{x^a}{f}\right)\right]^{\widetilde{\gamma}_a}$$

and, since the right-hand side is positive, we conclude that $w \in L(\phi)$. $\square$

LEMMA 2.4. *Suppose $\phi \in S$ is pure. Considering $L(\phi)$, the face $F(\phi)$ of $N(f)$ provided by lemma 2.3, and the $F(\phi)$-face $f_{F(\phi)}$ of f, there exists $\mu \in (0, \infty)^k$ such that for all $a \in L(\phi)$ we have*

$$\phi\left(\frac{x^a}{f}\right) = \frac{\mu^a}{f_{F(\phi)}(\mu)} .$$

We postpone the proof of lemma 2.4 until after that of the theorem. For $p \in R$ and $v \in \mathbb{R}^k$, let $\mathrm{in}_v(p)$ be the sum of $p_a x^a$ over those $a \in \mathrm{Log}(p)$ for which the dot product $a \cdot v$ is maximal. Observe that

$$\{\mathrm{in}_v(p) : v \in \mathbb{R}^k\} = \{p_F : F \in \mathcal{F}(p)\}.$$

Proof of Handelman's Theorem. It is clear that (ii) implies (i). Assume (i). Consider $F \in \mathcal{F}(p)$, and pick $v \in \mathbb{R}^k$ such that $\mathrm{in}_v(p) = p_F$. Then $\mathrm{in}_v(qp) = \mathrm{in}_v(q)\mathrm{in}_v(p) = \mathrm{in}_v(q)p_F$. Note that $\mathrm{in}_v(qp) \in R^+ \setminus \{0\}$ since

$qp \in R^+ \setminus \{0\}$. It follows that for $\alpha \in (0, \infty)^k$ the product $\mathrm{in}_v(q)(\alpha)p_F(\alpha)$ is nonzero. Therefore, $p_F(\alpha) \neq 0$, and (i) implies (iii).

Now suppose (iii) holds. Taking $F = N(p)$ in (iii), we see that one of p, $-p$ is numerically positive. First suppose p is numerically positive. Let $F \in \mathcal{F}(p)$. Pick $v = (v_1, \ldots, v_k) \in \mathbb{R}^k$ such that $\mathrm{in}_v(p) = p_F$, and put $m = \max\{\mathrm{Log}(p) \cdot v\}$. Consider $\alpha = (\alpha_1, \ldots, \alpha_k) \in (0, \infty)^k$. Since $p(\alpha_1 e^{v_1 t}, \ldots, \alpha_k e^{v_k t}) = \sum_{a \in \mathrm{Log}(p)} p_a \alpha^a e^{(a \cdot v)t}$, we have

$$\mathrm{in}_v(p)(\alpha) = \lim_{t \to \infty} e^{-mt} p(\alpha_1 e^{v_1 t}, \ldots, \alpha_k e^{v_k t}),$$

and it follows from the numerical positivity of p that $p_F(\alpha) = \mathrm{in}_v(p)(\alpha) \geq 0$. Using (iii), we conclude that $p_F(\alpha) > 0$; that is, each p_F is numerically positive. Now pick any $f \in R^+$ such that $\mathrm{Log}(f) = \mathrm{Log}(p)$. We will show that $f^n p \in R^+$ for large $n \in \mathbb{N}$. Consider the ordered ring S associated above with f and, for a pure state $\phi \in \mathcal{S}$, let $F(\phi)$ be the face of $N(f) = N(p)$ discussed in lemmas 2.3 and 2.4. Observe that $p/f \in S$ and, letting $\mu \in (0, \infty)^k$ be as in the conclusion of lemma 2.4, use Lemmas 2.2, 2.3, 2.4 to calculate $\phi(\frac{p}{f})$:

$$\phi\left(\frac{p}{f}\right) = \sum_{a \in \mathrm{Log}(p)} p_a \, \phi\left(\frac{x^a}{f}\right) = \sum_{a \in L(\phi)} p_a \, \phi\left(\frac{x^a}{f}\right)$$

$$= \sum_{a \in L(\phi)} \frac{p_a \, \mu^a}{f_{F(\phi)}(\mu)} = \frac{p_{F(\phi)}(\mu)}{f_{F(\phi)}(\mu)} > 0.$$

Combining this with lemma 2.1, we find that $\frac{p}{f} \in S^+$. Thus, in the case p is numerically positive, $f^n p \in R^+$ for large n. In the case $-p$ is numerically positive, apply this argument to $-p$ to see that $-f^n p \in R^+$ for large n. □

Note that the proof reveals that $\pm f^n p \in R^+$ for any $f \in R^+$ with $\mathrm{Log}(f) = \mathrm{Log}(p)$ and large enough n. It remains for us to prove lemmas 2.1 and 2.4.

3. Proof of Lemma 2.4. For pure $\phi \in \mathcal{S}$, list the elements of $L(\phi)$ as $a_0, a_1, \ldots, a_m$. If $m = 0$ then $F(\phi) = L(\phi) = \{a_0\}$ and we take $\mu = (1, \ldots, 1)$ to obtain

$$\frac{\mu^{a_0}}{f_{F(\phi)}(\mu)} = \frac{1}{f_{a_0}} = \frac{\phi(1)}{f_{a_0}} = \frac{1}{f_{a_0}} \phi\left(\sum_{a \in \mathrm{Log}(f)} f_a x^a / f\right) = \phi\left(\frac{x^{a_0}}{f}\right),$$

as desired. Now suppose $m \geq 1$. Put $b_i = a_i - a_0$ for $i = 1, \ldots, m$. Consider the $k \times m$ matrix M with b_i as its i-th column, $i = 1, \ldots, m$. Let w be the vector in $\mathbb{R}^m$ whose i-th coordinate equals $\log \phi(\frac{x^{a_i}}{f}) - \log \phi(\frac{x^{a_0}}{f})$. Regarding M and w as linear maps $M : \mathbb{R}^m \to \mathbb{R}^k$ and $w : \mathbb{R}^m \to \mathbb{R}$, we show that $\mathrm{Ker}(M) \subset \mathrm{Ker}(w)$: Since M has integral entries, $\mathrm{Ker}(M)$

has a basis consisting of vectors in $\mathbb{Q}^m$ and it suffices to check that $\mathbb{Q}^m \cap \mathrm{Ker}(M) \subset \mathrm{Ker}(w)$. Suppose $t = (t_1, \dots, t_m) \in \mathbb{Q}^m$ and $Mt = 0$. Then $\sum_{i=1}^m t_i b_i = 0$ and, putting $t_0 = \sum_{i=1}^m t_i$, we have $t_0 a_0 = \sum_{i=1}^m t_i a_i$. Find $N \in \mathbb{N}$ such that $Nt_i \in \mathbb{Z}$ for $i = 1, \dots, m$. Letting $I^+ = \{1 \leq i \leq m : t_i \geq 0\}$, $I^- = \{1 \leq i \leq m : t_i < 0\}$, $t^+ = \sum_{i \in I^+} t_i$ and $t^- = \sum_{i \in I^-} |t_i|$, we find that

$$\left(\frac{x^{a_0}}{f}\right)^{Nt^-} \prod_{i \in I^+} \left(\frac{x^{a_i}}{f}\right)^{Nt_i} = \left(\frac{x^{a_0}}{f}\right)^{Nt^+} \prod_{i \in I^-} \left(\frac{x^{a_i}}{f}\right)^{N|t_i|}.$$

Applying ϕ and taking N-th roots,

$$\prod_{i=1}^m \left[\phi\left(\frac{x^{a_i}}{f}\right)\right]^{t_i} = \left[\phi\left(\frac{x^{a_0}}{f}\right)\right]^{t_0}.$$

That is, $\sum_{i=1}^m t_i \log \phi(\frac{x^{a_i}}{f}) = \sum_{i=1}^m t_i \log \phi(\frac{x^{a_0}}{f})$, which means that $wt = 0$. This shows $\mathrm{Ker}(M) \subset \mathrm{Ker}(w)$. It follows from this inclusion that the map $w : \mathbb{R}^m \to \mathbb{R}$ factors through M; that is, there exists $s = (s_1, \dots, s_k) \in \mathbb{R}^k$ such that $w = sM$. Equivalently,

$$\log \phi\left(\frac{x^{a_i}}{f}\right) - \log \phi\left(\frac{x^{a_0}}{f}\right) = s \cdot a_i - s \cdot a_0.$$

Exponentiating and setting $\mu = (e^{s_1}, \dots, e^{s_k})$, we find that

$$\phi\left(\frac{x^{a_i}}{f}\right) / \phi\left(\frac{x^{a_0}}{f}\right) = \mu^{a_i - a_0}. \tag{*}$$

Hence

$$\frac{f_{F(\phi)}(\mu)}{\mu^{a_0}} = \sum_{i=0}^m f_{a_i} \mu^{a_i - a_0} = \left[\phi\left(\frac{x^{a_0}}{f}\right)\right]^{-1} \sum_{i=0}^m f_{a_i} \phi\left(\frac{x^{a_i}}{f}\right)$$

$$= \left[\phi\left(\frac{x^{a_0}}{f}\right)\right]^{-1} \sum_{a \in \mathrm{Log}(f)} f_a \phi\left(\frac{x^a}{f}\right) = \left[\phi\left(\frac{x^{a_0}}{f}\right)\right]^{-1} \phi(1),$$

and, since $\phi(1) = 1$, we have $\phi(\frac{x^{a_0}}{f}) = \mu^{a_0}/f_{F(\phi)}(\mu)$. We then use $(*)$ again to see that $\phi(\frac{x^{a_i}}{f}) = \mu^{a_i}/f_{F(\phi)}(\mu)$ for $i = 0, 1, \dots, m$. This completes the proof of lemma 2.4.

4. Proof of Lemma 2.1. We now extract from [2] the proof of lemma 2.1. As we mentioned, lemma 2.1 is a special case of a general result; see pp. 61–65, 81–86, 95–96 of [2] for further details and general results.

LEMMA 4.1. *Let G be a partially ordered Abelian group with order unit u and let H be a subgroup of G with $u \in H$. Let $g \in G$, $\theta \in \mathcal{S}(H, u)$, and set*

$$\alpha = \sup\{\theta(h)/n : h \in H, n \in \mathbb{N}, h \leq ng\},$$
$$\beta = \inf\{\theta(h)/n : h \in H, n \in \mathbb{N}, ng \leq h\}.$$

(a) *We have* $-\infty < \alpha \leq \beta < \infty$.

(b) *If* $\psi \in \mathcal{S}(H + \mathbb{Z}g, u)$ *and* ψ *extends* θ, *then* $\alpha \leq \psi(g) \leq \beta$.

(c) *If* $\alpha \leq \gamma \leq \beta$ *then there exists* $\psi \in \mathcal{S}(H + \mathbb{Z}g, u)$ *such that* ψ *extends* θ *and* $\psi(g) = \gamma$.

(d) *The state* $\theta \in \mathcal{S}(H, u)$ *extends to an element of* $\mathcal{S}(G, u)$.

Proof. Since u is an order unit, we can find $K, L \in \mathbb{N}$ with $-g \leq Ku$ and $g \leq Lu$. Then $-Ku \leq g \leq Lu$, which implies $-K \leq \alpha$, $\beta \leq L$. The rest of (a) and (b) are easily checked. To establish (c), first use the definition of α, β to observe that if $h \in H$ and $N \in \mathbb{Z}$ are such that $h + Ng \geq 0$ then $\theta(h) + N\gamma \geq 0$. It follows that if $h \in H$ and $N \in \mathbb{Z}$ satisfy $h + Ng = 0$ then both $\theta(h) + N\gamma \geq 0$ and $-\theta(h) - N\gamma \geq 0$ and, therefore, $\theta(h) + N\gamma = 0$. This means that θ extends to a group homomorphism $\psi : H + \mathbb{Z}g \rightarrow \mathbb{R}$ with $\psi(g) = \gamma$. The above observation shows that $\psi(h + Ng) \geq 0$ whenever $h + Ng \geq 0$. Finally, (d) follows from (c) by an application of Zorn's lemma. $\square$

PROPOSITION 4.1. *Let G be a partially ordered group with order unit u, let $g \in G$, and set*

$$\alpha = \sup\{m/n : m \in \mathbb{Z}, n \in \mathbb{N}, mu \leq ng\},$$
$$\beta = \inf\{m/n : m \in \mathbb{Z}, n \in \mathbb{N}, ng \leq mu\}.$$

(a) *We have* $-\infty < \alpha \leq \beta < \infty$.

(b) *If* $\phi \in \mathcal{S}(G, u)$ *then* $\alpha \leq \phi(g) \leq \beta$.

(c) *If* $\alpha \leq \gamma \leq \beta$ *then there exists* $\phi \in \mathcal{S}(G, u)$ *with* $\phi(g) = \gamma$.

Proof. (a) and (b) are easy to check. For the proof of (c), put $H = \mathbb{Z}u$ and note that $\mathcal{S}(H, u)$ consists of a single element, θ. Observing that

$$\alpha = \sup\{\theta(h)/n : h \in H = \mathbb{Z}u, n \in \mathbb{N}, h \leq ng\},$$
$$\beta = \inf\{\theta(h)/n : h \in H = \mathbb{Z}u, n \in \mathbb{N}, ng \leq h\},$$

apply lemma 4.1(c) to extend θ to $\psi \in \mathcal{S}(\mathbb{Z}u + \mathbb{Z}g, u)$ with $\psi(g) = \gamma$. Then apply lemma 4.1(d) to extend ψ to $\phi \in \mathcal{S}(G, u)$. $\square$

Proof of lemma 2.1. First observe that if $\phi(g) > 0$ for all $\phi \in \mathcal{S}$ then $g > 0$: If $\phi(g) > 0$ for all $\phi \in \mathcal{S}$ then, by the above proposition, there exist $m, n \in \mathbb{N}$ such that $0 < m1 \leq ng$. It follows that $g > 0$.

Now equip $\mathbb{R}^{\mathcal{S}}$ with the product topology, and $\mathcal{S} \subset \mathbb{R}^{\mathcal{S}}$ with the induced topology. It is easy to see that $\mathbb{R}^{\mathcal{S}}$ is then a locally convex topological vector space. For each $g \in \mathcal{S}$, find $K_g \in \mathbb{N}$ such that $-K_g 1 \leq g \leq K_g 1$. Then $\mathcal{S} \subset \prod_{g \in \mathcal{S}}[-K_g, K_g]$ and, using Tychonoff's theorem, we see that $\mathcal{S}$ is a compact convex subset of $\mathbb{R}^{\mathcal{S}}$. In addition, for any $g \in \mathcal{S}$, the map $\phi \mapsto \phi(g) : \mathcal{S} \rightarrow \mathbb{R}$ is affine and continuous. It is a well-known corollary of the Krein-Milman theorem (see pp. 81–86 of [2]) that an affine continuous map on a compact convex subset $\mathcal{S}$ of a locally convex topological vector space achieves its infimum at an extreme point of $\mathcal{S}$. So, if $\phi(g) > 0$ for all pure $\phi \in \mathcal{S}$ then the initial observation of the proof applies and shows $g > 0$. $\square$

REFERENCES

[1]   M. EINSIEDLER AND S. TUNCEL, When does a polynomial ideal contain a positive polynomial? To appear in J. Pure and Appl. Algebra.

[2]   K. GOODEARL, *Partially Ordered Abelian Groups with Interpolation*, Amer. Math. Soc., Providence, R.I., 1986.

[3]   D. HANDELMAN, Positive polynomials and product type actions of compact groups, *Mem. Amer. Math. Soc.* **320** (1985).

[4]   B. MARCUS AND S. TUNCEL, Matrices of polynomials, positivity, and finite equivalence of Markov chains, *J. Amer. Math. Soc.* **6** (1993), 131–147.

TOPOLOGICAL DYNAMICS OF CELLULAR AUTOMATA

PETR KŮRKA*

Abstract. This is an overview of some classical and recent results in topological dynamics of cellular automata on the space of twosided symbolic sequences. The concepts studied include surjectivity, transitivity, equicontinuity, closingness, openness, expansivity, attractors and the shadowing property.

Key words. Equicontinuity, sensitivity, transitity, expansivity.

AMS(MOS) subject classifications. Primary 54H20, 68D20.

1. Introduction. Cellular automata (CA) are dynamical systems with very rich and diversified behaviour. They are often used in physics and sciences as models of complicated behaviour. In computer science they serve as models of parallel processing and yield several complexity classes of languages.

Cellular automata have been introduced in the fifties by Ulam [28] and von Neumann [26]. The latter used them as models of self-reproduction and universal computation. A well-known example of a CA with surprising complexity and universal computation is Conway's Game of Life [14]. Deep mathematical theory of CA has been developed in Hedlund's pioneering paper [19]. Hedlund studied CA in the context of symbolic dynamics as homomorphisms of the shift dynamical system. Later, Wolfram [30, 31] studied computational aspects of CA. He performed extensive computer simulations and classified CA informally into four classes according to their behaviour on finite configurations. There have been several attempts to formalize Wolfram's classification, using computational, measure-theoretic and topological properties of CA, see e.g., Culik et al. [12], Gilman [16], Hurley [20] and Kůrka [21]. Simultaneously many results have been obtained interrelating these properties. While some classes of CA have been completely understood and some interesting examples have been thoroughly elucidated, many problems remain still open.

CA have been considered on many symbolic spaces: space of finite or periodic sequences, space of onesided or twosided infinite sequences, onesided or twosided mixing subshifts of finite type or configurations in n-dimensional grids. While these classes have some properties in common, there are also important differences. We concentrate here on one particular, if most familiar case, the space of twosided infinite symbolic sequences. With the product topology, this space is compact and metrizable and CA are continuous, so the concepts of topological dynamics apply. In the present study we disregard ergodic and computational aspects of CA

---

*Faculty of Mathematics and Physics, Charles University in Prague, Malostranské náměstí 25, CZ-118 00 Praha 1, CZECHIA.

and concentrate on topological dynamics. We present both classical and recent results and sometimes complete the classification with new theorems. The concepts studied include surjectivity, transitivity, equicontinuity, closingness, openness, expansivity, attractors and the shadowing property.

2. Topological dynamics. A **dynamical system** (DS) is a pair (X, F) where X is a compact metric space and $F : X \to X$ is a continuous map.

The n-th **iteration** of F is denoted by F^n, so $F^0 = \mathrm{Id}$ is the identity map and $F^{n+1} = F \circ F^n$.

A set $Y \subseteq X$ is **invariant**, if $F(Y) \subseteq Y$ and **strongly invariant** if $F(Y) = Y$. If Y is closed and invariant, (Y, F) is a DS which is called a **subsystem** of (X, F). The **orbit** of a point $x \in X$ is $\mathcal{O}(x) = \{F^n(x) : n > 0\}$. For every $x \in X$, the closure $\overline{\mathcal{O}(x)}$ of the orbit of x is a subsystem.

A **homomorphism** $\varphi : (X, F) \to (Y, G)$ is a continuous map $\varphi : X \to Y$ such that $\varphi \circ F = G \circ \varphi$.

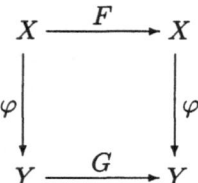

A **conjugacy** is a bijective homomorphism. The systems (X, F) and (Y, G) are **conjugate** if there exists a conjugacy between them. If φ is injective, $(\varphi(X), G)$ is a subsystem of (Y, G) as $\varphi(X) \subseteq Y$ is a closed invariant set. If φ is surjective, (Y, G) is a **factor** of (X, F).

A point $x \in X$ is **periodic** with period $n > 0$ if $F^n(x) = x$. If $F(x) = x$, x is a **fixed** point. A point $x \in X$ is **eventually periodic**, if $F^m(x)$ is periodic for some $m \geq 0$. If $m > 0$, x is **preperiodic** and m is its preperiod. A point is eventually periodic iff its orbit $\mathcal{O}(x)$ is finite.

A DS is **open**, if $F(U)$ is open for any open $U \subseteq X$.

A point $x \in X$ is a **transitive point**, if it has dense orbit, i.e., if $\overline{\mathcal{O}(x)} = X$.

A DS is **transitive**, if it has a transitive point. In this case the set of transitive points is **residual**, i.e., it contains a countable intersection of dense open sets. Equivalently, a DS is transitive, if for any nonempty open sets $U, V \subseteq X$ there exists $n > 0$ with $F^n(U) \cap V \neq \emptyset$.

A DS is **mixing**, if for any nonempty open sets $U, V \subseteq X$, $F^n(U) \cap V \neq \emptyset$ for all sufficiently large n. Every mixing DS is transitive.

A point $x \in X$ is an **equicontinuous point**, if

$$\forall \varepsilon > 0, \exists \delta > 0, \forall y \in B_\delta(x), \forall n \geq 0, d(F^n(y), F^n(x)) < \varepsilon .$$

Here d is the metric and $B_\delta(x) = \{y \in X : d(y, x) < \delta\}$ is the ball with center x and radius δ.

A DS is **equicontinuous**, if every its point is equicontinuous. In this case it is uniformly equicontinuous, i.e.,

$$\forall \varepsilon > 0, \exists \delta > 0, \forall x, y \in X, (d(x, y) < \delta \;\Rightarrow\; \forall n \geq 0, d(F^n(x), F^n(y)) < \varepsilon) \ .$$

A DS is **almost equicontinuous**, if the set of its equicontinuous points is residual.

A DS is **sensitive**, if

$$\exists \varepsilon > 0, \forall x \in X, \forall \delta > 0, \exists y \in B_\delta(x), \exists n \geq 0, d(F^n(y), F^n(x)) \geq \varepsilon \ .$$

Clearly, a sensitive system has no equicontinuous points. The converse is not true in general but holds in transitive systems. Every transitive system is either sensitive or almost equicontinuous (Akin et al. [2]). Every transitive almost equicontinuous system is **uniformly rigid** (Glasner and Weiss [18]). This means

$$\forall \varepsilon > 0, \exists n > 0, \forall x \in X, d(F^n(x), x) < \varepsilon \ .$$

We shall see that the dichotomy between sensitive and almost equicontinuous systems holds also for all CA. Moreover, transitive CA are sensitive.

A DS is **positively expansive**, if

$$\exists \varepsilon > 0, \forall x \neq y \in X, \exists n \geq 0, d(F^n(x), F^n(y)) \geq \varepsilon \ .$$

Every positively expansive DS on a perfect space (without isolated points) is sensitive.

A finite sequence $(x_i \in X)_{0 \leq i \leq n}$ is a **δ-chain** from x_0 to x_n, if

$$\forall i < n, d(F(x_i), x_{i+1}) < \delta.$$

A point x **ε-shadows** a sequence $(x_i)_{0 \leq i \leq n}$, if $d(F^i(x), x_i) < \varepsilon$ for all $0 \leq i \leq n$.

A DS has the **shadowing property** if for every $\varepsilon > 0$ there exists $\delta > 0$, such that every δ-chain is ε-shadowed by some point. It follows that every infinite δ-chain is ε-shadowed by some point.

A DS is **chain transitive** if for every $x, y \in X$ and for every $\varepsilon > 0$, there exists an ε-chain from x to y. Every transitive DS is chain transitive. Every chain transitive system with the shadowing property is transitive.

A point $x \in X$ is a **nonwandering point**, if for every open neighbourhood $U \subseteq X$ of x there exists $n > 0$ with $F^n(U) \cap U \neq \emptyset$. The set of nonwandering points is invariant and closed. A DS is **nonwandering**, if every its point is nonwandering.

The **omega limit** set of a closed invariant set $V \subseteq X$ is

$$\omega(V) = \bigcap_{n \geq 0} F^n(V) \ .$$

A set $Y \subseteq X$ is an **attractor** iff there exists a nonempty open set V such that $F(\overline{V}) \subseteq V$ and $Y = \omega(V)$. In a totally disconnected space, attractors are omega limit sets of clopen invariant sets. The **basin** of an attractor $Y \subseteq X$ is the set

$$B(Y) = \{x \in X; \ \lim_{n \to \infty} d(F^n(x), Y) = 0\}.$$

There exists always the largest attractor $\omega(X)$. The number of attractors is at most countable.

A periodic point $x \in X$ is **attracting** if its orbit $\mathcal{O}(x)$ is an attractor. Any attracting periodic point is equicontinuous.

A set $Y \subseteq X$ is a **minimal attractor**, if it is an attractor and no proper subset of Y is an attractor. An attractor is minimal iff it is chain transitive. A DS has unique attractor iff $\omega(X)$ is chain-transitive (Akin [1]).

A **quasi-attractor** is a countable intersection of attractors, which is not an attractor.

3. Subshifts. We call any finite set with at least two elements an **alphabet**. We frequently use alphabets $\mathbf{2} = \{0, 1\}$ and $\mathbf{3} = \{0, 1, 2\}$. The cardinality of a finite set A is denoted by $|A|$. A word over A is any finite sequence $u = u_0 \dots u_{n-1}$ of elements of A. The length of u is $|u| = n$. The **empty word** of length 0 is denoted by λ. The set of all words of length n is denoted by A^n, so $A^0 = \{\lambda\}$ and $A^1 = A$. The set of all nonzero words and the set of all words are

$$A^+ = \bigcup_{n > 0} A^n, \quad A^* = \bigcup_{n \geq 0} A^n.$$

A onesided infinite word over A is a map $x : \mathbb{N} \to A$, where $\mathbb{N} = \{0, 1, \dots\}$ is the set of non-negative integers. A twosided infinite word over A is a map $x : \mathbb{Z} \to A$, where $\mathbb{Z} = \{\dots, -1, 0, 1, \dots\}$ is the set of all integers. The position of the zeroth letter is denoted by a period, $x = \dots x_{-2} x_{-1} . x_0 x_1 \dots$. If u is a finite or infinite word and $I = \langle i, j \rangle$ is an interval of integers on which u is defined, put $u_{\langle i, j \rangle} = u_i \dots u_j$. Similarly for open or half-open integer intervals $u_{(i,j)} = u_{i+1} \dots u_{j-1}$. We say that v is a **subword** of u and write $v \sqsubseteq u$, if $v = u_I$ for some interval $I \subseteq \mathbb{Z}$.

If $u \in A^n$, then $u^\infty \in A^{\mathbb{Z}}$ is the infinite repetition of u, i.e., $(u^\infty)_{kn+i} = u_i$. If $v \in A^m$, then $x = v^\infty . u^\infty \in A^{\mathbb{Z}}$ is the word defined by

$$x_{km+i} = v_i \quad \text{for} \ k < 0, 0 \leq i < m$$
$$x_{kn+i} = u_i \quad \text{for} \ k \geq 0, 0 \leq i < n.$$

In $A^{\mathbb{N}}$ and $A^{\mathbb{Z}}$ we have metrics

$$d(x, y) = 2^{-n} \quad \text{where} \ n = \min\{i \geq 0 : x_i \neq y_i\}, \quad x, y \in A^{\mathbb{N}}$$

$d(x,y) = 2^{-n}$ where $n = \min\{i \geq 0 : x_i \neq y_i \vee x_{-i} \neq y_{-i}\}$, $x,y \in A^{\mathbb{Z}}$.

Both these metrics satisfy a stronger form of the triangle inequality

$$d(x,z) \leq \max\{d(x,y), d(y,z)\}.$$

Both $A^{\mathbb{N}}$ and $A^{\mathbb{Z}}$ are compact, perfect (do not contain isolated points) and totally disconnected, i.e., their clopen (closed and open) sets form a base of their topology. All compact, perfect and totally disconnected spaces are homeomorphic.

In $A^{\mathbb{N}}$ and $A^{\mathbb{Z}}$ the cylinder sets of a word $u \in A^n$ are

$$[u] = \{x \in A^{\mathbb{N}} : x_{\langle 0,n \rangle} = u\}$$

$$[u]_k = \{x \in A^{\mathbb{Z}} : x_{\langle k,k+n \rangle} = u\}, \ k \in \mathbb{Z}.$$

Cylinder sets are clopen and every clopen set is a finite union of cylinders.

The shift maps $\sigma : A^{\mathbb{N}} \to A^{\mathbb{N}}$ and $\sigma : A^{\mathbb{Z}} \to A^{\mathbb{Z}}$ are both defined by $\sigma(x)_i = x_{i+1}$ and they are continuous. Thus $(A^{\mathbb{Z}}, \sigma)$ and $(A^{\mathbb{N}}, \sigma)$ are DS which are called twosided and onesided **full shifts** respectively. While any twosided full shift is bijective, a onesided full shift is not; every point has $|A|$ preimages. Full shifts are mixing and have dense sets of periodic points.

A **onesided subshift** is any subsystem of a onesided full shift, so any closed set $\Sigma \subseteq A^{\mathbb{N}}$ with $\sigma(\Sigma) \subseteq \Sigma$. A DS on a totally disconnected space is conjugate to a onesided subshift iff it is positively expansive.

A **twosided subshift** is any closed set $\Sigma \subseteq A^{\mathbb{Z}}$ with $\sigma(\Sigma) = \Sigma$. We require strong invariance here, since the twosided subshift is bijective.

The language of a subshift Σ is the set of all subwords of points of Σ,

$$\mathcal{L}(\Sigma) = \{u \in A^* : \exists x \in \Sigma, u \sqsubseteq x\}.$$

A subshift Σ is a **subshift of finite type** (SFT) if there exists an integer $k \geq 2$, called its order, such that for every $x \in A^{\mathbb{N}}$, or for every $x \in A^{\mathbb{Z}}$,

$$x \in \Sigma \text{ iff } \forall n, x_{\langle n,n+k \rangle} \in \mathcal{L}(\Sigma).$$

A subshift is a SFT iff it has the shadowing property (Walters [29]). A onesided subshift is a SFT iff it is open (Parry [27]).

A subshift Σ is **sofic**, if it is a factor of some SFT.

Topological entropy of a subshift Σ is

$$h(\Sigma) = \lim_{n \to \infty} \frac{\ln P(n)}{n}$$

where $P(n) = |\mathcal{L}(\Sigma) \cap A^n|$ is the number of its words of length n.

4. Cellular automata.

DEFINITION 4.1. *A* **cellular automaton** *(CA) is a dynamical system* $(A^{\mathbb{Z}}, F)$, *such that* $F \circ \sigma = \sigma \circ F$.

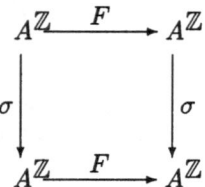

THEOREM 4.1 (Hedlund-Curtis-Lyndon [19]). *A dynamical system* $(A^{\mathbb{Z}}, F)$ *is a CA iff there exist integers* $m \leq a$ *(memory and anticipation) and a* **local rule** $f : A^{a-m+1} \to A$ *such that for every* $x \in A^{\mathbb{Z}}$,

$$F(x)_i = f(x_{\langle i+m, i+a \rangle}).$$

Proof. Let $(A^{\mathbb{Z}}, F)$ be a CA. For $\varepsilon = 1$ there exists $r \geq 0$ such that

$$d(x, y) < 2^{-r} \Rightarrow d(F(x), F(y)) < 1$$
$$x_{\langle -r, r \rangle} = y_{\langle -r, r \rangle} \Rightarrow F(x)_0 = F(y)_0 .$$

Thus there exists $f : A^{2r+1} \to A$ such that for every $x \in A^{\mathbb{Z}}$, $F(x)_0 = f(x_{\langle -r, r \rangle})$. Since F commutes with the shift,

$$F(x)_i = \sigma^i(F(x))_0 = F(\sigma^i(x))_0 = f(\sigma^i(x)_{\langle -r, r \rangle}) = f(x_{\langle i-r, i+r \rangle}) .$$

Thus we have a local rule with memory $m = -r$ and anticipation $a = r$. Conversely, if $F(x)_i = f(x_{\langle i+m, i+a \rangle})$ for some $m \leq a$, put $r = \max\{-m, a\}$. Since $-r \leq m \leq a \leq r$, for every $n \geq 0$,

$$d(x, y) < 2^{-n-r} \Rightarrow x_{\langle -n-r, n+r \rangle} = y_{\langle -n-r, n+r \rangle}$$
$$\Rightarrow x_{\langle -n+m, n+a \rangle} = y_{\langle -n+m, n+a \rangle}$$
$$\Rightarrow F(x)_{\langle -n, n \rangle} = F(y)_{\langle -n, n \rangle}$$
$$\Rightarrow d(F(x), F(y)) < 2^{-n}$$

so F is continuous and clearly commutes with the shift. $\square$

We can assume that the local rule is symmetric, so $F(x)_i = f(x_{\langle i-r, i+r \rangle})$, and $f : A^{2r+1} \to A$, where $r \geq \{-m, a\}$. Any r with this property is called a **radius** of F. Using a larger alphabet, it can be easily shown that any CA is conjugate to a CA with radius 1.

A **factor subshift** of a CA is any one-sided subshift which is its factor. If $I = \langle a, b \rangle$ is an integer interval, put $B = A^{b-a+1}$ and define a map $\varphi : A^{\mathbb{Z}} \to B^{\mathbb{N}}$ by $\varphi(x)_i = F^i(x)_{\langle a, b \rangle}$. The map φ is continuous and

$$\varphi(F(x))_i = F^{i+1}(x)_{\langle a, b \rangle} = \varphi(x)_{i+1} = \sigma(\varphi(x))_i .$$

Thus $\Sigma_I(F) = \varphi(A^{\mathbb{Z}}) \subseteq B^{\mathbb{N}}$ is a onesided subshift and $\varphi : (A^{\mathbb{Z}}, F) \rightarrow (\Sigma_I(F), \sigma)$ is a factor map

$$
\begin{array}{ccc}
A^{\mathbb{Z}} & \xrightarrow{\;\;F\;\;} & A^{\mathbb{Z}} \\
\varphi \downarrow & & \downarrow \varphi \\
\Sigma_I(F) & \xrightarrow{\;\;\sigma\;\;} & \Sigma_I(F)
\end{array}
$$

Let $d : A^{\mathbb{Z}} \rightarrow A^{\mathbb{Z}}$ be the homeomorphism defined by $d(x)_i = x_{-i}$. The **dual** of a CA $(A^{\mathbb{Z}}, F)$ is $(A^{\mathbb{Z}}, \overline{F})$, where $\overline{F} = d \circ F \circ d$. Clearly $d : (A^{\mathbb{Z}}, F) \rightarrow (A^{\mathbb{Z}}, \overline{F})$ is a conjugacy.

We show some typical examples of CA.

EXAMPLE 1 (P: product). $(2^{\mathbb{Z}}, P)$ *where* $P(x)_i = x_{i-1}x_ix_{i+1}$ *(Figure 1)*.   The radius is $r = 1$ and the local rule $f : 2^3 \rightarrow 2$ is defined by the table

000	001	010	011	100	101	110	111
0	0	0	0	0	0	0	1

FIG. 1. *P: the product CA.*

The behaviour of the product CA is displayed in the space time diagram in Figure 1. The n-th row represents a central part of the state at time n, i.e., the word $F^n(x)_{\langle -m, m \rangle}$, where $2m + 1$ is the width displayed. Ones are represented by black squares and zeros are left empty. We see that 0's propagate both to the left and to the right. If $x_i = 0$, then $F(x)_{\langle i-1, i+1 \rangle} = 000$ and $F^2(x)_{\langle i-2, i+2 \rangle} = 00000$. This implies that 0^∞ is an attracting fixed point. If $x_0 = 0$, i.e., if $d(x, 0^\infty) < 1$, then $d(F^n(x), 0^\infty) < 2^{-n}$. The CA has another fixed point 1^∞ which is not equicontinuous. If $d(x, 1^\infty) = 2^{-n} < 1$, then $d(F(x), 1^\infty) = 2^{-n+1}$. The product CA is not surjective:

$$
F(2^{\mathbb{Z}}) = \{x \in 2^{\mathbb{Z}} : 101 \not\sqsubseteq x \;\&\; 1001 \not\sqsubseteq x\}
$$
$$
F^n(2^{\mathbb{Z}}) = \{x \in 2^{\mathbb{Z}} : \forall m \in \langle 0, 2n \rangle, 10^m1 \not\sqsubseteq x\}
$$
$$
\omega(2^{\mathbb{Z}}) = \{x \in 2^{\mathbb{Z}} : \forall m > 0, 10^m1 \not\sqsubseteq x\}
$$

EXAMPLE 2 (M: majority). $(2^{\mathbb{Z}}, M)$ *where*

$$M(x)_i = \lfloor \frac{x_{i-1} + x_i + x_{i+1}}{2} \rfloor$$

000	001	010	011	100	101	110	111
0	0	0	1	0	1	1	1

FIG. 2. *M: the majority CA.*

The space time diagram of the majority CA is shown in Figure 2. We see that the cellular space is successively homogenized so that no isolated 0 or 1 remain. If $x_{\langle 0,1 \rangle} = 00$, then $F(x)_{\langle 0,1 \rangle} = 00$ no matter what value is in x_{-1} or x_2. Thus the cylinder set $[00]_0$, and every cylinder sets $[00]_i$ and $[11]_i$ are invariant. On the other hand, if $F(x)_{\langle -1,1 \rangle} = 010$, then necessarily $x_{\langle -2,2 \rangle} = 01010$. Indeed $x_{\langle -1,1 \rangle}$ contains at least two ones, but $x_{\langle 0,1 \rangle} = 11$ would imply $F(x)_{\langle 0,1 \rangle} = 11$. By induction we get

$$F^n(x)_{\langle -1,1 \rangle} = 010 \Rightarrow x_{\langle -n,n \rangle} = 01 \ldots 10$$
$$F^n(x)_{\langle -1,1 \rangle} = 101 \Rightarrow x_{\langle -n,n \rangle} = 10 \ldots 01$$

The clopen set $V = [00]_0 \cup [11]_0$ is invariant and

$$x \in V \Rightarrow \forall i \in \langle -n, n+1 \rangle, F^n(x)_{\langle i-1,i+1 \rangle} \notin \{010, 101\}$$

so

$$\omega(V) = \{x \in 2^{\mathbb{Z}} : \forall i \in \mathbb{Z} : x_{\langle i-1,i+1 \rangle} \notin \{010, 101\}\}$$

is a SFT whose every point is fixed. The set $\omega(V)$ is, however, not the largest attractor, since $\omega(A^{\mathbb{Z}})$ contains e.g., periodic points $(01)^{\infty}$, $(10)^{\infty}$. There are also many smaller attractors. If $u \in \mathcal{L}(\omega(V))$, i.e., if u does not contain neither 010 nor 101, then $[u]_i$ is an invariant set and $\omega([u]_i) = \omega(V) \cap [u]_i$ is an attractor.

EXAMPLE 3 (S: sum). $(2^{\mathbb{Z}}, S)$, *where* $S(x)_i = x_{i-1} + x_{i+1} \bmod 2$.

000	001	010	011	100	101	110	111
0	1	0	1	1	0	1	0

FIG. 3. *S: the sum CA.*

There is a formula for the n-th iteration. We regard **2** as a group with addition modulo 2.

$$S^2(x)_i = x_{i-2} + 2x_i + x_{i+2}$$

$$S^n(x)_i = \sum_{j=0}^{n} \binom{n}{j} x_{i-n+2j} \ .$$

Since $2x = 0$, we count only odd binomial coefficients. In particular for $n = 2^m$, all but two binomial coefficients are even, so

$$S^{2^m}(x)_i = x_{i-2^m} + x_{i+2^m} \ .$$

If the initial state is $0^\infty.10^\infty$ (Figure 3 top), then $F^{2^m}(x)$ contains exactly two 1's, one at 2^{-m} and another at 2^m. If the initial state is random (Figure 3 bottom), we see no pattern in the time development. We show that the sum CA is conjugate to the full shift on four symbols. In fact $\Sigma_{\langle 0,1 \rangle}(S) = B^{\mathbb{N}}$ where $B = A^2 = \{00, 01, 10, 11\}$ and $\varphi : (\mathbf{2}^{\mathbb{Z}}, S) \to (B^{\mathbb{N}}, \sigma)$ is a conjugacy. Given a point $y \in B^{\mathbb{N}}$, let us search a point $x \in A^{\mathbb{Z}}$ such that for every $n \geq 0$, $S^n(x)_{\langle 0,1 \rangle} = y_n$. We have

$$S^n(x)_0 + S^n(x)_2 = S^{n+1}(x)_1$$
$$S^n(x)_2 = S^n(x)_0 + S^{n+1}(x)_1$$

so the second column $(S^n(x)_2)_{n \geq 0}$ can be computed from the zeroth and the first columns. Similarly we compute the third and also (-1)-st columns. This shows that the map $\varphi : \mathbf{2}^{\mathbb{Z}} \to B^{\mathbb{N}}$ is bijective. Since the sum CA is conjugate to a full shift, it is mixing, surjective, positively expansive and has a dense set of periodic points.

In general, every CA has a periodic point and the eventually periodic points are dense in it.

PROPOSITION 4.1. *Every σ-periodic point of a CA $(A^{\mathbb{Z}}, F)$ is F-eventually periodic. Hence the set of eventually periodic points is dense.*

Proof. If $\sigma^n(x) = x$, then $\sigma^n(F(x)) = F(x)$. Since σ-periodic points with period n are in one-to-one correspondence with words of A^n, there is a finite number of them and for some preperiod $m \geq 0$ and period $p > 0$, $F^{m+p}(x) = F^m(x)$. $\square$

5. Equicontinuity. Almost equicontinuous systems are characterized by the presence of blocking words. In an equicontinuous system, every long enough word is blocking.

DEFINITION 5.1. *Let $s > 0$. A word $u \in A^+$ with $|u| \geq s$ is an s-blocking word for a CA $(A^{\mathbb{Z}}, F)$, if there exists an offset $p \in \langle 0, |u| - s \rangle$, such that*

$$\forall x, y \in [u]_0, \forall n \geq 0, F^n(x)_{\langle p, p+s \rangle} = F^n(y)_{\langle p, p+s \rangle} \ .$$

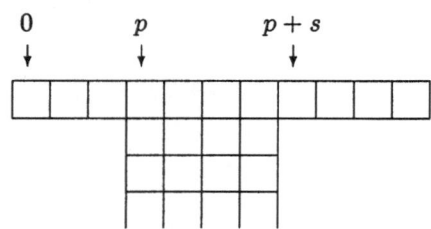

FIG. 4. *A blocking word.*

In the product CA of Example 1, 0 is a 1-blocking word (with offset 0), since $x_0 = 0$ implies $F^n(x)_0 = 0$ for all $n \geq 0$. In the majority CA of Example 2, 00 and 11 are 2-blocking words (with offset 0). Since $[00]_0$ is invariant, $x_{\langle 0,1 \rangle} = 00$ implies $F^n(x)_{\langle 0,1 \rangle} = 00$ for all $n \geq 0$.

THEOREM 5.1 (Kůrka [21]). *Let $(A^{\mathbb{Z}}, F)$ be a CA with radius $r > 0$. The following conditions are equivalent.*

1. $(A^{\mathbb{Z}}, F)$ is not sensitive.

2. $(A^{\mathbb{Z}}, F)$ has an r-blocking word.

3. $(A^{\mathbb{Z}}, F)$ is almost equicontinuous.

Proof. $1 \Rightarrow 2$: Suppose that F is not sensitive. Let m be an integer with $2m + 1 \geq r$. For $\varepsilon = 2^{-m}$ there exists $x \in A^{\mathbb{Z}}$ and $\delta = 2^{-m-p}$, $p \geq 0$, such that for all $y \in A^{\mathbb{Z}}$,

$$d(x, y) < \delta \ \Rightarrow \ \forall k \geq 0, d(F^k(x), F^k(y)) < \varepsilon \ .$$

Put $u = x_{\langle -m-p, m+p \rangle} \in A^{2m+2p+1}$. Then

$$y, z \in [u]_{-m-p} \Rightarrow \forall k, F^k(y)_{\langle -m,m \rangle} = F^k(z)_{\langle -m,m \rangle}$$

$$\Rightarrow \forall k, F^k(y)_{\langle -m,-m+r \rangle} = F^k(z)_{\langle -m,-m+r \rangle}$$

$$y, z \in [u]_0 \Rightarrow \forall k, F^k(y)_{\langle p,p+r \rangle} = F^k(z)_{\langle p,p+r \rangle}$$

$2 \Rightarrow 3$: Let $u \in A^q$ be an r-blocking word with offset $p \le q - r$. Let $Y_k \subseteq A^{\mathbb{Z}}$ be the set of points which contain at least k occurrences of u in both its positive and negative parts.

$$Y_k = \{y \in A^{\mathbb{Z}} : |\{j \le 0 : y_{\langle j,j+q \rangle} = u\}| \ge k \,\&$$
$$|\{j \ge 0 : y_{\langle j,j+q \rangle} = u\}| \ge k\}$$

Y_k is a dense open set, so $Y = \cap_{k>0} Y_k$ is residual. We show that every point of Y is equicontinuous. Let $x \in Y$ and $\varepsilon = 2^{-j}$. There exist $m \le -j - p$ and $n \ge j - p - r$ such that $x_{\langle m,m+q \rangle} = u = x_{\langle n,n+q \rangle}$.

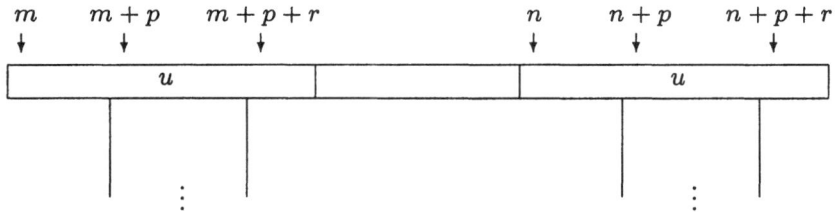

Put $\delta = \min\{2^m, 2^{-n-q}\}$ and suppose that $d(y, x) < \delta$. Then $x_{\langle m,n+q \rangle} = y_{\langle m,n+q \rangle}$. For $m + p + r < i < n + p$ we get

$$F(x)_i = f(x_{\langle i-r,i+r \rangle}) = f(y_{\langle i-r,i+r \rangle}) = F(y)_i$$

and by induction $F^k(x)_{\langle m+p,n+p+r \rangle} = F^k(y)_{\langle m+p,n+p+r \rangle}$ for every $k \ge 0$. Since $m + p \le -j < j \le n + p + r$, $d(F^k(x), F^k(y)) < 2^{-j}$. $3 \Rightarrow 1$ is clear. $\square$

In the product CA of Example 1, 0^∞ is an equicontinuous point but 1^∞ is not. In the majority CA, both 0^∞ and 1^∞ are equicontinuous but $(01)^\infty$ is not. Both product and majority CA are almost equicontinuous but not equicontinuous. The sum CA of Example 3 is sensitive as it is conjugate to a full shift. There are two trivial examples of equicontinuous CA:

EXAMPLE 4 (I: identity). $(A^{\mathbb{Z}}, I)$, where A is any alphabet and $I(x) = x$, is an equicontinuous CA.

EXAMPLE 5 (O: zero). $(A^{\mathbb{Z}}, O)$, where A is any alphabet, $a \in A$ and $O(x) = a^\infty$ for all $x \in A^{\mathbb{Z}}$, is an equicontinuous CA.

PROPOSITION 5.1. Every CA with radius zero is equicontinuous.

Proof. If $r = 0$, then the local rule is a map $f : A \to A$ and $F(x)_i = f(x_i)$. Thus

$$x_{\langle -n,n \rangle} = y_{\langle -n,n \rangle} \Rightarrow F(x)_{\langle -n,n \rangle} = F(y)_{\langle -n,n \rangle}$$
$$d(x, y) < 2^{-n} \Rightarrow d(F(x), F(y)) < 2^{-n}$$

□

However, not all equicontinuous CA have radius zero.

EXAMPLE 6 (E: An equicontinuous CA). $(2^{\mathbb{Z}}, E)$, where

$$E(x)_i = x_i + x_{i-1}x_{i+1} \bmod 2$$

000	001	010	011	100	101	110	111
0	0	1	1	0	1	1	0

FIG. 5. *E: An equicontinuous CA.*

We show first that 01110 is a 3-blocking word with offset 1. Considering all possibilities for the left and right extensions of 01110, we get

$$
\begin{array}{llll}
0011100 & 0011101 & 1011100 & 1011101 \\
0010100 & 0010111 & 1110100 & 1110111 \\
0011100 & 001110 & 011100 & 01110
\end{array}
$$

Thus $F^2[01110] \subseteq [01110]$, so if $x \in [01110]_{-2}$, then $F^{2n}(x)_{\langle -1,1 \rangle} = 111$ and $F^{2n+1}(x)_{\langle -1,1 \rangle} = 101$. By inspection we verify that every word in the set $B = \{00, 1111, 01110, 0110110, 010110, 011010, 10101\}$ is 2- blocking.

$$
\begin{array}{lllllll}
00 & 1111 & 01110 & 0110110 & 010110 & 011010 & 10101 \\
00 & 00 & 101 & 11111 & 1111 & 1111 & 11111 \\
00 & 00 & 01110 & 000 & 00 & 00 & 0000
\end{array}
$$

Every word v of length at least 10 contains at least one occurrence of some word from B, so it is also 2-blocking. It follows that the CA is equicontinuous.

THEOREM 5.2 (Kůrka [21]). *A CA $(A^{\mathbb{Z}}, F)$ is equicontinuous iff there exists a preperiod $m \geq 0$ and a period $p > 0$, such that $F^{m+p} = F^m$.*

Proof. For $\varepsilon = 1$ there exists $\delta = 2^{-k}$ such that for all $x, y \in A^{\mathbb{Z}}$, if $x_{\langle -k,k \rangle} = y_{\langle -k,k \rangle}$, then $F^n(x)_0 = F^n(y)_0$ for all $n > 0$. Let $u \in A^{2k+1}$ and let $x \in [u]_{-k}$ be the σ-periodic point with period $2k+1$. Then $(F^n(x)_0)_{n \geq 0}$ is an eventually periodic sequence with some preperiod $m_u \geq 0$ and period $p_u > 0$. For every $y \in [u]_{-k}$, $(F^n(y)_0)_{n \geq 0}$ has preperiod m_u and period p_u. Put

$$m = \max\{m_u : u \in A^{2k+1}\}, \quad p = \mathrm{lcm}\{p_u : u \in A^{2k+1}\} .$$

Here lcm is the least common multiple. For every $x \in A^{\mathbb{Z}}$, $F^m(x)_0 = F^{m+p}(x)_0$, so

$$F^m(x)_i = F^m(\sigma^i(x))_0 = F^{m+p}(\sigma^i(x))_0 = F^{m+p}(x)_i$$

and therefore $F^{m+p}(x) = F^m(x)$.

Conversely, suppose that there exist $m \geq 0$ and $p > 0$ such that for every x $F^{m+p}(x) = F^m(x)$. For $\varepsilon = 2^{-k}$ put $\delta = 2^{-k-r(m+p)}$. If $d(x,y) < \delta$, then

$$d(F(x), F(y)) < 2^{-k-r(m+p-1)} < 2^{-k}$$
$$i \leq m + p \quad \Rightarrow \quad d(F^i(x), F^i(y)) < 2^{-k-r(m+p-i)} \leq 2^{-k}$$

Since both $F^i(x)_{\langle -k,k \rangle}$ and $F^i(y)_{\langle -k,k \rangle}$ are eventually periodic with preperiod m and period p, and since the first $m + p$ their elements are equal, we get $d(F^n(x), F^n(y)) < 2^{-k}$ for every $n \geq 0$. $\square$

In Example 6, the preperiod and period are $m = p = 2$.

THEOREM 5.3. Let $(A^{\mathbb{Z}}, F)$ be a CA and $\Sigma \subseteq A^{\mathbb{Z}}$ an F- invariant subshift. If (Σ, F) is transitive then it is either sensitive or consists of a single periodic orbit.

Proof. Suppose that (Σ, F) is transitive and not sensitive. By a theorem of Glasner and Weiss [18], it is uniformly rigid, so for $\varepsilon = 1$ there exists n such that for all $x \in \Sigma$, $d(F^n(x), x) < 1$, i.e., $F^n(x)_0 = x_0$. For $y = \sigma^i(x)$ we get

$$F^n(x)_i = \sigma^i(F^n(x))_0 = F^n(\sigma^i(x))_0 = \sigma^i(x)_0 = x_i .$$

Thus F^n is identity and since (Σ, F) is transitive, it consists of a single periodic orbit. $\square$

COROLLARY 5.1 (Kůrka [21]). Every transitive CA is sensitive. The sum CA of Example 3 is transitive, and therefore sensitive.

EXAMPLE 7. $(2^{\mathbb{Z}} \times 2^{\mathbb{Z}}, \text{Id} \times \sigma)$ (i.e., $F(x,y)_i = (x_i, y_{i+1})$) is a surjective and sensitive CA which is not transitive.

6. Surjectivity. Let $(A^{\mathbb{Z}}, F)$ be a CA with radius $r \geq 0$ and a local rule $f : A^{2r+1} \to A$. We extend it to a function $f : A^* \to A^*$ by

$$f(u)_i = \begin{cases} f(u_{\langle i, i+2r \rangle}) & \text{if} \quad 0 \leq i < |u| - 2r \\ \lambda & \text{if} \quad i \geq |u| - 2r \end{cases} .$$

Thus $|f(u)| = \max\{0, |u| - 2r\}$. For example the successive images of a word 01100011 of the sum CA of Example 3 are

$$01100011 \mapsto 111011 \mapsto 0101 \mapsto 00 \mapsto \lambda \mapsto \lambda .$$

Observe that for every $n \geq 0$,

$$\sum_{u \in A^n} |f^{-1}(u)| = |A^{n+2r}| = |A|^{n+2r} .$$

The mean number of preimages of a word of length n is $|A|^{2r}$. We show that a CA is surjective iff every nonempty word has exactly $|A|^{2r}$ preimages.

THEOREM 6.1 (Hedlund [19]). *Let* $(A^{\mathbb{Z}}, F)$ *be a CA with local rule* $f :$ $A^{2r+1} \to A$. *Then* F *is surjective iff for every* $u \in A^+$, $|f^{-1}(u)| = |A|^{2r}$.

Proof. Let the condition be satisfied and $y \in A^{\mathbb{Z}}$. For $n \geq 0$ put

$$X_n = \{x \in A^{\mathbb{Z}} : f(x_{\langle -n-r, r+n \rangle}) = y_{\langle -n, n \rangle}\} .$$

By the assumption every X_n is nonempty, closed and $X_{n+1} \subseteq X_n$. By compactness there exists $x \in \bigcap_{n>0} X_n$ and $F(x) = y$, so F is surjective. Conversely, suppose that F is surjective and put

$$p = \min\{|f^{-1}(u)| : u \in A^+\} .$$

Since F is surjective, every word has at least one preimage, so $p > 0$. Indeed, if $y \in A^{\mathbb{Z}}$ contains u, then every preimage of y contains some preimage of u. We show

LEMMA 6.1. *If* $|f^{-1}(u)| = p$, *then for every* $a \in A$, $|f^{-1}(ua)| = p$.

Proof. By the assumption, for every $a \in A$, $f^{-1}(ua) \geq p$. Suppose that for some $a \in A$, $|f^{-1}(ua)| > p$. We have disjoint unions

$$\bigcup_{b \in A} \{vb : v \in f^{-1}(u)\} = \bigcup_{a \in A} f^{-1}(ua)$$

$$p|A| = \left| \bigcup_{b \in A} \{vb : v \in f^{-1}(u)\} \right| = \left| \bigcup_{a \in A} f^{-1}(ua) \right| > p|A|$$

and this is a contradiction. This proves Lemma 6.1. □

LEMMA 6.2. $p = |A|^{2r}$.

Proof. Let $u \in A^n$ be such that $|f^{-1}(u)| = p$. By Lemma 1, we have $|f^{-1}(uw)| = p$ for every $w \in A^*$. We have a disjoint union

$$\{vw : v, w \in f^{-1}(u)\} = \bigcup_{z \in A^{2r}} f^{-1}(uzu)$$

$$p^2 = |\{vw : v, w \in f^{-1}(u)\}| = \left| \bigcup_{z \in A^{2r}} f^{-1}(uzu) \right| = p|A|^{2r}$$

v		w	
u	z	u	

so $p = |A|^{2r}$. This proves Lemma 6.2. □

We finish now the proof of the theorem. Suppose that for some $n > 0$ and some $u \in A^n$, $|f^{-1}(u)| > |A|^{2r}$. Then

$$|A|^{n+2r} = \left| \bigcup_{v \in A^n} f^{-1}(v) \right| > |A|^n \cdot |A|^{2r}$$

and this is a contradiction. Thus for every $u \in A^*$, $|f^{-1}(u)| = |A|^{2r}$. $\square$

The condition $|f^{-1}(u)| = |A|^{2r}$ may be satisfied for $|u| = 1$ and fail for longer u. For example in the majority CA of Example 2, the condition works for $|u| = 1$ but does not work for $|u| = 2$, since there are six preimages of 00:

$$f^{-1}(00) = \{0000, 0001, 0010, 0100, 1000, 1001\} \,.$$

THEOREM 6.2 (Hedlund [19]). *Let $(A^{\mathbb{Z}}, F)$ be a CA with radius $r > 0$ which is not surjective. Then*
1. *There exists $w \in A^{2r}$ and distinct $u, v \in A^+$ with $|u| = |v|$ and $f(wuw) = f(wvw)$.*
2. *There exists a point $x \in A^{\mathbb{Z}}$ with an infinite (continuum) number of preimages.*

Proof. 1. By Theorem 6.1 there exists a word $z \in A^+$ such that $|f^{-1}(z)| = k > |A|^{2r}$. Choose a word $w \in f^{-1}(z)$, so $|w| = |z| + 2r \geq 2r$. For $m > 0$ consider sets

$$M_m = \{ww_0w_1 \ldots w_{m-1}w : w_i \in f^{-1}(z)\},$$
$$N_m = \{zv_0z \ldots zv_mz : v_i \in A^{2r}\}$$

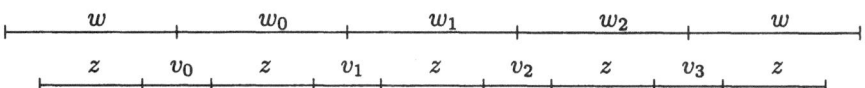

We have $f(M_m) \subseteq N_m$, $|M_m| = k^m$, $|N_m| = |A|^{2r(m+1)}$. Since $k > |A|^{2r}$, for m large enough, $k^m > |A|^{2r(m+1)}$, so there exist distinct $u', v' \in M_n$ with $f(u') = f(v')$. Since both u' and v' begin and end with w of length at least $2r$, the statement follows.
2. Let $w \in A^{2r}$, $u, v \in A^n$ be such that $u \neq v$ and $f(wuw) = f(wvw)$. Put $M = \{wu, wv\}^{\mathbb{Z}}$, i.e., $x \in M$ iff for all $i \in \mathbb{Z}$, $x_{\langle i(2r+n), i(2r+n)+2r \rangle} = w$ and $x_{\langle i(2r+n)+2r, (i+1)(2r+n)\rangle} \in \{u, v\}$. Then M is uncountable and all its elements have the same image. $\square$

A pair of words wuw, wvw with $f(wuw) = f(wvw)$ is called a diamond.

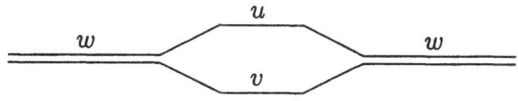

FIG. 6. *A diamond.*

COROLLARY 6.1. *Every injective CA is surjective and hence bijective.*

THEOREM 6.3. *Every surjective CA is nonwandering.*

Proof. We use the Poincaré recurrence theorem: Let $(A^{\mathbb{Z}}, F)$ be a CA with local rule $f : A^{2r+1} \to A$. For every clopen set U define its measure $\mu(U)$ as follows. If $u \in A^n$ and $k \in \mathbb{Z}$, then $\mu([u]_k) = |A|^{-n}$. If $U = V_1 \cup \cdots \cup V_m$ is a disjoint union of cylinders, put $\mu(U) = \mu(V_1) + \cdots + \mu(V_m)$. Clearly $\mu(A^{\mathbb{Z}}) = 1$. If $u \in A^n$, then $f^{-1}(u) \subseteq A^{n+2r}$ and $|f^{-1}(u)| = |A|^{2r}$, so

$$\mu(F^{-1}([u]_k)) = |A|^{2r} \cdot |A|^{-n-2r} = \mu([u]_k)$$

so for every clopen set U, $\mu(F^{-1}(U)) = \mu(U)$.

Let U be a clopen set and suppose that all $F^{-i}(U)$ are pairwise disjoint. Then for every $n > 0$,

$$n \cdot \mu(U) = \mu(U \cup \cdots \cup F^{-n+1}(U)) \leq \mu(A^{\mathbb{Z}}) = 1$$

and this is a contradiction. Thus for some $i < j$, $F^{-i}(U) \cap F^{-j}(U) \neq \emptyset$ and $U \cap F^{-j+i}(U) \neq \emptyset$. $\square$

THEOREM 6.4 (Blanchard and Tisseur [6]). *Any surjective almost equicontinuous CA has a dense set of periodic points.*

Proof. Let $u \in A^m$ be arbitrary and let $v \in A^p$ be an r-blocking word with offset j. Then $w = vuv$ is a q-blocking word with offset j, where $q = p + m + r$. Since $(A^{\mathbb{Z}}, F)$ is nonwandering, there exists $t > 0$ and $x \in [w]_0 \cap F^{-t}([w]_0)$. Put $y = (vu)^{\infty}$. Since $x, F^t(x), y \in [w]_0$,

$$F^t(y)_{\langle j, j+q \rangle} = F^t(x)_{\langle j, j+q \rangle} = x_{\langle j, j+q \rangle} = y_{\langle j, j+q \rangle} \ .$$

Since y and $F^t(y)$ are σ-periodic with period $p + m \leq q$, $F^t(y) = y$ and $y \in [u]_p$ is F-periodic with period t. $\square$

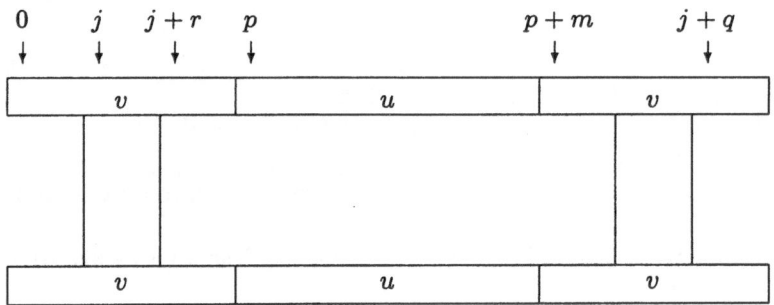

THEOREM 6.5 (Blanchard and Tisseur [6]). *If $(A^{\mathbb{Z}}, F)$ is a surjective equicontinuous CA, then there exists $p > 0$ such that $F^p = \mathrm{Id}$. In particular, F is bijective.*

Proof. By Theorem 5.2, there exists $m \geq 0$ and $p > 0$ such that $F^m = F^{p+m}$. Since every point is eventually periodic and nonwandering, it is periodic and $m = 0$. $\square$

7. Preimages. We study now the number of preimages of points in surjective CA's.

THEOREM 7.1 (Hedlund [19]). *Let* $(A^{\mathbb{Z}}, F)$ *be a surjective CA with radius* r. *Then for every* $y \in A^{\mathbb{Z}}$, $|F^{-1}(y)| \leq |A|^{2r}$.

Proof. Suppose that $x_0, x_1 \ldots x_{n-1}$ are distinct elements in $F^{-1}(y)$ and $n > |A|^{2r}$. There exists $m > r$, such that for all $i \neq j$, $(x_i)_{\langle -m,m \rangle} \neq (x_j)_{\langle -m,m \rangle}$, so $y_{\langle -m+r,m-r \rangle}$ has at least n preimages and this is a contradiction. $\square$

Although every finite word has exactly $|A|^{2r}$ preimages, infinite words may have strictly fewer preimages.

EXAMPLE 8 (L: A closing CA, Boyle et al. [7]). $(\mathbf{3}^{\mathbb{Z}}, L)$, *where*

$$L(x)_i = \begin{cases} x_i + x_{i+1} \bmod 2 & if \quad x_i \neq 2 \\ 2 & if \quad x_i = 2 \end{cases}$$

is a surjective CA with nonconstant number of preimages.

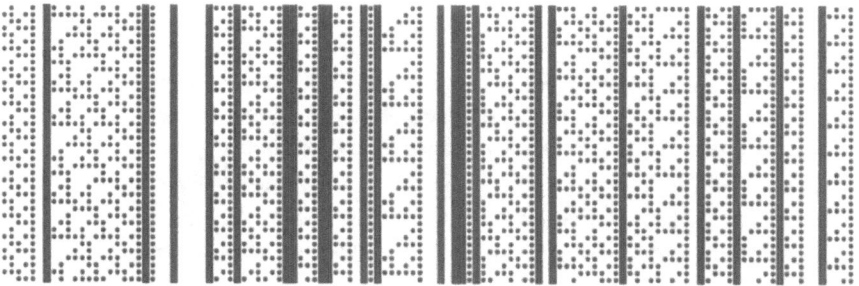

FIG. 7. *L: A closing CA.*

We will show later that $(\mathbf{3}^{\mathbb{Z}}, L)$ is a closing CA. In Figure 7, two's are represented by a black square, ones by a dot and zeros by empty space. While the only preimage of 2^{∞} is 2^{∞} itself, 1^{∞} has two preimages $(01)^{\infty}$ and $(10)^{\infty}$. We show that L is surjective. The radius is $r = 1$ so we show that every $v \in \mathbf{2}^n$ has exactly $|A|^{2r} = 9$ preimages. For every $u_{-1}, u_n \in A$ put successively for $i = n - 1, \ldots, 0$

$$u_i = \begin{cases} v_i + u_{i+1} \bmod 2 & if \quad v_i \neq 2 \\ 2 & if \quad v_i = 2 \end{cases} .$$

Then $f(u_{\langle -1,n \rangle}) = v_{\langle 0,n \rangle}$, so every pair $(u_{-1}, u_n) \in \mathbf{3}^2$ determines a distinct preimage and $|f^{-1}(v)| = |A|^2 = 9$.

We are going to characterize CA with constant number of preimages.

DEFINITION 7.1. *Let* $x, y \in A^{\mathbb{Z}}$ *and* $s \geq 0$.
1. x, y *are totally* s-*separated if* $\forall n \in \mathbb{Z}, x_{\langle n,n+s \rangle} \neq y_{\langle n,n+s \rangle}$.

2. x, y are left (right) s-separated if there exists $m \in \mathbb{Z}$, such that $x_{\langle n,n+s \rangle} \neq y_{\langle n,n+s \rangle}$ for all $n \leq m$ (for all $n \geq m$).

3. x, y are left (right) asymptotic, if there exists $m \in \mathbb{Z}$ such that $x_n = y_n$ for all $n \leq m$ (for all $n \geq m$).

4. A point $z \in A^{\mathbb{Z}}$ is σ-transitive, iff for every open $U \subseteq A^{\mathbb{Z}}$ there exist $n < 0 < m$ such that $\sigma^n(z), \sigma^m(z) \in U$.

LEMMA 7.1. Let $(A^{\mathbb{Z}}, F)$ be a surjective CA with radius r, let $x, y \in A^{\mathbb{Z}}$, $x \neq y$ be such that $F(x) = F(y)$ is a σ-transitive point. Then exactly one of the following three statements is true.

1. x and y are left $2r$-separated and right asymptotic.
2. x and y are both left and right $2r$-separated.
3. x and y are left asymptotic and right $2r$-separated.

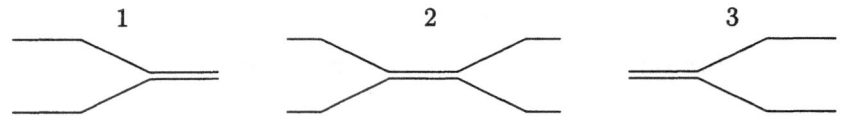

Proof. Assume that x and y are not right $2r$-separated, and that the statement (1) does not hold. Since $x \neq y$ there exists p with $x_p \neq y_p$. Since x and y are not right $2r$-separated, there exists $m > p$ such that $x_{\langle m,m+2r \rangle} = y_{\langle m,m+2r \rangle}$. Since (1) does not hold, there are two possibilities:

a. x and y are not left $2r$-separated, so there exists $n < p$ with $x_{\langle n,n+2r \rangle} = y_{\langle n,n+2r \rangle}$. In this case $u = x_{\langle n,m+2r \rangle}$, $v = y_{\langle n,m+2r \rangle}$ form a diamond and every point in $\{u,v\}^{\mathbb{Z}}$ would have the same image.

b. x and y are not right asymptotic. In this case there exists $q > m$ with $x_q \neq y_q$ and since x and y are not right $2r$- separated there exists $n > q$ with $x_{\langle n,n+2r \rangle} = y_{\langle n,n+2r \rangle}$. Again, $u = x_{\langle n,n+2r \rangle}$ and $v = y_{\langle m,n+2r \rangle}$ would form a diamond and this is not possible. We have proved that if x and y are not right $2r$-separated, then (1) holds. Similarly, if x and y are not left $2r$-separated then (3) holds. $\square$

LEMMA 7.2. Let $(A^{\mathbb{Z}}, F)$ be a surjective CA with radius r, $z \in A^{\mathbb{Z}}$ a σ-transitive point, and $x_1, \ldots x_k \in F^{-1}(z)$ pairwise right (left) $2r$-separated points. Then there exist $y_1, \ldots, y_k \in F^{-1}(z)$ which are pairwise totally $2r$-separated.

Proof. By the assumption there exists $p \in \mathbb{Z}$ such that for all $n \geq p$ and for every distinct $i, j \leq k$, $(x_i)_{\langle n,n+2r \rangle} \neq (x_j)_{\langle n,n+2r \rangle}$. For every $q > 0$ there exists $n_q > p + q$ such that $d(z, \sigma^{n_q}(z)) < 2^{-q}$. Choosing a subsequence if necessary, we can assume that for every $i \leq k$, there exist limits

$$\lim_{q \to \infty} \sigma^{n_q}(x_i) = y_i$$

Clearly $F(y_i) = z$. We show that $(y_i)_{i \leq k}$ are totally $2r$-separated. For $m \in \mathbb{Z}$ put $s = \max\{-m, m + 2r\}$. There exists $q \geq s$ such that for every $i \leq k$, $d(\sigma^{n_q}(x_i), y_i) < 2^{-s}$. Since $-s \leq m \leq m + 2r \leq s$,

$$(y_i)_{\langle m, m+2r \rangle} = \sigma^{n_q}(x_i)_{\langle m, m+2r \rangle} \neq \sigma^{n_q}(x_j)_{\langle m, m+2r \rangle} = (y_j)_{\langle m, m+2r \rangle}$$

so $(y_i)_{i \leq k}$ are totally $2r$-separated. $\square$

PROPOSITION 7.1 (Hedlund [19]). *Let $(A^{\mathbb{Z}}, F)$ be a surjective CA and $z \in A^{\mathbb{Z}}$ a σ-transitive point. Then its preimages are pairwise totally $2r$-separated.*

Proof. Let $k_0 > 0$ be the maximum number such that there exists points $x_1, \ldots, x_{k_0} \in F^{-1}(z)$, which are pairwise totally $2r$-separated. Assume that $F(y) = z$ and that y is different from all x_i. If y were right $2r$-separated from all x_i, by Lemma 7.2 there would exist $k_0 + 1$ pairwise totally $2r$-separated preimages of z. Thus there exists $j \leq k_0$ such that y and x_j are not right $2r$-separated and by Lemma 7.1 they are right asymptotic. Similarly we prove that there exists $l \leq k_0$ such that y and x_l are left asymptotic. Let $x_{k_0+1}, \ldots, x_{k_1}$ be all preimages of z different from $x_1, \ldots, x_{k_0}$. There exists $p > 0$ such that

$$\forall i \in (k_0, k_1), \exists j \leq k_0, (x_i)_{\langle p, \infty)} = (x_j)_{\langle p, \infty)}$$
$$\forall i \in (k_0, k_1), \exists l \leq k_0, (x_i)_{(-\infty, -p+2r\rangle} = (x_l)_{(-\infty, -p+2r\rangle}$$

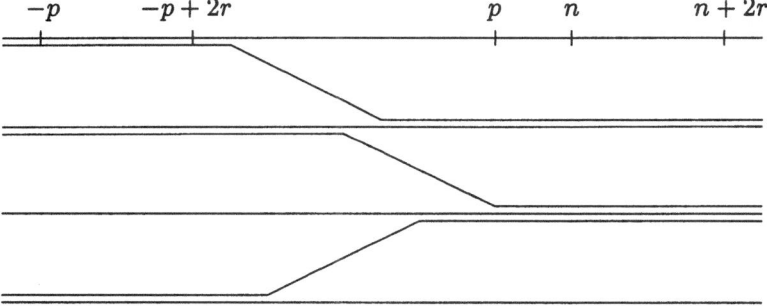

For every $q > 0$ there exists $n_q > p + q$ such that $d(\sigma^{-p}(z), \sigma^{n_q}(z)) < 2^{-q}$. By choosing a subsequence if necessary, we can assume that there exist limits

$$\lim_{q \to \infty} \sigma^{n_q}(x_i) = y_i, \ \ 1 \leq i \leq q_1$$

so $F(y_i) = \sigma^{-p}(z)$. Since $F^{-1}(\sigma^{-p}(z)) = \{\sigma^{-p}(x_i) : 1 \leq i \leq k_1\}$, for every $i \leq k_1$ there exists $j_i \leq k_1$ such that $y_i = \sigma^{-p}(x_{j_i})$. There exists $n > p$ such that for all $i \leq k_1$, $d(\sigma^{n+r}(x_i), y_i) < 2^{-r}$, so

$$(x_i)_{\langle n, n+2r \rangle} = (y_i)_{\langle 0, 2r \rangle} = (x_{j_i})_{\langle -p, -p+2r \rangle}$$

However, both sets $\{(x_i)_{\langle n,n+2r\rangle} : 1 \le i \le k_1\}$ and $\{(x_{j_i})_{\langle -p,-p+2r\rangle} : 1 \le i \le k_1\}$ contain exactly k_0 elements and they are equal. Consider a set of words $\{u_i = (x_i)_{\langle -p,n\rangle}; 1 \le i \le k_1\}$ and define a binary matrix $M = (M_{ij})_{1 \le i,j \le k_1}$ by

$$M_{ij} = 1 \iff (x_i)_{\langle n,n+2r\rangle} = (x_j)_{\langle -p,-p+2r\rangle} .$$

For every $i \le k_1$ there exists at least one $j \le k_1$ with $M_{ij} = 1$ and for some i there exist at least two j with $M_{ij} = 1$. Dually, for every j there exists at least one, and sometimes at least two i with $M_{ij} = 1$. Let $\Sigma \subseteq A^{\mathbb{Z}}$ be a SFT defined by $w \in \Sigma$ iff for every $i \in \mathbb{Z}$ there exists $j_i \le k_1$ such that

$$w_{\langle i(n+p),(i+1)(n+p)\rangle} = u_{j_i} \text{ and } M_{j_i,j_{i+1}} = 1 .$$

Then for every $x, y \in \Sigma$, $F(x) = F(y)$, but Σ is an infinite set and this is a contradiction. $\square$

PROPOSITION 7.2 (Hedlund [19]). *Let $(A^{\mathbb{Z}}, F)$ be a surjective CA and $y, z \in A^{\mathbb{Z}}$. Let z be σ-transitive and $|F^{-1}(z)| = m$. Then $|F^{-1}(y)| \ge m$ and there exist m pairwise totally $2r$-separated preimages of y. In particular every two σ-transitive points have the same number of preimages.*

Proof. Let $F^{-1}(z) = \{x_1, \dots, x_m\}$. By Proposition 7.1, $(x_i)_{i \le m}$ are pairwise totally $2r$-separated. For every q there exists n_q for which $d(\sigma^{n_q}(z), y) < 2^{-q}$. By choosing a subsequence if necessary, we can assume that $\sigma^{n_q}(x_i)$ converge to some w_i as $q \to \infty$. Then $(w_i)_{i \le m}$ are pairwise totally $2r$-separated and $F(w_i) = y$. $\square$

COROLLARY 7.1. *Let $(A^{\mathbb{Z}}, F)$ be a surjective CA and put*

$$m = \min\{|F^{-1}(x)| : x \in A^{\mathbb{Z}}\}.$$

The set $\{x \in A^{\mathbb{Z}} : |F^{-1}(x)| = m\}$ is residual and contains all σ-transitive points.

In the CA $(3^{\mathbb{Z}}, L)$ of Example 8, $m = 1$. Every point which has an infinite number of two's in $\langle 0, \infty)$, has only one preimage.

8. Openness. We characterize CA with constant number of preimages. Recall that a dynamical system (X, F) is open if for every open set $U \subseteq X$, $F(U)$ is open. Clearly every bijective DS is open. By a theorem of Parry [27], a one-sided subshift is open iff it is a SFT. Since the sum CA of Example 3 is conjugate to the full shift of four symbols, it is open. Moreover, every its point has exactly four preimages.

THEOREM 8.1 (Hedlund [19]). *Let $(A^{\mathbb{Z}}, F)$ be a CA. The following conditions are equivalent.*

1. *There exists $m > 0$ such that for all $x \in A^{\mathbb{Z}}$, $|F^{-1}(x)| = m$.*
2. *There exist continuous maps $F_1, \dots, F_m : A^{\mathbb{Z}} \to A^{\mathbb{Z}}$ such that for all $x \in A^{\mathbb{Z}}$, $F^{-1}(x) = \{F_1(x), \dots, F_m(x)\}$.*
3. *F is open.*

Proof. 1 ⇒ 2: By Proposition 7.2, the preimages of every point are pairwise totally $2r$-separated. Assume that $F^{-1}(y) = \{x_1, \ldots, x_m\}$. We show

$$\forall \varepsilon > 0, \exists \delta > 0, \forall y' \in B_\delta(y), \exists i \le m, \exists x' \in B_\varepsilon(x_i), F(x') = y' \ .$$

If this were not true, there would exist $\varepsilon > 0$ such that for every $\delta = 2^{-n}$ there would exist $y_n \in B_\delta(y)$, and its preimage $z_n \in F^{-1}(y_n)$, such that $d(z_n, x_i) \ge \varepsilon$ for every i. If z were a limit of a subsequence of $(z_n)_{n \ge 0}$, then $F(z) = y$, so y would have at least $m + 1$ preimages and this is a contradiction. Moreover, if $\varepsilon \le 2^{-r}$, then the m preimages of y' belong to distinct $B_\varepsilon(x_i)$, since they are pairwise totally $2r$-separated and therefore their distance is at least 2^{-r}. Thus we get a stronger condition

$$\forall \varepsilon > 0, \exists \delta > 0, \forall y' \in B_\delta(y), \forall i \le m, \exists! x_i' \in B_\varepsilon(x_i), F(x_i') = y' \ .$$

Suppose that δ has this property with $\varepsilon = 2^{-r}$. Let $\mathcal{V} = \{V_b : b \in B\}$ be a clopen partition of $A^{\mathbb{Z}}$ with $\mathrm{diam}(\mathcal{V}) < \delta$. Choose from every V_b a point $y_b \in V_b$ and denote its preimages $F^{-1}(y_b) = \{x_{b,1}, \ldots, x_{b,m}\}$. Define $F_i : A^{\mathbb{Z}} \to A^{\mathbb{Z}}$ by

$$x \in V_b \ \Rightarrow \ d(F_i(x), x_{b,i}) < \varepsilon, \ F(F_i(x)) = x \ .$$

Then F_i are continuous.

2 ⇒ 3: Assume that $U \subseteq A^{\mathbb{Z}}$ is an open set such that $F(U)$ is not open, so there exists $x \in U$ such that $F(U)$ is not a neighbourhood of $F(x)$. For some $i \le m$, $x = F_i(F(x))$. There exists a sequence $(y_n \notin F(U))_{n \ge 0}$ converging to $F(x)$ and $F_i(y_n) \notin U$. $(F_i(y_n))_{n \ge 0}$ has a subsequence converging to some $z \notin U$ and $F(z) = F(x)$. Since the preimages of every y_n are pairwise totally $2r$-separated, z differs from all $F_j(F(x))$ and this is a contradiction.

3 ⇒ 1: We show first that F is surjective. Suppose that $V = F(A^{\mathbb{Z}}) \ne A^{\mathbb{Z}}$, Since V is clopen and $(A^{\mathbb{Z}}, \sigma)$ is transitive, there exists $n > 0$ such that $\sigma^n(V) \cap (A^{\mathbb{Z}} \setminus V) \ne \emptyset$ and this is impossible since V is σ- invariant. Put $m = \min\{|F^{-1}(x)| : x \in A^{\mathbb{Z}}\}$ and assume that for some $y \in A^{\mathbb{Z}}$, $F^{-1}(y) = \{x_1, \ldots, x_p\}$, where $p > m$. Let $U_i \ni x_i$ be pairwise disjoint open sets. Then $V = F(U_1) \cap \cdots \cap F(U_p)$ is an open set containing x, and every element of V has at least p preimages. However, σ-transitive points are dense, so V contains a σ-transitive point which has only m preimages. □

The inverse maps F_i are in general not CA. In the sum CA of Example 3, the four inverse maps may be given by conditions

$$F_0(x)_{\langle 0,1 \rangle} = 00, \ \ F_1(x)_{\langle 0,1 \rangle} = 01, \ \ F_2(x)_{\langle 0,1 \rangle} = 10, \ \ F_3(x)_{\langle 0,1 \rangle} = 11 \ .$$

Only when $m = 1$, i.e., when F is bijective, the inverse map F^{-1} is a CA. This is the case of identity and of every almost equicontinuous open map.

THEOREM 8.2. *Every CA which is open and almost equicontinuous is bijective.*

Proof. Let $u \in A^m$ be a $2r$-blocking word with offset j. Suppose that some $z \in [u]_0$ has two distinct preimages x, y. Then $v = x_{(-r,m+r)}$, $w = y_{(-r,m+r)}$ are distinct (since x and y are totally $2r$-separated), and both $F([v]_{-r})$ and $F([w]_{-r})$ are contained in $[u]_0$. Moreover, v and w are $2r$-blocking words with offsets $j + r$. By Proposition 6.3, $(A^{\mathbb{Z}}, F)$ is nonwandering, so there exist $p, q > 0$ and $x' \in [v]_{-r} \cap F^{-p}([v]_{-r})$, $y' \in [w]_r \cap F^{-q}([w]_{-r})$. It follows that

$$x'_{\langle j, j+2r \rangle} = F^{pq}(x')_{\langle j, j+2r \rangle} = F^{pq-1}(z)_{\langle j, j+2r \rangle} = F^{pq}(y')_{\langle j, j+2r \rangle} = y'_{\langle j, j+2r \rangle}$$

$$x_{\langle j, j+2r \rangle} = x'_{\langle j, j+2r \rangle} = y'_{\langle j, j+2r \rangle} = y_{\langle j, j+2r \rangle}$$

so x and y are not $2r$-separated. Thus z has only one preimage, and since F is open, every point has exactly one preimage. □

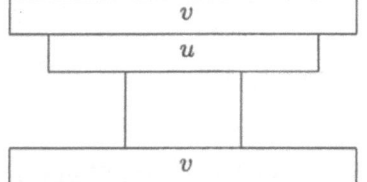 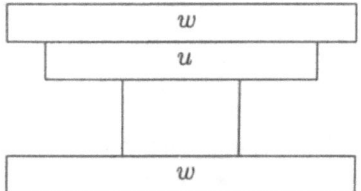

EXAMPLE 9 (B: A bijective almost equicontinuous CA).
$(A^{\mathbb{Z}}, B)$ *where* $A = \{000, 001, 010, 011, 100\} = \{0, 1, 2, 3, 4\}$, *and*

$$B(x, y, z)_i = (x_i, (1 + x_i)y_{i+1} + x_{i-1}z_i, (1 + x_i)z_{i-1} + x_{i+1}y_i)$$

(the addition is modulo 2).

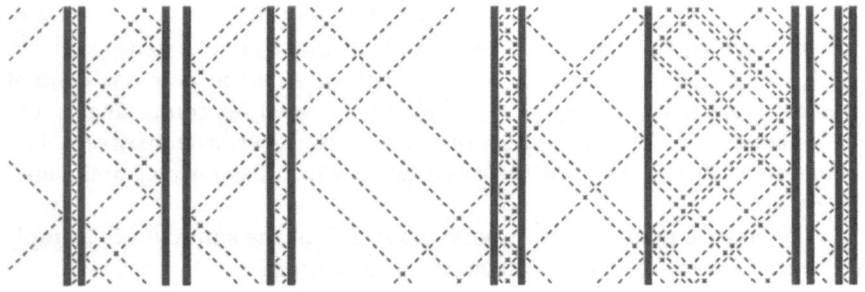

FIG. 8. *B: A bijective CA.*

The dynamics is conveniently described as movement of three types of particles, $1 = 001$, $2 = 010$ and $4 = 100$. Four is a wall which neither moves nor changes. A 1 goes to the left and when it hits a wall (4), it changes to 2. A two goes to the right and when it hits the wall, it changes to a one. One's and two's may cross, to form three's. Clearly 4 is a 1-blocking word, so the system is almost equicontinuous. It is bijective, as its inverse is

$$B^{-1}(x, y, z)_i = (x_i, (1 + x_i)y_{i-1} + x_{i+1}z_i, (1 + x_i)z_{i+1} + x_{i-1}y_i) .$$

9. Closingness.

DEFINITION 9.1. *A CA $(A^{\mathbb{Z}}, F)$ is right closing (left closing), if for every distinct left asymptotic (right asymptotic) $x, y \in A^{\mathbb{Z}}$, $F(x) \neq F(y)$. A CA is closing if it is either left or right closing.*
Clearly $(A^{\mathbb{Z}}, F)$ is right closing iff its dual $(A^{\mathbb{Z}}, \overline{F})$ is left closing.

The CA $(3^{\mathbb{Z}}, L)$ of Example 8 is left closing. If $L(x) = z$, then for every $i \in \mathbb{Z}$

$$x_i = \begin{cases} x_{i+1} + z_i \bmod 2 & \text{if } z_i \neq 2 \\ 2 & \text{if } z_i = 2 \end{cases} .$$

Thus if $L(x) = L(y)$ and if x and y are right asymptotic, then $x = y$. The CA is, however, not right closing. The points $x = 0^\infty.20^\infty$ and $y = 0^\infty.21^\infty$ are left asymptotic and their images coincide.

EXAMPLE 10 (R: a surjective CA). *The CA $(3^{\mathbb{Z}} \times 3^{\mathbb{Z}}, R)$, where $R = L \times \overline{L}$, is not closing but it is surjective.* The points $x = (0^\infty.20^\infty, 0^\infty)$ and $y = (0^\infty.21^\infty, 0^\infty)$ are left asymptotic but their images coincide. The points $x = (0^\infty, 0^\infty 2.0^\infty)$ and $y = (0^\infty, 1^\infty 2.0^\infty)$ are right asymptotic but their images coincide.

PROPOSITION 9.1. *Every closing CA is surjective.*

Proof. Let $(A^{\mathbb{Z}}, F)$ be a CA with radius r which is not surjective. By Theorem 6.2 there exists a diamond, i.e., words $w \in A^{2r}$ and $u \neq v \in A^+$ with $|u| = |v|$ and $f(wuw) = f(wvw)$. Put $x = w^\infty.uw^\infty$, $y = w^\infty.vw^\infty$. Then x, y are both left and right asymptotic and $F(x) = F(y)$, so F is neither right nor left closing. $\square$

PROPOSITION 9.2. *A CA $(A^{\mathbb{Z}}, F)$ is right closing iff there exists $m > 0$ such that*

$$x_{\langle -m, 0 \rangle} = y_{\langle -m, 0 \rangle} \ \& \ F(x)_{\langle -m, m \rangle} = F(y)_{\langle -m, m \rangle} \Rightarrow x_0 = y_0 .$$

Proof. Suppose that F is right closing. For $m > 0$ let X_m consist of all pairs $(x, y) \in A^{\mathbb{Z}} \times A^{\mathbb{Z}}$ such that

$$x_{\langle -m, 0 \rangle} = y_{\langle -m, 0 \rangle}, \quad x_0 \neq y_0, \quad F(x)_{\langle -m, m \rangle} = F(y)_{\langle -m, m \rangle} .$$

Every X_m is closed and $X_{m+1} \subseteq X_m$. If all X_m were nonempty, their intersection would contain a pair (x, y) of distinct left asymptotic words with $F(x) = F(y)$, so F would not be right closing. Thus for some m,

$X_m = \emptyset$.

Conversely, suppose that F is not right closing, so there exist distinct left asymptotic words x, y with $F(x) = F(y)$. Taking the shifts of x and y if necessary, we can assume $x_{(-\infty,0)} = y_{(-\infty,0)}$ and $x_0 \neq y_0$. Thus for every $m > 0$, the condition is not satisfied. $\square$

PROPOSITION 9.3. *Let* $(A^{\mathbb{Z}}, F)$ *be a right closing CA. There exists* $m > 0$ *such that if* $u \in A^m$, $v \in A^{2m}$ *and* $F([u]_{-m}) \cap [v]_{-m} \neq \emptyset$, *then*

$$\forall b \in A, \exists! a \in A, F([ua]_{-m}) \cap [vb]_{-m} \neq \emptyset.$$

Proof. By Proposition 9.2, for every b there exists at most one a with $F([ua]_{-m}) \cap [vb]_{-m} \neq \emptyset$. Put

$$B = \{(u,v) \in A^m \times A^{2m} : F([u]_{-m}) \cap [v]_{-m} \neq \emptyset\} .$$

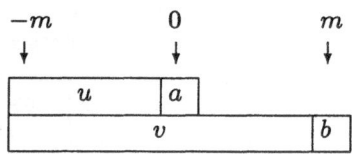

Consider an oriented graph with vertices B. For every pair of words $u \in A^{m+1}$, $v \in A^{2m+1}$ such that $F([u]_{-m}) \cap [v]_{-m} \neq \emptyset$, put an oriented edge

$$(u_{\langle 0, m-1 \rangle}, v_{\langle 0, 2m-1 \rangle}) \to (u_{\langle 1, m \rangle}, v_{\langle 1, 2m \rangle}) .$$

This graph defines a SFT $\Sigma \subseteq B^{\mathbb{Z}}$ of order two, and there is a conjugacy $\varphi : (A^{\mathbb{Z}}, \sigma) \to (\Sigma, \sigma)$ defined by $\varphi(x)_i = (x_{\langle i, i+m-1 \rangle}, F(x)_{\langle i, i+2m-1 \rangle})$. The topological entropy is $h(\Sigma, \sigma) = h(A^{\mathbb{Z}}, \sigma) = \ln |A|$. In the graph of Σ, from every vertex $(u_{\langle 0, m-1 \rangle}, v_{\langle 0, 2m-1 \rangle})$ there are at most $|A|$ outgoing edges (for every v_{2m} at most one). If there are fewer than $|A|$ outgoing edges from some vertex, then the topological entropy $h(\Sigma, \sigma)$ would be smaller than $\ln |A|$ (Lind and Marcus, [22], Corollary 4.4.9.). $\square$

THEOREM 9.1. *If* $(A^{\mathbb{Z}}, F)$ *is a right closing CA, then for some* $p > 0$, $(A^{\mathbb{Z}}, \sigma^p \circ F)$ *is a factor of a two-sided full shift.*

Proof. Let $m > 0$ be the number from Proposition 9.3. We can assume $m \geq r$, where r is the radius. Put $p = 3m + 1$ and

$$B = \{(u,v) \in A^{2m+1} \times A^{2m+1} : F^{-1}([u]_{-m}) \cap [v]_0 \neq \emptyset\} .$$

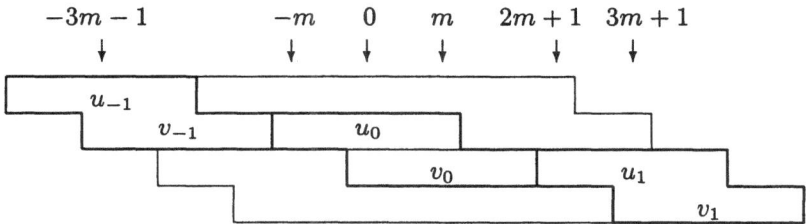

If (u_0, v_0) and (u_1, v_1) belong to B, then there exists $y \in [u_1]_{2m+1} \cap F^{-1}([v_1]_{3m+1})$. We can change y in interval $\langle 0, 2m \rangle$ to obtain a point $x \in [v_0 u_1]_0 \cap F^{-1}([v_1]_{3m+1})$. By Proposition 9.3, x has a preimage in $[u_0]_{-m}$, so

$$F^1([u_0]_{-m}) \cap [v_0 u_1]_0 \cap F^{-1}([v_1]_{3m+1}) \neq \emptyset .$$

If $(u, v) \in B^{\mathbb{Z}}$, then $X_1 = F^1([v_{-1} u_0]_{-3m-1}) \cap [v_0 u_1]_0 \cap F^{-1}([v_1 u_2]_{3m+1}) \neq \emptyset$, and similarly, for every $q > 0$,

$$X_q = \bigcap_{n=-q}^{q} F^{-n}([v_n u_{n+1}]_{(3m+1)n}) \neq \emptyset .$$

By compactness, $\bigcap_{q>0} X_q$ is nonempty. If $x, y \in X_q$, then

$$x_{\langle -q(3m+1)+mq, q(3m+1)-mq \rangle} = y_{\langle -q(3m+1)+mq, q(3m+1)-mq \rangle} .$$

Thus there exists a unique $\varphi(u, v) \in \bigcap_{q>0} X_q$ and $\varphi : (B^{\mathbb{Z}}, \sigma) \to (A^{\mathbb{Z}}, \sigma^p \circ F)$ is a factor map. $\square$

THEOREM 9.2 (Boyle and Kitchens [8]). *Every closing CA $(A^{\mathbb{Z}}, F)$ has a dense set of points which are both F-periodic and σ-periodic.*

Proof. We can assume that F is right closing. Since $(A^{\mathbb{Z}}, \sigma^p \circ F)$ is a factor of a two-sided full shift, it has a dense set of periodic points. We can assume $p > m$ where m is the integer from Proposition 9.3. We show that if $x \in A^{\mathbb{Z}}$ is a $(\sigma^p \circ F)$-periodic point, then it is both σ-periodic and F-periodic. Put $B = A^{2p+1}$ and construct a map $\varphi : A^{\mathbb{Z}} \to B^{\mathbb{N}}$ by

$$\varphi(x)_i = (\sigma^p \circ F)^i(x)_{\langle -p, p \rangle} = F^i(x)_{\langle pi-p, pi+p \rangle}$$

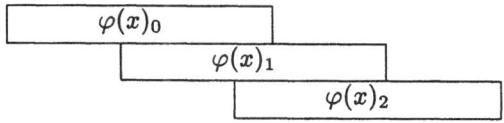

Because F is right closing and $p > m$, $\varphi(\sigma(x))$ is uniquely determined by $\varphi(x)$, in fact, $\varphi(\sigma(x))_i$ depends only on $\varphi(x)_i$ and $\varphi(x)_{i+1}$. If $x \in A^{\mathbb{Z}}$ is $(\sigma^p \circ F)$-periodic, i.e., if $(\sigma^p \circ F)^q(x) = x$, then $\varphi(x)$ is σ- periodic since

$$\varphi(x)_{i+q} = (\sigma^p \circ F)^{i+q}(x)_{\langle -p,p \rangle} = (\sigma^p \circ F)^i(x)_{\langle -p,p \rangle} = \varphi(x)_i .$$

It follows that $\varphi(\sigma(x))$ is σ-periodic with period q too. In $B^{\mathbb{N}}$, there are at most $|B|^q$ σ-periodic points with period q, so there exist $0 \leq k < k + s < |B|^q$ such that $\varphi(\sigma^{k+s}(x)) = \varphi(\sigma^k(x))$ and for all $l \geq k$, $\varphi(\sigma^{l+s}(x)) = \varphi(\sigma^l(x))$. Applying this argument to $\sigma^{-j}(x)$ where $j = |B|^q$, we get $\varphi(\sigma^{l+s}(x)) = \varphi(\sigma^l(x))$ for all $l \geq 0$. In particular, $\varphi(\sigma^s(x)) = \varphi(x)$, so $\sigma^s(x) = x$ and

$$x = (\sigma^p \circ F)^{qs}(x) = F^{qs} \circ \sigma^{pqs}(x) = F^{qs}(x) .$$

Thus x is both σ-periodic and F-periodic. $\square$

THEOREM 9.3. *A CA is open iff it is both left closing and right closing.*

Proof. If $(A^{\mathbb{Z}}, F)$ is open and $z \in A^{\mathbb{Z}}$, then every two preimages of z are totally $2r$-separated, so they are neither left nor right asymptotic. This means that F is both right and left closing.

Conversely, Suppose that $(A^{\mathbb{Z}}, F)$ is a CA with radius r which is both left and right closing. Let $m > 0$ be an integer with property of Proposition 9.3 and $u \in A^{2k+1}$. Then $F([u]_{-k})$ is open, since it is the union of all cylinders $[v]_{-k-m}$ such that $v \in A^{2k+2m+1}$ and $F([u]_{-k}) \cap [v]_{-k-m} \neq \emptyset$. $\square$

10. Expansivity.

PROPOSITION 10.1. *Every positively expansive CA $(A^{\mathbb{Z}}, F)$ with radius r is conjugate to $(\Sigma_{\langle -r,r \rangle}(F), \sigma)$.*

Proof. Suppose that there exist $x \neq y$ such that for all $n \geq 0$, $F^n(x)_{\langle -r,r \rangle} = F^n(y)_{\langle -r,r \rangle}$. Let $x_k \neq y_k$, so $|k| > r$. Assume $k > 0$ and define $z \in A^{\mathbb{Z}}$ by $z_{(-\infty,0)} = x_{(-\infty,0)}$, $z_{\langle 0,\infty)} = y_{\langle 0,\infty)}$. Then $x \neq z$ and for all $n \geq 0$, $F^n(x)_{(-\infty,r)} = F^n(z)_{(-\infty,r)}$. It follows that for every $k \geq 0$

$$F^n(\sigma^{-k}(x))_{(-\infty,r+k)} = F^n(\sigma^{-k}(z))_{(-\infty,r+k)}$$

and this is a contradiction. Thus $\varphi : (A^{\mathbb{Z}}, F) \to (\Sigma_{\langle -r,r \rangle}(F), \sigma)$ is injective, hence a conjugacy. $\square$

THEOREM 10.1 (Kůrka [21], Nasu [25]). *Every positively expansive CA is open and conjugate to a onesided SFT.*

Proof. We show that if $(A^{\mathbb{Z}}, F)$ is positively expansive, then it is both left and right closing. Let $\delta = 2^{-m}$ be the constant of expansivity, so

$$x \neq y \Rightarrow \exists n \geq 0, F^n(x)_{\langle -m,m \rangle} \neq F^n(y)_{\langle -m,m \rangle} .$$

Suppose that x, y are distinct left asymptotic points with $F(x) = F(y)$. By taking a shift of x and y if necessary, we can assume $x_{(-\infty,m)} = y_{(-\infty,m)}$, so

$d(F^n(x), F^n(y)) < \delta$ for all $n \geq 0$ and this is a contradiction. By Theorem 9.3, $(A^{\mathbb{Z}}, F)$ is open, so $(\Sigma_{\langle -r,r \rangle}(F), \sigma)$ is open too. By a Theorem of Parry, [27], a one sided subshift is a SFT iff it is open, so $\Sigma_{\langle -r,r \rangle}(F)$ is a SFT. $\square$

THEOREM 10.2 (Blanchard and Maass [3]). *Every positively expansive CA is mixing.*

Proof. Let $(A^{\mathbb{Z}}, F)$ be a positively expansive CA with radius r, put $B = A^{2r+1}$ and let k be the order of SFT $\Sigma_{\langle -r,r \rangle}(F)$, so $w \in \Sigma_{\langle -r,r \rangle}(F)$ iff for all $n \geq 0$, $w_{\langle n, n+k \rangle} \in \mathcal{L}(\Sigma_{\langle -r,r \rangle}(F))$. Let $u, v \in A^{nr}$. There exists $p > 0$ such that for every x, y,

$$\forall i < p, F^i(x)_{\langle -r,r \rangle} = F^i(y)_{\langle -r,r \rangle} \;\Rightarrow\; x_{\langle -(k+n)r, (k+n)r \rangle} = y_{\langle -(k+n)r, (k+n)r \rangle}.$$

We will show that for all $m \geq p$, $F^m([u]_0) \cap [v]_0 \neq \emptyset$. Choose any $x \in A^{\mathbb{Z}}$, such that $x_{\langle kr+1, (k+n)r \rangle} = u$. Since F is surjective, there exists $y \in A^{\mathbb{Z}}$ such that

$$F^m(y)_{\langle -kr, kr \rangle} = F^m(x)_{\langle -kr, kr \rangle}, \; F^m(y)_{\langle kr+1, (k+n)r \rangle} = v$$

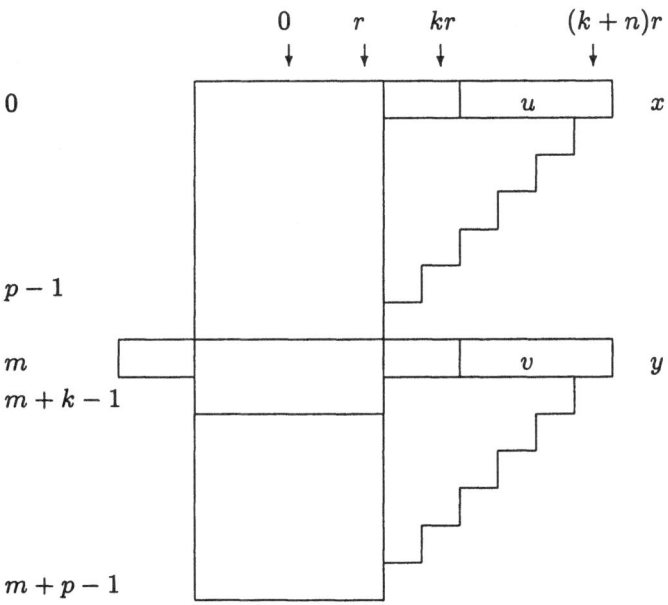

It follows that for $m \leq i < m+k$, $F^i(x)_{\langle -r,r \rangle} = F^i(y)_{\langle -r,r \rangle}$. Construct $w \in B^{m+p}$ by

$$w_i = \begin{cases} F^i(x)_{\langle -r,r \rangle} & \text{if } 0 \leq i \leq m+k-1 \\ F^i(y)_{\langle -r,r \rangle} & \text{if } m \leq i \leq m+p-1 \end{cases}.$$

Since both $w_{\langle 0,m+k\rangle}$ and $w_{\langle m,m+p\rangle}$ belong to the language of the subshift, w belongs to this language too and there exists $z \in A^{\mathbb{Z}}$ such that for $i < m+p$, $F^i(z)_{\langle -r,r\rangle} = w_i$. Thus $z_{\langle kr+1,(k+n)r\rangle} = u$ and $F^m(z)_{\langle kr+1,(k+n)r\rangle} = v$, so $\sigma^{kr+1}(z) = [u]_0 \cap F^{-m}([v]_0)$. $\square$

There is a large class of permutive local rules which yield expansive, or at least open CA.

DEFINITION 10.1. *Let $s \geq 1$ and $f : A^s \to A$ be a local rule.*
1. *f is left permutive iff $\forall u \in A^{s-1}, \forall b \in A, \exists! a \in A, f(au) = b$.*
2. *f is right permutive iff $\forall u \in A^{s-1}, \forall b \in A, \exists! a \in A, f(ua) = b$.*

PROPOSITION 10.2. *Let $(A^{\mathbb{Z}}, F)$ be a CA and let $f : A^{a-m+1} \to A$ be its local rule. Here $m \leq a$ is the memory and anticipation respectively.*
1. *If f is left permutive, then F is left closing.*
2. *If f is right permutive, then F is right closing.*
3. *If f is both left and right permutive and $m < 0 < a$, then F is expansive.*

The sum CA of Example 3 is both left and right permutive with $m = -1$, $a = 1$. The closing CA of Example 8 is left permutive with $m = 0$ and $a = 1$. It is not right permutive. However, the bijective and hence open CA of Example 9 is neither left nor right permutive.

11. Shadowing property. If $(x_i)_{i \geq 0}$ is a 2^{-m}-chain in a CA $(A^{\mathbb{Z}}, F)$, then for all i, $F(x_i)_{\langle -m,m\rangle} = (x_{i+1})_{\langle -m,m\rangle}$, so $u_i = (x_i)_{\langle -m,m\rangle}$ satisfy $F([u_i]_{-m}) \cap [u_{i+1}]_{-m} \neq \emptyset$. Conversely, if a sequence $(u_i \in A^{2m+1})_{i \geq 0}$ satisfies this property and $x_i \in [u_i]_{-m}$, then $(x_i)_{i \geq 0}$ is a 2^{-m}-chain.

THEOREM 11.1 (Kůrka [21]). *Let $(A^{\mathbb{Z}}, F)$ be a CA. If for every sufficiently large n, $\Sigma_{\langle -n,n\rangle}(F)$ is a SFT, then F has the shadowing property.*

Proof. For a given $\varepsilon = 2^{-n} \leq 2^{-r}$, let $m + 1 > 1$ be the order of $\Sigma_{\langle -n,n\rangle}(F)$, so $w \in \Sigma_{\langle -n,n\rangle}(F)$ iff for every $i \geq 0$, $w_{\langle i,i+m\rangle} \in \mathcal{L}(\Sigma_{\langle -n,n\rangle}(F))$. Put $k = n + rm$ and $\delta = 2^{-k}$. We show that every δ-chain is ε-shadowed by some point. Let $(x_i)_{i \geq 0}$ be a δ-chain, so $F(x_i)_{\langle -k,k\rangle} = (x_{i+1})_{\langle -k,k\rangle}$. We get

$$F^2(x_i)_{\langle -k+r,k-r\rangle} = F(x_{i+1})_{\langle -k+r,k-r\rangle} = (x_{i+2})_{\langle -k+r,k-r\rangle}$$

and $\forall j \leq m, F^j(x_i)_{\langle -n,n\rangle} = (x_{i+j})_{\langle -n,n\rangle}$. Since $\Sigma_{\langle -n,n\rangle}(F)$ is a SFT of order $m+1$, the sequence $((x_i)_{\langle -n,n\rangle})_{i \geq 0}$ belongs to $\mathcal{L}(\Sigma_{\langle -n,n\rangle}(F))$, so there exists a point $x \in A^{\mathbb{Z}}$, such that for all $i \geq 0$, $F^i(x)_{\langle -n,n\rangle} = (x_i)_{\langle -n,n\rangle}$, so x ε-shadows the sequence $(x_i)_{i \geq 0}$. $\square$

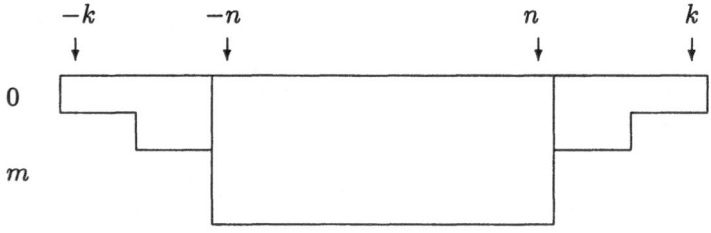

Every expansive CA has the shadowing property since it is conjugate to a SFT. It is easy to see that the factor subshifts of the product CA and the majority CA are SFT, so these CA have the shadowing property.

THEOREM 11.2 (Kůrka [21]). *If $(A^{\mathbb{Z}}, F)$ is a CA with the shadowing property, then every its factor subshift is sofic.*

Proof. Let $(A^{\mathbb{Z}}, F)$ be a CA. We show that for every $n \geq 0$, $\Sigma_{\langle -n,n\rangle}(F)$ is a sofic subshift. For $\varepsilon = 2^{-n}$ there exists $\delta = 2^{-k}$ such that every δ-chain is ε-shadowed by some point. We can assume $k \geq n$. Put $B = A^{2k+1}$, $C = A^{2n+1}$ and define a SFT $\Sigma \subseteq B^{\mathbb{N}}$ of order 2 so that for $u, v \in B$,

$$uv \in \mathcal{L}(\Sigma) \quad \Leftrightarrow \quad F([u]_{-k}) \cap [v]_{-k} \neq \emptyset \ .$$

Define a map $\varphi : \Sigma \to C^{\mathbb{N}}$ by $\varphi(u)_i = (u_i)_{\langle k-n,k+n\rangle}$. For $u \in \Sigma$ consider any sequence of points $(x_i \in A^{\mathbb{Z}})_{i \geq 0}$ with $x_i \in [u_i]_{-k}$. Then $(x_i)_{i \geq 0}$ is a δ-chain so there exists a point $x \in A^{\mathbb{N}}$ which ε-shadows it. We get

$$F^i(x)_{\langle -n,n\rangle} = (x_i)_{\langle -n,n\rangle} = (u_i)_{\langle k-n,k+n\rangle}$$

so $\varphi(u) \in \Sigma_{\langle -n,n\rangle}(F)$. Thus $\varphi(\Sigma) \subseteq \Sigma_{\langle -n,n\rangle}(F)$. Conversely, if $v \in \Sigma_{\langle -n,n\rangle}(F)$, then there exists $x \in A^{\mathbb{Z}}$ with $v_i = F^i(x)_{\langle -n,n\rangle}$ and $v = \varphi(u)$, where $u_i = F^i(x)_{\langle -k,k\rangle}$. Thus $\Sigma_{\langle -n,n\rangle}(F) = \varphi(\Sigma)$ is sofic. $\square$

PROPOSITION 11.1. *Every equicontinuous DS on $A^{\mathbb{Z}}$ has the shadowing property.*

Proof. By the assumption for every $\varepsilon = 2^{-n}$ there exists $\delta = 2^{-k}$ such that if $x_{\langle -k,k\rangle} = y_{\langle -k,k\rangle}$, then $F^i(x)_{\langle -n,n\rangle} = F^i(y)_{\langle -n,n\rangle}$ for all $i \geq 0$. Let $(x_i \in A^{\mathbb{Z}})_{i \geq 0}$ be a δ- chain. Then $F(x_i)_{\langle -k,k\rangle} = (x_{i+1})_{\langle -k,k\rangle}$, so for every $m > 0$,

$$F^m(x_0)_{\langle -n,n\rangle} = F^{m-1}(x_1)_{\langle -n,n\rangle} = \cdots F(x_{m-1})_{\langle -n,n\rangle} = (x_m)_{\langle -n,n\rangle} \ .$$

Thus x_0 ε-shadows $(x_i)_{i \geq 0}$. $\square$

12. Attractors.

THEOREM 12.1 (Hurley [20]). *Let $(A^{\mathbb{Z}}, F)$ be a CA and $x \in A^{\mathbb{Z}}$ an attracting periodic point. Then $\sigma(x) = x$ and $F(x) = x$.*

Proof. Let p be the period of x, so $F^p(x) = x$. By the assumption, there exists a clopen set U with $\mathcal{O}(x) = \omega(U)$. There exists a nonempty open set $U_0 \subseteq U$ such that if $y \in U_0$, then $\lim_{n \to \infty} F^{np}(y) = x$. Since $(A^{\mathbb{Z}}, \sigma)$ is mixing, for k large enough, both $\sigma^k(U_0) \cap U_0$ and $\sigma^{k+1}(U_0) \cap U_0$ are nonempty. For $y \in \sigma^k(U_0) \cap U_0$ and $z \in \sigma^{k+1}(U_0) \cap U_0$ we get

$$\sigma^k(x) = \lim_{n \to \infty} F^{np}(y) = x = \lim_{n \to \infty} F^{np}(z) = \sigma^{k+1}(x) \ .$$

It follows $\sigma(x) = x$, so $x = a^{\infty}$ for some $a \in A$. Then $F(x) = b^{\infty}$ for some $b \in A$ and $a^{\infty} = \lim_{n \to \infty} F^{np}(a^{\infty}.b^{\infty}) = b^{\infty}$. Thus $a = b$ and therefore $p = 1$. $\square$

In the product CA of Example 1, 0^∞ is an attracting fixed point. There is also another larger attractor $\omega(2^{\mathbb{Z}})$.

THEOREM 12.2 (Hurley [20]). *If a CA has two disjoint attractors, then every its attractor contains two disjoint attractors and a continuum of quasi-attractors.*

Proof. Let Y_1 and Y_2 be two disjoint attractors and U_1, U_2 clopen invariant sets with $\omega(U_1) = Y_1$, $\omega(U_2) = Y_2$. If $U_1 \cap U_2$ were nonempty, then $\omega(U_1 \cap U_2)$ would be an attractor contained in both Y_1 and Y_2. Thus $U_1 \cap U_2 = \emptyset$. Let Y be an attractor and U a clopen invariant set with $\omega(U) = Y$. Since $(A^{\mathbb{Z}}, \sigma)$ is mixing, there exists $n > 0$ such that $V_1 = \sigma^n(U_1) \cap U \neq \emptyset$ and $V_2 = \sigma^n(U_2) \cap U \neq \emptyset$ are clopen invariant sets. It follows that $\omega(V_1)$ and $\omega(V_2)$ are disjoint attractors contained in Y. $\square$

In the majority CA of Example 2, $[00]_0$ and $[11]_0$ are disjoint invariant sets, so their omega-limits are disjoint attractors. While $\omega(2^{\mathbb{Z}})$ and $Y = \omega([00]_0 \cup [11]_0)$ are subshifts, other attractors are not.

THEOREM 12.3 (Hurley [20]). *If a CA has a minimal attractor then it is a subshift, it is contained in every other attractor and its basin of attraction is a dense open set.*

Proof. Let Y be a minimal attractor and let Z be another attractor. If $Y \cap Z$ were empty, then Y would contain two disjoint attractors. Since Y is a minimal attractor, $Y \subseteq Z$. Let V be a clopen invariant set with $\omega(V) = Y$. Then $\sigma^{-1}(V)$ is a clopen invariant set, so $Y \subseteq \omega(\sigma^{-1}(V)) = \sigma^{-1}(Y)$. Thus $\sigma(Y) \subseteq Y$ and similarly, $\sigma^{-1}(Y) \subseteq Y$, so Y is a subshift. The basin $\mathcal{B}(Y)$ is an open σ-invariant set. Since $(A^{\mathbb{Z}}, \sigma)$ is transitive, for every nonempty open $U \subseteq A^{\mathbb{Z}}$ there exists $n > 0$ with

$$\emptyset \neq \sigma^n(\mathcal{B}(Y)) \cap U = \mathcal{B}(Y) \cap U$$

and $\mathcal{B}(Y)$ is dense. $\square$

Of course an attracting fixed point is an example of a minimal attractor. To see an example of a minimal attractor which is not a fixed point consider

EXAMPLE 11 (A minimal attractor). $(2^{\mathbb{Z}} \times 2^{\mathbb{Z}}, P \times \sigma)$, *has a minimal attractor* $\{0^\infty\} \times 2^{\mathbb{Z}}$.

COROLLARY 12.1 (Hurley [20]). *For any CA exactly one of the following statements holds.*
1. *There exists two disjoint attractors and a continuum of quasi-attractors.*
2. *There exists a unique minimal quasi-attractor. It is a subshift and it is contained in every attractor.*
3. *There exists a unique minimal attractor which is contained in every other attractor.*

Proof. If there do not exist disjoint attractors, then every two attractors are comparable by inclusion. The number of attractors is at most countable, so the intersection of all attractors is either a quasi-attractor or an attractor and it is σ-invariant. $\square$

EXAMPLE 12 (Q : A quasi-attractor, Hurley [20]). $(2^{\mathbb{Z}}, Q)$ *where*

$$Q(x)_i = x_i x_{i+1}$$

has a quasi-attractor $\{0^\infty\}$.

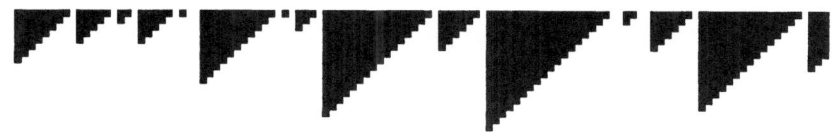

FIG. 9. *Q: A quasi-attractor.*

The point 0^∞ is equicontinuous but not attracting. As in Example 1, we have

$$Y = \omega(2^{\mathbb{Z}}) = \{x \in 2^{\mathbb{Z}} : \forall k > 0, 10^k 1 \not\sqsubseteq x\}.$$

For every $m \in \mathbb{Z}$, $[0]_m$ is a clopen invariant set and

$$Y_m = \omega([0]_m) = Y \cap \{x \in 2^{\mathbb{Z}} : \forall i \leq m, x_i = 0\}$$

is an attractor. For example Y_0 consists of all points $0^\infty.1^n 0^\infty$ where $0 \leq n \leq \infty$. We have $Y_{m+1} \subset Y_m$ and $\bigcap_{m \geq 0} Y_m = \{0^\infty\}$ is the unique minimal quasi-attractor.

13. Unique attractor. Recall that a DS (X, F) has a unique attractor iff $\omega(X)$ is chain transitive (Akin [1], p.66).

THEOREM 13.1. *An equicontinuous CA has either two disjoint attractors or an attracting fixed point, which is the unique attractor.*

Proof. Suppose that $(A^{\mathbb{Z}}, F)$ is an equicontinuous CA which has not disjoint attractors. By Theorem 5.2, there exists a preperiod $m \geq 0$ and a period $p > 0$, such that $F^{m+p} = F^m$. Assume that there exist two distinct attractors $Z \subset Y \subseteq A^{\mathbb{Z}}$. For $y_1 \in Y \setminus Z$ construct a sequence $(y_n \in Y)_{n \geq 1}$ with $F(y_{n+1}) = y_n$. This is possible, since $F(Y) = Y$, and clearly, $y_n \in Y \setminus Z$. There exists a converging subsequence $x = \lim_{i \to \infty} y_{n_i} \in Y$. It follows that x is a nonwandering point. Since it is eventually periodic, it must be periodic and $F^p(x) = x$. There exists $\varepsilon > 0$ such that $B_\varepsilon(F^i(x)) \cap Z = \emptyset$ for all $i < p$. Since x is equicontinuous, there exists $\xi > 0$ such that if $d(z, x) < \xi$, then $d(F^n(z), F^n(x)) < \varepsilon$ for all n. By Proposition 11.1, $(A^{\mathbb{Z}}, F)$ has the shadowing property, so there exists $\delta > 0$ such that every δ-chain which starts in x, is ξ-shadowed by some point, so it remains ε-close to the orbit of x. Let $C_\delta(x)$ be the set of points which can be reached by δ-chains from x, and $V = \overline{C_\delta(x)}$ be its closure. Then $V \cap Z = \emptyset$. We show that V is inward, i.e., $F(V) \subseteq V^\circ$. There exists $\eta > 0$ such that for all $v, w \in A^{\mathbb{Z}}$,

$$d(v, w) < \eta \Rightarrow d(F(v), F(w)) < \delta/2.$$

Given $z \in V$, there exists $w \in C_\delta(x)$ with $d(z, w) < \eta$ and a δ-chain $x = x_0, \ldots, x_{n-1}, w$. If $d(v, F(z)) < \frac{\delta}{2}$, then

$$d(F(w), v) \leq d(F(w), F(z)) + d(F(z), v) < \delta$$

so $x = x_0, \ldots, x_{n-1}, v, w$ is a δ-chain, and $v \in V$. Thus $B_{\frac{\delta}{2}}(F(z)) \subseteq V$, so $F(V) \subseteq V^\circ$ and $\omega(V)$ is an attractor disjoint from Z and this is a contradiction. Thus $(A^{\mathbb{Z}}, F)$ has unique attractor $Y = \omega(A^{\mathbb{Z}})$. Since Y is a minimal attractor it is chain transitive. By Proposition 11.1, (Y, F) has the shadowing property, so it is transitive. By Theorem 5.3, Y consists of a single periodic orbit, which must be a fixed point by Theorem 12.1. □

The zero CA of example 5 is equicontinuous and has a unique attractor. The identity CA of Example 4 is equicontinuous and has disjoint attractors. There is also an example which is not surjective.

EXAMPLE 13 (J: disjoint attractors). $(3^{\mathbb{Z}}, J)$ where $J(x)_i = \lfloor \frac{x_i + 1}{2} \rfloor$, is equicontinuous, has disjoint attractors and is not surjective.

THEOREM 13.2. If a CA has an attracting fixed point which is a unique attractor, then it is equicontinuous.

Proof. Let a^∞ be the attracting fixed point, so $\omega(A^{\mathbb{Z}}) = \{a^\infty\}$. If $b \in A$, $b \neq a$, then $b \notin \mathcal{L}(\{a^\infty\})$, so for some $n > 0$, $b \notin F^n(A^{\mathbb{Z}})$. Taking the maximum of these n for all $b \in A \setminus \{a\}$, we get $F^n(A^{\mathbb{Z}}) = \{a^\infty\}$ and clearly F is equicontinuous. □

PROPOSITION 13.1. Let $(A^{\mathbb{Z}}, F)$ be a surjective CA. If $U \subseteq A^{\mathbb{Z}}$ is an invariant clopen set, then $F^{-1}(U) = U$ and $F(U) = U$.

Proof. If $F(U) \subset U$, then $U \subseteq F^{-1}F(U) \subseteq F^{-1}(U)$. Suppose that there exists $x \in F^{-1}(U) \setminus U$. There exists an open set $V \ni x$ with $V \cap U = \emptyset$ and $F(V) \subseteq U$. There exists $n > 0$ and words $u_0, \ldots, u_{m-1}, u_m \in A^{2n+1}$ such that

$$U = [u_0]_{-n} \cup \cdots \cup [u_{m-1}]_{-n}, \quad [u_m]_{-n} \subseteq V.$$

Define an integer matrix of the type $(m+1) \times m$ by

$$M_{ij} = |\{w \in A^{2n+2r+1} : w_{\langle r, r+2n \rangle} = u_i, \ w \in f^{-1}(u_j)\}|.$$

Here r is the radius and $f : A^{2r+1} \to A$ the local rule. Since F is surjective, for every $j < m$, $|f^{-1}(u_j)| = |A|^{2r}$ by Theorem 6.1, so

$$\sum_{i=0}^{m} M_{ij} \leq |A|^{2r} \quad \text{and} \quad \sum_{i=0}^{m} \sum_{j=0}^{m-1} M_{ij} \leq m|A|^{2r}.$$

Since $F(U) \subseteq U$ and $F(V) \subseteq U$, for every $i \leq m$,

$$\sum_{j=0}^{m-1} M_{ij} = |A|^{2r} \text{ and } \sum_{i=0}^{m} \sum_{j=0}^{m-1} M_{ij} = (m+1)|A|^{2r}$$

and this is a contradiction. Thus $F^{-1}(U) = U$ and $F(U) = U$. $\square$

THEOREM 13.3. *A surjective CA has either a unique attractor or a pair of disjoint attractors.*

Proof. Suppose that a surjective CA $(A^{\mathbb{Z}}, F)$ has at least two attractors. There exists a nonempty clopen invariant set $U \neq A^{\mathbb{Z}}$ and by Proposition 13.1, $V = A^{\mathbb{Z}} \setminus U$ is a clopen invariant set too. Thus $\omega(U)$ and $\omega(V)$ are disjoint attractors. $\square$

EXAMPLE 14 (U: A unique attractor, Gilman [15]). $(2^{\mathbb{Z}}, U)$ *where*

$$U(x)_i = x_{i+1}x_{i+2} \; .$$

FIG. 10. *U: A unique attractor.*

Similarly as in Example 1 we show

$$\omega(2^{\mathbb{Z}}) = \{x \in 2^{\mathbb{Z}} : \forall n > 0, 10^n1 \not\sqsubseteq x\}$$

$\omega(2^{\mathbb{Z}})$ is not transitive. If $x \in [10]_0 \cap \omega(2^{\mathbb{Z}})$, then $x_{\langle 1,\infty \rangle} = 0^\infty$, so for every $n > 0$, $F^n(x) \notin [11]_0$. The following chains show that $\omega(2^{\mathbb{Z}})$ is chain transitive

0000	0001	1111	01	1111	0
0001	11	1110	0	1100	0
0011	11	1000	0	0000	
0111	11	0000			
1111					

It follows that the CA does not have the shadowing property and by Theorem 11.2, its factor subshifts $\Sigma_{\langle -n,n \rangle}(U)$ are not sofic.

EXAMPLE 15 (C: Coven's CA [9]). $(2^{\mathbb{Z}}, C)$ *where*

$$C(x)_i = x_i + x_{i+1}(x_{i+2} + 1) \bmod 2$$

FIG. 11. *Coven's CA.*

is almost equicontinuous, has a unique attractor and does not have the shadowing property (Blanchard and Maass [4]). It is left closing but not open.

The CA is left closing since it is left permutive. It is not right closing, since it has not constant number of preimages. 0^∞ has the only preimage itself, while 1^∞ has two preimages $(01)^\infty$ and $(10)^\infty$. To show other properties, observe first that for every $a, b \in 2$, $f(1a1b) = 1c$ where $c = a + b + 1$ (here f is the local rule and the addition is modulo 2). Define a CA $(2^{\mathbb{Z}}, G)$ by $G(x)_i = x_i + x_{i+1} + 1 \bmod 2$ and a map $\varphi : 2^{\mathbb{Z}} \to 2^{\mathbb{Z}}$ by

$$\varphi(x)_{2i} = 1, \quad \varphi(x)_{2i+1} = x_i .$$

Then $\varphi : (2^{\mathbb{Z}}, G) \to (2^{\mathbb{Z}}, C)$ is an injective homomorphism, so $(2^{\mathbb{Z}}, G)$ is a transitive subsystem of $(2^{\mathbb{Z}}, C)$.

LEMMA 13.1. *For every $n \geq 0$,*

$$x_{\langle 0,1 \rangle} = 10 \Rightarrow \forall i \leq n, C^i(x)_{\langle 0,1 \rangle} \in \{10, 11\} .$$

Proof. The statement clearly holds for $n = 0$. Assume that it holds for n and let us prove it for $n + 1$. Let $m > 0$ be the first integer for which $C^m(x)_{\langle 0,1 \rangle} \neq 10$, so $C^m(x)_{\langle 0,1 \rangle} = 11$ and $C^{m-1}_{\langle 0,3 \rangle} = 1010$. By the induction hypothesis

$$C^n(x)_{\langle 0,3 \rangle} \in \{1010, 1011, 1110, 1111\}$$

so $C^{n+1}(x)_{\langle 0,1 \rangle} \in \{10, 11\}$. $\square$

LEMMA 13.2. *The word 000 is a 2-blocking word with offset 0.*

Proof. Let $x_{\langle 0,2 \rangle} = 000$ and let $n > 0$ be the first integer with $C^n(x)_{\langle 0,2 \rangle} \neq 000$ so $C^n(x)_{\langle 0,2 \rangle} = 001$ and $C^{n-1}(x)_{\langle 0,4 \rangle} = 00010$. By Lemma 13.1, $C^m(x)_3 = 1$ for all $m \geq n$, so $C^k(x)_{\langle 0,1 \rangle} = 00$ for all $k \geq 0$. $\square$

We show now that for every $n > 0$ there are chains $1^{n+1} \to 01^n \to 1^{n+1}$. This follows from the fact that a transitive system $(2^{\mathbb{Z}}, G)$ is a subsystem of $(2^{\mathbb{Z}}, C)$. The chains differ slightly when n is even or odd.

11111	10	01111	10	111111	0	011111	0
11110	0	01110	0	111101	11	011101	11
11010	10	01010	10	110101	0	010101	0
01111		11111		011111		111111	

Since $(2^{\mathbb{Z}}, C)$ is almost equicontinuous, it is not transitive. Since it is chain transitive, it does not have the shadowing property. However, its factor subshifts are sofic, so the converse of Theorem 11.2 is not true. For example

$$\Sigma_{\langle 0,1 \rangle}(C) = \{10, 11\}^{\mathbb{N}} \cup \{11, 01\}^{\mathbb{N}} \cup \{01, 00\}^{\mathbb{N}} .$$

14. Classification. We have seen that many topological properties of CA are closely interrelated while others are independent. The following two tables summarize some of the results obtained.

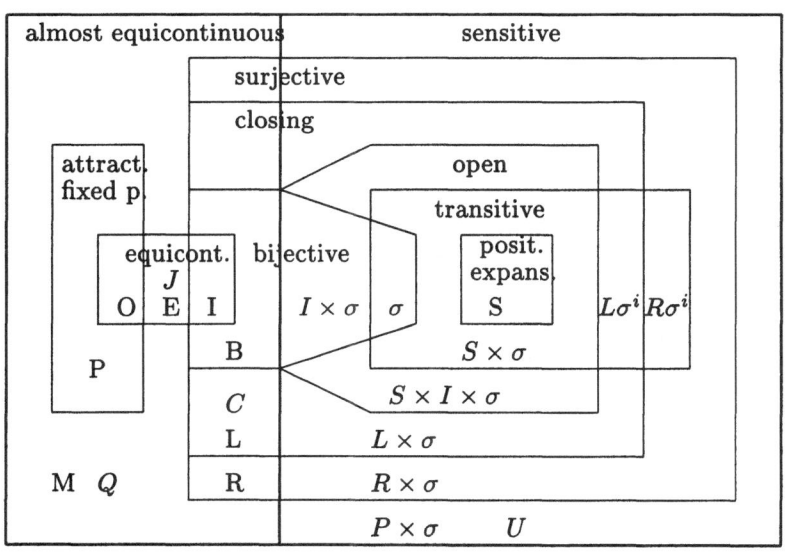

FIG. 12. *Equicontinuity classes.*

COROLLARY 14.1.

1. *Every positively expansive CA is open and transitive (Theorems 10.1, 10.2).*
2. *Every transitive CA is surjective and sensitive (Corollary 5.1).*
3. *Every open CA is closing (Theorem 9.3).*
4. *Every closing CA is surjective (Proposition 9.1).*
5. *Every open and almost equicontinuous CA is bijective (Theorem 8.2).*

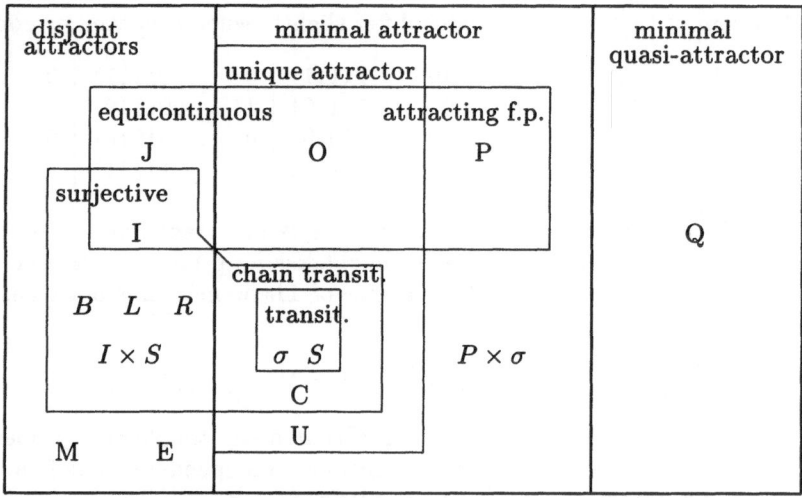

FIG. 13. *Attractor classes.*

6. *Every surjective equicontinuous CA is bijective (Theorem 6.5).*

7. *Every CA is either sensitive or almost equicontinuous (Corollary 5.1).*

8. *Every CA with an attracting fixed point is almost equicontinuous.*

9. *Every transitive system is chain transitive.*

10. *Every equicontinuous CA has either two disjoint attractors or an attracting fixed point, which is the unique attractor (Theorem 13.1).*

11. *Every CA whose unique attractor is an attracting fixed point is equicontinuous (Theorem 13.2).*

12. *Every surjective CA has either a unique attractor or a pair of disjoint attractors (Theorem 13.3).*

13. *Every CA has either two disjoint attractors or a unique minimal attractor or a unique minimal quasi-attractor (Corollary 12.1)*

The nonemptiness of various CA classes is shown by following examples:

P: The product CA of Example 1 has an attracting fixed point, it is almost equicontinuous but not equicontinuous.

M: The majority CA of Example 2 is almost equicontinuous. It is not equicontinuous and it does not have an attracting fixed point.

S: The sum CA of Example 3 is positively expansive.

I: The identity CA of Example 4 is equicontinuous, bijective and has disjoint attractors.

O: The zero CA of Example 5 is equicontinuous and has an attracting fixed point.

E: Example 6 is equicontinuous, not bijective and does not have an attracting fixed point.

L: Example 8 is right closing but not left closing, so it is not open.

B: Example 9 is bijective and almost equicontinuous but not equicontinuous.

R: Example 10 is neither left nor right-closing but it is surjective

J: Example 13 has disjoint attractors, it is equicontinuous and not surjective.

U: Example 14 has a unique attractor, it is not surjective.

C: Coven's CA of Example 15 is almost equicontinuous, has a unique attractor, it is surjective but not transitive.

σ: The shift CA is bijective, transitive, surjective and sensitive.

$L \circ \sigma^i$: is transitive (for high enough i) and not open.

$R \circ \sigma^i$: is transitive (for high enough i) and not closing.

Q: Example 12 has a quasi-attractor.

15. Open problems. The relationships among some CA classes are still unknown. One interesting class form CA which are chain transitive without being transitive. Coven's CA of Example 15 belongs to this class and it is closing. Does there exist an open or even bijective CA which is chain transitive without being transitive?

A long standing problem is density of periodic points in surjective CA. For two particular classes this property has been recently established: almost equicontinuous surjective CA (Theorem 6.4, Blanchard and Tisseur [6]) and closing CA (Theorem 9.2, Boyle and Kitchens [8]). For general surjective CA, the problem seems to be still open.

Another hard problem has been considered in Nasu [25] as an analogue of Theorem 10.1 which says that every positively expansive CA is conjugate to a onesided SFT. A bijective DS (X, F) is said to be **expansive**, if

$$\exists \varepsilon > 0, \forall x \neq y \in X, \exists n \in \mathbb{Z}, d(F^n(x), F^n(y)) \geq \varepsilon .$$

Every bijective expansive DS on $A^{\mathbb{Z}}$ is conjugate to a twosided subshift. Is any bijective expansive CA conjugate to a twosided SFT? Or is it at least transitive?

Finally there are many particular CA whose topological properties are unknown. One of the most interesting CA is Coven's CA of Example 15 generalized in Blanchard and Maass [4] to a class of aperiodic Coven CA. The classification of periodic Coven's automata seems to be even more difficult. While they are closing, it is unknown whether they are transitive, sensitive or chain transitive. One particular case is

EXAMPLE 16 (V: Periodic Coven's CA). $(2^{\mathbb{Z}}, V)$ *where*

$$V(x)_i = x_i + x_{i+1}x_{i+2} \bmod 2$$

FIG. 14. *V: Periodic Coven's CA.*

REFERENCES

[1] E. AKIN, *The General Topology of Dynamical Systems*, AMS, 1991.

[2] E. AKIN, J. AUSLANDER, AND K. BERG, *When is a transitive map chaotic ?* Conference in Ergodic Theory and Probability (eds. Bergelson, March & Rosenblatt) (1996), de Gruyter and Co., pp. 25–40.

[3] F. BLANCHARD AND A. MAASS, *Dynamical properties of expansive one-sided cellular automata*, Israel Journal of Mathematics **99** (1997), pp. 149–174.

[4] F. BLANCHARD AND A. MAASS, *Dynamical behaviour of Coven's cellular automata*, Theoret. Computer Sci. **163** (1996), pp. 291–302.

[5] F. BLANCHARD, P. KŮRKA, AND A. MAASS, *Topological and measure-theoretic properties of one-dimensional cellular automata*, Physica D **103** (1997), pp. 86–99.

[6] F. BLANCHARD AND P. TISSEUR, *Some properties of cellular automata with equicontinuous points*, to appear in Annalles de l'IHP, serie de Probabilités et Statistiques.

[7] M. BOYLE, D. FIEBIG, U. FIEBIG, *A dimension group for local homeomorphisms and endomorphisms of one-sided shifts of finite type*, J. Reine Angew. Math. **497** (1997), pp. 27–59.

[8] M. BOYLE AND B. KITCHENS, *Periodic points for onto cellular automata*, Indagationes Math., 1999.

[9] E.M. COVEN AND G.A. HEDLUND, *Periods of some non-linear shift registers*, J.Combinatorial Th. **A 27** (1979), pp. 186197.

[10] E.M. COVEN, *Topological entropy of block maps*, Proc. Amer. Math. Soc. **78** (1980), pp. 590–594.

[11] E.M. COVEN AND M. PAUL, *Endomorphisms of irreducible shifts of finite type*, Math. System Th. **8** (1974), pp. 167–175.

[12] K. CULIK II., L.P. HURD, AND S. YU, *Computation theoretic aspects of cellular automata*, Physica D **45** (1990), pp. 357–378.

[13] K. CULIK II., J. PACHL, AND S. YU, *On the limit sets of cellular automata*, SIAM J.Comput. **18** (1989), pp. 831–842.

[14] M. GARDNER, *Mathematical games*, Scientific American, October 1970, February 1971.

[15] R.H. GILMAN, *Notes on cellular automata*, manuscript 1988.

[16] R.H. GILMAN, *Classes of cellular automata*, Ergod. Th. & Dynam. Sys. **7** (1987), pp. 105–118.

[17] R.H. GILMAN, *Periodic behavior of linear automata*, in Dynamical Systems, J.C. Alexander, (ed.), Lecture Notes in Mathematics 1342, Springer-Verlag, Berlin 1988.

[18] E. GLASNER AND B. WEISS, *Sensitive dependence on initial conditions*, Nonlinearity **6** (1993), pp. 1067–1075.

[19] G.A. HEDLUND, *Endomorphisms and automorphisms of the shift dynamical system*, Math. Sys. Th. **3** (1969), pp. 320–375.

[20] M. HURLEY, *Attractors in cellular automata*, Ergod. Th. & Dynam. Sys. **10** (1990), pp. 131–140.

[21] P. KŮRKA, *Languages, equicontinuity and attractors in cellular automata*, Ergod. Th. & Dynam. Sys. **17** (1997), pp. 417–433.

[22] D. LIND, B. MARCUS, *An Introduction to Symbolic Dynamics and Coding*, Cambridge University Press, Cambridge, 1995.

[23] D.A. LIND, *Applications of ergodic theory and sofic systems on cellular automata*, Physica **D 10** (1984), pp. 36–44.

[24] A. MAASS, *On the sofic limit sets of cellular automata*, Ergod. Th. & Dynam. Syst. **15** (1995), pp. 663–684.

[25] M. NASU, *Textile systems for endomorphisms and automorphisms of the shift*, Mem. Amer. Math. Soc. **546**, 1995.

[26] J. VON NEUMANN, *The general and logical theory of automata*, Cerebral Mechanics of Behaviour, L.A. Jeffress (ed.), Wiley, New York 1951.

[27] W. PARRY, *Symbolic dynamics and transformations of the unit interval*, Trans. Amer. Math. Soc. **122** (1966), pp. 368–378.

[28] S. ULAM, *Random processes and transformations*, Proc. Int. Congress of Math. **2** (1952), pp. 264–275.

[29] P. WALTERS, *On the pseudo-orbit tracing property and its relationship to stability*, Lecture Notes in Mathematics 668, pp. 231–244, Springer-Verlag, 1987.

[30] S. WOLFRAM, *Computation theory of cellular automata*, Comm. Math. Phys. **96** (1984), pp. 15–57.

[31] S. WOLFRAM, *Theory and Applications of Cellular Automata*, World Scientific, Singapore 1986.

A SPANNING TREE INVARIANT FOR MARKOV SHIFTS

DOUGLAS LIND* AND SELIM TUNCEL*

Abstract. We introduce a new type of invariant of block isomorphism for Markov shifts, defined by summing the weights of all spanning trees for a presentation of the Markov shift. We give two proofs of invariance. The first uses the Matrix-Tree Theorem to show that this invariant can be computed from a known invariant, the stochastic zeta function of the shift. The second uses directly the definition to show invariance under state splitting, from which all block isomorphisms can be built.

Key words. Markov shift, block isomorphism, spanning tree, Matrix-Tree Theorem.

AMS(MOS) subject classifications. Primary: 37A35, 37A50, 37B10, 60J10.

1. Introduction. Invariants of dynamical systems typically make use of recurrent or asymptotic behavior. Examples include entropy, mixing, and periodic points. Here we define a quantity for stochastic Markov shifts that is invariant under block isomorphism, and which has a different flavor. For a given presentation of the Markov shift, we add up the weights of all spanning trees for the graph. Since spanning trees are maximal subgraphs without loops, this is in some sense an operation that is orthogonal to recurrent behavior.

We prove invariance of the spanning tree quantity under block isomorphism in two ways. The first shows that it can be computed from the stochastic zeta function of the Markov shift, an invariant introduced in [6]. The second is a more "bare-hands" structural approach, using only the definition to show that it is invariant under the elementary block isomorphisms corresponding to state splittings.

2. The Matrix-Tree Theorem. In this section we give a brief account of the Matrix-Tree Theorem for directed graphs. See [1, II.3] for more details.

Let G be a (finite, directed) graph. We suppose that the vertex set of G is $\mathcal{V} = \{1, 2, \ldots, v\}$. We sometimes call vertices *states*. Let $\mathcal{E}$ be the edge set of G. Denote the subset of edges from state i to state j by $\mathcal{E}_i^j$. Put $\mathcal{E}_i = \bigcup_j \mathcal{E}_i^j$, the set of all edges starting at state i, and $\mathcal{E}^j = \bigcup_i \mathcal{E}_i^j$, the set of all edges ending at state j.

A *tree in G rooted at $r \in \mathcal{V}$* is a subgraph T of G such that every vertex in T except r has a unique outgoing edge in T, there is no outgoing edge in T at r, and from every vertex in T except r there is a unique path ending at r. See Figure 1(a). We abbreviate this by saying that T is a *tree in G*

---

*Department of Mathematics, University of Washington, Box 354350, Seattle, WA 98195. The authors were supported in part by NSF Grant DMS-9622866.

at r. A tree is *spanning* if it contains every state. Let $\mathcal{S}_r$ denote the set of spanning trees at r, and $\mathcal{S} = \bigcup_r \mathcal{S}_r$ be the set of all spanning trees in G.

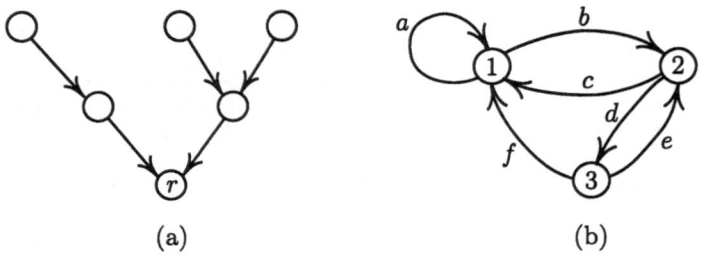

(a)   (b)

FIG. 1. *A typical tree at* r, *and a graph.*

Consider the elements of $\mathcal{E}$ to be commuting abstract variables, and form the ring $\mathbb{Z}[\mathcal{E}]$ of polynomials in the variables from $\mathcal{E}$ with integer coefficients. For any subgraph H of G define the *weight* of H to be $\prod_{e \in H} e \in \mathbb{Z}[\mathcal{E}]$, where the product is over the edges in H. For a subset $\mathcal{F} \subset \mathcal{E}$ put $\Sigma(\mathcal{F}) = \sum_{e \in \mathcal{F}} e \in \mathbb{Z}[\mathcal{E}]$. The *Kirchhoff matrix* K of G is the $v \times v$ matrix $K = [K_{ij}]$ defined by

$$K_{ij} = \Sigma(\mathcal{E}_i)\delta_{ij} - \Sigma(\mathcal{E}_i^j),$$

where $\delta_{ij} = 1$ if $i = j$ and 0 otherwise. Notice that no self-loops occur in K. Let $K^{(r)}$ denote the rth principal minor of K, that is the determinant of the matrix formed by removing the rth row and rth column from K. Let adj K be the adjoint matrix of K, and let tr denote the trace of a matrix.

THEOREM 2.1 (Matrix-Tree Threorem [1, II.3]). *Using the notations above,*

$$\sum_{S \in \mathcal{S}_r} w(S) = K^{(r)}, \quad \text{and so} \quad \sum_{S \in \mathcal{S}} w(S) = \text{tr}[\text{adj } K] .$$

EXAMPLE 1. For the graph in Figure 1(b),

$$K = \begin{bmatrix} b & -b & 0 \\ -c & c+d & -d \\ -f & -e & e+f \end{bmatrix}.$$

Then $K^{(1)} = ce + cf + df$ enumerates the spanning trees at 1, and similarly for $K^{(2)} = be + bf$ and $K^{(3)} = bd$.

3. Markov shifts. Let $P = [p_{ij}]$ be a $v \times v$ stochastic matrix, so that $p_{ij} \geq 0$ and $\sum_j p_{ij} = 1$ for every i. We assume from now on that P is irreducible. Let $G(P)$ be the directed graph with vertex set $\mathcal{V} = \{1, \ldots, v\}$, and with exactly one edge from state i to state j if $p_{ij} > 0$, and no such edge if $p_{ij} = 0$. Let $\mathcal{E}$ denote the resulting edge set for $G(P)$.

The *shift of finite type* determined by $G(P)$ is the subset $X_{G(P)}$ of $\mathcal{E}^{\mathbb{Z}}$ defined by

$$X_{G(P)} = \{\ldots e_{-1}e_0e_1\cdots \in \mathcal{E}^{\mathbb{Z}} : e_{n+1} \text{ follows } e_n \text{ in } G(P) \}.$$

See [2, Chap. 2] for further details.

By the irreducibility assumption, there is a unique Markov probability measure μ_P on $X_{G(P)}$ with transition probabilities p_{ij}. Let σ_P denote the left shift on $X_{G(P)}$, so that μ_P is σ_P-invariant. The measure-preserving system $(X_{G(P)}, \mu_P, \sigma_P)$ is the *Markov shift determined by* P.

Let Q be another stochastic matrix, of possibly different dimension. We say that the Markov shifts determined by P and by Q are *block isomorphic* if there is a shift-commuting measure-preserving homeomorphism between them. In other words, a block isomorphism from $(X_{G(P)}, \mu_P, \sigma_P)$ to $(X_{G(Q)}, \mu_Q, \sigma_Q)$ is a homeomorphism $\psi : X_{G(P)} \rightarrow X_{G(Q)}$ such that $\sigma_Q \circ \psi = \psi \circ \sigma_P$ and $\mu_Q = \mu_P \circ \psi^{-1}$.

4. The spanning tree invariant. As in the previous section, let $P = [p_{ij}]$ be an irreducible stochastic matrix and $G = G(P)$ be its associated directed graph. If $e \in \mathcal{E}$ goes from i to j put $p(e) = p_{ij}$. For any subgraph H of G define the *P-weight* (or simply the *weight*) of H to be $w_P(H) = \prod_{e \in H} p(e)$.

DEFINITION 4.1. *Let P be an irreducible stochastic matrix, and let $\mathcal{S}$ denote the set of spanning trees for $G(P)$. Define the* spanning tree invariant *of P to be*

$$\tau(P) = \sum_{S \in \mathcal{S}} w_P(S).$$

REMARK 4.1. Sums of weights of spanning trees arise naturally in probability theory on graphs and networks, and are an important tool (see [4] and [3] for example).

EXAMPLE 2. (1) If $P = \begin{bmatrix} p & 1-p \\ q & 1-q \end{bmatrix}$ then $\tau(P) = 1 - p + q$. In particular, if $p = q = 1/2$ then $\tau(P) = 1$.

(2) If

$$P = \begin{bmatrix} 0 & 1/2 & 1/2 \\ 1/2 & 0 & 1/2 \\ 1/2 & 1/2 & 0 \end{bmatrix}, \quad \text{then } \tau(P) = \frac{9}{4}.$$

Note that there is a uniformly three-to-one measure-preserving factor map from this Markov shift onto the Bernoulli shift in part (1) with $p = q = 1/2$. Thus τ is not in general preserved by such factor maps.

To justify its name, we will prove that τ is invariant under block isomorphism.

THEOREM 4.1. *If P and Q are irreducible stochastic matrices whose associated Markov shifts are block isomorphic, then $\tau(P) = \tau(Q)$.*

We will give two proofs of invariance. The first computes $\tau(P)$ in terms of a known invariant, the stochastic zeta function of P. The second is more "structural," showing proving invariance of τ for each of the basic building blocks of a block isomorphism.

5. First proof of invariance. Define $\phi_P \colon \mathbb{Z}[\mathcal{E}] \to \mathbb{R}$ on the variables e by $\phi_P(e) = p(e)$, and extend it to a ring homomorphism. Applying ϕ_P to the Matrix-Tree Theorem for $G = G(P)$ gives

$$
\tau(P) = \sum_{S \in \mathcal{S}} w_P(S) = \sum_{S \in \mathcal{S}} \phi_P\left(w(S)\right) = \phi_P\left(\sum_{S \in \mathcal{S}} w(S)\right)
$$
$$
= \phi_P\left(\text{tr}[\text{adj}\, K]\right) = \text{tr}\left[\text{adj}\, \phi_P(K)\right].
$$

Now $\phi_P(K) = I - P$ since P is stochastic. Hence $\tau(P) = \text{tr}\left[\text{adj}(I - P)\right]$.

Let the eigenvalues of P be $\lambda_1 = 1$, λ_2, ..., λ_v, where $\lambda_j \neq 1$ and $|\lambda_j| \leq 1$ for $2 \leq j \leq v$. Since formation of the adjoint commutes with conjugation and trace is invariant under conjugation, conjugating P to its Jordan form shows that

$$
(5.1) \qquad \tau(P) = \text{tr}\left[\text{adj}(I - P)\right] = \prod_{j=2}^{v}(1 - \lambda_j).
$$

Recall the *stochastic zeta function* $\zeta_P(t)$ of P, defined in [6] as

$$
\zeta_P(t) = \exp\left[\sum_{n=1}^{\infty} \frac{t^n}{n} \sum_{C \in \mathcal{C}_n} w_P(C)\right],
$$

where $\mathcal{C}_n$ is the set of all cycles in $G(P)$ of length n. The stochastic zeta function is invariant under block isomorphism. It can be computed in terms of P as

$$
\zeta_P(t) = \frac{1}{\det[I - tP]}.
$$

Hence

$$
(1/\zeta_P)(t) = \det[I - tP] = \prod_{k=1}^{v}(1 - \lambda_k t),
$$

so that

$$
(1/\zeta_P)'(t) = \sum_{k=1}^{v} -\lambda_k \prod_{j \neq k}(1 - \lambda_j t).
$$

Thus

$$(1/\zeta_P)'(1) = - \prod_{j=2}^{v}(1 - \lambda_j) = -\tau(P).$$

This shows that $\tau(P)$ can be computed from ζ_P, and hence is an invariant of block isomorphism.

6. Invariance under in-splitting. Every block isomorphism between Markov shifts is a composition of basic block isomorphisms obtained from state splitting and permuting states. This was a fundamental discovery of R. Williams [7]. For further background on state splitting and the decomposition of block isomorphisms, the reader is referred to [6] as well as §2.4 and Theorem 7.1.2 of [2]. Permuting states clearly preserves τ, so we focus on the behavior of τ under state splitting.

Let k be a fixed state in $G = G(P)$. There are two types of state splitting at k: in-splitting from a partition of the incoming edges to k, and out-splitting from a partition of the outgoing edges from k. These are handled by separate arguments, in-splitting in this section and out-splitting in the next. As might be expected from the directional nature of shifts of finite type, in-splitting is easier to handle that out-splitting.

It is sufficient, as well as notationally simpler, to consider in-splitting k into just two states. For this we partition $\mathcal{E}^k$ into the sets $\mathcal{F}_1$ and $\mathcal{F}_2$. Form a new graph G' as follows. Replace state k with two new states k_1 and k_2. Every edge in G from k to $j \neq k$ is duplicated as two edges in G', one from k_1 to j and one from k_2 to j. An edge f from i to k lies in either $\mathcal{F}_1$ or $\mathcal{F}_2$. If $f \in \mathcal{F}_1$, then in G' put a corresponding edge from i to k_1 and no edge from i to k_2 (if $i = k$, then in G' there are edges from both k_1 and k_2 to k_1); similarly if $f \in \mathcal{F}_2$. Figure 2 depicts such an in-splitting.

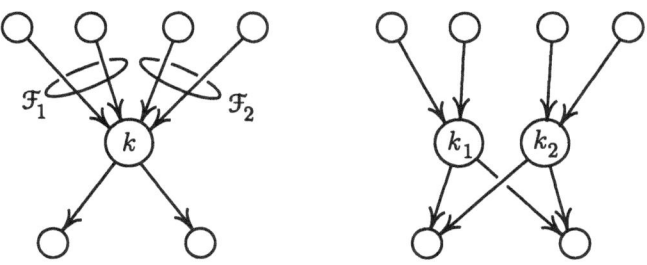

FIG. 2. *In-splitting a state.*

Preservation of measure shows that under such an in-splitting the transition matrix P becomes P' on G' defined as follows. For notational convenience we use $p'(i, j)$ instead of p'_{ij}. If $i, j \neq k_m$ ($m = 1, 2$) then $p'(i, j) = p(i, j)$. If $i \neq k_1, k_2$ and the edge from i to k is in $\mathcal{F}_m$ then $p'(i, k_m) = p(i, k)$. Finally, if the edge from k to k is in $\mathcal{F}_m$, then $p'(k_n, k_m) = p(k, k)$ for

$n = 1, 2$. For example, if $k = 1$, $\mathcal{F}_1 = \{1, \ldots, \ell\}$, and $\mathcal{F}_2 = \{\ell + 1, \ldots, v\}$, then P' has the form

$$P' = \begin{bmatrix} p_{11} & 0 & p_{12} & \cdots & p_{1v} \\ p_{11} & 0 & p_{12} & \cdots & p_{1v} \\ p_{21} & 0 & p_{22} & \cdots & p_{2v} \\ \vdots & \vdots & \vdots & & \vdots \\ p_{\ell 1} & 0 & p_{\ell 2} & \cdots & p_{\ell v} \\ 0 & p_{\ell+1,1} & p_{\ell+1,2} & \cdots & p_{\ell+1,v} \\ \vdots & \vdots & \vdots & & \vdots \\ 0 & p_{v1} & p_{v2} & \cdots & p_{vv} \end{bmatrix}.$$

We next construct a correspondence between certain sets of spanning trees in G and similar sets in G'. Consider a triple of the form (T, T_1, T_2), where T is a tree in G at some vertex r, each T_m is a tree in G at k using only edges from $\mathcal{F}_m$ ($m = 1, 2$), the three trees are disjoint except for the common vertex k, and they span all vertices of G. We specifically allow the possibility of the empty tree (with no vertices or edges), and also the tree consisting of a single vertex and no edges. In particular, if $k = r$ then T is empty.

Each such triple (T, T_1, T_2) in G corresponds to a triple (T', T'_1, T'_2) in G', where T is copied over to T' verbatim, and T'_m is the tree at k_m obtained from T_m ($m = 1, 2$). Figure 3 illustrates this correspondence. This triple has the property that T' is a tree in G' at r, T'_m is a tree in G' at k_m for $m = 1, 2$, all three trees are disjoint, and they span the vertices of G'. There is clearly a one-to-one correspondence between the set of such triples in G and those in G'.

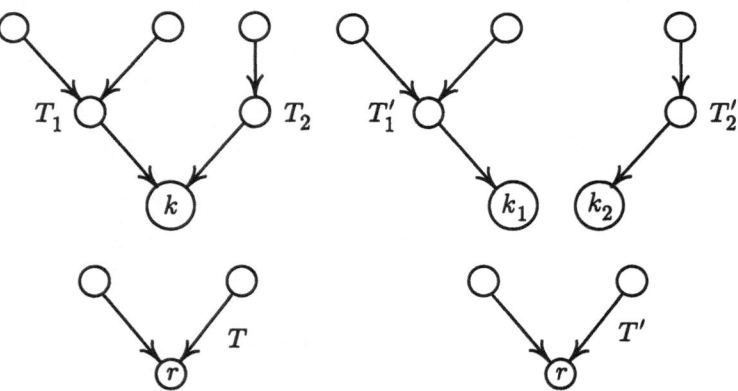

FIG. 3. *Correspondence of trees under an in-splitting.*

Let $\mathcal{S}(T, T_1, T_2)$ be the set of all spanning trees in G at r containing T, T_1, and T_2. Similarly define $\mathcal{S}(T', T'_1, T'_2)$ in G'. Clearly the set $\mathcal{S}(G)$ of spanning trees in G is the disjoint union of the $\mathcal{S}(T, T_1, T_2)$ over all possible triples, and similarly $\mathcal{S}(G')$ is the disjoint union of the $\mathcal{S}(T', T_1', T_2')$. Hence to prove that $\tau(P) = \tau(P')$, it suffices to show that

$$(6.1) \qquad \sum_{S \in \mathcal{S}(T, T_1, T_2)} w_P(S) = \sum_{S' \in \mathcal{S}(T', T_1', T_2')} w_{P'}(S').$$

Fix a triple (T, T_1, T_2). The only way to create a spanning tree in G at r containing these trees is to add an edge from k to T. Thus if $p(k, T)$ denotes the sum of the transition probabilities from k to the vertices of T, it follows that

$$\sum_{S \in \mathcal{S}(T, T_1, T_2)} w_P(S) = w_P(T) w_P(T_1) w_P(T_2) p(k, T).$$

Consider the corresponding triple (T', T_1', T_2') in G'. There are now three ways to form a spanning tree at r in G' containing these trees: (1) join k_1 to T_2' and k_2 to T', (2) join k_2 to T_1' and k_1 to T', and (3) join both k_1 and k_2 to T'. The contribution of adding these two edges to the total weight is, respectively, $p'(k_1, T_2') p'(k_2, T')$, $p'(k_2, T_1') p'(k_1, T')$, and $p'(k_1, T') p'(k_2, T')$. Hence

$$\sum_{S' \in \mathcal{S}(T', T_1', T_2')} w_{P'}(S') = w_{P'}(T') w_{P'}(T_1') w_{P'}(T_2') \times \big[p'(k_1, T_2') p'(k_2, T')$$
$$+ p'(k_2, T_1') p'(k_1, T') + p'(k_1, T') p'(k_2, T') \big]$$

Let us assume that if there is an edge from k to itself in G, then this edge lies is $\mathcal{F}_1$. Now $w_P(T) = w_{P'}(T')$, $w_P(T_1) = w_{P'}(T_1')$, and $w_P(T_2) = w_{P'}(T_2')$. Furthermore, $p'(k_1, T') = p'(k_2, T') = p(k, T)$, and $p'(k_1, T_2) = p(k, T_2) - p(k, k)$, $p'(k_2, T_1) = p(k, T_1)$. Hence

$$p'(k_1, T_2') p'(k_2, T') + p'(k_2, T_1') p'(k_1, T') + p'(k_1, T') p'(k_2, T')$$
$$= p(k, T) \big[p(k, T_1) + p(k, T_2) - p(k, k) + p(k, T) \big] = p(k, T)$$

since T_1 and T_2 are disjoint except for the common vertex k. This proves (6.1), and completes the proof that τ is invariant under in-splitting.

7. Invariance under out-splitting. To consider out-splittings, fix a state k in G. Partition the set $\mathcal{E}_k$ of outgoing edges from k into two sets $\mathcal{F}_1$ and $\mathcal{F}_2$. Form the out-split graph G' as follows. Replace k with two new states k_1 and k_2. Each incoming edge from a state $i \neq k$ to k is duplicated to two edges, one from i to k_1 and one from i to k_2. An edge $f \in \mathcal{F}_1$ from k to j induces a corresponding edge from k_1 to j in G' (if $j = k$, then include edges from k_1 to both k_1 and k_2), and similarly for $\mathcal{F}_2$. Figure 4 depicts a typical out-splitting at k.

The matrix P' on G' corresponding to P is defined as follows. Let $q = \sum_{e \in \mathcal{F}_1} p(e)$, so that $1 - q = \sum_{e \in \mathcal{F}_2} p(e)$. If $i, j \neq k_m$ put $p'(i, j) = p(i, j)$.

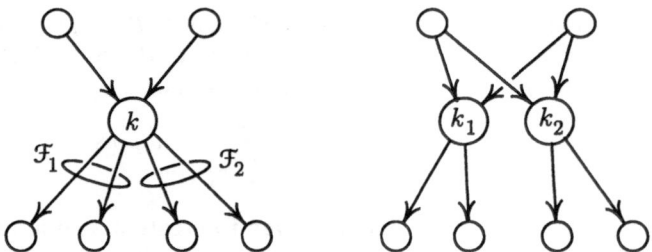

FIG. 4. *Out-splitting a state.*

If $j \neq k_m$ put $p'(k_1, j) = p(k, j)/q$ and $p'(k_2, j) = p(k, j)/(1 - q)$. If $i \neq k_m$ put $p'(i, k_1) = q\,p(i, k)$ and $p'(i, k_2) = (1 - q)p(i, k)$. Finally, if there is a loop at k, assume that it is contained in $\mathcal{F}_1$ (the alternative case is similar). Then put $p'(k_1, k_1) = q\,p(k, k)/q = p(k, k)$ and $p'(k_1, k_2) = (1-q)p(k, k)/q$. For example, if $k = 1$, $\mathcal{F}_1 = \{1, 2, \dots, \ell\}$, and $\mathcal{F}_2 = \{\ell + 1, \dots, v\}$, then

$$
P' = \begin{bmatrix}
\dfrac{q}{q}p_{11} & \dfrac{1-q}{q}p_{11} & \dfrac{1}{q}p_{12} & \cdots & \dfrac{1}{q}p_{1\ell} & 0 & \cdots & 0 \\[2mm]
0 & 0 & 0 & \cdots & 0 & \dfrac{p_{1,\ell+1}}{1-q} & \cdots & \dfrac{p_{1v}}{1-q} \\[2mm]
q\,p_{21} & (1-q)p_{21} & p_{22} & \cdots & p_{2\ell} & p_{2,\ell+1} & \cdots & p_{2v} \\[1mm]
\vdots & \vdots & \vdots & \vdots & \vdots & \vdots & & \vdots \\[1mm]
q\,p_{v1} & (1-q)p_{v1} & p_{v2} & \cdots & p_{v\ell} & p_{v,\ell+1} & \cdots & p_{vv}
\end{bmatrix}.
$$

Next, consider pairs (T, U) of subgraphs of G such that T is a tree at some vertex r, U is a tree at k, and T and U are disjoint and contain all vertices of G. For each such pair (T, U) let $\mathcal{B}$ denote the set of immediate predecessor states of k in U, so that $i \in \mathcal{B}$ if and only if the edge from i to k is in U. Each subset $B \subset \mathcal{B}$ induces two subtrees $U_1(B)$ and $U_2(B)$ rooted at k and which together span U, where $U_1(B)$ is the subtree of U including all predecessors in U of states in B, and $U_2(B)$ is defined similarly using $B^c = \mathcal{B} \setminus B$.

Each $B \subset \mathcal{B}$ then yields a triple $\big(T', U'_1(B), U'_2(B)\big)$ in G', where T' is copied directly from T, $U'_1(B)$ is the tree in G' at k_1 using the edges of $U_1(B)$, and $U'_2(B)$ is the tree in G' at k_2 using the edges of $U_2(B)$. Thus each pair (T, U) corresponds to the collection of triples $\{\big(T', U'_1(B), U'_2(B)\big) : B \subset \mathcal{B}\}$. Figure 5 illustrates this construction.

Let $\mathcal{S}(T, U)$ denote the set of spanning trees in G at r containing T and U, and $\mathcal{S}\big(T', U'_1(B), U'_2(B)\big)$ be the set of spanning trees in G' containing T', $U'_1(B)$, and $U'_2(B)$. Then $\mathcal{S}(G)$ is the disjoint union of the $\mathcal{S}(T, U)$ and $\mathcal{S}(G')$ is the disjoint union of the $\mathcal{S}\big(T', U'_1(B), U'_2(B)\big)$. Therefore it suffices to show that for each pair (T, U) we have that

$$
(7.1) \qquad \sum_{S \in \mathcal{S}(T, U)} w_P(S) = \sum_{B \subset \mathcal{B}} \ \sum_{S' \in \mathcal{S}(T', U'_1(B), U'_2(B))} w_{P'}(S').
$$

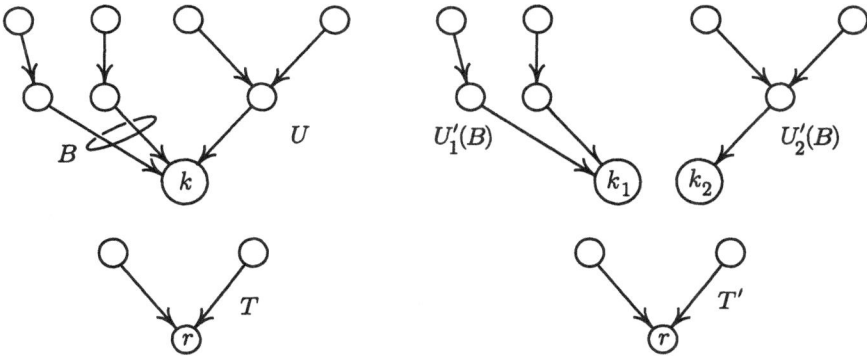

FIG. 5. *Correspondence of trees under an out-splitting.*

Fix a pair (T, U). Let

$$a_1 = p(\mathcal{F}_1, T) = \sum \{p(e) : e \in \mathcal{F}_1 \text{ and } e \text{ terminates in } T\},$$

and similarly $b_1 = p(\mathcal{F}_1, U)$, $a_2 = p(\mathcal{F}_2, T)$, and $b_2 = p(\mathcal{F}_2, U)$. Then $a_1 + b_1 = q$ and $a_2 + b_2 = 1 - q$.

The only additional edge needed to form a spanning tree from (T, U) is an edge from k to T. Hence

$$\sum_{S \in \mathcal{S}(T,U)} w_P(S) = w_P(T) w_P(U) p(k, T) = w_P(T) w_P(U) [a_1 + a_2].$$

Next, let $B \subset \mathcal{B}$, and form the triple $\big(T', U_1'(B), U_2'(B)\big)$. There are now three ways to form a spanning tree in $\mathcal{S}\big(T', U_1'(B), U_2'(B)\big)$: (1) join k_1 to $U_2'(B)$ and k_2 to T', (2) join k_2 to $U_1'(B)$ and k_1 to T', and (3) join both k_1 and k_2 to T'. Hence

$$\sum_{S' \in \mathcal{S}\big(T', U_1'(B), U_2'(B)\big)} w_{P'}(S') = w_{P'}(T') w_{P'}(U_1'(B)) w_{P'}(U_2'(B)) \Phi(B),$$

where

$$\Phi(B) = p'(\mathcal{F}_1, U_2'(B)) p'(k_2, T') + p'(\mathcal{F}_2, U_1'(B)) p'(k_1, T') \\ + p'(k_1, T') p'(k_2, T').$$

Note that $w_{P'}(T') = w_P(T)$. Let $n = |\mathcal{B}|$. Since $U_1'(B)$ uses $|B|$ incoming edges each of whose weight has been multiplied by the factor q, and $U_2'(B)$ uses $n - |B|$ edges each of whose weight is multiplied by a factor $1 - q$, we have that

$$w_{P'}(U_1'(B)) w_{P'}(U_2'(B)) = q^{|B|} (1 - q)^{n - |B|} w_P(U).$$

Cancelling the common term $w_P(T)w_P(U)$ reduces (7.1) to proving that

$$(7.2) \qquad a_1 + a_2 = \sum_{B \subset \mathcal{B}} q^{|B|}(1-q)^{n-|B|}\Phi(B).$$

Now $p'(k_1, T') = a_1/q$ and $p'(k_2, T') = a_2/(1-q)$. Let $\mathcal{F}_1(U)$ denote the set of edges in $\mathcal{F}_1$ ending in U. Then by interchanging the order of summation see that

$$\sum_{B \subset \mathcal{B}} q^{|B|}(1-q)^{n-|B|}p'(\mathcal{F}_1, U_2'(B)) = \sum_{e \in \mathcal{F}_1(U)} p'(e) \sum_{e \in B^c} q^{|B|}(1-q)^{n-|B|}$$

$$= \sum_{e \in \mathcal{F}_1(U)} \frac{p(e)}{q} \sum_{k=0}^{n-1} \binom{n-1}{k} q^k (1-q)^{n-k}$$

$$= \frac{1-q}{q} \sum_{e \in \mathcal{F}_1(U)} p(e) = \frac{1-q}{q} b_1.$$

Similarly,

$$\sum_{B \subset \mathcal{B}} q^{|B|}(1-q)^{n-|B|}p'(\mathcal{F}_2, U_1'(B)) = \frac{q}{1-q} b_2.$$

Since $b_1 = 1 - a_1$ and $b_2 = 1 - q - a_2$, we obtain that

$$\sum_{B \subset \mathcal{B}} q^{|B|}(1-q)^{n-|B|}\Phi(B) = \frac{1-q}{q}b_1 \frac{a_2}{1-q} + \frac{q}{1-q}b_2 \frac{a_1}{q} + \frac{a_1}{q}\frac{a_2}{1-q}$$

$$= \frac{a_2(q-a_1)}{q} + \frac{a_1(1-q-a_2)}{1-q} + \frac{a_1 a_2}{q(1-q)}$$

$$= a_1 + a_2.$$

This establishes (7.2), and completes the proof.

8. Concluding remarks. (1) The possibility of using spanning trees to define an invariant was first observed experimentally using *Mathematica*.

(2) It is possible to obtain finer invariants by use of the matrix of powers $P^t = [p_{ij}^t]$ as in [5].

(3) If P is $v \times v$, then (5.1) shows that $\tau(P) \le 2^{v-1}$. Thus $1 + \log_2 \tau(P)$ is a lower bound on the size of any irreducible Markov shift that is block isomorphic to P.

(4) Using elementary matrix operations, one can show directly that $\tau(P) = \tau(P')$, where P' is derived from P using in-splitting or out-splitting as above. This shows that τ is an invariant of block isomorphism without use of the stochastic zeta function.

(5) Graphs with positive weights can be interpreted as electrical resistance networks, and the use of spanning trees to compute total resistance goes back to Kirchhoff. It may be possible to use ideas from electrical networks to find other invariants of Markov shifts.

REFERENCES

[1] BÉLA BOLLOBÁS, *Modern Graph Theory*, Springer, New York, 1998.

[2] DOUGLAS LIND AND BRIAN MARCUS, *An Introduction to Symbolic Dynamics and Coding*, Cambridge Univ. Press, 1995.

[3] RUSSELL LYONS, AND YUVAL PERES, *Probability on Trees and Networks*, to appear.

[4] PHILIPPE MARCHAL, *Loop erased random walks, spanning trees, and Hamiltonian cycles*, Electronic Communications in Probability, to appear.

[5] WILLIAM PARRY AND SELIM TUNCEL, *On the stochastic and topological structure of Markov Chains*, Bull. London Math. Soc. **14** (1982), 16–27.

[6] W. PARRY AND R. WILLIAMS, *Block-coding and a zeta function for finite Markov chains*, Proc. London Math. Soc. **35**(3) (1977), 483–495.

[7] R. F. WILLIAMS, *Classification of subshifts of finite type*, Annals of Math. **98** (1973), 120–153; erratum, Annals of Math. **99** (1974), 380–381.

LIST OF WORKSHOP PARTICIPANTS

- Scot Adams, Department of Mathematics, University of Minnesota
- Roy Adler, IBM Watson Research Center
- Dakshi Agrawal, Coordinated Science Laboratory, University of Illinois at Urbana-Champaign
- Srinivas Aji, Department of Electrical Engineering, Caltech
- Brian Allen, Department of Mathematics, University of Notre Dame
- Venkat Anantharam, Department of Electrical Engineering and Computer Science, University of California - Berkeley
- John Anderson, Department of Information Technology, University of Lund
- Dieter Arnold, Laboratory for Signal & Information Processing, ETH Zurich
- Amir Banihashemi, Systems and Computer Engineering, Carleton University
- Louay Bazzi, Laboratory for Information & Decision Systems, MIT
- Marie-Pierre Beal, Institut Gaspard Monge, Universite de Marne-la-Vallee
- Ezio Biglieri, Departimento di Elettronica, Politecnico di Torino
- James Bond, SAIC
- Nigel Boston, Department of Mathematics, University of Illinois at Urbana-Champaign
- Mike Boyle, Department of Mathematics, University of Maryland
- Roger Brockett, DEAS, Harvard University
- Karen Brucks, Department of Mathematics, University of Wisconsin - Milwaukee
- Randy Bryant, Carnegie Mellon University
- Jorge Campello, IBM Almaden Research Center
- James P. Carr, Department of Mathematical Science, University of Wisconsin
- Olivier Carton, Institut Gaspard Monge, University de Marne-la-Vallee
- Rong-Rong Chen, University of Illinois at Urbana-Champaign
- Pyo Dong Chi, Department of Mathematics, Seoul National University
- Zhipei Chi, Electrical and Computer Engineering, University of Minnesota

- Sae-Young Chung, Electrical Engineering & Computer Science, MIT
- Ethan Coven, Department of Mathematics, Wesleyan University
- Ajay Dholakia, IBM Zurich Research Laboratory
- Changyan Di, Department of Mathematics, University of Notre Dame
- Rich Echard, Naval Research Laboratory
- Evangelos Eleftheriou, Zurich Research Laboratory, IBM Research Division
- Fabio Fagnani, Dip. Matematica Politecnico di Torino
- John Fan, Department of Electrical Engineering, Stanford University
- Patrick Fitzpatrick, Department of Mathematics, National University of Ireland
- David Forney, Laboratory for Information and Decision Systems, MIT
- Bill Freeman, Mitsubishi Electric Research Laboratories
- Brendan Frey, Computer Science, University of Waterloo
- Christiane Frougny, LIAFA
- Paul Fuhrmann, Department of Mathematics and Computer Science, Ben Gurion University
- Lijun Gao, Electrical and Computer Engineering, University of Minnesota
- Javier Garcia-Frias, Electrical Engineering Department, UCLA
- Paul Garrett, Department of Mathematics, University of Minnesota
- Elizabeth Gumustop, Department of Mathematics, University of Wisconsin
- Fernando Guzman, Department of Mathematics, SUNY-Binghamton
- Mark Hagen, Western Digital Corporation
- Jonathan I. Hall, Department of Mathematics, Michigan State University
- Masayuki Hattori, Information & Network Technologies Lab Sony Corporation
- Fred Heller, Department of Mathematical Sciences, University of Wisconsin at Milwaukee
- Uwe Helmke, Department of Mathematics, University of Wuerzburg
- Kjell Jorgen Hole, Department of Informatics, University of Bergen

- Gavin Horn, Department of Electrical Engineering, Caltech
- Danrun Huang, Department of Mathematics, St. Cloud State
- Stefen Hui, Mathematical Sciences, San Diego State University
- Tommi Jaakkola, Electrical Engineering and Computer Science, MIT
- Heera Lal Janwa, Mathematics and Computer Science, University of Puerto Rico
- Hui Jin, Electrical and Computer Engineering, University of Minnesota
- Hui Jin, Department of Electrical Engineering, Caltech
- Rolf Johannesson, Department of Information Theory, University of Lund
- Kimberly Johnson, Department of Mathematics, University of North Carolina
- Natasha Jonoska, Department of Mathematics, University of South Florida
- Jorn Justesen, Department of Telecommunication, Denmark Technical University
- Hiroshi Kamabe, Information Science, Gifu University
- Alex Kavcic, Harvard University
- Aamod Khandekar, Department of Electrical Engineering, Caltech
- Saejoon Kim, Department of Electrical Engineering, Cornell University
- Bruce Kitchens, IBM Watson Research Center
- Kevin Kochanek, Department of Applied Mathematics, Brown University
- Ralf Koetter, University of Illinois
- Frank Kschischang, Department of Electrical & Computer Engineering, University of Toronto
- Margreet Kuijper, Department of Electrical & Electronic Engineering, University of Melbourne
- John Lafferty, School of Computer Science, Carnegie Mellon University
- E.B. Lee, ECE Department, University of Minnesota
- Yu Liao, Electrical and Computer Engineering, University of Minnesota
- Samuel J. Lightwood, Erwin Schroedinger Institute
- Douglas Lind, Department of Mathematics, University of Washington
- Hans-Andrea Loeliger, Endora Tech AG
- Sergio Lopez-Permouth, Department of Mathematics, Ohio University

- Chung-Chin Lu, Department of Electrical Engineering, Princeton University
- Jun Ma, Electrical and Computer Engineering, University of Minnesota
- David J.C. MacKay, Department of Physics, University of Cambridge
- Yongyi Mao, System and Computer Engineering, Carleton University
- Brian Marcus, IBM Almaden Research Center
- Clyde Martin, Department of Mathematics, Texas Tech University
- Jim Massey, ETH Zurich and Lund University
- Robert McEliece, Department of Electrical Engineering, Caltech
- Jamie McGauhey, Department of Mathmatics, University of South Florida
- Peyman Meshkat, Deparment of Electrical Engineering, UCLA
- Thomas Mittelholzer, IBM Zurich Research Laboratory
- Sanjoy Mitter, Laboratory for Information & Decision Systems, MIT
- Dharmendra Modha, IBM Almaden Research Center
- Christopher Monico, Department of Mathematics, University of Notre Dame
- Mos Kaveh, Electrical and Computer Engineering, University of Minnesota
- Radford Neal, Department of Statistics, University of Toronto
- Nikolai Nefedov, Communication Laboratory, Helsinki University of Technology, Nokia
- Mike O'Sullivan, Department of Mathematics, National University of Ireland
- Travis Oenning, Electrical & Computer Engineering, University of Minnesota
- Geir Egil Oien, Department of Telecommunications, Norwegian University of Science & Technology
- Nicholas Ormes, Department of Mathematics, University of Texas at Austin
- Payam Pakzad, University of California - Berkeley
- Jongseung Park, Department of Electrical & Computer Engineering, University of Minnesota
- Alan Parks, Department of Mathematics, Lawrence University
- Dominique Perrin, Universite de Marne la Vallee
- Karl Petersen, Department of Mathematics, University of North Carolina

- Harish Kumar Pillai, Department of Mathematics, University of Groningen
- Tom Posbergh, AEM, University of Minnesota
- M.S. Ravi, Department of Mathematics, East Carolina University
- Tom Richardson, Lucent, Bell Labs
- Joachim Rosenthal, Department of Mathematics, University of Notre Dame
- Ronny Roth, Department of Computer Science, Technion Israel Institute of Technology
- Pierre-Paul Sauve, Communications Research Centre, Industry Canada
- Klaus Schmidt, Mathematics Institute, University of Vienna
- Amin Shokrollahi, Bell Labs
- Paul H. Siegel, Department of Electrical and Computer Engineering, University of California - San Diego
- Roxana Smarandache, Department of Mathematics, University of Notre Dame
- Emina Soljanin, Bell Laboratories
- Baldur Steingrimsson, Deparment of Electrical Engineering, University of Minnesota
- Jai Narayan Subrahmanyam, Advanced Projects & Staging Lab, Western Digital Corporation
- Allen Tannenbaum, Department of Electrical Engineering and Computer Science, University of Minnesota
- Michael Tanner, Department of Computer Science, University of California - Santa Cruz
- Thomas Taylor, Department of Mathematics, Arizona State University
- Jean-Pierre Tillich, Computer Science Department, University Paris-Sud
- Paul Trow, Department of Mathematical Sciences, University of Memphis
- Dewey Tucker, Electrical Engineering & Computer Science, MIT
- Selim Tuncel, Department of Mathematics, University of Washington
- Ruediger Urbanke, Bell Laboratories, Lucent Techologies
- Alexander Vardy, Department of Eletrical and Computer Engineering, University of California - San Diego
- Peter Vasiliev, VLSI, Seagate Technology
- Petr Vojtechovsky, Department of Mathematics, Iowa State University

- Pascal Olivier Vontobel, Laboratory for Signal & Information Processing, ETH Zurich
- Simon Waddington, NDS Limited
- Martin Wainwright, Electrical Engineering and Computer Science, Massachusetts Institute of Technology
- Zhe-Xian Wan, Department of Information Technology, Lund University
- Peter Webb, School of Mathematics, University of Minnesota
- Paul Weiner, Department of Mathematics, Saint Mary's University of Minnesota
- Christian Weiss, Institute for Communications Engineering, Munich University of Technology
- Yair Weiss, Computer Science Division, University of California - Berkeley
- Jan Willems, Mathematics Institute, University of Groningen
- Meina Xu, Department of Electrical Engineering, Caltech
- Byung K. Yi, Orbital Science Co.
- Sandro Zampieri, Dipartimento di Elettronica e Informatica, Universitá di Padova
- Gilles Zemor, Departement Reseau, E.N.S.T.
- Jia Zeng, Department of Electrical Engineering, University of Minnesota

IMA SUMMER PROGRAMS

1987 Robotics
1988 Signal Processing
1989 Robust Statistics and Diagnostics
1990 Radar and Sonar (June 18–29)
      New Directions in Time Series Analysis (July 2–27)
1991 Semiconductors
1992 Environmental Studies: Mathematical, Computational, and
      Statistical Analysis
1993 Modeling, Mesh Generation, and Adaptive Numerical Methods
      for Partial Differential Equations
1994 Molecular Biology
1995 Large Scale Optimizations with Applications to Inverse Problems,
      Optimal Control and Design, and Molecular and Structural
      Optimization
1996 Emerging Applications of Number Theory (July 15–26)
      Theory of Random Sets (August 22–24)
1997 Statistics in the Health Sciences
1998 Coding and Cryptography (July 6–18)
      Mathematical Modeling in Industry (July 22–31)
1999 Codes, Systems, and Graphical Models (August 2–13, 1999)
2000 Mathematical Modeling in Industry - A Workshop for Graduate
      Students (July 19–28)
2001 Geometric Methods in Inverse Problems and PDE Control
      (July 16–27)

IMA "HOT TOPICS" WORKSHOPS

- Challenges and Opportunities in Genomics: Production, Storage,
  Mining and Use, April 24–27, 1999
- Decision Making Under Uncertainty: Energy and Environmental
  Models, July 20–24, 1999,
- Analysis and Modeling of Optical Devices, September 9–10, 1999
- Decision Making under Uncertainty: Assessment of the Reliability
  of Mathematical Models, September 16–17, 1999
- Scaling Phenomena in Communication Networks, October 22–24,
  1999
- Text Mining, April 17–18, 2000
- Mathematical Challenges in Global Positioning Systems (GPS),
  August 16-18, 2000
- Modeling and Analysis of Noise in Integrated Circuits and Systems,
  August 29–30, 2000
- Mathematics of the Internet: E-Auction and Markets, December
  3–5, 2000
- Analysis and Modeling of Industrial Jetting Processes, January
  10–13, 2001

SPRINGER LECTURE NOTES FROM THE IMA:

The Mathematics and Physics of Disordered Media
 Editors: Barry Hughes and Barry Ninham
 (Lecture Notes in Math., Volume 1035, 1983)

Orienting Polymers
 Editor: J.L. Ericksen
 (Lecture Notes in Math., Volume 1063, 1984)

New Perspectives in Thermodynamics
 Editor: James Serrin
 (Springer-Verlag, 1986)

Models of Economic Dynamics
 Editor: Hugo Sonnenschein
 (Lecture Notes in Econ., Volume 264, 1986)

The IMA Volumes in Mathematics and its Applications

Current Volumes:

FORTHCOMING VOLUMES

1997–1998: *Emerging Applications of Dynamical Systems*
     Multiple–Time–Scale Dynamical Systems

1998–1999: *Mathematics in Biology*
     Pattern Formation and Morphogenesis
     Endocrinology: Mechanism of Hormone Secretion and Control
     Membrane Transport and Renal Physiology
     Mathematical Approaches for Emerging and Reemerging Infectious Disease

1999 Summer Program: *Codes, Systems, and Graphical Models*

1999–2000: *Reactive Flow and Transport Phenomena*
     Fire